Consulting Editor

George A. Anastassiou
Department of Mathematical Sciences
University of Memphis

Handbook of Computational and Numerical Methods in Finance

Svetlozar T. Rachev
Editor

Springer Science+Business Media, LLC

Svetlozar T. Rachev
Department of Statistics and Applied Probability
University of California
Santa Barbara, CA 93106
U.S.A.
and
Department of Economics and Business Engineering
Universität Karlsruhe
D-76128 Karlsruhe
Germany

Library of Congress Cataloging-in-Publication Data
Handbook of computational and numerical methods in finance / Svetlozar T. Rachev, editor.
   p. cm.
   Includes bibliographical references.
   ISBN 978-1-4612-6476-7   ISBN 978-0-8176-8180-7 (eBook)
   DOI 10.1007/978-0-8176-8180-7
    1. Finance–Mathematical models. I. Rachev, S. T. (Svetlozar Todorov)

HG106.H36 2004
332'.01'51-dc22   2004043735
                                              CIP

AMS Subject Classifications: Primary: 62P05, 62P20, 65C05; Secondary: 60E07, 62L20, 62M20

ISBN 978-1-4612-6476-7   Printed on acid-free paper.

© 2004 Springer Science+Business Media New York
Originally published by Birkhäuser Boston in 2004
Softcover reprint of the hardcover 1st edition 2004

9 8 7 6 5 4 3 2 1   SPIN 10923511

www.birkhasuer-science.com

Contents

Preface

Numerical Methods in Finance have recently emerged as a new discipline at the intersection of probability theory, finance and numerical analysis. They bridge the gap between financial theory and computational practice and provide solutions to problems where analytical methods are often non-applicable.

Numerical methods are more and more used in several topics of financial analysis: computation of complex derivatives; market, credit and operational risk assessment, asset liability management, optimal portfolio theory, financial econometrics and others.

Although numerical methods in finance have been studied intensively in recent years, many theoretical and practical financial aspects have yet to be explored. This volume presents current research focusing on various numerical methods in finance. The contributions cover methodological issues, Genetic Algorithms, Neural Networks, Monte-Carlo methods, Finite Difference Methods, Stochastic Portfolio Optimization as well as the application of other numerical methods in finance and risk management.

As editor, I am grateful to the contributors for their fruitful collaboration. I would particularly like to thank Stefan Trueck and Carlo Marinelli for the excellent editorial assistance received over the progress of this project. Thomas Plum did a splendid word-processing job in preparing the manuscript. I owe much to George Anastassiou (Consultant Editor, Birkhäuser) and Ann Kostant Executive Editor, Mathematics and Physics, Birkhäuser for their help and encouragement.

Svetlozar Rachev
Santa Barbara and Karlsruhe, 2003

Preface

*Handbook of Computational
and Numerical Methods
in Finance*

1

Skewness and Kurtosis Trades

Oliver J. Blaskowitz

Wolfgang K. Härdle

Peter Schmidt

ABSTRACT   In this paper we investigate the profitability of 'skewness trades' and 'kurtosis trades' based on comparisons of implied state price densities versus historical densities. In particular, we examine the ability of SPD comparisons to detect structural breaks in the options market behaviour. While the implied state price density is estimated by means of the Barle and Cakici Implied Binomial Tree algorithm using a cross section of DAX option prices, the historical density is inferred by a combination of a non-parametric estimation from a historical time series of the DAX index and a forward Monte Carlo simulation.

1.1   Introduction

From a trader's point of view, implied state price densities (SPDs) may be used as market indicators and thus constitute a good basis for advanced trading strategies. Deviations of historical SPDs from implied SPDs have led to skewness and kurtosis trading strategies [1]. Such strategies were investigated in [4] for the period from 04/97 until 12/99. The trades applied to European options on the German DAX index generated a positive net cash flow.

However, it is market consensus that option markets behavior changed as a consequence of the stock market bubble that burst in March 2000, Figure 1. The purpose of this paper is to examine the trading profitability and the informational content of both the implied and the historical SPD for the extended period from 04/97 to 07/02. Our analysis focuses on the ability of SPD skewness and kurtosis comparisons to detect structural breaks.

For this purpose we use EUREX DAX option settlement prices and DAX closing prices. All data is included in MD*Base (http://www.mdtech.de), a database located at CASE (Center for Applied Statistics and Economics, http://www.case.hu-berlin.de) of Humboldt-Universität zu Berlin.

We start by explaining skewness and kurtosis trades in Section 1.2. In Section 1.3 we motivate the transition from Black–Scholes implied and historical volatility

FIGURE 1. DAX from 01/97 to 01/03.

comparisons to implied and historical SPD comparisons. The SPD estimation techniques are discussed in Section 1.4, and Section 1.5 presents the estimation results. Section 1.6 investigates the trading performance, Section 1.7 concludes.

1.2   What are Skewness and Kurtosis Trades ?

In derivatives markets option strategies such as risk-reversals and strangles, [16], are used to exploit asymmetric and fat-tailed properties of the underlying's risk-neutral distribution. A risk-reversal is a portfolio of two European options with time to maturity τ. More precisely, it consists of a short position in a put with strike K_1 and a long position in a call with strike K_2, where $K_1 < K_2$. Its payoff profile at maturity as shown in Figure 2 suggests that an investor in this portfolio considers high prices of the underlying to be more likely to occur than low prices. Similarly, an investor believing that large moves of the underlying are likely to occur will buy a long strangle, which consists of a long position in a European put with strike K_1 and time to maturity τ and a long position in a European call with strike K_2 and time to maturity τ.

In our study we will use a risk-reversal and a modified strangle portfolio to exploit differences in two risk-neutral SPDs. To motivate the SPD comparison we recall the general pricing equations for European put and call options. From option pricing theory it follows that:

$$P = e^{-r\tau} \int_0^\infty \max(K_1 - S_T, 0)q(S_T)dS_T, \qquad (1.1)$$

$$C = e^{-r\tau} \int_0^\infty \max(S_T - K_2, 0)q(S_T)dS_T,$$

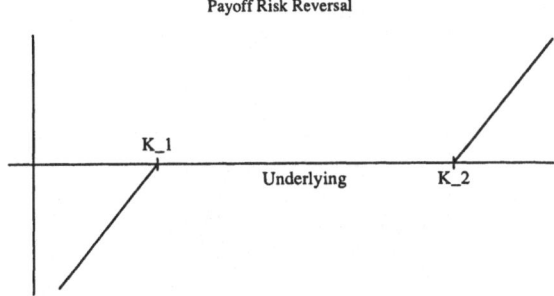

FIGURE 2. Payoff Risk Reversal

where P and C are put (respectively call) prices, r is the risk-free interest rate, S_T is the price of the underlying at maturity T and q is a risk-neutral density [9]. Consider two risk-neutral densities denoted f^* and g^* as in Figure 3 where density f^* is more negatively skewed than g^*. Then equation (1.1) implies that the price of a European call option with strike K_2 computed with density f^* is lower than the price computed with density g^*. The reason for this is that f^* assigns less probability mass to prices $S_T > K_2$ than g^*. If the call is priced using f^* but one regards density g^* as a better approximation of the underlying's distribution, one would buy the option. Along these lines one would sell a put option with strike K_1, which finally results in a risk-reversal portfolio or, as we will call it, a skewness 1 trade.

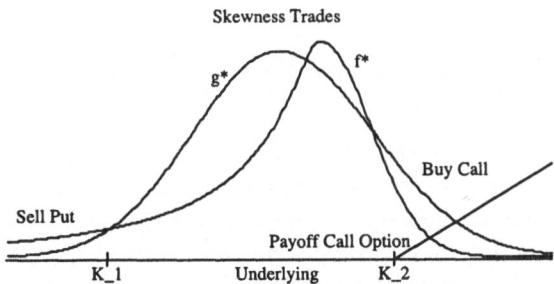

FIGURE 3. Skewness Trade

The same probability mass reasoning leads to kurtosis trades. We buy and sell calls and puts of different strikes as shown in Figure 4. The payoff profile at maturity is given in Figure 5. In Section 1.6 we will specify the regions in which to buy or sell options in terms of the moneyness $K/S_t e^{r\tau}$.

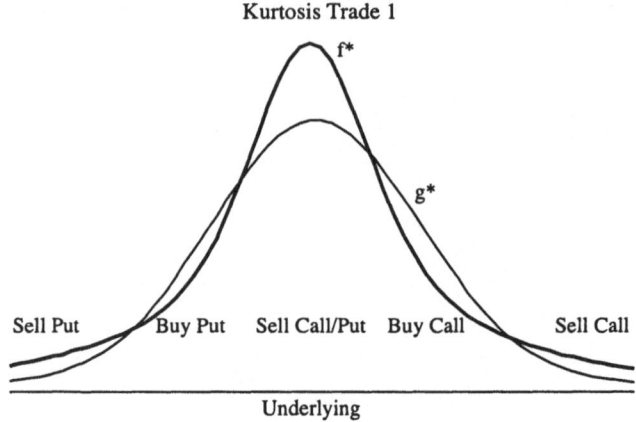

FIGURE 4. Kurtosis Trade 1

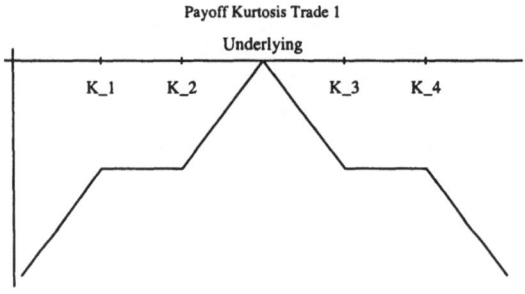

FIGURE 5. Payoff Kurtosis Trade 1

Note, in a complete market model admitting no arbitrage opportunities, there exists exactly *one* risk-neutral density. If markets are not complete, for example when the volatility is stochastic, there are in general many risk-neutral measures. Comparing two risk-neutral densities, as we do, amounts rather to comparing two different models, and trades are initiated depending on the model in which one believes more. The next section will discuss briefly how this approach is implemented in practice.

1.3   Skewness and Kurtosis in a Black–Scholes World

Black–Scholes' assumption that the underlying's process S_t follows a geometric Brownian motion

$$dS_t = \mu S_t dt + \sigma S_t dW_t,$$

where μ, σ are constants and dW_t is a Wiener Process, implies the underlying to be log-normally distributed with mean $exp(\mu dt)$ and variance $exp(2\mu dt)\{exp(\sigma^2) -1\}$. Skewness and kurtosis solely depend on the volatility parameter σ. As σ increases, skewness and kurtosis increase as well. If there is only one implied volatility (IV) for all options, trading differences in Black–Scholes implied and historical volatilities, σ_{imp} respectively σ_{hist}, amounts to a comparison of two log-normal distributions. More precisely, such traders compare two Black–Scholes models with constant parameters σ_{imp} and σ_{hist}. Within this framework of two log-normals and a constant volatility, one would buy all options if the historical volatility is higher than the IV and if one believes in such a 'historical volatility model'.

This way traders can trade volatility differences, but the assumption of log-normality does not allow the implementation of skewness and kurtosis trades. Comparing skewness and kurtosis requires the information contained in the Black–Scholes IV smile that is observed on option markets. Traders often use this smile to assess the market's view on the underlying's risk-neutral probabilistic behavior. Applying the inverted Black–Scholes option pricing formula to prices obtained from a model with a left-skewed risk-neutral distribution, for example, entails out-of-the-money (OTM) calls to have a higher IV than at-the-money (ATM) calls or puts and the latter to have higher IVs than OTM puts. If the unknown risk-neutral distribution has more kurtosis than the log-normal distribution, then OTM calls and OTM puts have higher Black–Scholes IVs than ATM options. Depending on the smile's location, slope and curvature, a skewness or a kurtosis trade is set up.

Relaxing the assumption that neither the implied nor the historical risk-neutral distribution is log-normal it is possible to extend trading rules based on Black–Scholes IVs to a framework where comparisons within a more general class of risk-neutral distributions are possible. In light of this, the approach we follow in this paper amounts to a generalization of a comparison of implied and historical volatility. In the following section we will briefly motivate the notion of implied and historical SPDs and describe the methods we used to extract them.

1.4   Implied and Historical DAX SPDs

Modern options markets have a high degree of market liquidity, i.e., prices on these markets are determined by supply and demand. This is particularly true for DAX options traded on the EUREX, the world's leading market for the trading of futures and options on stocks and stock indices. Some 425 participants in 17 countries traded more than 106 million options and futures contracts in March 2003. DAX options belong to the most frequently traded contracts (www.eurexchange.com).

Given a set of market option prices, an implied distribution q is the distribution that simultaneously satisfies the pricing equation for all observed options in the set. As we work with European options, prices are given by equation (1.1). The implied state price density of an asset should be viewed as a way of characterizing the prices of derivatives contingent upon this asset. It is the density used to price options and has therefore a 'forward looking character' [5].

We will later see that the historical SPD is inferred from a time series of past underlying's prices without involving option prices at all. Since we will use this distribution to compare it to the implied SPD, we call it an SPD too, a 'historical SPD'.

1.4.1   Extracting the Options Implied SPD

In recent years a number of methods have been developed to infer implied SPDs from cross-sectional option prices, see [5] and [13] for an overview. As Binomial Trees are widely used and accepted by practitioners, we use Implied Binomial Trees (IBT) in order to obtain a proxy for the option implied SPD, which is denoted by f^* from now on. The IBT algorithm is a modification of the Cox–Ross–Rubinstein (CRR) algorithm. The numerous IBT techniques proposed by [15], [6], [7] and [2] represent discrete versions of a continuous time and space diffusion model

$$\frac{dS_t}{S_t} = \mu(S_t, t)\, dt + \sigma(S_t, t) dW_t.$$

Whereas the classical CRR binomial tree assumes the instantaneous local volatility function to be constant, i.e., $\sigma(S_t, t) = \sigma$, the IBT allows $\sigma(S_t, t)$ to be dependent on time and space.

Relying on the work of [12] we decided to work with Barle & Cakici's method for two reasons. First, the authors provide interactive XploRe quantlets to compute the IBTs proposed by Derman & Kani and Barle & Cakici. Second, according to the authors the latter method proved to be more robust.

The procedure works as follows: From a cross-section of two weeks of options data the XploRe quantlet **volsurf.xpl** estimates the IV surface over 'forward' moneyness and time to maturity, which we measure assuming 250 trading days per year. The quantlet **IBTbc.xpl** computes the IBT assuming a flat yield curve, a constant time to maturity of three months and taking the IV surface as input.

Furthermore, the IBT consists of three trees, the tree of stock prices, the tree of transition probabilities and finally the tree of Arrow–Debreu prices. If the tree is discretised by N time steps of length $\Delta t = \tau/N$, the tree consists of $N + 1$ final stock prices $S_{N+1,i}, i \in \{1, 2, \ldots, N + 1\}$, and $N + 1$ final Arrow–Debreu prices $\lambda_{N+1,i}, i \in \{1, 2, \ldots, N + 1\}$. Compounding the Arrow–Debreu prices to maturity $e^{r\tau}\lambda_{N+1,i}, i \in \{1, 2, \ldots, N + 1\}$, and associating them to annualized stock returns

$$u_{N+1,i} = \left\{\log(S_{N+1,i}) - \log(S_t)\right\} \tau^{-1}, \; i \in \{1, 2, \ldots, N + 1\},$$

we obtain the option implied risk-neutral SPD f^* over log–returns. A more detailed description of the procedure is given in [4].

Figure 6 displays the implied SPD on Monday, June 23, 1997, and $N = 10$ time steps. This is the fourth Monday in June 1997. On that day, the DAX index S_t was at 3748.79 and the risk–free three month rate r was at 3.12. The plot shows the three

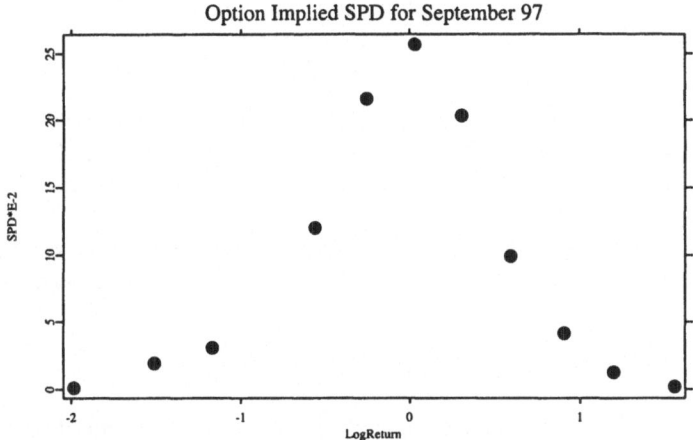

FIGURE 6. Option Implied SPD on Monday, June 23, 1997, for Friday, September 19, 1997, with $S_t = 3748.79, r = 3.12, N = 10$.

months ahead risk-neutral SPD for Friday, September 19, 1997. This is the third Friday of September 1997, the expiration day of September 97 options. The SPDs standard deviation is 0.5, its skewness is -0.45 and its kurtosis is 4.17.

We are interested in the SPD on the third Friday of the expiry month since later we will design the trading strategies such that we set up skewness and kurtosis portfolios on the 4th Monday of each month. These portfolios will consist of long and short positions in call and put options expiring on the 3rd Friday three months later.

1.4.2   Extracting the Historical SPD

The risk-neutral historical SPD g^* is estimated by assuming that the underlying S_t follows a continuous diffusion process:

$$dS_t = \mu(S_t)dt + \sigma(S_t)dW_t.$$

If we assume, as above, a flat yield curve and the existence of a bank account which evolves according to $B_t = B_0 e^{rt}$, then from Itô's formula and Girsanov's theorem we obtain the risk-neutral dynamics:

$$dS_t^* = rS_t^* dt + \sigma(S_t^*)dW_t^*. \tag{1.2}$$

Note, since here the underlying is the DAX performance index, we do not take dividend yields into account.

The instantaneous local volatility function is identical under both the actual and the risk-neutral dynamics. It is estimated by means of Härdle and Tsybakov's (1997) non-parametric version of the minimum contrast estimator:

$$\hat{\sigma}^2(S) = \frac{\sum_{i=1}^{N^*-1} K_1(\frac{S_i-S}{h_1})N^*\{S_{(i+1)/N^*} - S_{i/N^*}\}^2}{\sum_{i=1}^{N^*} K_1(\frac{S_i-S}{h_1})},$$

where K_1 is a kernel function, h_1 is a bandwidth parameter, S_i are discretely observed daily DAX closing prices and N^* is the total number of observed daily DAX closing prices. In the model specified in equation (1.2) $\hat{\sigma}^2(S)$ is an unbiased estimator of $\sigma^2(S)$.

Using three months of past daily DAX closing prices, we estimate $\sigma^2(S)$ and then simulate $M = 10000$ paths of the diffusion process for a time period of three months:

$$dS_t^* = rS_t^*dt + \hat{\sigma}_a(S_t^*)dW_t^*,$$

with $\hat{\sigma}_a(S) = \hat{\sigma}(S)\tau^{-1}$ being the estimated annualized diffusion coefficient. As the DAX is a performance index, the continuous dividend yield is 0.

Collecting the endpoints of the simulated paths, we compute annualized log-returns:

$$u_{m,t} = \{\log(S_{m,T}) - \log(S_t)\}\tau^{-1}, m = 1, \ldots, M.$$

Using the notation $u = \log(S_T/S_t)$ and knowing that

$$P(S_T \leq S) = P(u \leq \log(S/S_t)) = \int_{-\infty}^{\log(S/S_t)} p_t^*(u)du,$$

g^* is obtained by

$$g^*(S) = \frac{\partial}{\partial S}P(S_T < S) = \frac{\hat{p}^*\{\log(S/S_t)\}}{S},$$

where $\hat{p}^*$ is a non-parametric kernel density estimation of the continuously compounded log-returns. $\hat{p}^*$ is given by

$$\hat{p}^*(u) = \frac{1}{Mh_2}\sum_{m=1}^{M} K_2\left(\frac{u_{m,t} - u}{h_2}\right),$$

with K_2 being a kernel function and h_2 a bandwidth parameter. g^* is $\sqrt{N^*}$-consistent for $M \to \infty$ even though $\hat{\sigma}^2$ converges at a slower rate [1].

In order to satisfy the condition that under the absence of arbitrage the mean of the underlying's risk-neutral density is equal to the futures price, we translate the Monte Carlo simulated historical density:

$$\bar{S} = S - E(S) + S_t e^{r_{t,\tau}}.$$

As for the SPD comparison later on, we are only interested in the standard deviation, skewness and kurtosis measures of g^*. Because the annualized log-returns contain already all the necessary information, we finally compute only these statistics for the simulated log-returns.

Consider the period in the example above. On Monday, June 23, 1997, we used past daily DAX closing prices of the period of time between Monday, March 23,

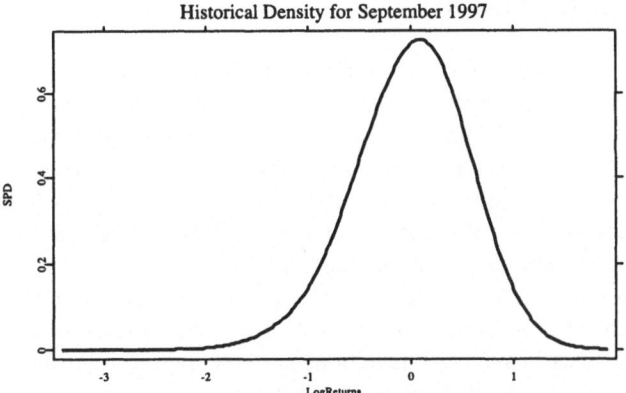

FIGURE 7. Historical SPD on Monday, June 23, 1997, for Friday, September 19, 1997, with $S_t = 3748.79, r = 3.12$.

1997, and Friday, June 20, 1997, to estimate σ^2. Following, on Monday, June 23, 1997, we simulate $M = 10000$ paths to obtain the three months ahead SPD, shown in Figure 7, whose standard deviation is 0.52, skewness is -0.39 and kurtosis is 3.23. Figure 8 illustrates both procedures.

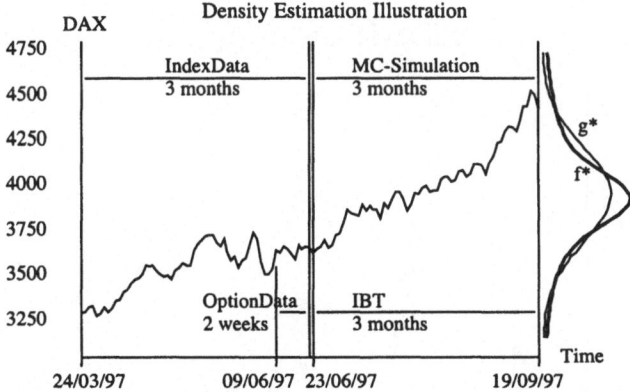

FIGURE 8. Comparison of procedures to estimate historical and implied SPD of Friday, 19/09/97. SPDs estimated on Monday, 23/06/97, by means of three months of index data respectively two weeks of option data.

1.5   SPD Comparison

In order to compare both SPDs, we computed the three month implied and historical densities every three months. More precisely, in March we compute implied and historical densities for June. Following, we compute in June the densities for September etc. The reason is that DAX options maturing in March, June, September and December are most liquid, thus containing most information. Starting in June 1997 we estimate the first SPDs for September. We compare both SPDs by looking at the standard deviation, skewness and kurtosis.

Figure 9 shows the standard deviations of the implied (blue line with triangles) and the historical SPD (red line with circles). Although difficult to interpret, it appears that differences in standard deviations are less significant at the end of 1997 and in 1998. It seems that deviations in the dispersion become more pronounced from 1999 on.

In contrast, skewness and kurtosis measures of implied and historical SPD as shown in Figures 10 and 11 give a less unambiguous picture. Whereas the skewness signal changes in the beginning of 2001, the kurtosis comparison yields in almost all periods a one-sided signal, ignoring the outlier in September 2001.

Given that market participants agree upon a structural break occurring in March 2000, the methodology applied above does not seem to provide useful information about such a break in the options markets behavior.

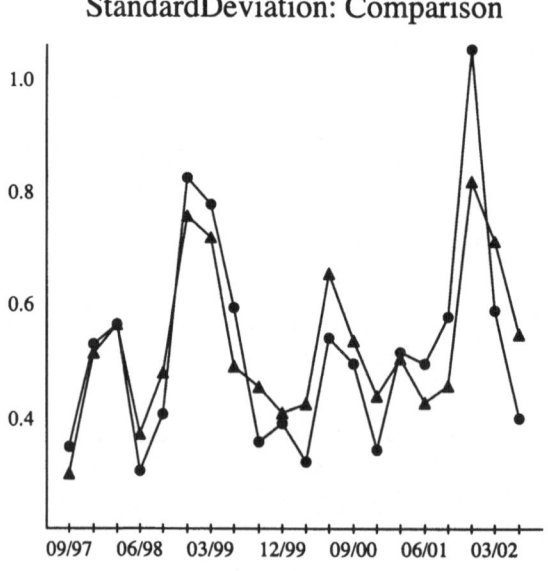

FIGURE 9. Comparison of Standard Deviations of Implied and Historical Densities. Historical and implied SPD are denoted by a circle, respectively a triangle.

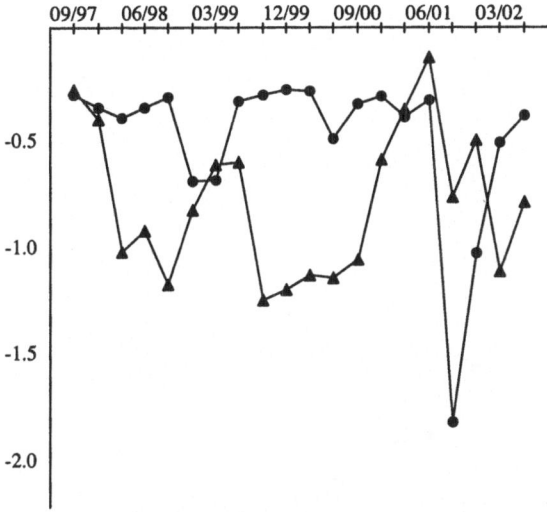

Skewness: Comparison

FIGURE 10. Comparison of Skewness of Implied and Historical Densities. Historical and implied SPD are denoted by a circle, respectively a triangle.

1.6   Skewness and Kurtosis Trades

The results from the previous section indicate that the implied SPD is more negatively skewed and has a higher kurtosis than the historical SPD. In this section, we investigate the skewness 1 and kurtosis 1 trade performance for the periods from June 1997 to March 2000, from June 2000 to March 2002, and for the overall period from June 1997 to March 2002. Each period consists of non–overlapping three month subperiods, in which we set up portfolios of calls and puts with a time to maturity of three months. For each subperiod we measure the portfolio return by:

$$\text{portfolio return} = \frac{\text{net cash flow at maturity}}{\text{net cash flow at initiation}} - 1.$$

The investment to set up the portfolio comprises the net cash flow from buying and selling calls and puts. Whenever the options sold are worth more than the options bought, entailing a positive cash inflow, it is not possible to compute a return measure. To ensure that the net cash flow at initiation is negative, we buy one share of the underlying for each call option sold and deposit in a bank account the cash value of each sold put options strike. Such an approach amounts to a very careful performance measurement. Applying margin deposits required by EUREX, for example, would lower the cash outflow at initiation and thus increase the profitability. Since a DAX option contract on EUREX consists of five options and one index point has a value of 5 EUR, we 'charge' 1 EUR for an index point. At maturity, we sum up

Kurtosis Comparison

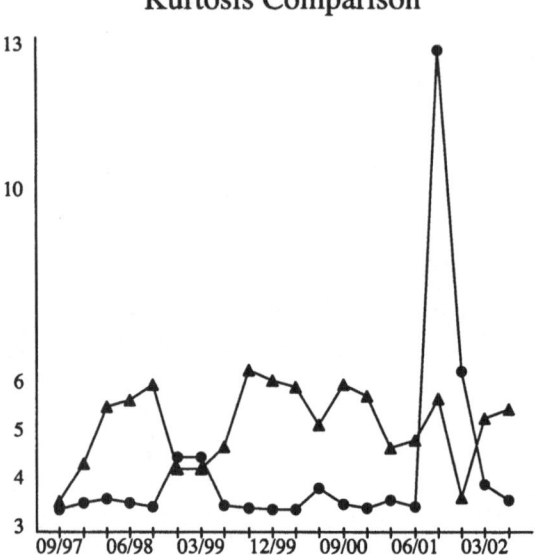

FIGURE 11. Comparison of Kurtosis of Implied and Historical Densities. Historical and implied SPD are denoted by a circle, respectively a triangle.

all option payoffs, the bank account balance and the DAX value. For simplicity, we assume that the bank account's interest rate is zero.

As for the skewness trade, we consider put options with a moneyness, $K/S_t e^{r\tau}$, of less than 0.95 as OTM. We sell all OTM put options available in the market. We buy all available ITM call options, i.e., call options with a moneyness of more than 1.05, see Table 1. In our trading simulation one call or put is traded on each moneyness. As Table 2 shows, the performance for the two subperiods reversed. The annualized total returns as well as Sharpe ratios turned from positive to negative.

A kurtosis 1 portfolio is set up by selling and buying puts and calls as given in Table 1. The kurtosis trade performed similarly to the skewness trade. In the first period it was highly profitable. In the second period it turned out to be a bad strategy compared to a risk-free investment.

1.7 Conclusion

Given that the trading performance of the skewness as well as the kurtosis trade differ significantly in both subperiods, it is disappointing that the SPD comparison does not reveal a similar pattern. One could argue that what we see within the two subperiods is just a feature of the risk premium as pointed out by [1]. However, as market participants agree that options markets behave differently since March 2000,

TABLE 1. Skewness 1 Trade: Definitions of moneyness regions.

Skewness 1 Trade		Kurtosis 1 Trade	
Position	Moneyness	Position	Moneyness
short puts	< 0.95	short puts	< 0.90
		long puts	0.90 − 0.95
		short puts	0.95 − 1.00
		long calls	1.00 − 1.05
		short calls	1.05 − 1.10
long calls	> 1.05	long calls	> 1.10

TABLE 2. Skewness 1 Trade Performance. Only Total Return is annualized. Returns are given in percentages.

	Skewness 1 Trade		
Period	06/97–03/00	06/00–03/02	Overall
Number of Subperiods	12	8	20
Total Return	4.85	-8.53	-2.05
Return Volatility	3.00	9.79	6.78
Minimum Return	-3.66	-25.78	-25.78
Maximum Return	7.65	7.36	7.65
Sharpe Ratio (Strategy)	0.10	-0.46	-0.24
Sharpe Ratio (DAX)	0.38	-0.35	0.02

TABLE 3. Kurtosis 1 Trade Performance. Only Total Return is annualized. Returns are given in percentages.

	Kurtosis 1 Trade		
Period	06/97–03/00	06/00–03/02	Overall
Number of Subperiods	12	8	20
Total Return	14.49	-7.48	2.01
Return Volatility	3.87	13.63	9.33
Minimum Return	-4.54	-28.65	-28.65
Maximum Return	8.79	18.14	18.14
Sharpe Ratio (Strategy)	0.55	-0.32	-0.05
Sharpe Ratio (DAX)	0.38	-0.35	0.02

we believe that there is more to exploit from an SPD comparison. In light of this, a topic for future research will be to investigate different methodologies with respect to their potential to improve and fine tune such an SPD comparison.

References

[1] Ait–Sahalia, Y., Wang, Y., Yared, F. 2001. Do Option Markets correctly Price the Probabilities of Movement of the Underlying Asset?, *Journal of Econometrics* **102**: 67–110.

[2] Barle, S., Cakici, N., 1998. How to Grow a Smiling Tree, *The Journal of Financial Engineering* **7**: 127–146.

[3] Black, F., Scholes, M., 1973. The Pricing of Options and Corporate Liabilities, *Journal of Political Economy* **81**: 637–659.

[4] Blaskowitz, O., Schmidt, P. 2002. Trading on Deviations of Implied and Historical Density, in: *Applied Quantitative Finance*, W. Härdle, T. Kleinow, G. Stahl, eds., Springer-Verlag, Heidelberg.

[5] Cont, R. 1998. Beyond Implied Volatility: Extracting Information from Options Prices, in: *Econophyiscs*, Kluwer, Dodrecht.

[6] Derman, E., Kani, I. 1994. The Volatility Smile and Its Implied Tree, http://www.gs.com/qs/

[7] Dupire, B. 1994. Pricing with a Smile, *Risk* **7**: 18–20.

[8] Florens–Zmirou, D. 1993. On Estimating the Diffusion Coefficient from Discrete Observations, *Journal of Applied Probability* **30**: 790–804.

[9] Franke, J., Härdle, W., Hafner, C. 2001. *Einführung in die Statistik der Finanzmärkte*, Springer-Verlag, Heidelberg.

[10] Härdle, W., Simar, L. 2003. *Applied Multivariate Statistical Analysis*, Springer-Verlag, Heidelberg.

[11] Härdle, W., Tsybakov, A., 1997. Local Polynomial Estimators of the Volatility Function in Nonparametric Autoregression, *Journal of Econometrics*, **81**: 223–242.

[12] Härdle, W., Zheng, J. 2002. How Precise Are Price Distributions Predicted by Implied Binomial Trees?, in: *Applied Quantitative Finance*, W. Härdle, T. Kleinow, G. Stahl, eds., Springer-Verlag, Heidelberg.

[13] Jackwerth, J.C. 1999. Option Implied Risk Neutral Distributions and Implied Binomial Trees: A Literature Review, *The Journal of Derivatives* **Winter**: 66–82.

[14] Kloeden, P., Platen, E., Schurz, H. 1994. *Numerical Solution of SDE Through Computer Experiments*, Springer-Verlag, Heidelberg.

[15] Rubinstein, M. 1994. Implied Binomial Trees, *Journal of Finance* **49**: 771–818.

[16] Willmot, P. 2002. *Paul Willmot Introduces Quantitative Finance*, Wiley.

2

Valuation of a Credit Spread Put Option: The Stable Paretian model with Copulas

Dylan D'Souza

Keyvan Amir-Atefi

Borjana Racheva-Jotova

2.1   Introduction

Financial institutions are making a concerted effort to measure and manage credit risk inherent in their large defaultable portfolios. This is partly in response to regulatory requirements to have adequate capital to meet credit event contingencies, but risk managers are also concerned about the sensitivity of the value of their portfolios to potential deteriorating credit quality of issuers. These changes in portfolio value can be quite significant for financial institutions such as commercial banks, insurance companies and investment banks, exposed to credit risk inherent in their large bond and loan portfolios. Credit derivatives are instruments used to manage financial losses due to credit risk, but unlike derivatives to manage market risk they are relatively less liquid and are more complicated to price because of the relative illiquidity of the underlying reference assets.

Credit derivatives are over-the-counter contracts used by investors to manage credit risk, allowing investors to achieve any desired risk/return profile. The global market for credit derivatives is growing exponentially as these derivatives facilitate market completion and provide for efficient risk sharing among market participants, specifically, these derivatives permit the transfer of credit risk to other investors without the need to trade the reference asset. The main credit derivatives traded in financial markets are credit default swaps, credit spread options and total return swaps. This paper focuses on valuing a credit spread put option written on a Baa-rated coupon bond as a reference asset.[1]

---

* The analysis and conclusions set forth in this paper represent the views of the authors and do not indicate concurrence by other members of the institutions they represent.

[1] D'Souza et al [2002] value a credit default swap referenced to a Baa coupon bond using a trinomial tree and a stable Paretian model for the default-free spot interest rate process.

Credit spread options are contracts with payoffs that depend on a defined reference spread over an asset with no credit risk, e.g., a default-free asset such as a Treasury security or a LIBOR security. The reference asset can be a single asset or a portfolio of defaultable assets. An investor insuring against adverse credit spread movements pays an option premium upfront in return for a bullet payment, contingent on the event that the reference asset's credit spread exceeds a certain prespecified value, called the strike or exercise spread.

Our paper focuses on the valuation of a credit spread put option with a noncallable defaultable Baa-rated coupon bond as the reference asset. The reference asset is priced using the term structure of Baa credit spreads obtained from the daily yields of Moody's aggregate Baa bond index. This aggregate index is a well diversified bond portfolio that includes bonds from corporate and utilities industry classifications of different maturities. The yield of the Baa index is a weighted average yield of all the corporate and utility bonds in the index, weighted by the value of each bond in the portfolio. The credit spread of the Baa index is defined as the difference between the weighted average yield of the Baa index and the yield of an otherwise equivalent Treasury security. Since we have no information about the portfolio's duration, we assume the Baa index has a duration of 30 years since it consists of long-term bonds. Therefore, the credit spread of the Baa index is computed using the 30-year constant maturity Treasury security as the default-free reference asset.

An individual bond in the Baa index will not significantly affect the volatility of the index credit spread changes as the portfolio is reasonably well diversified. Idiosyncratic risk is diversified away but systematic risk remains. Baa index credit spread movements arise from movements in systematic state variables that influence all individual bond's credit spread movements in the portfolio. Theoretically and empirically, macroeconomic variables affect each bond's credit spread which consequently impacts the Baa index credit spread. Such changes in credit spreads are termed systematic credit spread changes, as Pedrosa and Roll [1998] point out, since systematic state variables influence the variations, the Baa index credit spreads. The default-free spot interest rate is a state variable that in theory should affect the credit spread of every defaultable bond. Default-free spot rate changes induce investors to modify their assessments of the probability of default of all defaultable bonds. When the spot rate moves, investors revise the present values of firms' debt obligations which directly affects the ability to pay back bondholders, causing investors to re-estimate the probability of default, thus credit spreads change as the spot rate changes. This correlation between the spot rate and the credit spread represents correlation between market risk and credit risk which is empirically observed and theoretically justified.[2]

An investor holding the Baa index portfolio of defaultable bonds would be concerned about large movements in the spot rate that affects the Baa credit spread, which could potentially affect the index value adversely. Financial institutions concerned about insuring against large portfolio losses could hedge these losses by buying a credit spread put option that protects them from large adverse movements in the

---

[2] Jarrow and Turnbull [2000] discuss the importance of integrating market risk and credit risk for pricing models and risk management.

credit spread reflecting deteriorating credit quality. In the option valuation problem, the option's price would then directly depend on the joint probability distribution of changes in the spot rate and Baa credit spread. This joint distribution would be important to risk managers who are concerned about extreme movements in the defaultable portfolio's asset return process. For example, in VaR computations, tail probabilities of an asset's return distribution are of central importance. The tail probabilities of a defaultable asset's return distribution would depend upon the nature of the joint distribution of spot rate and credit spread changes. It would be a worthwhile exercise to examine the empirical joint distribution of the spot and credit spread movements to compute a fair price for the credit spread put option as the option's contingent payoffs directly depend on this joint distribution. Modeling as accurately as possible the empirical distribution of the spot rate and credit spread processes that incorporate relevant market information pertinent to the option valuation makes the framework empirically consistent.

In the credit derivative valuation literature, Brownian motion is assumed to drive the reference asset's return process. The discrete time counterpart of Brownian motion is a random walk with Gaussian innovations. However, empirical evidence suggests the distribution of asset returns are non-Gaussian. Many financial asset return processes exhibit heavy tails, skewness, stochastic volatility and volatility clustering. As a result of the non-Gaussian nature of asset returns, Mandelbrot [1963,1967] and Fama [1965a,b] were among the first to propose the use of stable Paretian distributions as an alternative to model these empirically observed properties.[3] The advantage of employing stable Paretian distributions over other non-Gaussian distributions is that sums of stable random variables are also stable distributed, which is convenient for portfolio choice theory. Stable distributions have heavy tails with a higher probability mass around the mean. Figure 1 depicts the probability density function for various indices of stability: $\alpha = 2$ (i.e., Gaussian distribution), $\alpha = 0.5$, 1 and $\alpha = 1.5$. The smaller the index of stability the heavier the tail. Figure 2 shows the flexibility of stable non-Gaussian distributions in modeling various levels of skewness. This flexibility permits stable non-Gaussian distributions to provide a better fit to the empirically observed distributions of asset returns.[4] The Gaussian distribution is a special case of stable distributions. It is the only stable distribution with a finite second moment.

Derivative valuation models fundamentally depend upon the assumptions made about which stochastic process drives the uncertainty of the underlying risk factors being priced. In the academic literature, Gaussian-based Brownian diffusion processes are the risk drivers. In this paper we examine the empirical distributions of the spot rate and Baa credit spread index changes and use it to formulate a model that accurately incorporates their joint distributional behavior. The empirical distributions display a significant departure from Gaussian properties, which violate the assumptions of classical option pricing theory.

---

[3] Rachev and Mittnik [2000], Rachev, Schwartz and Khindanova [2000] are among others that have applied stable Paretian models to financial asset returns.

[4] Mittnik and Rachev [2000] discuss this in detail.

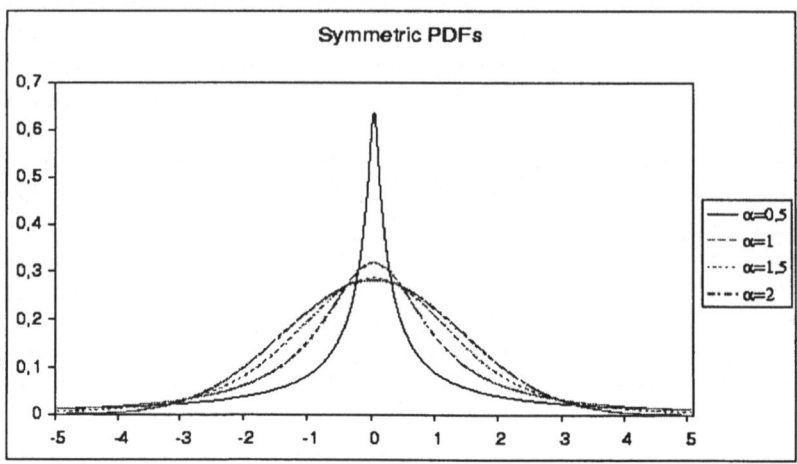

FIGURE 1. The probability density function for a standard symmetric alpha-stable random variable, varying the index of stability, alpha

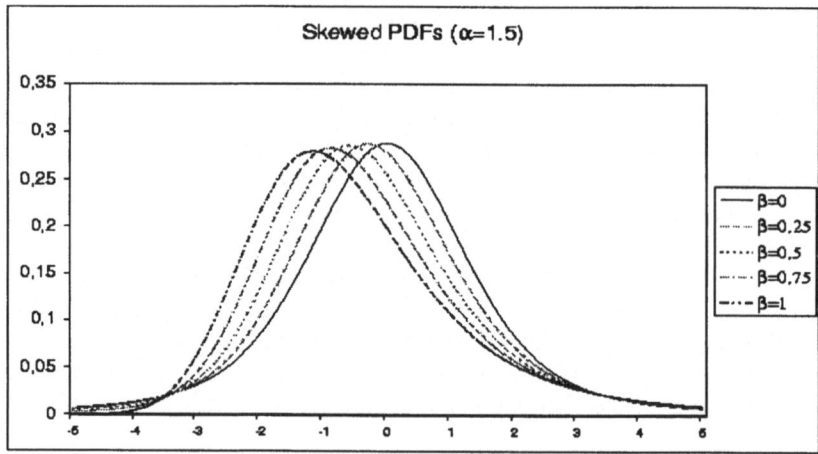

FIGURE 2. Probability density functions for skewed stable random variables, varying the skewness parameter beta, for a given index of stability, alpha=1.5

We value a one-year credit spread put option referenced on a 10-year Baa non-callable coupon bond that makes annual coupon payments. We incorporate the empirical stochastic properties of the spot rate and the credit spread processes by employing a truncated stable Paretian model and a copula function that models their tail dependence, enabling us to provide a more realistic valuation framework.

Section 2 provides a brief literature review on credit risk modeling. Section 3 discusses the credit risk methodology we use to value the defaultable reference asset. The dynamic models for the spot and intensity rate processes are discussed in Sections 4 and 5. The stable Paretian model as the alternative to the Gaussian model is discussed in Section 6 and in Section 7 we describe copulas used to model dependence between the two non-Gaussian marginal processes. In Section 8 we explain the computational procedures used to implement our valuation framework and in Section 9 we describe the option's payoff on the trinomial tree. We provide the numerical results in Section 10 and Section 11 concludes.

2.2   Literature review of credit risk models

Credit risk models address two sources of uncertainty associated with a defaultable asset. They are the probability of default and the loss rate given default. The two broad classes of models in the academic literature[5] that address these two sources of risks are called structural and reduced form models. Structural models derive the probability of default and the recovery given default endogenously by explicitly modeling the underlying assets of the firm that issues the defaultable asset. However, reduced form models treat the default process as exogenous and models it as a point process with a stochastic intensity or hazard rate governing the probability of default where the recovery rate of the defaultable asset is also exogenous.

2.2.1   Structural models

Structural models, also called firm value models, pioneered by the work of Black and Scholes [1973] and Merton [1974], was the first approach to valuing defaultable assets in modern continuous-time finance. In this class of contingent claim models the value of a firm's liability depends on the stochastic evolution of the firm's asset value or return process, the dynamics of the interest rate, payouts to claimants (i.e. dividends, coupons) and the specifics of the firm that relates to reorganization of the firm in the case of bankruptcy. This option theoretic approach views the defaultable asset as a contingent claim depending on the market value of a firm's assets. The stock of the firm is a contingent claim on the market value of the firm's assets, specifically, it is a call option on the firm's asset value with a strike price equal to the face value of the bond outstanding, with the option maturity equal to the maturity of the defaultable debt. The probability of default and the recovery given default is derived by modeling the firms's assets relative to its liabilities. The firm defaults when its asset's value is insufficient to cover the payments owed to bond holders.

Merton [1974] provides a simple model of the firm's capital structure that yields useful economic insights into defaultable asset valuation and the model is easily amenable to extensions that give a more realistic valuation methodology. In Merton's model a firm issues a single zero-coupon bond with a promised payoff F, equal to

---

[5] For a survey on the literature of credit risk models see Cooper and Martin [1996], Cossin [1997], Lando [1998] and Schönbucher [1998]. Crouhy, Galai and Mark [2000] provides an excellent exposition on credit risk methologies used by financial institutions.

the face value of the bond that matures in time T. From Black and Scholes [1973], when equity holders issue debt they sell the firm's assets to the bond holders, keeping a call option to buy them back. Equivalently, the equity holders own the firm's assets and buy a put option on the firm's assets. At maturity T of the zero-coupon bond, if the value of the firm's assets is greater than the amount owed to the bond holders, F, the equity holders pay off the bond holders and retain ownership of the firm's assets. However, if the value of the firm's assets is less than F, then equity holders default on their bond obligations. The model assumes no costs of default and that the absolute priority rule is obeyed, allowing bond holders to take control of the firm's assets, leaving equity with no value assuming limited liability of equity. In Merton's [1974] model framework, the value of the defaultable zero-coupon bond, $\bar{P}(t, T)$, can be shown to be given by

$$\bar{P}(t, T) = FP(t, T) - P_{put}[V(t)]$$

where $P(t, T)$ is the value of a default-free zero-coupon bond at time t, $V(t)$ is the time-t value of the firm's assets and $P_{put}[V(t)]$ is the value of a European put option on the assets of the firm with strike price F and maturity T. The following assumptions were made to obtain an explicit valuation formula for the defaultable bond. In addition to the standard assumptions of perfect capital markets, the other assumptions are: the riskless interest rate is constant over time, the firm pays no dividends over the life of the debt and the asset's value process is lognormally distributed.

The benefit of using the Merton [1974] model to price defaultable bonds is that the bond's value is derived directly from the economic fundamentals of the firm, the value of the firm's assets. Default is the economic mechanism of change of ownership of the firm's assets when the value of the firm's assets is insufficient to pay back the current value of the debt outstanding. This modeling approach is useful for examining the relative powers of equity holders and bond holders and optimal capital structure design. The structural approach can incorporate correlations between firms' asset values, useful for analyzing portfolios of defaultable bonds. CreditMetrics, JP Morgan's credit risk management software, uses precisely the correlation between firms' asset values proxied by the firms' equity correlations to compute a credit value at risk of a portfolio of defaultable bonds.

Merton's [1974] model has practical limitations and its oversimplified assumptions make it inadequate to value complicated debt structures. Merton's option valuation formulae are derived on the basis that the asset value's evolution can be described by a geometric Brownian motion diffusion process. In order to define the diffusion process we need estimates of the parameters that characterize the value process, obtained from historical market values of the firm's assets. This is problematic given that the value of a firm is not a tradeable asset and not easily observable. A firm typically has complex debt contracts that are relatively illiquid and therefore make it difficult to obtain a market price for the firm value, much less the asset's return process. Additionally, the volatility of the asset's return process would need to be estimated, which is harder given that we do not have observations on the firms' asset's return process. In Merton's framework, it is necessary to value all the different classes of liabilities senior to debt prior to valuing debt, which is challenging from a computational standpoint, given that in practice most firms have complicated liability structures. Most firms have many classes of equity and debt in their capital structure. Debt products can have callability and convertibility features in their indentures that can complicate pricing. Furthermore, in this framework default occurs only at the time of the principal repayment, however in practice default can be triggered at any time since other liabilities that are not explicitly modeled can bring about bankruptcy.

The diffusion models of Black and Scholes [1973] and Merton [1974] have not been very satisfactory in explaining empirically observed credit spreads. Jones, Mason and Rosenfeld [1984] find corporate bond credit spreads are much higher than implied by the diffusion models of the structural approach. This is because the firm's diffusion value process is a continuous process where a sudden drop in firm value is not permissible, a firm never defaults unexpectedly, i.e., by surprise. So, if a firm does not default unexpectedly and is not currently in financial distress, then the probability that it will default in a short period of time is zero, implying short term credit spreads should be equal to zero. Furthermore, the diffusion approach in structural models imply that the term structure of credit spreads should always start at zero and slope upwards for firms that are not currently in financial distress. However, in the empirical literature, Fons [1994] and Sarig and Warga [1989] document empirical credit spreads that are sometimes flat and downward sloping. A reasonably parameterized firm's value diffusion process does an inadequate job explaining the nature of short term credit spreads. There are many extensions to the classical firm-value approach of Merton in the literature, but the problems associated with observing the firm's value remain, which led to the development of the reduced form model.

2.2.2 Reduced Form models

Reduced form models, also called intensity rate or hazard rate models, attempt to avoid the practical difficulties associated with structural form models. Motivated by the idea that the default event arrives unexpectedly, the reduced form approach models exogenously the infinitesimal likelihood of default and the model is calibrated to market data. In other words, it is possible from market credit spreads to deduce the probabilities of default of bond issuers. While structural form models explicitly model the dynamics of a firm's value process, to deduce the probability of default and the recovery given default, reduced form models implicitly reference this process. Instead default is treated as an exogenous process and is modeled with a Poisson process governed by a stochastic intensity rate process. The default time is a totally inaccessible stopping time with a stochastic intensity rate and recovery is exogenous as well.

The default event is viewed as an unpredictable event where the probability of arrival of this event is given by the instantaneous intensity rate representing the frequency of defaults that can occur in an instantaneous time interval. The analytical tractability and the ease with which this model can be calibrated to market data makes it a popular approach to credit derivative valuation. In the academic literature papers by Artzner and Delbaen [1992], Jarrow and Turnbull [1995], Lando [1994;1998],

Jarrow, Lando and Turnball [1997], Madan and Unal [1998], Flesaker et al. [1994], Duffie and Singleton [1997;1999], Duffie, Schroder and Skiadas [1996], Duffie and Huang [1996] and Duffie [1994] among others represent the reduced form approach.

In the reduced form framework, default is modeled as a point process. Over a small time interval $[t, t + \Delta t]$, the probability of default conditional on no default before time t is approximately equal to $h(t)\Delta t$ where $h(t)$ is called the intensity(hazard) rate function. The intensity function $h(t)$ can be deterministic or an arbitrary stochastic process. From the term structure of credit spreads for each rating class, it is possible to infer the expected loss of a defaultable bond over a small time interval $[t, t + \Delta t]$, which turns out to be the product of the intensity rate $h(t)$ and the loss rate $L(t)$ where the expectation is under the equivalent martingale measure, also called the "risk-neutral" measure. The market's assessment of the default process implicit in market credit spreads is used to extract relevant and important information to value credit derivatives. Lando [1998] models the default counting process with a Cox process, a process where the intensity function can in general be stochastic depending on a vector of state variables. Intuitively, a Cox process is a process which when conditioned on the realization of state variables behaves as an inhomogeneous Poisson process. The intensity function is ad hoc, i.e., the function can depend on state variables to model systematic risk and can also depend on idiosyncratic factors to model firm specific risk.

Consider a defaultable contingent claim that pays a random value X at time T, conditional on no default at that time. Duffie and Singleton [1999] derive an intuitive result in a recovery framework where upon default, the defaultable bond's value is some fraction of its value just prior to default. They show time-t value of the defaultable claim as

$$V_t = E_t^Q \left[e^{\left(-\int_t^T \bar{r}(s)ds\right)} X \right] \tag{2.1}$$

where $\bar{r}(t) = r(t) + h(t)L(t)$. The loss rate function $L(t) = 1 - \delta(t)$, where $\delta(t)$ is the recovery rate function. The term $h(t)L(t)$ is called the "risk-neutral mean loss rate", attributable to default. Thus, when valuing defaultable claims, future cash flows are discounted by a default-adjusted spot rate $\bar{r}(t)$ and under certain technical conditions the term structure methodology developed for default-free assets are directly applicable to defaultable assets.

Any model of recovery for defaultable assets has a trade-off between analytic tractability and practical applicability. This is because it may be very difficult to accurately model the bankruptcy recovery process which involves a complex and long drawn out process of litigation and negotiation. In the academic literature there are three main recovery models. They are called the recovery of market value (RMV), recovery of face value and accrued interest (RFV) and recovery of treasury value (RT).

RMV: $\varphi(t) = (1 - L(t))\bar{P}(t-, T)$, in this recovery model the creditor receives an exogenous and possible random fraction of the pre-default value of the defaultable bond, $\bar{P}(t-, T)$.

RFV: $\varphi(t) = (1 - L(t))$, the creditor recovers $\delta(t) = 1 - L(t)$, in general a random fraction of face value ($1) of the bond immediately upon default.

RT: $\varphi(t) = (1 - L(t))P(t, T)$, where $L(t)$ is an exogenously specified and possible random fractional loss process and $P(t, T)$ is the price at time t of an otherwise equivalent default-free bond.[6]

Duffie and Singleton [1999] discuss the relative merits of each recovery specification. We employ the RMV formulation, which allows a simple application of a default-free term structure model to compute the value of a defaultable discount bond which is used as a reference asset for the credit spread put option.

2.2.3   Influence of macro-economic variables on credit spreads

In the empirical literature on credit risk, Duffee [1998] shows that credit spread changes on an aggregate bond index is negatively correlated with changes in default-free spot interest rates and a term spread. For different maturities, Duffee [1998] demonstrates lower rated bonds have credit spread changes that are more negatively related to changes in the spot rate and a term spread than higher rated bonds. Duffee's results are for bonds that do not have a call provision or sinking fund feature. Das and Tufano [1996] demonstrate a similar result. Alessandrini [1999] finds long-term interest rates are mainly responsible for business cycle effects on credit spreads. Longstaff and Schwartz [1995a,b] find that changes in credit spreads are negatively related to changes in 30-year treasury yields for investment grade corporate bonds with Moody's rating Aaa, Aa, A and Baa industrials. However, their data set includes bonds with embedded options which as Duffee [1998] points out can significantly impact the estimated coefficients.

Shane [1994] finds high yield bonds have a higher correlation with the return on an equity index compared to low yield bonds, but have lower correlation with the return on a Treasury bond index. Wilson [1997a,b] estimates default rates using macroeconomic variables as explanatory variables. He reports a high R-squared statistic for variables like GDP growth rate, unemployment rate, long-term interest rates, foreign exchange rates and the aggregate saving rate. In Altman [1983, 1990], first order differences of percentage changes in real GNP, money supply, Standard & Poor index and new business formation are used to explain aggregate default rates. He finds that changes in the aggregate number of business failures is negatively related to these macro-economic variables. All of these empirical studies show macro-economic variables clearly influence credit spreads and therefore our modeling framework attempts to parametrize this dependence, which has implications for our credit spread put option pricing model.

---

[6] See Jarrow and Turnbull [1995] for this recovery specification.

2.3   Credit risk model framework

In the reduced form model we employ, the default process is driven by a Cox process. We assume an arbitrage-free and complete frictionless financial market. In this market, default-free and defaultable bonds trade and there exits a default-free money market account. A filtered probability space $(\Omega, \mathcal{F}, \mathcal{Q})$ characterizes uncertainty, where Ω denotes the state space, $\mathcal{F}$ is a σ-algebra of measurable events and $\mathcal{Q}$ denotes the risk-neutral measure. The information structure is given by the filtration $\mathcal{F}(t)$. In this economy all securities are priced in terms of the default-free spot interest rate process. The default-free spot interest rate is a non-negative, bounded, $\mathcal{F}(t)$ adapted process, $r(t)$, which defines the money market account, where the value of the account at time t is given as

$$B(t) = e^{\int_0^t r(s)ds}$$

2.3.1   Default Time Model

In this reduced form intensity-based framework, the default time corresponds to the first jump of a Poisson process N which is characterized by a general intensity process, denoted as $h(t)$. For a constant intensity rate, the default time $\tau = T_1$ is an exponential random variable with a parameter h and the probability of default in the time interval $[0, T]$ is given by

$$F(T) = \Pr[\tau \leq T]$$
$$= 1 - e^{hT}.$$

The intensity or hazard rate $h(t)$ at time t, is the conditional probability of default, given no prior default,

$$\Pr[\tau \in (t + dt) \mid \tau > t] = h(t)dt + o(dt).$$

We assume that the default counting process $N(t)$

$$N(t) = \sum_{i=1}^{\infty} 1_{\{\tau_i \leq t\}}$$

is a Cox process with the intensity process $h(t)$. Here τ_i is the time of the i-th default. The increasing sequence of defaults are modeled by the Cox process. Recall that $N = (N(t), t \geq 0)$ is called a Cox process, if there is a non-negative $\mathcal{F}_t$- adapted stochastic process $h(t)$, the intensity of the Cox process, with $\int_0^t h(s)ds < \infty$, $\forall t$, and conditional on the realization $\{h(t)\}_{\{t>0\}}$ of the intensity, $N(t)$ is a time-inhomogeneous Poisson process with intensity $h(t)$. The events of default occur at the times when N jumps in value. For an inhomogeneous Poisson process, given the realization of h, the probability of having exactly n jumps is

$$\Pr[N(T) - N(t) = n \mid \{h(s)\}_{\{T \geq s \geq t\}}] = \frac{1}{n!} \left(\int_t^T h(s)ds \right)^n e^{\left\{ -\int_t^T h(s)ds \right\}}.$$

In a Cox process the intensity of the process is independent of previous defaults and therefore the time of default becomes a totally inaccessible stopping time. When the intensity rate is formulated as a stochastic process, it allows for rich dynamics of the credit spread process and is flexible to capture the empirically observed stochastic credit spreads.

2.3.2 Fractional Recovery and Multiple Default Model

For the recovery process specification we use the Duffie and Singleton [1999] recovery of market value (RMV) model, where bond holders in the event of default recover a fraction of the pre-default discount bond price. With this assumption the value of a defaulted discount bond is

$$\bar{P}(t, T) = \delta(t) \bar{P}(t^-, T)$$

where $\delta(t) = 1 - \bar{L}(t)$ is the exogenous fractional recovery rate process, $\bar{P}(t^-, T)$ represents the price of the defaultable discount bond just prior to default and $\bar{P}(t, T)$ denotes the value of the defaultable discount bond at default.

In practice, a default event does not terminate the debt contract, firms instead reorganize and re-float their debt.[7] Bond holders can recover a positive value either in the form of cash or in terms of newly floated debt. This framework allows for subsequent defaults and hence multiple defaults are possible with the debt restructuring at each default event. This is a more realistic assumption, since in practice firms reorganize their debt instead of liquidating each time a default-triggering event occurs. Liquidating a firm's assets may be a very costly alternative or may not be an option at all, sovereign debt is an example of an impossible liquidation. Therefore, the RMV model is extended to allow for multiple defaults following Schönbucher [1996, 1998, 1999]. In this framework the defaulted security is worth a fraction of its pre-default value, however it continues to trade at a fraction of its pre-default price.

The increasing sequence of default times $\{\tau_i\}_{i \in N}$ is driven by a Cox process. At each default τ_i the defaultable discount bond's face value is reduced by a factor l_i, the loss rate of which can be a random variable. A defaultable discount bond promising to pay \$1 at maturity T has a final payoff at time T under the RMV formulation and multiple default given by

$$R(T) := \prod_{\tau_i \leq T} (1 - l(\tau_i)) \tag{2.2}$$

where $R(T)$ is a product of the face value reductions after all defaults until maturity T. We assume the loss rate associated with the defaultable bond is a constant,

---

[7] See Franks and Torous [1994].

$l = 50\%$. In this paper we have made the assumption that the recovery rate is constant, for example, this constant recovery rate can represent the historical average for bonds of this Baa rating class. Empirically, recovery rates are uncertain and stochastic recovery rates can be incorporated in our framework. Modeling recovery rate uncertainty is important as the recovery rate process fundamentally influences the credit spread of defaultable assets referenced by credit derivatives.[8] However, we choose to keep the recovery rate constant to identify our credit risk model given that the only data we have available are defaultable bond prices.[9] Our exogenous specification of recovery rates is consistent with violation of absolute priority rules. Absolute priority rules are rarely upheld in distressed firms, found empirically in Franks and Torous [1994] among others.

2.3.2.1   Valuing discount bonds

In a financial market that is complete with no arbitrage opportunities, standard arbitrage pricing theory can be applied to value default-free and defaultable discount bonds that trade using the unique equivalent martingale or the risk-neutral measure, Q.

2.3.2.2   Valuing a default-free discount bond

The price of a default-free discount bond is obtained by discounting the bond's future cash flow using the default-free spot rate and then taking an expectation with respect to the martingale measure Q. Let $P(t, T)$ denote the price of a default-free discount bond, then its price under the pricing measure Q is given as

$$P(t, T) = E_t^Q[e^{(-\int_t^T r(u)du)}],$$ (2.3)

where $E_t^Q[\cdot]$ is the risk-neutral conditional expectation, conditional on available information up to time t and $r(t)$ is the instantaneous default-free spot rate at time t.

2.3.2.3   Valuing a defaultable discount bond

Valuing a defaultable discount bond requires a knowledge of the default characteristics of the bond, the default-free term structure of interest rates, the promised payments and the bonds recovery, its payment in the event of default. A firm that issues a discount bond that *promises* to pay the bond's face value ($1) at maturity T has some finite probability of defaulting on the final principal repayment. Let $\bar{P}(t, T)$

---

[8] We refer the interested reader to Bakshi, Madan and Zhang [2001], Altman et al [2002] and Gupton and Stein [2002] and references therein for papers on recovery uncertainty modeling.

[9] See Duffie and Singleton [1999].

denote the price of the defaultable discount bond at time t that promises to pay \$1 at time T. Then the defaultable discount bond's price under the RMV/multiple default specification can be written as

$$\bar{P}(t, T) = E_t[R(t)\bar{P}(\tau-, T)e^{-\int_t^\tau r(u)du} 1_{\{t \leq \tau \leq T\}} + e^{-\int_t^T r(u)du} 1_{\{T \leq \tau\}}] \quad (2.4)$$

where $E_t[\bullet]$ denotes the conditional expectation with respect to risk-neutral measure Q given all the information $\mathcal{F}_t$ until time t, and τ is the time of default. There are two components in the defaultable discount bond's value. The first component is the present value of the promised payment if default occurs and the second component is the present value of the promised payment if default does not occur. Duffie and Singleton [1999] derive an expression for a defaultable discount bond under RMV assumption which is applicable to multiple defaults following Schönbucher [1998,1999]. The value of a defaultable discount bond at time t promising a dollar at time T in the risk-neutral measure is given by

$$\bar{P}(t, T) = 1(\tau > t)E_t^Q \left[e^{(-\int_t^T (r(s)+h(s)L(s))ds)} \right]$$

where the loss function $L(t) \equiv 1-\delta(t)$ and $\delta(t)$ is the recovery rate function and $h(t)$ is the risk-neutral default intensity rate. The indicator function, $1(\tau > t)$, indicates that the default event occurs after time t. The recovery rate process determines the form of the discount bond price which is important for both pricing and estimation. The above defaultable discount price can be expressed as

$$\bar{P}(t, T) = 1(\tau > t)E_t[e^{-\int_t^T \bar{r}(s)ds}] \quad (2.5)$$

where $1(\tau > t)$ is an $\mathcal{F}_t$-adapted process and denotes the survival probability of the bond up to time t.[10] The process $\bar{r}$ is called the default-adjusted spot rate which can be decomposed into a default-free interest rate and a credit spread written as

$$\bar{r}(t) = r(t) + h(t)L(t) \quad (2.6)$$

where $r(t)$ is the instantaneous default-free spot rate, $h(t)$ denotes the intensity rate of default and $L(t)$ is the fractional loss rate each time the bond defaults. The credit spread is just the product of the risk-neutral instantaneous conditional probability of default and the instantaneous loss rate. The valuation of a defaultable bond is similar to the valuation of a default-free discount bond except the spot rate is adjusted for default risk. In this framework, standard term structure models can be used to parameterize the default-free spot rate and the credit spread process to value the defaultable discount bond. It is possible to infer the market's perception of the term structure of default probabilities using Equation (2.5) and observed market prices

---

[10] Duffie and Singleton [1999] discuss the technical conditions when this can be done.

of default-free and defaultable discount bonds. The key components that need to be modeled in the defaultable discount bond valuation are the spot, intensity and recovery rate stochastic dynamics. We model the term structure of defaultable bond prices by parameterizing the spot rate and the intensity rate processes separately using a two-factor model that permits empirically observed dependence between these processes consistent with dependence between market and credit risk.

2.3.3   Valuing a defaultable coupon bond

This section provides an expression for the value of a defaultable coupon bond with the RMV and multiple default specification of Duffie and Singleton [1999] and Schönbucher [1998,1999] respectively. Consider a defaultable coupon bond that promises to make a sequence of coupon payments $\{c_i\}$ at each time $\{t_i\}$, for $i = 1, \ldots, n$ where n is the number of promised payments corresponding to the maturity of the bond and a principal repayment equal to the face value F at time t_n. If $\bar{P}_c(t, T)$ denotes the value of the coupon bond time t maturing at time T conditional upon no default at time t, the value of the defaultable coupon bond in our RMV/multiple default framework is given as

$$\bar{P}_c(t, T) = E_t^Q \left\{ \sum_{i=1}^n c_i e^{\left(-\int_t^{t_i}(r(s)+h(s)L(s))ds\right)} + F e^{\left(-\int_t^{t_n}(r(s)+h(s)L(s))ds\right)} \right\}$$

(2.7)

$$= \sum_i^n c_i \bar{P}(t, t_i) + FV\bar{P}(t, t_n).$$

(2.8)

The defaultable coupon bond comprises a portfolio of defaultable discount bonds with portfolio weights equal to the coupon payments c_i at each payment date t_i and the face value F at time t_n. This expression for a defaultable coupon bond is useful particularly when valuing short term credit derivatives with long term defaultable coupon bonds as reference assets.

2.3.4   Credit spread put option on a defaultable discount bond

Let $P(t, T_2)$ and $\bar{P}(t, T_2)$ be the time-t price of a default-free and defaultable discount bond maturing at time T_2. Let T_1 denote the exercise date of the credit spread put option. The holder of the option has the right to sell the defaultable discount bond at a spread $\bar{s}$ over that of the yield of an equivalent treasury security. If the option is in-the-money at the exercise date, then the option holder has the right to exchange a single defaultable discount bond valued at $\bar{P}(T_1, T_2)$ for $\bar{S}$ default-free bonds which is valued at $P(T_1, T_2)$, where $\bar{S} := e^{-\bar{s}(T_2-T_1)}$ is the exchange ratio. The payoff function

$$\max\left[\bar{S}P(T_1, T_2) - \bar{P}(T_1, T_2), 0\right].$$

(2.9)

TABLE 1. Statistical summary of the 1-year Treasury bill series

Mean	6.17 %
Median	5.6 %
Maximum	15.21 %
Minimum	2. 49 %
Standard Deviation	2.44 %
Skewness	1.100804
Kurtosis	4.239598
Jarque-Bera	2706.148
Probability	0.000000

We assume the option contract does not terminate at default, instead the option holder gets the face value of the bond minus the fractional recovery of the bond's pre-default price. This contract partially protects the buyer of the option from losses on the defaultable discount bond when the spread is higher than $\bar{s}$ and from default, but the buyer of the option contract is still exposed to interest rate risk. The price of the option referenced on a Baa coupon bond is simply a portfolio of options referenced on Baa discount bonds.

2.4 Model for the default-free spot interest rate process

The default-free spot interest rate is a key financial economic variable as it is the only state variable affecting the prices of all other assets in the economy. The spot rate movements influence the Baa credit spread movements which directly affects the option's payoff. Therefore any model for the dynamics of the spot rate needs to capture its stochastic properties as precisely as possible. Table 1 provides a statistical summary of the 1-year Treasury bill series. Figure 3 depicts the evolution of the spot rate time series and Figure 4 plots the first difference of the spot rate series. In Figure 5 we have a histogram of the changes in the spot rate time series. It is clear from these plots and the histogram that the spot rate clearly deviates from Gaussian statistical properties. The change in the spot rate has a kurtosis of 19.63 compared to a kurtosis of 3 for a Gaussian distribution. Furthermore, Figure 4 shows that the spot rate exhibits stochastic volatility.

Despite possessing non-Gaussian statistical properties, as a base case, we look at the spot rate using a Gaussian model. In the Markovian economy we model the spot rate dynamics with a simplified Hull–White [1994a,1994b] no-arbitrage term structure model. This is a mean-reverting, Ornstein–Uhlenbeck process with risk-neutral dynamics described by the stochastic differential equation

$$dr(t) = [\theta(t) - ar(t)] dt + \sigma dW(t), 0 \le t \le T \qquad (2.10)$$

where $r(0) = r_0$, a constant, $\theta \in \mathcal{L}^1$ is a deterministic process, a and σ are constants and $a \neq 0$. The standard Wiener process, $dW(t)$ is under the risk-neutral measure

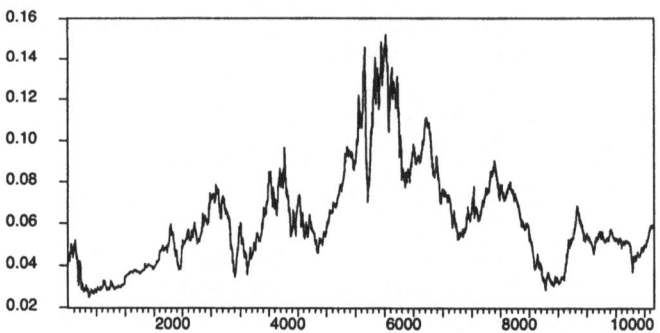

FIGURE 3. The evolution of the 1-year Treasury bill series

and it drives the evolution of the term structure of default-free discount bonds. This continuous spot rate process is a conditionally Gaussian process with a constant initial value. The Hull–White [1994a,1994b] arbitrage-free model provides us with analytic tractability, we get explicit solutions for default-free discount bond prices and the calibration of parameters to observed bond prices is fast and easy to compute. The model is consistent with the current term structure when

$$\theta(t) = f_t(0, t) + af(0, t) + \frac{\sigma^2}{2a}(1 - e^{-2at}), \ t \ge 0 \tag{2.11}$$

where $f(t, T)$ is the forward rate of the default-free discount bonds.

In this exponential affine term structure model the price of the discount bond can now be expressed as

$$P(t; T) = e^{-A(t,T)-B(t,T)r(t)}. \tag{2.12}$$

Equation (2.12) defines the price of a discount bond at time t with maturity T in terms of the spot rate at time t, $r(t)$ with $B(t, T)$ as

$$B(t, T) = \frac{1}{a}(1 - e^{-a(T-t)})$$

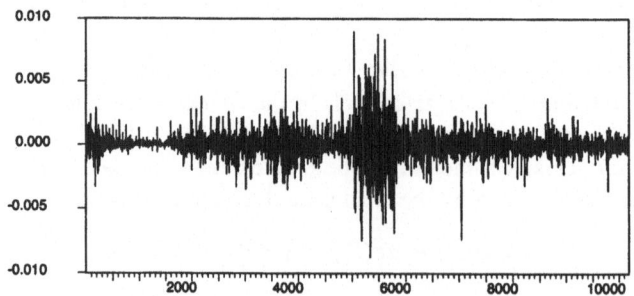

FIGURE 4. Plot of the first difference in the 1-year Treasury bill series

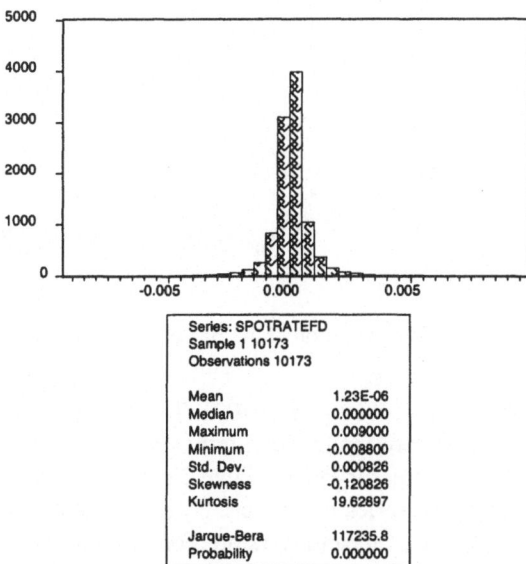

FIGURE 5. Histogram of the first difference in the 1-year Treasury bill series

and $A(t, T)$ as

$$A(t, T) = \frac{1}{2} \int_t^T \sigma^2(s) B(t, s)^2 ds - \int_t^T B(t, s) \theta(s) ds.$$

2.5   Model for the Baa credit spread index

Longstaff and Schwartz [1995] examine the empirical properties of Moody's credit spread indices for investment grade bonds with Aaa and Baa ratings. They find evidence of mean-reversion behavior in the logarithm of the credit spreads. Pedrosa and Roll [1998] propose a Gaussian mixture to model credit spread changes in a study of corporate bond indices that are pooled by rating and sectors. In their study of Moody's Aaa and Baa corporate bond indexes, Prigent, Renault and Scaillet [2001] propose an Ornstein–Uhlenbeck model with jumps to model credit spread dynamics for the investment grade indices. We examine the empirical properties of the Baa credit spread process to define a dynamic model with the dual objective of making our model empirically consistent and analytically tractable.

2.5.1   Data on the Baa credit spread index

A sample of daily yields of Moody's investment grade Baa corporate bond index and 30-year constant maturity treasury bonds were obtained from the Federal Reserve

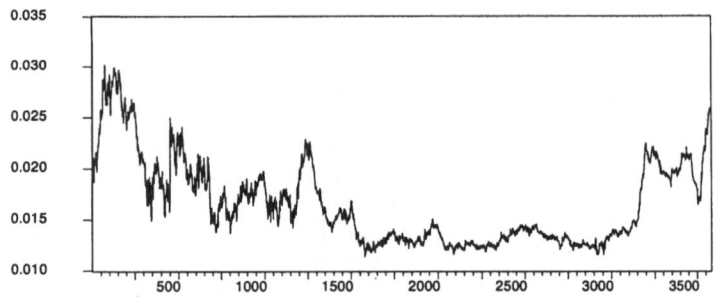

FIGURE 6. The evolution of the Baa credit spread index

Board for the period January 1986 to April 2001. The credit spread of the Baa index is defined as the difference between the long-term (assumed to be 30 year) Baa bond index yield and the 30-year treasury yield. Table 2 provides a summary of the statistics of the Baa credit spread series and Figure 6 depicts the evolution of this series which displays some degree of mean reversion. Formalizing these observations statistically, following Longstaff and Schwartz [1995], we regress daily credit spread changes on the credit spread of the previous day. Let CS_t denote the credit spread at time t; we estimate the regression

$$\Delta CS_{t+1} = b_0 + b_1 CS_t + \varepsilon_t \tag{2.13}$$

where b_0 and b_1 are the intercept and slope coefficient respectively and $\Delta CS_{t+1} = CS_{t+1} - CS_t$. The estimated coefficients are given in Table 3. The slope coefficient is significantly negative suggesting mean reversion behavior in the credit spread time series. A look at Figure 7 that plots the first difference of the time series indicates this process has a random volatility. In Figure 8 the plot of the histogram of first differences in the credit spread times series unambiguously indicates non-Gaussian statistical properties. This time series has an excessive kurtosis of 22 compared to a Gaussian kurtosis of 3. Despite the departure from Gaussian properties, as our base case, we model the Baa credit spread dynamics with a mean-reverting conditional Gaussian process.

2.5.2   Model for the credit spread/intensity rate process

We model the term structure of credit spreads and its dynamics by modeling the intensity of the Cox process, $h(t)$, which is the instantaneous conditional probability of default at time t, given all the available information at this time. The intensity rate process is modeled since the credit spread process is just the intensity process adjusted by a constant fractional loss rate, $l = 0.5$. Consistent with our empirical findings, we use the mean-reverting Hull–White [1994a,1994b] term structure model for the dynamics of the intensity process of the Baa index. Under the risk-neutral measure the stochastic intensity, $h(t)$ has the differential

$$dh(t) = \left[\bar{\theta}(t) - \bar{a}h(t)\right]dt + \bar{\sigma}d\bar{W}(t), \ 0 \leq t \leq T. \tag{2.14}$$

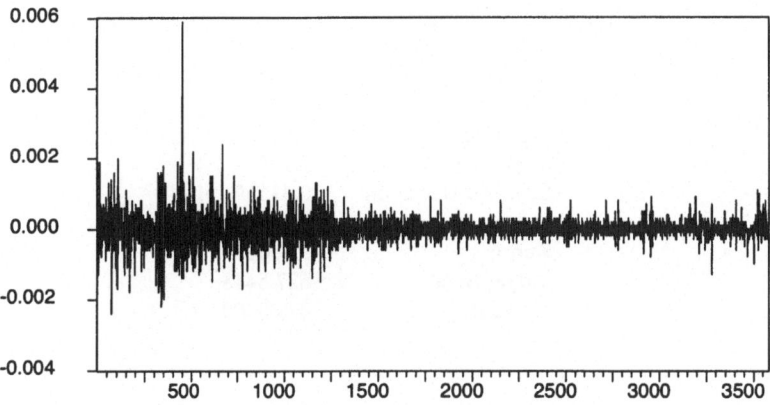

FIGURE 7. Plot of the first difference in the Baa credit spread time series

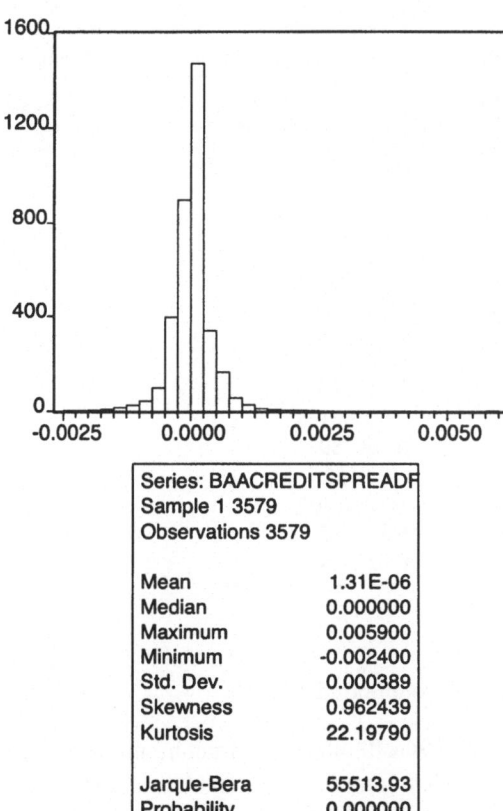

Series: BAACREDITSPREADF	
Sample 1 3579	
Observations 3579	
Mean	1.31E-06
Median	0.000000
Maximum	0.005900
Minimum	-0.002400
Std. Dev.	0.000389
Skewness	0.962439
Kurtosis	22.19790
Jarque-Bera	55513.93
Probability	0.000000

FIGURE 8. Histogram of the first difference in the credit spread time series

TABLE 2. Statistical summary of the Baa credit spread time series

Mean	1.6581%
Median	1.52%
Maximum	3.02%
Minimum	1.14%
Standard Deviation	0.4155%
Skewness	1.010487
Kurtosis	3.350239
Jarque-Bera	627.5448
Probability	0.000000

TABLE 3. Parameters estimation for the regression of daily changes in credit spread on the previous days credit spread

Variable	Coefficient	Standard Error	t-Statistic	Prob
Creditspread(t)	-0.003878	0.001566	-2.476210	0.0133
Constant	0.0000656	0.0000268	2.451077	0.0143

R-squared	0.001711
Adjusted R -squared	0.001432
S.E of regression	0.000389
Sum squared of resid	0.000541
Log likelihood	23024.55
Durbin-Watson stat	2.225314
Mean dependent var	0.00000131
S.D depéndent var	0.000389
Akaike info criterion	-12.86535
Schwarz criterion	-12.86190
F-statistic	6.131617
Prob (F – Statistic)	0.013324

2.5.2.1   Defaultable discount bond formula

From the Hull–White [1994a,1994b] term structure model for the intensity rate process, $h(t)$, we have explicit formulas for defaultable discount bonds. In this exponential affine term structure model the time t price of a defaultable discount bond maturing at time T is

$$\bar{P}(t, T) = P(t, T)e^{[\bar{A}(t,T) - \bar{B}(t,T)h(t)]}$$

(2.15)

where

$$\bar{B}(t, T) = \frac{1}{\bar{a}}\left[1 - e^{(-\bar{a}(T-t))}\right], \tag{2.16}$$

$$\bar{A}(t, T) = \frac{1}{2}\int_t^T \sigma^2(s)\bar{B}(t, s)^2 ds - \int_t^T \bar{B}(t, s)\tilde{\theta}(s)ds \tag{2.17}$$

and

$$\tilde{\theta}(s) = \bar{\theta}(t) + \rho\bar{\sigma}(t)\sigma(t)B(s, T) \tag{2.18}$$

where $\bar{a}$ is the mean reversion rate and $\bar{\sigma}$ is the instantaneous standard deviation of the mean-reverting continuous-time intensity process. $d\bar{W}(t)$ is a standard Wiener process, under the risk-neutral measure, driving the evolution of the term structure of credit spreads.

In the two-factor model for the yield of a defaultable discount bond we allow for dependence between the spot rate and the intensity rate processes, represented by a correlation ρ, between the two Wiener processes, $dW(t)$ and $d\bar{W}(t)$.

2.6 A Stable Paretian model

In theoretical and empirical financial analysis important implications are drawn from the distributional assumptions made about asset return processes. Essential to all derivative valuation models are the assumptions made about which stochastic process drives the underlying risk factors that are being traded. In the classical finance literature it is assumed that asset return processes are Gaussian because Gaussian-based models are both parsimonious and analytically tractable. In the Black–Scholes option pricing model, the underlying asset price process is lognormally distributed. However, in the empirical finance literature it is well known that Gaussian models are limited in their ability to explain anomalies such as leptokurtosis, skewness, stochastic volatility and volatility clustering observed in asset return processes.

Furthermore, the Black–Scholes model cannot explain the "smile effect" observed in market option prices. If the Black–Scholes model were the "true" model, then the implied volatility for the asset return process should be constant according to the model's assumptions. However, in practice the implied volatility resembles a "smile", the volatility is a convex function of the strike price for different expiration dates. In addition to this anomaly it is an empirically observed phenomena that high frequency data exhibits higher kurtosis than low frequency data. For example daily return time series have a higher kurtosis than monthly or yearly time series.

In the empirical finance literature many methods have been employed to modify the Black–Scholes model to account for asymmetric leptokurtic behavior in asset return processes and to account for the "smile" effect. Chaos theory, fractal brownian motion, stable processes, time-changed brownian motion, generalized hyperbolic models, to name a few models, attempt to capture the asymmetric leptokurtic behavior in asset returns. Stochastic volatility models such as GARCH, jump-diffusions, affine stochastic-volatility and affine jump diffusions and models based on Lévy processes try to explain the "smile" volatility shape implicit in empirically observed option prices. Kou [2002] provides a good exposition on the advantages and disadvantages in these various approaches.

Mandelbrot [1963,1967] and Fama [1965a,b] were among the first to observe asset return processes displaying "heavy" tails instead of "light" tails implied by Gaussian distributions. They studied the application of stable non-Gaussian distributions to model financial asset returns as an alternative to Gaussian distributions. Their seminal work has spawned a large body of literature in empirical finance that applies stable processes in finance.[11]

The Hull–White [1994] term structure model we employ as the base case for the spot rate and intensity rate dynamics assumes these processes are conditionally Gaussian. However, an empirical examination of their daily movements clearly indicates significant departure from the Gaussian statistical properties. If the data generating process for daily changes in the two processes were Gaussian, then we would not observe excessive kurtosis of 19.68 and 22.2 for the spot and intensity processes seen in Figures 5 and 8. Additionally, Figures 4 and 7 indicate these processes exhibit stochastic volatility. It is this stochastic volatility that generates the excess kurtosis or heavy-tailedness observed in the unconditional distributions of the spot and intensity rate daily movements.

Defaultable asset returns typically have distributions characterized by positive skewness and values that are negatively skewed with heavy downside tails. This heavy-tailed nature of defaultable asset returns implies a greater probability for large outlier events to occur than thin-tailed distributions would predict. Defaultable bonds typically have returns that are left skewed. The reason that these returns are left skewed is because the probability of a defaultable bond earning a substantial price appreciation is relatively small. However, there is large probability of receiving a small profit through interest rate earnings. The distribution tends to be skewed around a positive value with a very small positive tail reflecting the limited upside potential. Adverse movements in credit quality occur with small probability but can have a significant negative impact on the value of an asset. These credit quality migrations have the ability to generate significant losses, thus producing skewed distributions with large downside tails. Additionally these skewed returns with heavy downside tails are characteristic of portfolios of defaultable bonds as well.

We propose the use of a stable Paretian model for several reasons. The stable Paretian model possesses the desirable statistical properties of heavy tails, high peaks, skewness, stochastic volatility and volatility clustering.[12] This model can flexibly capture the asymmetric leptokurtic behavior observed in empirical asset returns

---

[11] See Rachev and Mittnik [2000] and the references therein.

[12] In the empirical finance literature Rachev and Mittnik [2000] model financial assets returns which exhibit properties such as heavy-tails, asymmetries, volatility clustering, temporal dependence of tail behavior and short and long range dependence.

and the "smile" effect in option prices with a parsimonious set of parameters. The stable Paretian model is a generalization of the Gaussian distribution that possesses a property very desirable in portfolio choice theory. Stable distributions are the only limiting distribution of properly normalized and centered partial sum processes for i.i.d. random variables X_i.[13] It is stable with respect to the operation of addition and is in the class of infinitely divisible distributions. A distribution in the domain of attraction of a stable distribution will have properties close to that of a stable distribution.

A stable non-Gaussian distribution has an infinite second moment with power decaying tails. However, tail truncation guarantees the existence of all moments and specifically exponential moments. We employ a truncated stable law for the spot and intensity processes ensuring bounded second moments but with a truncation deep in the tails of the distribution. The truncated stable Paretian model allows us to use the explicit bond price formulas for default-free and defaultable discount bonds as in the Gaussian model and also captures extreme co-movements in the spot and intensity processes,unachievable with the Gaussian model. The truncated stable law satisfies the Lévy pre-central limit theorems discussed in Rachev and Mittnik [2000].[14]

The truncated stable Paretian model is a stochastic volatility model which is responsible for generating the excessive kurtosis observed in spot and intensity rate's unconditional distributions. The stochastic volatility model implies markets are incomplete with no unique equivalent martingale measure. In an incomplete market it may not be possible to determine option prices from riskless hedging arguments per se as there may not be a dynamic self-financing strategy of replicating assets that can reproduce an option's payoffs almost surely. Intuitively, when the volatility of an underlying asset's return process is stochastic, it introduces another source of randomness that cannot be completely hedged unless the volatility is the price of a traded asset that is perfectly correlated to one of the replicating assets used to construct the self-financing strategy. An additional feature of our truncated stable Paretian model is its capability of generating more pronounced leptokurtic behavior with higher frequency data which is consistent with empirical observations. GARCH models following Bollerslev [1986] are competing stochastic volatility models applied to option pricing, however we feel the truncated stable model is appropriate since there are many financial time series that exhibit heavy tails even after examining their GARCH filtered residuals and this is well documented in Rachev and Mittnik [2000].

A paper by Pedrosa and Roll [1998] point out that excessive kurtosis observed in the unconditional distributions of credit spread index changes can be generated by various mixtures of Gaussian distributions. A priori, it is reasonable to assume the spot rate and Baa credit spread processes have increments that are conditionally Gaussian. This is because the independent actions of many market participants influence the price of the 1-year Treasury bill and the value of the Baa index. Since there are many market participants, the central limit theorem dictates the limiting distri-

---

[13] See Embrechts et al. [1997] and Rachev and Mittnik [2000].

[14] See Rachev and Mittnik [2000], pages 77–81.

bution for sums of independent random increments with a finite variance to be a symmetric bell-shaped Gaussian distribution. However, our view is that the spot rate and the Baa credit spread processes are attracted by a truncated stable law, namely the treasury bill and Baa index values result from the sum of independent random increments in their prices, but these increments have very large variances with a stable limiting distribution. We posit the stable distribution is truncated at ten standard deviations from the mean of the corresponding base case Gaussian distribution. Therefore, non-Gaussian stable processes represent an important class of stochastic processes that are used to model relatively infrequent extreme events combined with frequent events that have returns close to the mean. The Gaussian model is inadequate for modeling financial market crashes and upturns because these extreme market conditions are ignored by Gaussian models.

Figures 9 and 10 show heavy tails in empirical distributions for the spot and intensity rate daily movements, indicating that extreme values occur with relatively higher probabilities than Gaussian distributions. The empirical densities are obtained via kernel density estimation (dashed lines), together with the fitted Gaussian distribution and fitted α-stable distributions for daily changes in the spot and intensity rates respectively. The plots show the α-stable model provides a much closer approximation to the empirical distributions; the number of extreme observations in the spot and intensity rates are too many relative to what would be observed if these empirical distributions were Gaussian.

The degree to which a stable distribution is heavy tailed depends on a parameter called the index of stability. As the index of stability, α, becomes smaller than 2 ($\alpha = 2$, represents a Gaussian distribution) the distribution gets more peaked around its

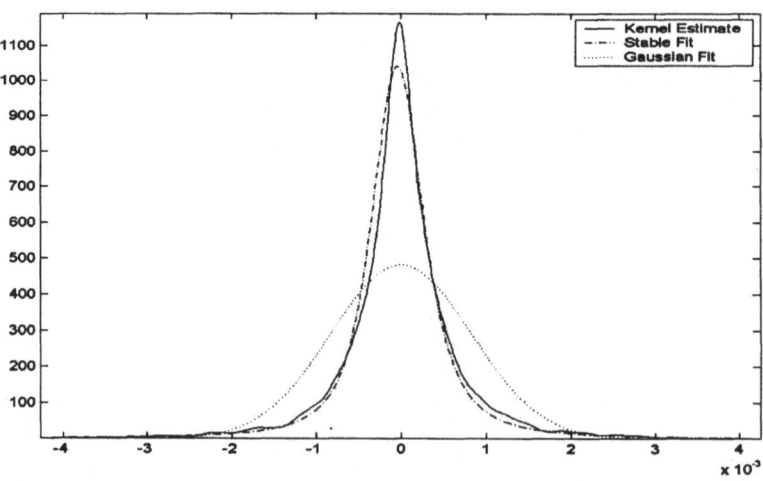

FIGURE 9. Comparison of the Gaussian and stable non-Gaussian fit for the residuals of the daily changes in the 1-year Treasury bill return

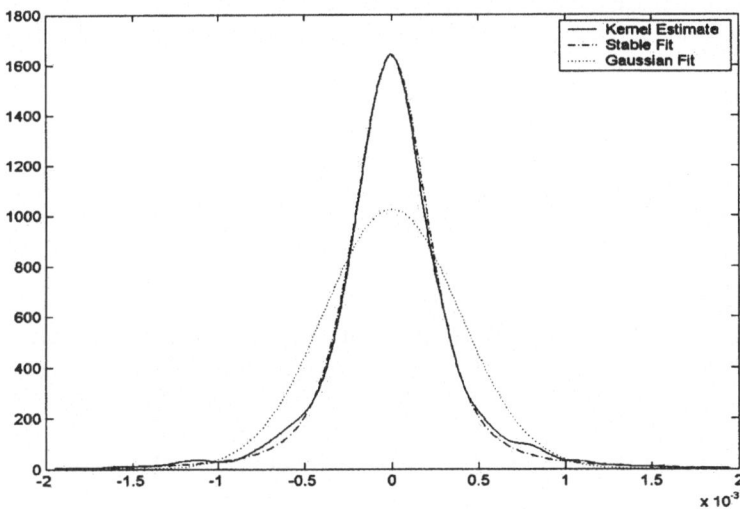

FIGURE 10. Comparison of the Gaussian and stable non-Gaussian fit for the residuals of the daily changes in Baa credit spread time series

mean and heavy tailed. High frequency financial time series have indices smaller than 2 while low frequency time series have indices close to 2, consistent with empirical findings. Determining whether a distribution is in the domain of attraction of a stable distribution requires an examination of the tails of the distribution, which influences the properties of the distribution.

We model the spot interest rate and the intensity rate with symmetric α-stable distributions, denoted by $S\alpha S$. In the following sections we briefly discuss the properties of stable non-Gaussian distributions, the estimation method used to obtain stable parameters and the properties of symmetric α-stable random variables.

2.6.1 Description of Stable Distributions

Stable distributions are characterized by the equality in distribution between a random variable and the normalized sum of any number of i.i.d. replicas of the same variable. The Gaussian distribution is a special case of the stable distribution, it is the only stable distribution with finite second moments, with an index of stability, α, equal to 2.

Definition 1. A random variable X (or a distribution function, F_X) is called stable if it satisfies

$$c_1 X_1 + c_2 X_2 \stackrel{d}{=} b(c_1, c_2)X + a(c_1, c_2) \tag{2.19}$$

for all non-negative numbers c_1, c_2 and appropriate real numbers $b(c_1, c_2) > 0$ and $a(c_1, c_2)$, where X_1 and X_2 are independent random variables, $X_i \stackrel{d}{=} X_1, i = 1, 2$.

$\stackrel{d}{=}$ denotes equality in distribution. Consider the sum S_n of i.i.d. stable random variables. From Equation (2.19) we have for some real constants a_n and $b_n > 0$ and $X = X_1$,

$$S_n = X_1 + \cdots + X_n \stackrel{d}{=} b_n X + a_n, \; n \geq 1 \qquad (2.20)$$

which can be rewritten as

$$b_n^{-1}(S_n - a_n) \stackrel{d}{=} X. \qquad (2.21)$$

Then it can be concluded that if a distribution is stable, it is the unique limit distribution for sums of i.i.d. random variables. In general the stable distribution has no explicit form for the distribution function but can be expressed with its characteristic function which is

$$\Phi_X(t) = \exp\left\{-\sigma^\alpha \mid t \mid^\alpha (1 - i\beta sign(t) \tan\frac{\pi\alpha}{2}) + i\delta t\right\}, \; \text{if } \alpha \neq 1, \quad (2.22)$$

$$\Phi_X(t) = \exp\left\{-\sigma \mid t \mid (1 - i\beta\frac{2}{\pi}sign(t)\ln t) + i\delta t\right\} \; \text{if } \alpha = 1.$$

The characteristic function is described by the following parameters: $\alpha \in (0, 2]$, called the index of stability, $\beta \in [-1, 1]$, the skewness parameter, $\delta \in \Re$, is the location parameter and $\sigma \in [0, \infty)$, is the scale parameter. A stable random variable is represented as $X \sim S_{\alpha,\beta}(\delta, \sigma)$. The Gaussian distribution is a special case of the stable distribution with $\alpha = 2$. As this stability index gets smaller the distribution becomes more leptokurtic, i.e., with a higher peak and a fatter tail. If $\beta = 0$ the distribution is symmetric like the Gaussian. When $\beta > 0(\beta < 0)$ the distribution is skewed to the right(left). The distribution is called symmetric α-stable when $\delta = 0$ and $\beta = 0$. The scale parameter generalizes the definition of standard deviation. The stable analog of variance is variation, denoted by σ^α. The pth absolute moment of X , $E|X|^p = \int_0^\infty P(|X| > y)dy$, is finite if $0 < p < \alpha$ and infinite otherwise. Hence when $\alpha \leq 1$ the first moment is infinite and when $\alpha < 2$ the second moment is infinite. The Gaussian distribution is the only stable distribution that has finite second moments.

2.6.1.1   Maximum Likelihood Estimation

Mittnik et al. [1996] use an approximate conditional maximum likelihood (ML) procedure to estimate the parameters of the stable distribution. The unconditional (ML) estimate $\Theta = (\alpha, \beta, \mu, \sigma)$ is obtained by maximizing the logarithm of the likelihood function

$$L(\Theta) = \prod_{t=1}^{T} S_{\alpha,\beta}\left(\frac{r_t - \mu}{\sigma}\right)\sigma^{-1}. \qquad (2.23)$$

The estimation is similar to DuMouchel [1973a,b] but differs in that the stable density function, is numerically approximated using Fast Fourier Transforms (FFT) of the stable characteristic function in equation (2.22). Since the stable density function is approximated, the estimated stable parameters are approximate. For further details on stable maximum likelihood estimation see Mittnik et al. [1996], Paulauskas and Rachev [1999] and Rachev and Mittnik [2000]. The stable non-Gaussian parameters estimated for the daily changes in the 1-year treasury bills return series and the intensity rate series of the Baa index are presented in Table 4.

TABLE 4. Gaussian and stable non-Gaussian parameters for 1-year Treasury bill returns and the credit spread of the Baa–rated bond index

	Gaussian parameters		Stable Non-Gaussian parameters			
	mu	sigma	alpha	beta	sigma	delta
1-year Treasury bill returns	0.00	0.00083	1.3145	0.0591	0.000306	0.00
Credit spread on Baa bond index	0.00	0.00078	1.6070	0.00	0.003	0.00

A symmetric α-stable random variable Z that we construct belongs to the sub-Gaussian class where $Z = A^{1/2}G$, G is a zero-mean Gaussian random variable, i.e., $G \sim N(0, \sigma)$ and A is an $\alpha/2$-stable random variable totally skewed to the right and independent of G. A has the distribution $A \sim S_{\alpha/2}(1, 0, \sigma)$; A is called the $\alpha/2$ stable subordinator. A symmetric α-stable random variable can be interpreted as a random rescaling transformation of a Gaussian random variable. The random rescaling variable is the stable subordinator, A. The stable random variable can be defined for a two-dimensional case[15] where the two stable random variables in our analysis are the default-free spot interest rate and the default intensity rate. A two-dimensional vector $X = (X_1, X_2)$ is defined by

$$X = (A_1^{1/2}G_1, A_2^{1/2}G_2). \tag{2.24}$$

The vector X is called a sub-Gaussian[16] $S\alpha S$ vector in $\Re^2$ under the Gaussian vector $G = (G_1, G_2)$.

---

[15] In general it can be extended to the d-dimenisonal case.

[16] A_i are α_i-stable subordinators, $i = 1, 2, 1 < \alpha_i < 2$, and independent of the Gaussian vector $G = (G_1, G_2)$. If $A_i = A, i = 1, 2$, then X is sub-Gaussian. In general $X = (X_1, X_2)$, $X_i = A_i^{1/2}G_i$, is only a bivariate infinitely divisible vector. In fact $X = (X_1, X_2)$ is (α_1, α_2)-stable. See Rachev and Mittnik [2000], page 729. The class of infinitely divisible random variables is larger than the class of stable random variables.

2.6.1.2   Stable Non-Gaussian Dependence

Dependence between stable random variables when $\alpha = 2$ is defined by a covariance function or a correlation function. However when $1 < \alpha < 2$, the covariance is replaced by a covariation. We first introduce the signed power function before defining covariation between joint $S\alpha S$ random variables. The signed power $a^{\langle p \rangle}$ is defined as

$$a^{\langle p \rangle} = \left\{ \begin{array}{l} a^p \text{ if } a \geq 0 \\ -|a|^p \text{ if } a < 0 \end{array} \right\}. \tag{2.25}$$

We define covariation as follows.[17] Let X_1 and X_2 be jointly $S\alpha S$ with $\alpha > 1$ and let Γ be the spectral measure of random vector (X_1, X_2). The covariation of X_1 on X_2 is the real number

$$[X_1, X_2]_\alpha = \int_{S_2} s_1 s_2^{\langle \alpha - 1 \rangle} \Gamma(ds). \tag{2.26}$$

Let $(G_1, \dots, G_n)$ denote zero-mean jointly Gaussian random variables with covariance $R_{ij} = EG_i G_j$, $i, j = 1, \dots, n$ and let $A \sim S_{\alpha/2}((\cos \frac{\pi\alpha}{4})^{2/\alpha}, 1, 0)$ be independent of $(G_1, \dots, G_n)$. Then the sub-Gaussian random vector $X = (X_1, \dots, X_n)$ where $X_k = A^{1/2} G_k$, $k = 1, \dots, n$, $1 < \alpha < 2$ has the covariation,

$$[X_i, X_j]_\alpha = 2^{-\alpha/2} R_{ij} R_{jj}^{(\alpha-2)/2}. \tag{2.27}$$

If $R_{ii} = R_{jj}$, then $[X_i, X_j]_\alpha = [X_j, X_i]_\alpha$. However the covariation is not symmetric in its arguments (in contrast to the Gaussian covariance). The variation function only exists for Gaussian random variables. It is replaced by variation for stable random variables where $1 < \alpha < 2$. The variation of a symmetric α-stable random variable X is given as

$$Var(X) = [X, X]_\alpha. \tag{2.28}$$

As the covariance of stable distributed random variables with $\alpha < 2$ is infinite, the sub-Gaussian $S\alpha S$ vector can be employed to incorporate the Gaussian dependence structure among different stable random variables. The Gaussian dependence is easy to compute and can easily be transferred to the sub-Gaussian case.

We generate independent and dependent $S\alpha S$ random variables that use truncated subordinators. Generating two-dimensional random vector X is performed by simulating a two-dimensional Gaussian random vector G with correlated elements G_1 and G_2 and a stable subordinator vector $A = (A_1, A_2)$ that is independent of $G = (G_1, G_2)$.[18]

---

[17] See Samorodnitsky and Taqqu [1994].

[18] See Rachev, Schwartz and Khindanova [2000] for more details.

2.6.2   Subordinated asset price model

The benchmark paradigm in option pricing assumes asset return processes are driven by Wiener processes, which are continuous functions of physical time. However, more general asset return dynamics can be specified. For example, in models with subordinators, asset returns have dynamics driven by Wiener processes that are subordinated to an intrinsic time process. The physical time directing the Wiener process is replaced by an intrinsic time process $\{T(t), t \geq 0\}$. In fact, the intrinsic time process is a stochastic rescaling transformation of the physical time scale. The process $\{T(t), t \geq 0\}$ is a non-decreasing stochastic directing process with stationary independent increments. Let $S(t)$ be the price of an asset at time t. In the general subordinated asset price model the price of the asset at time $t + s$ is given by

$$S(t + s) = S(t)e^{\{\mu s + \sigma[W(T(t+s)) - W(T(t))]\}}, \; t, s \geq 0 \qquad (2.29)$$

where $W(T(t))$ is a Wiener process which is subordinated to the intrinsic time process $T(t)$.[19] Transforming the process by taking the natural logarithm of the above equation leads to a process that can be written as

$$\Delta L(t, s) = L(t + s) - L(t) = \mu s + \sigma[W(T(t + s)) - W(T(t))].$$

This intrinsic time $T(t)$, represents *trading time*, which can be interpreted as volume of trades executed at time t.[20] This intrinsic time process is the random variable responsible for generating heavy-tails observed in empirical distributions of asset returns. When the time scale is stochastic, modeled with a stochastic intrinsic time process, $T(t)$, the subordinated model allows for the possibility of trading time periods marked with high volume trading and volatility and also periods with low volume trading and volatility. Therefore, "regime" changes in financial markets, manifested in changing volumes and volatility, are well described by this stochastic intrinsic time process $\{T(t)\}$. Another way to interpret the directing time process $\{T(t), t \geq 0\}$ is to think of it as an information flow. If on average the intrinsic time process moves faster than physical time, then market participants are responding to fast unexpected changes in information and the market is characterized by high volatility. If on the other hand the average directing process is slower than physical time, then information flow arrives slowly and the market responds with low volatility.

We model both the spot rate and intensity rate with a symmetric α-stable Lévy process denoted by Z. The probabilistic attributes of these return processes can be analyzed by combining the constant volatility parameter with the subordinated process $W(T(t))$ to get a new process

$$Z = \tilde{W}(T(t)) = \sigma W(T(t))$$

where $\tilde{W}(t)$ is a Wiener process with stationary independent increments

---

[19] See Mittnik and Rachev [2000] and references therein.
[20] See Marinelli, Rachev, Roll and Göppl [1999].

$$\Delta \tilde{W}(t, s) = \tilde{W}(t+s) - \tilde{W}(t) \tilde{} N(0, \sigma^2 s), \quad s, t \geq 0.$$

The expected value, variance and the kurtosis of the increments $\Delta Z(t, s) = Z(t + s) - Z(t)$, $s, t \geq 0$ are

$$\mu_{Z,s} = E(\Delta Z(t, s)) = 0, \qquad (2.30)$$
$$\sigma^2_{Z,s} = E(\Delta Z(t, s)) - \mu_{Z,s})^2 = \sigma^2 \mu_{T,s}$$

where $\mu_{T,s} = E(\Delta T(t, s))$ is the expected value of an intrinsic time in the physical time interval $[t, t+s]$ of length s. The kurtosis of the stable random variable $Z(t, s)$ is given by

$$\kappa_{Z,s} = \frac{E\left[(\Delta Z(t, s) - \mu_{Z,s})^4\right]}{\sigma^4_{Z,s}} = 3\left(1 + \frac{\sigma^2_{T,s}}{\mu^2_{T,s}}\right) \qquad (2.31)$$

where $\sigma^2_{T,s} = E\left[(\Delta T(t, s) - \mu_{T,s})^2\right]$ is the variance of the intrinsic time process. The skewness of the distribution of $\Delta Z_{t,s}$ is given as

$$\beta_{Z,s} = \frac{E\left[(\Delta Z_{t,s} - \mu_{Z,s})^3\right]}{\sigma^3_{Z,s}} = 0.$$

The distribution for $\Delta Z(t, s)$ is symmetric around the expected value of zero. From Equation (2.30), the increments of Z has a finite variance if the expected values $\mu_{T,s}$ of the intrinsic time increments are finite. When the increments of the intrinsic time process is non-deterministic and the expected value and variance of the intrinsic time increments are finite, the kurtosis seen in equation (2.31) is always higher than 3. The intrinsic time scale as a directing process for a Wiener process enables us to generate leptokurtic return processes with probability densities that are heavy-tailed and have a higher peak around the mode relative to a Gaussian distribution. The asset returns generated from a subordinated model depends on the probabilistic characteristics of the intrinsic time process. We use the log-stable model for the intrinsic time process.

2.6.2.1   The Log-Stable Model

In Mandelbrot [1963,1967] and Fama [1965], the asset return processes follow a symmetric α-stable Lévy process, a process where the intrinsic time increments $\Delta T(t, s)$ is distributed stable with characteristic exponent $\alpha/2$,

$$\Delta T(t, s) \tilde{} S_{\alpha/2}(cs^{2/\alpha}, 1, 0), \quad 0 < \alpha < 2, \quad c > 0.$$

The characteristic function of $\Delta Z(t, s)$ is obtained using a Laplace transform (see Rachev and Mittnik [2000], pp. 605–608) and is expressed as

$$\Phi_{Z,s} = e^{\left\{-\left(\frac{\sigma\sqrt{\frac{1}{2}c}}{\cos(\frac{\pi\alpha}{4})^{1/\alpha}}\right)^\alpha |\theta|^\alpha s\right\}}$$
$$= e^{(-\tilde{c}^\alpha s |\theta|^\alpha)}$$

with

$$\tilde{c} = \frac{\sigma\sqrt{c/2}}{\cos(\frac{\pi\alpha}{4})^{1/\alpha}}.$$

Since the characteristic function of a symmetric α-stable random variable X with location parameter equal to zero is defined by

$$\phi_X(\theta) = e^{(-c_X^{\alpha}|\theta|^{\alpha})},$$

it follows that the increments $\Delta Z(t, s)$ are α-stable distributed with scale parameter $\tilde{c}s^{1/\alpha}$ and scale and location parameter equal to zero:

$$\Delta Z(t, s)^\sim S_\alpha(\tilde{c}s^{1/\alpha}, 0, 0).$$

The process Z is a symmetric α-stable Lévy process. If $\alpha \to 2$, then the intrinsic time process converges to the deterministic physical time and the classical lognormal model is obtained as a special case.

In the log-stable model we employ, the mean of the intrinsic time increments is truncated, i.e., the variance of Z is truncated and we obtain a truncated stable law. We refer the reader to Mittnik and Rachev [2000] and references therein for an exposition of pre-limit theorems used as an alternative to Lévy type central limit theorems which are applied to truncated stable laws.

2.7 Modeling non-Gaussian dependence with Copulas

The dependence structure for multivariate distributions is completely described by the correlation in the class of elliptical distributions of which the multivariate Gaussian is a special case. Outside of this class, Embrechts et al [2001] point out that the dependence structure is only partially expressed with a correlation function. Correlation does not give us any information about the degree of tail dependence since it captures only the first two moments of the joint distributions and ignores higher moments that are useful for describing tail dependencies. For example, the bi-variate Gaussian has a weak tail dependence but we need a need a function that allows for strong tail dependence since the stable Paretian model we employ allows for extreme co-movements in the spot and intensity processes. This tail dependence is a property of a copula function.[21] This section briefly discusses properties of the copula function and we describe how to implement a copula in the section on computational procedures.

---

[21] We refer the reader to a survey paper by Embrechts et al. [2001] that applies copulas for risk management.

2.7.1   Copula functions

The mathematical tool used to capture the underlying *dependence structure* of a general multivariate distribution is the copula function. The copula refers to a class of multivariate distribution functions supported on a unit hypercube with standard uniform marginals. Copula functions are concerned with probability integrals and quantile transforms. This section explains the general idea of copula and begins with a common definition of copulas.[22] We use the definitions and theorems following Mashal and Zeevi [2002].

Definition 2. (Copula) A function $C : [0, 1]^d \longmapsto [0, 1]$ is a d-dimensional copula if it satisfies the following properties:

1. For all $u_i \in [0, 1], C(1, \dots, 1, u_i, 1, \dots, 1) = u_i$.
2. For all $u \in [0, 1]^d$, $C(u_1, \dots, u_d) = 0$ if at least one of the coordinates, u_i, equals zero.
3. C is grounded and d-decreasing, i.e., the C-measure of every box whose vertices lie in $[0, 1]^d$ is non-negative.

Copulas are important because they capture the dependence structure of a multivariate distribution which can be seen from Sklar's theorem, (see Sklar [1959] and Embrechts et al. [2001]). The primary purpose of the copula method is that a joint distribution of random variables can be expressed as a function of the marginal distributions.

Theorem 1. *(Sklar's Theorem) Given a d-dimensional distribution function H with continuous marginal cumulative distributions $F_1, \dots, F_d$, then there exists a unique d-dimensional copula $C : [0, 1]^d \longmapsto [0, 1]$ such that*

$$H(x_1, \dots, x_d) = C(F_1(x_1), \dots, F_d(x_d)). \tag{2.32}$$

The function C is called the copula of F. Sklar's theorem is general as a copula can be used to represent any joint distribution. If the marginals of H are continuous, then F has a unique copula, but if any one of the marginals are discontinuous, then there is more than one copula for F. From Sklar's theorem we see that for continuous multivariate distribution functions, the univariate margins and the multivariate dependence structure can be separated, and the dependence structure can be represented by a copula. Sklar's theorem is important since it shows us how to apply multivariate models with any prescribed marginals. We simply apply any copula, a multivariate distribution function with uniform marginals, to those marginals. To show how this unique copula is related to the cumulative distribution function, we need the following definition. Let F be a univariate distribution function. The *generalized inverse of F* is defined as

$$F^{-1}(t) = \inf\{x \in \Re : F(x) \geq t\}$$

for all $t \in [0, 1]$, using $\inf\{\emptyset\} = \infty$.

---

[22] See Schweizer [1991] and Nelsen [1999] for the mathematical treatment on copulas.

Corollary 1. *Let H be a d-dimensional function with continuous marginals $F_1, \dots,$ F_d and copula C (where C satisfies Equation (2.32). Then for any $u \in [0, 1]^d$,*

$$C(u_1, \dots, u_d) = H(F_1^{-1}(u_1), \dots, F_d^{-1}(u_d)).$$

Without the continuity assumption, this may not hold [see, e.g., Nelsen [1999]]. Note, unlike correlation that captures the full dependence structure in multivariate Gaussian distributions (and more generally in the class of elliptic distributions), the copula summarizes this dependence structure for any multivariate distribution (with continuous marginals).

Theorem 2. *(Copula invariance) Consider d continuous random variables $(X_1,$ $\dots, X_d)$ with copula C. If $g_1, \dots, g_d : \Re \longmapsto \Re$ are strictly increasing on the range of $X_1, \dots, X_d$, then $(g_1(X_1), \dots, g_d(X_d))$ also have C as their copula.*

The copula is invariant under (strictly) increasing transformations, that is marginal distributions can change but the copula is unchanged. This theorem does not hold for correlation which is only invariant under linear transformations. We proceed to define the Gaussian copula as we employ this copula to model the dependence between the spot and intensity rate processes.

2.7.1.1   Gaussian copula

The Gaussian multivariate copula like other copulas allows any marginal distributions. In our case the marginal distribution for the spot and intensity rate daily movements are the stable non-Gaussian distributions. It is called the Gaussian copula because like the multivariate Gaussian distribution it constructs dependence using pairwise correlations among variables that have arbitrary marginals.

Definition 3. (Gaussian-copula) Let Φ denote the standard normal cumulative distribution function and let Φ_Σ^d denote the standard multivariate normal cumulative distribution function with zero mean, unit variance for each marginal and linear correlation matrix $\Sigma \in \Re^{d \times d}$, i.e., for $x \in \Re^d$,

$$\Phi_\Sigma^d(x) = \int_{-\infty}^{x} \frac{1}{(2\pi)^{d/2} |\Sigma|^{1/2}} e^{(-\frac{1}{2} y^T \Sigma^{-1} y)} dy$$

where $|\Sigma|$ is the determinant of Σ, x^T denotes the transpose of the vector $x \in \Re^d$ and the above integral denotes componentwise integration. Then, for $u = (u_1, \dots, u_d) \in [0, 1]^d$,

$$C^G(u_1, \dots, u_d; \Sigma) = \Phi_\Sigma^d \left(\Phi^{-1}(u_1), \dots, \Phi^{-1}(u_d) \right)$$

$$= \int_{-\infty}^{\Phi^{-1}(u)} \frac{1}{(2\pi)^{d/2} |\Sigma|^{1/2}} e^{(-\frac{1}{2} y^T \Sigma^{-1} y)} dy$$

is the Gaussian or normal copula, parameterized by Σ. Here $\Phi^{-1}(u) := (\Phi^{-1}(u_1),$
$\ldots, \Phi^{-1}(u_d))$, and $\Phi(\cdot)$ is the standard normal cumulative distribution function.

The Gaussian copula's flexibility and analytical tractability make it a popular approach to modeling dependence.

2.8   Computational Procedures

2.8.1   Model Inputs

The Hull–White [1994a, 1994b] extended Vasicek model describes the dynamics of the entire term structure in a manner that is consistent with the initial or observed market data. The Hull–White [1994a, 1994b] trinomial tree algorithm is a discrete time approximation of the two mean reverting continuous processes for the spot and intensity rate. The trinomial tree algorithm is able to describe the discrete-time dynamics of the spot rate and the intensity rate process that exactly fits the term structures of default-free and defaultable discount bond prices. Uncertainty in the trinomial tree is captured through branching. The tree is three-dimensional with one time dimension and two state space dimensions representing the states for $r(t)$, the spot rate and $h(t)$, the intensity rate of default. To price the credit spread put option we require default-free and defaultable discount bond prices on each node of the tree.

We calibrate the Hull–White [1994a, 1994b] model to market data using a least squares approach together with the assumption that the fractional loss rate in the event of default is constant, set at 50 %, i.e., the defaultable bond loses 50 % of its pre-default value at each default event. We obtain estimates of the parameters that characterize the two mean-reverting spot and intensity rate processes using the closed form expressions for discount bonds given in earlier sections and using 1-year Treasury bill and Baa index values assuming the index has a 30 year duration. Additionally, the correlation coefficient is estimated using this calibration procedure. Given the estimated parameters we obtain default-free and defaultable discount bond prices at each node of the tree from the explicit formulas of the Hull–White model. These calibrated parameters are given in Table 5.

TABLE 5. Estimated parameter values for the default-free spot rate and intensity rate mean reverting processes

α	0.89
σ	0.26
θ	0.055
$\bar{\alpha}$	0.92
$\bar{\sigma}$	0.01
$\bar{\theta}$	0.06
ρ	0.308

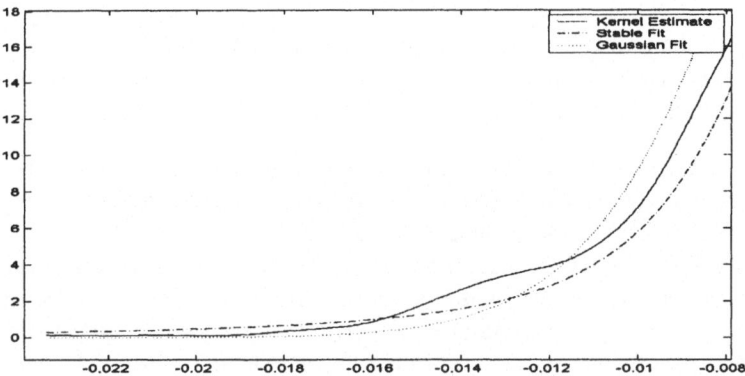

FIGURE 11. Comparison of the Gaussian and stable non-Gaussian left tail fit for residuals of the daily change in 1-year Treasury bill return

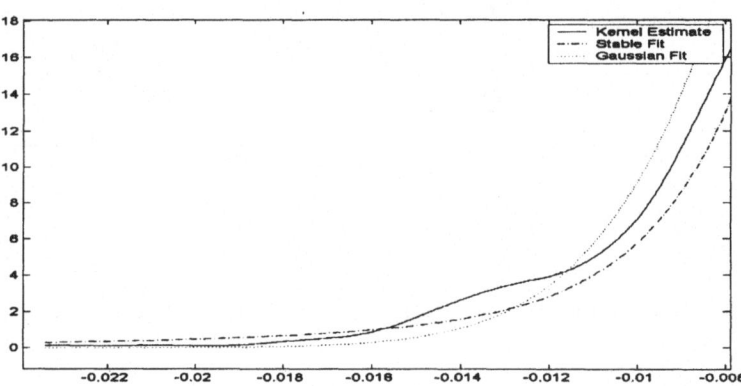

FIGURE 12. Comparison of the Gaussian and stable non-Gaussian left tail fit for residuals of the daily change in Baa credit spread time series

2.8.2 Default Branching

The next two sections discuss the default branching and recovery mechanism on the tree. From explicit bond price formulas we have default-free and defaultable discount bond prices at each node of the tree, therefore we know the current default intensity $h(t)$ at each node. The survival probability in the period from t to $t + \Delta t$ is given by

$$1 - p = E_t[e^{(-\int_t^{t+\Delta t} h(s)ds)}].\tag{2.33}$$

Over this small time interval $[t, t + \Delta t)$, the default intensity rate is constant, the survival probability can then be written as

$$1 - p = e^{-h(t)\Delta t} \tag{2.34}$$

where p is the probability of default over the time interval $[t, t + \Delta t)$. If the time step Δt is not large, we can assume with little loss in accuracy that if default occurs, it occurs at the left end of the time interval, i.e., default occurs at $\tau = t$.

As Figure 13 indicates, an additional branch is added to the each node of the tree to incorporate the default event. If the bond survives, the tree continues with the evolution of the spot rates and default intensities.

Using Figure 13, we see that the conditional probability of surviving and reaching nodes u, m, and d, the up, middle and down nodes, over the time interval Δt are $(1 - p) \cdot p_u$, $(1 - p) \cdot p_m$ and $(1 - p) \cdot p_d$ respectively. A security with survival contingent payoffs x_u, x_m and x_d in the up, middle and down nodes and zero at default, will have an expected value $x' = x_u p_u + x_m p_m + x_d p_d$, if default is not considered. However in the case of default the expected value is $x = (1 - p) \cdot x'0$.

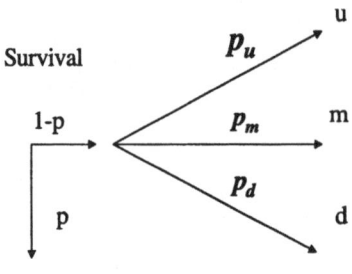

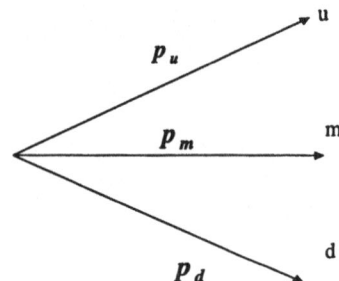

FIGURE 13. Default branching at each node of the trinomial tree

FIGURE 14. Scheme A : Standard trinomial tree branching

2.8.3   Recovery Modelling in the Default Branch

Similar to Schönbucher [June 1999] we permit multiple defaults.[23] In the discrete-time fractional recovery model, we can approximate multiple defaults in two alternative ways.[24] The number of defaults is restricted to a single default in each time interval $[t, t + \Delta t)$ or multiple defaults are allowed in this time interval. If V_n denotes the value of the defaultable asset at time $t = n \Delta t$, and V_n^* its value if it survived until time $t = n \Delta t$ and if only a single default is allowed each period, the following recursion will hold for V_n (ignoring discounting by default-free interest rates):

$$V_n = e^{(-h_n \Delta t)} V_{n+1}^* + (1 - e^{(-h_n \Delta t)})(1 - l)V_{n+1}^*$$
$$= (1 - l(1 - e^{(-h_n \Delta t)}))V_{n+1}^*. \tag{2.35}$$

---

[23] See Schönbucher [1996,1998] for more details on multiple defaults.
[24] See Schönbucher [June 1999].

The above equation states that the value of the defaultable security at time period $n \Delta t$ is the expected value of the defaultable security in the next time period $(n + 1)\Delta t$ depending on survival or default at period $(n + 1)\Delta t$. Alternatively, allowing for multiple defaults in the interval $[t, t + \Delta t)$ and assuming $l * \Delta t$ is small, the value is given by,

$$V_n = e^{(-lh_n \Delta t)} V_{n+1}^*. \tag{2.36}$$

The value given in equation (2.35) converges to the value in equation (2.36) as $\Delta t \to 0$, and for reasonably small time step sizes the difference is negligible. If the time step is large (e.g., larger than $\frac{1}{12}$) the approach in (2.36) is more appropriate. We do not consider stochastic recovery rates in this paper, however it can be incorporated in this framework[25].

2.8.4   Independence between the spot and intensity rate processes

In the first case that we examine, we assume the spot rate is independent of the intensity rate process of the defaultable bond. The price of the defaultable discount bond is given by

$$\bar{P}(t, T) = E_t[R(T) e^{(-\int_t^T r(s)ds)}]$$

$$= E_t[e^{(-\int_t^T r(s)ds)}] E_t[R(T)] = P(t, T) \Pr(t, T), \tag{2.37}$$

conditional on no default up to time t, where $R(T)$ is the face value reductions up to time t and

$$\Pr(t, T) = E_t[e^{(-\int_t^T lh(s)ds)}] \tag{2.38}$$

where $\Pr(t, T)$ is the survival probability from time t to T. When the spot and intensity process are independent we can model them separately.

2.8.4.1   Hull–White trinomial tree algorithm for the spot and intensity rate processes

The Hull and White [1994a, 1994b] numerical method constructs a discrete-time trinomial tree that approximates the mean-reverting spot rate process as given in the equation

$$dr(t) = [\theta(t) - ar(t)]dt + \sigma dW(t), 0 \le t \le T.$$

The Hull–White algorithm is a robust and efficient numerical procedure that reproduces in discrete-time the mean reverting property of the continuous spot rate

---
[25] Schönbucher [June 1999] discusses stochastic recovery rates.

process and is capable of exactly fitting the current term structure of the default-free discount bond prices and its volatility. The trinomial tree describes the spot rate suitably because it is capable of duplicating the expected drift and the instantaneous standard deviation of the continuous-time process at each node of the tree. The spot rate is in terms of an unknown function of time, $\theta(t)$, which is implicitly obtained to exactly match the current term structure of default-free discount bond prices. The spot rate tree is constructed using the estimated parameters of the Hull–White [1994a, 1994b] term structure model.

The trinomial tree represents movements in the state variable r, by approximating the continuous-time model using discrete time intervals, Δt. The time interval $[0, T]$ is divided into n sub-intervals. The interest rate at each time interval is $r_0 + j\Delta r$ where j is the index indicating the position or the node of the tree, j can take positive and negative integers. Therefore, each node on the spot rate trinomial tree is summarized by the coordinates (n, j). From each node there are three alternative branching processes as shown in Figures 14, 15 and 16. The geometry of the tree is constructed so that the middle branch corresponds with the expected value of r, making tree construction faster and permitting more accurate pricing and better values for hedge parameters.

The tree is constructed in two stages. In the first stage the tree is constructed assuming $\theta(t) = 0$ and the initial value of r, $r(0)$ is zero. With this assumption the dynamics for r^* is given by the SDE

$$dr^* = -ar^*dt + \sigma dW(t). \tag{2.39}$$

In discrete time, r^* corresponds to the continuously compounded Δt period interest rate. $r^*(t + \Delta t) - r^*(t)$ is normally distributed with mean $r^*(t)M = r^*a\Delta t$ and variance $\sigma^2\Delta t$. The central node of the tree at each time step corresponds to $r_0^* = 0$. Error minimization in the numerical procedures suggests the size of the interest rate step or the vertical distance between nodes at each time step be $\Delta r^* = \sqrt{3V} = \sigma\sqrt{3\Delta t}$. At each time step Δt, the variance in r^* is σ^2. Therefore node (n, j) represents the node where $t = n\Delta t$ and $r^* = j\Delta r^*$. Let p_u, p_m and p_d denote the probabilities of the up, middle and down branches respectively emanating from a node. These probabilities are chosen to match the expected change and variance of the change in r^* over the time interval Δt and they should add up to one. These conditions give us three equations in three unknowns, given as follows,

$$E[r^{*n+1} - r^{*n}] = p_u\Delta r_u^* + p_m\Delta r_m^* + p_d\Delta r_d^* = -ar_j^{*n}\Delta t, \tag{2.40}$$

$$E[(r^{*n+1} - r^{*n})^2] = p_u(\Delta r_u^*)^2 + p_m(\Delta r_m^*)^2 + p_d(\Delta r_d^*)^2 = \sigma^2\Delta t + a^2(r_j^{*n})^2\Delta t^2 \tag{2.41}$$

and

$$p_u + p_m + p_d = 1. \tag{2.42}$$

The tree is constructed so that at each node the first two moments of the discrete-time process coincide with the continuous-time process up to terms of order Δt^2. Changes in the spot rate from node (n, j) are given by Δr_u, Δr_m and Δr_d where the spot rate goes to an upper, middle and lower node respectively. The solutions to these equations depends on the branching from node (n, j). If the branching from node (n, j) is as in Scheme A (Figure 14) then the solution to the above equations is

$$p_u = \frac{1}{6} + \frac{1}{2}(j^2 M^2 + jM),$$

$$p_m = \frac{2}{3} - j^2 M^2,$$

$$p_d = \frac{1}{6} + \frac{1}{2}(j^2 M^2 - jM).$$

If the branching from node (n, j) is as in Scheme B (Figure 15), then the solution is given by

$$p_u = \frac{1}{6} + \frac{1}{2}(j^2 M^2 - jM), \tag{2.43}$$

$$p_m = -\frac{1}{3} - j^2 M^2 + 2jM, \tag{2.44}$$

$$p_d = \frac{7}{6} + \frac{1}{2}(j^2 M^2 - 3jM). \tag{2.45}$$

If the branching from node (n, j) is as in Scheme C (Figure 16), then the probabilities are given as

$$p_u = \frac{7}{6} + \frac{1}{2}(j^2 M^2 + 3jM), \tag{2.46}$$

$$p_m = -\frac{1}{3} - j^2 M^2 - 2jM, \tag{2.47}$$

$$p_d = \frac{1}{6} + \frac{1}{2}(j^2 M^2 + jM). \tag{2.48}$$

The trinomial branch in scheme A is appropriate most of the time. However to ensure the probabilities are positive, branching switches from scheme A to scheme C when j is large and when $a > 0$. For the same reason it is necessary to switch from branching as shown in scheme A to scheme B when j is small. Switching from one branching alternative to another truncates the tree to keep the tree from growing very

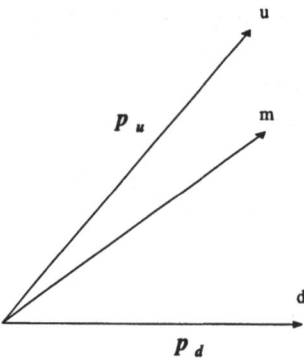

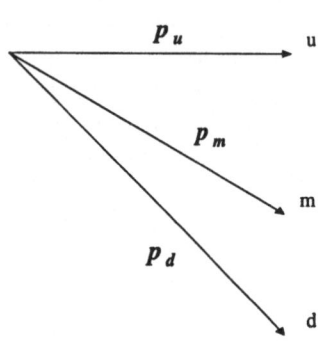

FIGURE 15. Scheme B, "Up" branching in the trinomial tree

FIGURE 16. Scheme C, "Down" branching in the trinomial tree

large as the number of time periods increases. Let j_{max} be the value of j where the branching is switched from scheme A to scheme C and j_{min} be the value of j when the branching is switched from scheme A to scheme B. The probabilities p_u, p_m and p_d are always positive when j_{max} is between $-0.184/M$ and $-.816/M$ provided $a > 0$ and $M > 0$. The probabilities at each node are a function of j, additionally the tree structure branching is symmetrical.

In the second stage of tree construction we employ forward induction. Starting from time period zero to the end of the tree, each node of the tree is displaced by an amount α_i, in order to match the initial term structure and simultaneously preserving the probabilities. The displacement magnitude is the same for all nodes at a particular time $n\Delta t$. The new tree for r will be for

$$r(t) = \alpha(t) + r^*(t)$$

and the initial discount curve is obtained,

$$P(0, T) = E[e^{(-\int_0^T r(s)ds)}].$$

The new displaced tree models the mean-reverting stochastic process

$$dr(t) = [\theta(t) - ar(t)]dt + \sigma dW(t).$$

If $\hat{\theta}(t)$ is defined to be the discrete-time estimate of $\theta(t)$ between time t and $t + \Delta t$, and the drift in r at time $i \Delta t$ at the midpoint of the tree is $\hat{\theta}(t) - a\alpha_i$ such that

$$[\hat{\theta}(t) - a\alpha_n] \Delta t = \alpha_n - \alpha_{n-1},$$

then

$$\hat{\theta}(t) = \frac{\alpha_n - \alpha_{n-1}}{\Delta t} + a\alpha_n.$$

In the limit as $\Delta t \to 0$, $\hat{\theta}(t) \to \theta(t)$.[26]

Let $Q_{i,j}$ denote the present value of a security that pays off \$1 if node (n, j) is reached and zero otherwise. $Q_{i,j}$ is then the state price of node (n, j). The α_i's and $Q_{i,j}$'s are calculated using forward induction. The initial value of the state price, $Q_{0,0}$ is one. The value of α_0 is chosen to match the discount bond price that matures in time Δt. So α_0 is set to the initial Δt period interest rate given as

$$\alpha_0 = -\frac{1}{\Delta t} \ln P(0, 1). \tag{2.49}$$

The next step is to compute $Q_{1,1}$, $Q_{1,0}$, $Q_{1,-1}$. This procedure is expressed formally as the following. (This section is taken from Hull [2000], page 583). Suppose the $Q_{i,j}$'s have been determined up to time $n \leq m$ ($m \geq 0$). The next step is to determine α_m so the tree correctly prices a discount bond maturing at time $(m + 1)\Delta t$. The interest rate at (m, j) is $\alpha_m + j\Delta r$. The price of a discount bond maturing at time $(m + 1)\Delta t$ is given by

$$P(0, m + 1) = \sum_{j=-n_m}^{j=n_m} Q_{m,j} e^{[-(\alpha_m + j\Delta r)\Delta t]} \tag{2.50}$$

where n_m is the number of nodes on each side of the central node $m\Delta t$. Then α_m can be written as

$$\alpha_m = \frac{\ln \sum_{j=-n_m}^{j=n_m} Q_{m,j} e^{[-j\Delta r * \Delta t]} - \ln[P(0, m + 1)]}{\Delta t}. \tag{2.51}$$

Once α_m has been determined, the $Q_{i,j}$ for $n = m + 1$ can be calculated using

$$Q_{m+1,j} = \sum_k Q_{m,k} p(k, j) \exp\left[-(\alpha_m + k\Delta r)\Delta t\right] \tag{2.52}$$

where $p(k, j)$ is the probability of moving from node (m, k) to node $(m + 1, j)$ and the summation is taken for all values of k which are non-zero.

A trinomial tree is also built for the risk-neutral mean-reverting stochastic intensity process $h(t)$ which is given as

$$dh(t) = \left[\bar{\theta}(t) - \bar{a}h(t)\right] dt + \bar{\sigma} d\bar{W}(t), \ 0 \leq t \leq T.$$

The process of building the intensity rate tree is analogous to the construction of the spot rate tree. We build the tree specifically for the credit spread hl by building a tree for intensity rate $h(t)$, since one is a multiple of the other. The tree is then fitted to the initial discount curve of credit spreads.

---

[26] See Hull and White [1994a, b]

TABLE 6. Joint branching probabilities for the default-free spot rate and the intensity rate movements for a single period of time given independence between their dynamics

		r move			Marginal
		Down	Middle	Up	
	Up	$p'_u p_d$	$p'_u p_m$	$p'_u p_u$	p'_u
h move	Middle	$p'_m p_d$	$p'_m p_m$	$p'_m p_u$	p'_m
	Down	$p'_d p_d$	$p'_d p_m$	$p'_d p_u$	p'_d
	Marginal	p_d	p_m	p_u	1

$$\Pr(0, T) = E[e^{(-\int_0^T lh(s)ds)}] = \frac{\bar{P}(0, T)}{P(0, T)}. \tag{2.53}$$

Once the intensity rate tree is constructed, default branches are added to each node of the tree as described in Section 8.2. The trinomial intensity rate tree has at each node the probability of default and survival in the next time interval Δt. When combining the trees it is required that both trees should have the same time step size Δt but do not need to have the same space step size. The trees are combined after constructing the spot rate and the intensity rate trees. The independence between the spot rate and the intensity rate processes permitted us to fit the h-tree separately. The combined tree is a three-dimensional tree, with two space dimensions (r and hl) and a third time dimension. Table 6 shows the branching probabilities associated with moves in the spot rate and the intensity rate under assumption of independence. The combined tree can now be used to numerically compute the value of the credit spread put option.

At each time level n, the tree has $(i_{max} - i_{min}) * (j_{max} - j_{min})$ survival nodes and the same number of default nodes. From node (n, i, j) there are ten different branch possibilities. Both rates r and intensities h have three possible combinations, and there is a tenth branch to default. The combined tree is constructed to fit the term structure of the default-free bond prices and defaultable bond prices and then used to value the credit spread put option.

2.8.5   Correlated processes with Gaussian innovations

In practice the spot rate and the intensity rate of default are correlated with each other. The procedure for constructing trinomial trees with correlation is different than the case of independence. First the spot rate tree is built to fit the term structure of default-free discount bond prices. Then the tree for the credit spread is built but not fitted to the term structure of the credit spread. Instead the two trees are combined and the correlation between the two mean reverting processes are incorporated. Default branches are then added to the combined correlated tree. The combined tree is now fitted to the term structure of defaultable bond prices while preserving the fit

TABLE 7. Joint branching probabilities for the default-free spot rate and the intensity rate movements for a single period give a positive correlation between their dynamics

| | | r move | | | Marginal |
		Down	Middle	Up	
	Up	$p'_u p_d - \varepsilon$	$p'_u p_m - 4\varepsilon$	$p'_u p_u + 5\varepsilon$	p'_u
h move	Middle	$p'_m p_d - 4\varepsilon$	$p'_m p_m + 8\varepsilon$	$p'_m p_u - 4\varepsilon$	p'_m
	Down	$p'_d p_d + 5\varepsilon$	$p'_d p_m - 4\varepsilon$	$p'_d p_u - \varepsilon$	p'_d
	Marginal	p_d	p_m	p_u	1

to default-free bond prices. This resulting combined tree which is fitted to the term structure defaultable discount bond prices can now be used to numerically value the credit spread put option given dependence between the spot and intensity rate processes.

TABLE 8. Joint branching probabilities for the default-free spot rate and the intensity rate movements for a single period, given a negative correlation between their dynamics

| | | r move | | | Marginal |
		Down	Middle	Up	
	Up	$p'_u p_d + 5\varepsilon$	$p'_u p_m - 4\varepsilon$	$p'_u p_u - \varepsilon$	p'_u
h move	Middle	$p'_m p_d - 4\varepsilon$	$p'_m p_m + 8\varepsilon$	$p'_m p_u - 4\varepsilon$	p'_m
	Down	$p'_d p_d - \varepsilon$	$p'_d p_m - 4\varepsilon$	$p'_d p_u + 5\varepsilon$	p'_d
	Marginal	p_d	p_m	p_u	1

We use the Hull–White [1994b] procedure that considers correlation in a two-dimensional tree.[27] For a correlation $\rho \neq 0$ the procedure adjusts the transition probabilities given in Table 6 to the probabilities given in Table 7 and Table 8 depending on whether correlation is positive or negative using a correction parameter ϵ,

$$\epsilon = \begin{matrix} \frac{1}{36}\rho \text{ for } \rho > 0, \\ -\frac{1}{36}\rho \text{ for } \rho < 0. \end{matrix} \qquad (2.54)$$

Table 7 and 8 give the probabilities of the indicated combined movements of r and h in the combined tree for a given positive or negative correlation $\rho = \pm 36\epsilon$. Default

---

[27] Correlation between the spot rate and the intensity rate process is also discussed in Schönbucher [June, 1999]. For a more detailed discussion of the procedure used to incorporate correlation, see Hull and White [1994b].

and survival are ignored in these tables, to reach the probabilities for the movements and survival over the next time interval, the probabilities are multiplied by $(1 - p)$. The original probabilities are: r: up p_u; middle p_m; down p_d. h: up p'_u; middle p'_m; down p'_d. The adjustment for correlation given in Tables 7 and 8 only work if ϵ is not too large. Thus there is a maximum value for the correlation that can be implemented for a given time step size Δt. As the refinement is increased ($\Delta t \to 0$) this restriction becomes weaker and the maximum correlation approaches one[28].

To match the terms structure of the defaultable bond yield given correlation, the tree building procedure is similar to the case when there is independence except the shift of the tree occurs in the h-dimension without affecting the tree built for the spot rate, thus preserving the spot rate tree.

2.8.6   Computation procedure with Stable innovations

This section describes the tree construction when the spot and intensity rate processes are driven by stable non-Gaussian innovations. We first estimate the stable parameters for the 1-year Treasury bills and the Baa credit spread time series using maximum likelihood described earlier in Section 6.1.1.

In Section 6.1 we defined a symmetric α-stable random variable Z, constructed as $Z = A^{1/2}G$, where G is a zero-mean Gaussian random variable, i.e., $G \sim N(0, \sigma^2)$ and A is an $\alpha/2$-stable random variable totally skewed to the right and independent of G. A is distributed as, $A \sim S_{\alpha/2}(1, 0, 0)$; A is called the $\alpha/2$ stable subordinator. We model daily changes in the spot and intensity rate as stable non-Gaussian random variables, incorporating covariation between them. A (α_r, α_h)-stable vector X is defined by

$$X = (A_r^{1/2}G_r, A_h^{1/2}G_h). \tag{2.55}$$

The elements of the vector X are daily changes in the spot rate, $\Delta r \stackrel{d}{=} A_r^{1/2}G_r$ and the daily changes in the intensity rate, $\Delta h \stackrel{d}{=} A_h^{1/2}G_h$ in one unit of time. The covariation between the changes in the spot rate and the intensity rate of the defaultable bond is achieved through copula functions that construct a dependence structure between the truncated subordinators for the spot rate and intensity rate.

In discrete time, when spot rate changes are Gaussian, we have $r(t+\Delta t) - r(t) \sim N(0, \sigma_r^2 \Delta t)$, where $\Delta t = 1$. These daily changes in the spot rate are distributed Gaussian with mean zero and variance σ^2. However, when daily changes in the spot rate are modelled employing the stable subordinator we obtain, $r(t + \Delta t) - r(t) \sim A_r^{1/2}N(0, \sigma_r^2)$. A_r is a stable random variable totally skewed to the right with stable parameters $\Theta_{A_r} = (\alpha_r/2, 1, 0)$, where α_r represents the index of stability, the (right) skewness parameter is equal to 1 and with location parameter equal to zero. Note the stable parameter α_r was estimated for the daily return series of 1-year Treasury bills using the ML procedure. When the stable subordinator $A_r^{1/2}$, is multiplied by the

---

[28] For a detailed discussion see Hull and White [1994b].

Gaussian random variable with mean zero and variance σ^2, we get a sub-Gaussian symmetric α-stable random variable $Z \sim N(0, A_r * \sigma_r^2)$. The Gaussian random variable has its variance rescaled to $A_r * \sigma_r^2$. This rescaled variance is stochastic since the stable subordinator A_r is a stable random variable. The variance σ_r^2 of the Gaussian random variable is perturbed with a stable subordinator, A_r. Since A_r is a potentially large multiplicative factor we truncate it to take on a maximum value of 10 in absolute value terms.[29] In other words, we allow for events that are ten standard deviations from the mean to occur in the economy. The stable subordinator is truncated to model termination in trading of an asset in an exchange if the asset return changes significantly in an adverse direction. Thus we have the spot rate change, $r(t + \Delta t) - r(t) \sim N(0, A_r * \sigma_r^2)$, as a stable random variable, a truncated Gaussian random variable with a rescaled variance. The subordinator has introduced a stochastic volatility term and increases the probability of outlier events, allowing for extreme spot rates to occur, thus permitting us to represent the empirical data with a greater degree of accuracy. We construct a stable sub-Gaussian random variable for daily changes in the intensity rate $\Delta h \overset{d}{=} A_h^{1/2} G_h$. Note that $G = (G_r, G_h)$ are dependent. A_r and A_h while independent of G might be dependent. This is achieved by generating A_r and A_h by a single uniform random variable or from using a copula function. For maximum and minimum association between the stable non-Gaussian spot and intensity rate processes we use a uniform distribution and for all other degrees of association we use a Gaussian copula function.

In the trinomial tree procedure, employing Gaussian innovations, we note the size of the spot rate step between nodes at each time step is $\Delta r = \sqrt{3V_r} = \sigma_r * \sqrt{3\Delta t}$. This is also the case for the intensity rate step, i.e., $\Delta h = \sigma_h * \sqrt{3\Delta t}$. However employing stable innovations changes the vertical distance between nodes for the spot rate and the intensity rate to $\Delta r = A_r^{1/2} * \sigma_r * \sqrt{3\Delta t}$ and $\Delta h = A_h^{1/2} * \sigma_h * \sqrt{3\Delta t}$. For each realization of the stable subordinator for the spot and intensity rate, A_r and A_h, we generate a new three-dimensional trinomial tree for the spot rate and the intensity rate. Dependence between the spot and intensity stable processes is achieved by modeling dependence between the stable subordinators A_r and A_h. In the next section we discuss the Gaussian copula methodology that is used to generate dependence between the stable subordinators, A_r and A_h. We model maximum and minimum association between the two stable subordinators using standard uniform marginal distributions and a Gaussian copula function for other intermediate associations.

2.8.7 Procedure for the copula implementation

We apply the Gaussian copula for the stable non-Gaussian marginals of the daily changes in the spot and intensity rates for its flexibility, analytical tractability and

---

[29] Indeed, truncating the subordinator results in Z being in the domain of attraction of a Gaussian law. However, because the truncation is deep in the tail of the subordinator the random variable Z is statistically indistinguishable from the symmetric α-stable law and satisfies the Lévy pre-limit theorem discussed in Rachev and Mittnik [2000], pages 77–81.

its computational simplicity. We adopt the procedure outlined by Bravo Risk Management Group for the application of a Gaussian copula with stable non-Gaussian marginals. This copula provides a good degree of tractability for multi-dimensional stable random vectors even for skewed stable marginals.

Unlike the sub-Gaussian approach discussed earlier the copula approach does not require the elements of the stable random vector to be decomposed into a symmetric dependent and skewed independent part. The implementation procedure for Gaussian copulas for stable random vectors is outlined as follows:[30]

From an I-dimensional vector of random variables, X, where $X_{i,j}$ is the j-th observation of the i-th element of X, where $i = 1, \ldots, I$ and $j = 1, \ldots, J$, each X_i has stable non-Gaussian marginal distributions. Their dependence will be described by a Gaussian copula. The copula is developed as follows:

1. For each observation $j = 1, \ldots, J$ of a random-vector element X_i, the value of the cumulative density function is estimated: $U_{i,j} = S_i(X_{i,j})$, $i = 1, \ldots, I$ where S_i is the fitted stable cumulative distribution function (CDF) for the i-th random variable. $U_{i,j} = S_i(X_{i,j}) \in U(0, 1)$

2. For each set of observations j an I-dimensional multivariate vector N_j of a multivariate normal distribution is constructed with components $N_{i,j}$. The i-th sample has a normal distribution $N(m_i, v_i)$ with mean m_i and variance v_i. m_i and v_i are the sample mean and sample variance estimated from the sample $X_{i,j}$, $j = 1, \ldots, J$. The I-dimensional multivariate vector N_j with Gaussian marginals is obtained by transforming the I-dimensional vectors U_j. $N_j = (\Phi^{-1}(U_{1,j}), \ldots, \Phi^{-1}(U_{I,j}))$, where Φ_i is a CDF of $N(m_i, v_i)$. Assuming that the real dependence is described by a Gaussian copula, this vector will be a multivariate normal vector.

3. With the N_j, $j = 1, \ldots, J$, a multivariate normal distribution is fitted, $N(\bullet, \sum)$.

4. To simulate scenarios, samples are drawn from the multivariate $N(\bullet, \sum)$

5. The current Gaussian marginals have to be transformed into stable ones. Thus, each coordinate of each draw is converted to $U(0, 1)$ by $W_j = (\Phi_1(N_{1,j}), \ldots, \Phi(N_{I,j}))$. The transformed simulations are denoted by $W_{i,j}$.

6. For each $i = 1, \ldots, I$, $S_i^{-1}(W_{i,j})$ is constructed. The resultant vectors are simulated multivariate random vectors with stable marginals and Gaussian copula.

We use the Gaussian copula for its computational simplicity especially for high-dimensional random vectors. In the academic literature, T-copulas are more popular for modeling joint heavy-tailed distributions than Gaussian copulas, however from a computational standpoint, T-copulas are more demanding to model multivariate dependencies.

---

[30] For a detailed desciption the reader is referred to the technical document of Cognity, a software package of the Bravo Risk Management Group [2002].

2.9 Valuation of the credit spread put option on the tree

Once the tree is constructed and fitted to market discount bond prices it can be used to value defaultable coupon bonds and credit derivative securities written on them. A credit derivative security is characterized by its payoff given the occurrence of the predefined credit event that is linked to the referenced asset. In our case the credit event is the movement of the Baa term credit spread beyond a pre-specified threshold, the strike credit spread. We briefly describe the characteristics of the credit spread put option whose value we compute using the constructed three-dimensional trinomial tree.

The credit spread put option expires in one year and has two counterparties to the contract that are both default-free. In this contract the protection buyer A, pays an option premium upfront to a protection seller B in exchange for a contingent payment, contingent on the reference asset's credit spread exceeding the strike credit spread which is set at $\bar{s}$ =200 basis points. There are two securities underlying the credit spread put option. They are the default-free coupon bond denoted by P_c and the defaultable Baa coupon bond $\bar{P}_c$. The payoffs of the option at the future time $T = N\Delta t$ is computed for various possible interest rates $r = j\Delta r$ at the option expiration date. To obtain the payoff of the option we need the default-free coupon bond price P_c and the strike price of the option, the price K_j^N that is equivalent to the price of the defaultable coupon bond at a credit spread of $\bar{s} = 200$ bps over the value of the default-free bond, P_c. Since we assume the option is not knocked out in the event of default the writer of the option, in the event of default, pays the difference between the face value of the bond and the fractionally recovered pre-default price to the buyer.

The notation used here follows Schönbucher [1999]. Let f_{ij}^n denote the payoff the protection seller makes to the protection buyer if the bond defaults. The protection buyer pays an option premium at the option initiation date for the protection. Let V_{ij}^N denote the value of the option at node (N, i, j), where N is the maturity date of the option. If default has not occurred at this terminal level of the tree, then the value of the option is

$$V_{ij}^N = F_{ij}^N \tag{2.56}$$

where F_{ij}^N is the payoff of the option when the actual credit spread exceeds the strike credit spread $\bar{s}$ at node (N, i, j) and is given as

$$F_{ij}^N = \max{(K_j^N - \bar{P}_{c_{ij}}^N, 0)}.$$

To compute the value of the option at the initial date we use backward induction along the tree. Conditional on survival at time $n\Delta t$, the value of the option is the discounted expected value of the option's value at time $(n + 1)\Delta t$, which can be expressed as,

$$V_{ij}^{||n} = \sum_{k,l \in Succ(n,i,j)} p_{kl}^n e^{-r_j^n \Delta t} V_{kl}^{n+1}. \tag{2.57}$$

The value of the option at time period $n\Delta t$ at the survival node is the discounted expected value of the option from each of its emanating branches. $Succ(n, i, j)$ represent all the nodes $(n+1, k, l)$ at time period $(n+1)\Delta t$ that are connected from node (n, i, j) at time period $n\Delta t$, except for the default node. The value of the option at node (n, i, j), at time period n considering both survival and default is given as

$$V_{ij}^{in} = e^{-h_i^n \Delta t} V_{ij}^{iin} + (1 - e^{-h_i^n \Delta t}) f_{ij}^n. \tag{2.58}$$

The above expression states that the value of the credit spread put option at time period $n\Delta t$ is the expected value of the option given its payoffs in survival and default.

In the default node the protection seller pays the par value of the bond less the fractional recovery of the reference coupon bond in the event of default. The protection sellers' payment to the buyer can be expressed as

$$f_{ij}^n = 1000 - (1 - l)\bar{P}_{c_{ij}}^n \tag{2.59}$$

where $\bar{P}_{c_{ij}}^n$ is the price of the Baa-rated coupon bond at time period $n\Delta t$ at node (n, i, j) just prior to default and $(1 - l)$ is the recovery rate of this coupon bond. The payoff of the option depends on the Baa coupon bond's price at each node on the tree, which is obtained using explicit formulas for discount bond prices and noting that the defaultable coupon bond is a portfolio of defaultable discount bonds.

2.9.1 Data and Computational Results

We use daily yields of the 1-year Treasury bill for the period July/15/1959: April/20/2001 for the spot rate time series and daily yields of the long-term Baa index and 30-year constant maturity Treasury bonds for the period January/02/1986:April/20/2001 to compute the intensity rate time series.

The estimated Gaussian and stable non-Gaussian parameters for the 1-year treasury bill and Baa credit spread series are shown in Table 4. The estimated parameters for the spot and intensity mean reverting processes are given in Table 5.

In this numerical exercise we compute the value of a credit spread put option maturing in one year referenced on a 10-year non-callable Baa coupon bond with an 8.5 per cent coupon rate with annual coupon payments. The value of the option is computed on the trinomial tree when the spot and intensity rate dynamics have Gaussian and stable non-Gaussian innovations. In both these cases we compute the value of the option assuming dependence between these processes for various levels of association. We examine maximum, minimum and intermediate association between the processes. Maximum association is perfect positive dependence and minimum association is perfect negative dependence. In the stable non-Gaussian case we compute a value for the option by averaging the results generated from 1000 realizations of the stable subordinators for the spot and intensity rate, A_r and A_h.

In Table 9 we compare the value of the credit spread put option, for different degrees of association between the spot and intensity rate processes when these processes are driven by Gaussian and stable non-Gaussian innovations. A possible reason for a much larger option price in the truncated stable Paretian model might be

TABLE 9. The price of a credit spread put option referenced on a 10-year non-callable Baa coupon bond with a strike credit spread of 200 basis points. The coupon rate is 8.5% and coupons are paid annually. The maturity of the option is one year. Prices are given for different degrees of association between the spot and intensity rate processes under both the Gaussian base model and the stable Paretian model

Association	Gaussian model	Stable Paretian model
Minimum association	36.82	67.90
Rho = -0.5	39.68	67.92
Rho = 0	42.76	67.92
Rho = 0.5	46.05	67.92
Maximum association	49.13	67.93

because we have assumed that the stable subordinator vector $A = (A_1, A_2)$ is independent of the Gaussian vector $G = (G_1, G_2)$.[31]. Modeling this dependence is computationally more demanding.

2.10   Conclusion

We value a credit spread put option referenced on a Baa-rated coupon bond. The coupon bond's value is constructed from a term structure of credit spreads using Moody's Baa index yields. Default is modeled with a Cox process and recovery rate uncertainty is modeled using the RMV approach of Duffie and Singleton [1999] extended for multiple defaults following Schönbucher's [1996,1999]. In this framework the key variables that need to be dynamically modeled in the credit spread put option valuation problem are the default-free spot interest rate and the Baa credit spread processes. The instantaneous credit spread of a defaultable discount bond turns out to be the product of the instantaneous intensity and the loss rate. The credit spread process is represented by an intensity rate process since we assume the loss rate is fixed at fifty per cent. Standard term structure models developed for default-free bonds are directly applicable to defaultable bonds except the defaultable bond's instantaneous spot rate is adjusted for default risk. This default-adjusted spot rate has a default-free spot rate component and a credit spread component, representing the compensation for bearing credit risk. We parameterize the spot and intensity rate processes separately with a term structure model for the short rate of the defaultable bond. We use the no-arbitrage two-factor Hull–White [1994a,1994b] term structure model for the dynamics of the spot and intensity rate processes. Apart from empirical consistency, the Hull–White [1994a,1994b] model provides us with a great deal of analytical tractability and computational efficiency in calibrating the parameter to observed market prices.

---

[31] Research by Bravo Risk Management Group shows that A and G are actually dependent.

We numerically compute the value of a credit spread put option using the trinomial tree procedure outlined in Schönbucher [June 1999]. This algorithm constructs a trinomial tree, which is a three-dimensional tree with two space dimensions for the spot rate and the intensity rate, with time as the third dimension. The trinomial tree is a discrete-time approximation of the continuous-time spot and intensity rate processes. Each node on the tree has an absorbing state of default.

The Hull–White [1994a,1994b] term structure model for the spot and intensity processes is a Gaussian model. However, the empirical spot and intensity processes have distributions that are leptokurtic and exhibit stochastic volatility. This significant departure from Gaussian statistical properties suggest the application of a truncated stable Paretian model, a model with many advantages over other competing asymmetric leptokurtic and stochastic volatility models. In the valuation of the credit spread put option, accurately modelling the underlying risk factors provides an empirically consistent valuation framework.

Furthermore, it is important from a risk management perspective as well to accurately model the empirical distributions of the spot rate and Baa credit spread changes as VaR computations are sensitive to these empirical tail distributions. Our truncated stable model for the spot and intensity rate processes account for their extreme co-movements, and VaR computations in this framework provide better estimates of the likelihood of large losses reflecting extreme adverse market conditions when compared to a Gaussian based model. A risk manager that overlooks the leptokurtic nature and other anomalies of the empirical distributions of assets' returns will significantly underestimate the likelihood of large losses and undermine the effectiveness of risk management efforts.

In the truncated stable Paretian model the stochastic volatility model is generated from the stable subordinator. This stable subordinator is a stable non-Gaussian random variable totally skewed to the right. The stochastic volatility model captures daily regime shifts observed in financial markets. A priori we assume these daily regime shifts are random and independent from one another. In other words there is no intertemporal dependence in the volatility of the spot and credit spread processes observed in financial markets. Daily regime shifts that are independent across time are modeled by generating independent and identically distributed stable subordinator random variables for the spot and intensity rate processes.

When the spot rate and the default intensity rate are modeled with stable non-Gaussian distributions, we need a more accurate measure for the dependence structure between these non-Gaussian processes than correlation coefficient. The correlation coefficient only accounts for the first two moments of the distributions but ignores the higher moments. In fact, the tail dependence would be a more representative model for dependence accounting for the possibility of extreme co-movements. Copula functions are a popular approach to model dependence structures when the marginal distributions of the spot and intensity rate processes are non-Gaussian. We employ a Gaussian copula function to model the dependence structure between the stable non-Gaussian spot and intensity rate processes.

It is clear that the credit spread put option price is highly sensitive to the assumptions made about the distribution governing the default-free spot rate and intensity rate processes. For all degrees of association between the spot rate and the intensity rate processes we notice that the option price is significantly higher when the spot rate and intensity rate processes are modeled with a stable non-Gaussian model than with a Gaussian model. This is because employing stable non-Gaussian distributions results in extreme outcomes for the changes in the spot rate and intensity rate and assigns higher probabilities for these extreme outcomes to occur. Modeling the spot rate and intensity rate with a stable subordinator random variable makes our framework a stochastic volatility model. It is this stochastic volatility that reproduces the excess kurtosis observed in the default-free spot rate and the Baa credit spread distributions. We also notice that the variation in the option prices for different degrees of association between the spot rate and the intensity rate is much higher in the Gaussian model than the stable Paretian model even after modeling the dependence structure with copulas. A possible reason for the small variation in option prices might be because the credit spread put option price is bounded.

In credit derivative valuation models of the reduced form, Brownian-based term structure models are typically used for their analytic tractability but from an empirical standpoint are grossly inadequate. These Gaussian models fail to capture empirically observed stochastic properties of heavy tails, excessive kurtosis, skewness and stochastic volatility. We replace the traditional Brownian-based model with a truncated stable Paretian model to capture these empirical anomalies. Generalizing the distributional assumptions made about the default-free spot interest rate and Baa index default intensity rate processes, we obtain values for the credit spread put option that better embody market information about the reference asset's return process, thus enabling us to provide a more empirically grounded framework than what the Brownian-based models can offer and this has significant implications for option pricing as well as for risk management.

References

[1] Alessandrini, F. (1999). Credit risk, interest rate risk and the Business cycle. *The Journal of Fixed Income*, 1(9):42–53.
[2] Altman, E.I. (1983/1990). *Corporate Financial Distress*. Wiley, New York.
[3] Altman, E.I., Brady, B., Resti, A., and Sironi, A. (2002). The link between default and recovery rates: Implications for credit risk models and procylicality.
[4] Ammann, M. (1998). *Pricing Derivative Credit Risk*. Springer-Verlag, New York.
[5] Anson, M. J.P. (1999). *Credit Derivatives*. Frank J. Fabozzi Associates, New Hope, PA.
[6] Artzner, P., and Delbaen, F. (1992). *Credit risk and prepayment option.* 22: 81–96.
[7] Bakshi, G., Madan, D., and Zhang, F. (2001). Recovery in default risk modeling: Theoretical foundations and empirical applications. University of Maryland and the Federal Reserve Board.
[8] Black, F., and Cox., J. (1976). Valuing corporate securities: Some effects of bond indenture provisions. *Journal of Finance*, 31(2):351–367.
[9] Black, F., and Scholes, M. (1973). The pricing of options and corporate liabilities. *Journal of Political Economy*, 81: 637–654.

[10] Bollerslev, T. (1986). Generalized autoregressive conditional heteroskedasticity. *Journal of Econometrics*, **31**: 307–327.

[11] Bollerslev, T. (1987). A conditional heteroskedastic time series model for speculative prices and rates of return. *Review of Economics and Statistics*, **69**:542–547.

[12] Brémaud, P. (1981). *Point Processes and Queues*. Springer-Verlag, Berlin, Heidelberg, New York.

[13] Campbell, J.Y., Lo, A.W., and MacKinlay, A.C. (1997). *The Econometrics of Financial Markets*. Princeton University Press, Princeton, NJ.

[14] Clewlow, L., and Strickland, C. (1998). *Implementing Derivatives Models*. Wiley Series in Financial Engineering.

[15] Cooper, I., and Martin, M. (1996). Default risk and derivative securities. *Applied Mathematical Finance*, **3**(1): 109–126.

[16] Cossin, D. (1997). Credit risk pricing: A literature survey. *Finanzmarkt und Portfolio Management*, **11**(4): 398–412.

[17] Cossin, P., and Pirotte, H. (2001). *Advanced Credit Risk Analysis: Financial approaches and mathematical models to assess, price and manage credit risk*. John Wiley and Sons Ltd.

[18] Crouhy, M., Galai, D., and Mark, R. (2000). A comparative analysis of current credit risk models. *Journal of Banking and Finance*, **24** (1–2): 59–117.

[19] Das, S.R. (1995). Credit risk derivatives. *Journal of Derivatives*, **2**(3): 7–23.

[20] Das, S.R. (1999). Pricing credit derivatives. *Handbook of Credit Derivatives*. McGraw-Hill.

[21] Das, S.R., and Tufano, P. (1996). Pricing credit — sensitive debt when interest rates, credit ratings and credit spreads are stochastic. *Journal of Financial Engineering*, 5(2):161–198.

[22] D'Souza, D.M., Amir-Atefi, K., and Racheva-Jotova, B. (2002). Valuation of a credit default swap: The stable non-Gaussian versus the Gaussian approach. to appear in the *Proceedings of the 8th Econometrics Conference on Credit Risk Management*.

[23] Duffie, D. (1998). Defaultable term structure models with fractional recovery of par. Working paper, Graduate School of Business, Stanford University.

[24] Duffie, D. (1994). Forward rate curves with default risk. Working paper, Graduate School of Business, Stanford University.

[25] Duffie, D. (1999). Credit swap valuation. Working paper, Graduate School of Business, Stanford University.

[26] Duffie, D. (1999). Credit swap valuation. Association for Investment Management and Research, p. 73–86.

[27] Duffie, D., and Huang, M. (1996). Swap rates and credit quality. *Journal of Finance*, **51**: 921–949.

[28] Duffie, D., Schroder, M., and Skiadas, C. (1996). Recursive valuation of defaultable securities and the timing of resolution of uncertainty. *Annals of Probability*, **6**:1075–1090.

[29] Duffie, D., and Singleton. K. (1997). An econometric model of the term structure of interest rate swap yields. *Journal of Finance*, **52**(4): 1287–1322.

[30] Duffie, D., and K. Singleton. (1999). Modeling term structures of defaultable bonds. *The Review of Financial Studies*, **12**(4): 687–720.

[31] DuMouchel, W. (1973a). Stable distributions in statistical inference: Symmetric stable distribution compared to other symmetric long-tailed distributions. *Journal of the American Statistical Association*, **68**:469–477.

[32] DuMouchel, W. (1973b). On the asymptotic normality of the maximum-likelihood estimate when sampling from a stable distribution. *Annals of Statistics*, **3**:948–957.

[33] Embrechts, P., Klüppelberg, C., and Mikosch, T. (1997). *Modelling Extremal Events for Insurance and Finance*. Springer-Verlag, Berlin.

[34] Embrechts, P., McNeil, A., and Straumann, D. (1999). Correlation: Pitfalls and Alternatives. Department of Mathematics, ETH Zentrum, CH-8092 Zürich.

[35] Embrechts, P., Lindskog, F., and McNeil, A. (2001). *Modeling dependence with copulas and applications to risk management*.

[36] Fama, E. (1965a). The behavior of stock market prices. *Journal of Business*, **38**:34–105.

[37] Fama, E. (1965b). Portfolio analysis in a stable Paretian market. *Management Science*, **11**:404–419.

[38] Flesaker, B., Houghston, L., Schreiber, L., and Sprung, L. (1994). Taking all the credit. *Risk Magazine*, **7**:105–108.

[39] Fons, J. (1994). Using default rates to model the term structures of defaultable bonds. *Financial Analysts Journal*, September–October, 25–32.

[40] Franks, J.R. and Torous, W.N. (1994). A comparison of financial re-contracting in distressed exchanges and Chapter 11 reorganizations. *Journal of Financial Economics*, **35**:349–370.

[41] Geske, R. (1977). The valuation of corporate liabilities as compound options. *Journal of Financial and Quantitative Analysis*, **12**:541–552.

[42] Gupton, G.M. and Stein, R.M. (2002). LossCalc^{TM} Moody's Model for predicting loss given default (LGD), *Global Credit Research*, Moody's Investor Services.

[43] Heath, D., Jarrow, R., and Morton, A. (1992). Bond pricing and the term structure of interest rates: A new methodology for contingent claims valuation. *Econometrica*, **60**:77–105.

[44] Hull, J.C. (2000). *Options, Futures, and Other Derivatives*. Prentice-Hall International.

[45] Hull, J. and A. White. (1990). Pricing Interest-Rate-Derivative Securities. *Review of Financial Studies*, **3**:573–592.

[46] Hull, J. and A. White. (1994a). Numerical procedures for implementing term structure models I : Single-Factor models. *Journal of Derivatives*, **2**:7–16.

[47] Hull, J. and A. White. (1994b). Numerical procedures for implementing term structure models II : Two-Factor models. *Journal of Derivatives*, **2**:37–48.

[48] Jarrow, R.A., Lando, D., and Turnbull, S.M. (1997). A Markov model of the term structure of credit risk spreads. *Review of Financial Studies*, **10**(2), Summer.

[49] Jarrow, R.A., and S.M. Turnbull. (1995). Pricing derivatives on financial securities subject to credit risk. *Journal of Finance*, **50**:53–85.

[50] Jarrow, R.A. and Turnbull, S.M. (2000). The intersection of market and credit risk. *Journal of Banking and Finance*, **24**:271–299.

[51] Jarrow, R.A. and Yildirim, Y. (2002). Valuing default swaps under market and credit risk correlation. *The Journal of Fixed Income*, 7–19.

[52] Jones, E.P., Mason, S.P., and Rosenfeld, E. (1984). Contingent claim analysis of corporate capital structures: An empirical investigation. *Journal of Finance*, **39**(3).

[53] Kou, S.G. (2002). A jump-diffusion model for option pricing. *Management Science*, **48**(8):1086–1101.

[54] Lando, D. (1994). Three essays on contingent claims pricing. Ph.D. thesis, Graduate School of Management, Cornell University.

[55] Lando, D. (1998). On Cox processes and credit risky securities. *Review of Derivatives Research*, **2**(2/3):99–120.

[56] Longstaff, F.A., and Schwartz, E. (1995a). A simple approach to valuing risky fixed and floating rate debt. *Journal of Finance* **50**(3): 789–819.

[57] Longstaff, F.A., and Schwartz, E. (1995b). The pricing of credit derivatives. *Journal of Fixed Income*, **5**(1): 6–14.

[58] Madan, D and H. Unal. (1998). Pricing the risks of default. *Review of Derivatives Research*, 2(2/3): 121–160.

[59] Mandelbrot, B.B., (1963). The variation of certain speculative prices. *Journal of Business*, 26:394–419.

[60] Mandelbrot, B.B., (1967). The variation of some other speculative prices. *Journal of Business*, 40:393–413.

[61] Marinelli, C., Rachev, S.T., Roll, R., and Göppl, H. (1999). Subordinated stock price models: Heavy tails and long-range dependence in the high-frequency Deutsche bank price record. Technical Report, Department of Statistics and Mathematical Economics, University of Karlsruhe.

[62] Martin, B., Rachev, S.T., Schwartz, E.S. (2000). Stable non-Gaussian models for credit risk management. Working paper, University of Karlsruhe, Germany.

[63] Mashal, R., and Zeevi, A. (2002). *Beyond correlation: Extreme co-movements between financial assets.* Columbia, Graduate School of Business, Columbia University.

[64] Merton, R.C. (1974). On the pricing of corporate debt: The risk structure of interest rates. *Journal of Finance*, 29:449–470.

[65] Mittnik, S., Rachev, S.T., Doganoglu, T. and Chenyao, D. (1996) Maximum likelihood estimation of stable Paretian models. Working paper, Christian Albrechts University, Kiel.

[66] Mittnik, S., and Rachev, S.T., (2000). Diagnosing and treating the fat tails in financial returns data. *Journal of Empirical Finance* 7:389–416.

[67] Nelsen, R.B. (1999). *An Introduction to Copulas.* Springer-Verlag, New York.

[68] Nielsen, L. (1999). *Pricing and Hedging of Derivative Securities.* Oxford University Press.

[69] Paulauskas, V. and Rachev, S.T. (1999). Maximum likelihood estimators in regression models with infinite variance innovation. Working paper, Vilnius University, Lithuania

[70] Pedrosa, M., and Roll, R. (1998). Systematic risk in corporate bond credit spreads. *The Journal of Fixed Income*, 7–26.

[71] Prigent, J.L., Renault, O., Scaillet, O. (2001). *An empirical investigation in credit spread indices.*

[72] Rachev, S.T., (1991). *Probability Metrics and the Stability of Stochastic Models.* John Wiley & Sons, Chichester, New York.

[73] Rachev, S.T., and Mittnik, S. (2000). *Stable Paretian Models in Finance.* Wiley & Sons, New York.

[74] Rachev, S.T., Racheva-Jotova, B., Hristov, B., I. Mandev (1999). Software Package for Market Risk (VaR) Modeling for Stable Distributed Financial Distributed Returns Mercury 1.0 Documentation.

[75] Rachev, S.T., Schwartz, E., and Khindanova, I., (2000). Stable modeling of credit risk. Working paper UCLA.

[76] Rachev, S.T., and Tokat, Y. (2000) Asset and liability management: recent advances. *CRC Handbook on Analytic Computational Methods in Applied Mathematics.*

[77] Samorodnitsky, G., Taqqu, M.S. (1994). *Stable Non-Gaussian Random Variables.* Chapman and Hall, New York.

[78] Sarig, O., and Warga, A. (1989). Some empirical estimates of the risk structure of interest rates. *Journal of Finance*, 44(5), 1351–1360.

[79] Schönbucher, P. (August 1996). The term structure of defaultable bond prices. Discussion Paper B-384, University of Bonn, SFB 303.

[80] Schönbucher, P. (1998). Term structure modelling of defaultable bonds. Discussion Paper B-384,University of Bonn, SFB 303. Review of Derivatives Studies, Special Issue: *Credit Risk* 2(2/3): 161–192.

[81] Schönbucher, P. (1999). Credit Risk Modelling and Credit Derivatives. Ph.D.-thesis, Faculty of Economics, Bonn University.

[82] Schönbucher, P. (June 1999). A tree implementation of a credit spread model for credit derivatives. Department of Statistics, Bonn University.

[83] Schönbucher, P. (2002). A tree implementation of a credit spread model for credit derivatives. *The Journal of Computational Finance*, 6(2), Winter 2002/3.

[84] Schweizer, B. (1991). Thirty Years of Copulas. G. Dall'Aglio, S. Kotz, G. Salinetti, eds. *Advances in Probability Distributions with Given Marginals*, Kluwer, Dordrecht, Netherlands. 13–50.

[85] Shane, H. (1994). Comovements of low grade debt and equity returns of highly levered firms. *Journal of Fixed Income* 3(4):79–89.

[86] Sklar, A. (1959). Fonctions de Répartition à n Dimensions et Leurs Marges. Publications de l'Institut Statistique de l'Université de Paris, 8:229–231.

[87] Tavakoli, Janet M. (1998). *Credit Derivatives: A guide to instruments and applications*. John Wiley & Sons.

[88] Tokat, Y., Rachev, S.T., Schwartz, E.S. (2001). Asset Liability Management: A review and some new results in the presence of heavy tails. Ziemba, W.T. (Ed) *Handbook of Heavy Tailed Distributions in Finance*, Handbook of Finance.

[89] Vasicek, O. (1977). An equilibrium characterization of the term structure. *Journal of Financial Economics*, 5: 177–188.

[90] Wilson, T. (1997a). Portfolio credit risk (1). *Risk* 10(9): 111–116.

[91] Wilson, T. (1997b). Portfolio credit risk (2). *Risk* 10(10): 56–61.

3

GARCH-Type Processes in Modeling Energy Prices

Irina Khindanova

Zauresh Atakhanova

Svetlozar Rachev

ABSTRACT   High price volatility is a long-standing characteristic of world oil markets and, more recently, of natural gas and electricity markets. However, there is no widely accepted answer to what the best models and measures of price volatility are because of the complexity of distribution of energy prices. Complex distribution patterns and volatility clustering of energy prices have motivated considerable research in energy finance. Such studies propose dealing with the non-normality of energy prices by incorporating models of time-varying conditional volatility or using stochastic models. Several GARCH models have been developed and successfully applied to modeling energy prices. They represent a significant improvement over models of unconditionally normally distributed energy returns. However, such models may be further improved by incorporating the Pareto stable distributed error term. The article compares the performance of normal GARCH models with the statistical properties of unconditional distribution models of energy returns. We then present the results of estimation of energy GARCH based on the stable distributed error term and compare the performance of normal GARCH and stable GARCH.

3.1   Introduction

High price volatility is a long-standing characteristic of world oil markets and, more recently, of natural gas and electricity markets. However, there is no widely accepted answer to what the best models and measures of price volatility are because of the complexity of distribution of energy prices. Complex distribution patterns and volatility clustering of energy prices have motivated considerable research in energy finance. Many analysts employ the normal distribution for modeling returns on energy assets. However, Khindanova and Atakhanova (2002) demonstrate that the normality assumption is incapable of incorporating the asymmetry and heavy tails of the empirical distribution of energy returns. They suggest modeling of energy prices based on the assumption of independently and identically distributed Pareto stable distribution. Some studies propose dealing with the non-normality of energy prices by incorporating models of time-varying conditional volatility or using stochastic models. In this paper, we focus on the former approach. We describe generalized autoregressive conditional heteroskedasticity (GARCH) models, which have been successfully applied to modeling energy prices. Applications of energy GARCH models include a study of efficiency of crude oil markets, an investigation of the effect of resource price uncertainty on rates of economic growth, studies of the links between related products, testing the theory of storage and the leverage effect, analysis of the effects of regulatory changes and the introduction of futures trading, and forecasting of energy prices. Most of the energy GARCH models assume the normally distributed disturbances. They represent a significant improvement over models of unconditionally normally distributed energy returns. However, such models may be further improved by incorporating the Pareto stable distributed error term. Following Rachev and Mittnik (2000), we use GARCH based on the assumption of Pareto stable distributed error term to model energy prices. Such a model allows us to better understand the nature of volatility in the energy markets. We demonstrate the advantages of the GARCH models in the area of energy risk management, particularly the Value-at-Risk (VaR). We show that for those series that are characterized by a stationary volatility process, stable GARCH VaR yields accurate estimates of losses, especially at high confidence levels.

The paper is organized as follows. The next section discusses numerical aspects of GARCH processes. Section 3 reviews applications of GARCH models in energy economics and finance. Section 4 contains the results of our estimation of GARCH models based on a normally distributed error term. We compare the performance of normal GARCH models with the statistical properties of the unconditional distribution models of energy returns. Then we present the results of estimation of energy GARCH based on the stable distributed error term and compare the performance of normal GARCH and stable GARCH. Section 5 concludes.

3.2   Numerical Aspects of GARCH Processes

It is well known that the volatility of asset returns is characterized by clustering. In other words, large price changes of either sign tend to be followed by large price changes of either sign, while small price changes are followed by small price changes. To model such pattern of asset returns, Engle (1982) proposed Autoregressive Conditional Heteroskedasticity (ARCH (q)) and Bollerslev (1986) generalized Autoregressive Conditional Heteroskedasticity and proposed GARCH (p, q) process. This section describes numerical aspects of the GARCH processes, including specifications of the error term and associated estimation of parameters.

The GARCH model can be defined as follows. Let the conditional mean process y_t be described by a function of past values of y, current and past values of the error term, and exogenous variables, i.e., $\mathbf{x}_t = \{y_{t-1}, y_{t-2}, \ldots, y_{t-r}, \varepsilon_t,$

$\varepsilon_{t-1}, \varepsilon_{t-2}, \ldots, \varepsilon_{t-s}, x_1, x_2, \ldots, x_m$}. Then, $E[y_t|y_{t-1}] = \mathbf{x}_t\theta$ and Var $(y_t|y_{t-1}) = E_{t-1}[(y_t - \mathbf{x}_t\theta)^2] = E_{t-1}[\varepsilon_t^2] = \sigma_{t-1}^2$.

The error process is given by

$$\varepsilon_t = v_t\sigma_t,$$
$$\sigma_t^2 = \omega + \alpha_1\varepsilon_{t-1}^2 + \alpha_2\varepsilon_{t-2}^2 + \ldots + \alpha_p\varepsilon_{t-p}^2 + \beta_1\sigma_{t-1}^2 + \beta_2\sigma_{t-2}^2 + \cdots$$
$$\cdots + \beta_q\sigma_{t-q}^2, \tag{3.1}$$

where $v_t \sim$ iid $(0, 1)$, $\omega > 0$, $\alpha_i \geq 0$ for $i = 1, \ldots, p$, and $\beta_i \geq 0$ for $i = 1, \ldots, q$. If $\beta_i = 0$ for all i, then y_t follows the Autoregressive Conditional Heteroskedasticity or ARCH (p) process introduced by Engle (1982). The requirement for stationarity is $\sum_{i=1}^{m} \alpha_i + \sum_{i=1}^{m} \beta_i < 1$, where $m = \max(p, q)$. If $\sum_{i=1}^{m} \alpha_i + \sum_{i=1}^{m} \beta_i = 1$, then the process for ε_t has a unit root and such a process is known as the integrated GARCH (Engle and Bollerslev, 1986 and Nelson, 1990) or **IGARCH**. If the variance has d unit roots, then it is integrated of order d (usually an integer).

GARCH (p, q) parameters are obtained by maximizing the log likelihood function. The sequence $\{\sigma_t^2\}_{t=1}^{T}$ can be used to evaluate the log likelihood as follows:

$$L(\theta) = \sum_{t=1}^{T} \log f(y_t|\mathbf{x}_t, y_{t-1}, \theta, \omega, \alpha, \beta). \tag{3.2}$$

For a given density function $f(y_t|\mathbf{x}_t, \mathbf{Y}_{t-1})$, the expression given by (3.2) can be maximized numerically with respect to parameters $\theta, \omega, \alpha_1, \alpha_2, \ldots, \alpha_p, \beta_1, \beta_2, \ldots, \beta_q$ of a GARCH (p, q) process. For example, let us consider a GARCH (1,1) process with normally distributed error term:

$$y_t = \beta x_t + \varepsilon_t$$
$$\varepsilon_t = v_t(\omega + \alpha_1\varepsilon_{t-1}^2 + \beta_1\sigma_{t-1}^2)^{0.5},$$

where $v_t \sim$ iid N(0,1). Then the conditional variance of ε_t is given by

$$\sigma_t^2 = \omega + \alpha_1\varepsilon_{t-1}^2 + \beta_1\sigma_{t-1}^2 = \omega + \alpha_1(y_{t-1} - \beta x_{t-1})^2 + \beta_1\sigma_{t-1}^2$$

and the log likelihood function of this GARCH (1,1) process is given by

$$\log L = -\frac{T}{2}\ln 2\pi - \frac{1}{2}\sum_{t=1}^{T}\ln\sigma_t^2 - \frac{1}{2}\sum_{t=1}^{T}\sigma_t^{-2}(y_t - \beta x_t)^2.$$

The log likelihood function of a GARCH (1,1) process can be maximized with respect to β, ω, α_1, and β_1 by recursive iterations. In general, GARCH (1, 1) is found to be the most parsimonious GARCH (p, q) representation and is the most commonly used process in modeling economic and financial series.

Empirical studies often find that standardized GARCH residuals v_t from model (3.1) have heavier than normal tails. Therefore, several researchers propose modifying GARCH models by allowing v_t to follow a non-normal distribution. Here we describe three types of GARCH models that are based on non-normal distributions of the error term: the generalized error distribution, the Student's t-distribution, and the stable Paretian distribution.

Nelson (1991) proposed the *generalized error distribution* to describe the process for v_t in $\varepsilon_t = v_t \sigma_t$:

$$f(v_t) = \frac{v \exp\left(-\frac{1}{2}\left|\frac{v_t}{\lambda}\right|^v\right)}{\lambda \cdot 2^{\frac{v+1}{v}} \Gamma\left(\frac{1}{v}\right)}.$$

Here $\Gamma(\cdot)$ is the gamma function and λ is a constant given by

$$\lambda = \left\{\frac{2^{-\frac{2}{\lambda}}\Gamma\left(\frac{1}{v}\right)}{\Gamma\left(\frac{3}{v}\right)}\right\}^{\frac{1}{2}},$$

where $v > 0$ is the parameter governing the thickness of the tails of the distribution of v_t. If $v = 2$, $\lambda = 1$, then $v_t \sim N(0,1)$. If $v < 2$, then the tails are thicker and if $v > 2$, they are thinner than in the normal distribution case. $E|v_t| = \frac{\lambda \cdot 2^{\frac{1}{v}}\Gamma\left(\frac{2}{v}\right)}{\Gamma\left(\frac{1}{v}\right)}$ and if $v = 2$, then $E|v_t| = \sqrt{(2/\pi)}$ (Hamilton, 1994, 668–669).

Another possible specification of the error process is to assume that it follows *Student's t-distribution* with v degrees of freedom (Bollerslev 1987):

$$f(v_t) = \Gamma\left(\frac{v+1}{2}\right)\Gamma\left(\frac{v}{2}\right)^{-1}((v-2)\sigma_t^2)^{-0.5}(1+\varepsilon_t^2\sigma_t^{-2}(v-2)^{-1})^{-(v+1)/2}.$$

The *t* distribution is symmetric around zero and it allows for fatter than normal tails.

Liu and Brorsen (1995) and Rachev and Mittnik (2000) propose an alternative distribution for v_t and define the **stable GARCH** process as

$$\begin{aligned} y_t &= \mu_t + v_t \sigma_t, \\ \sigma_t &= \kappa + \delta\sigma_{t-1} + \gamma|y_{t-1} - \mu_{t-1}|, \end{aligned} \tag{3.3}$$

where v_t is an independently distributed *stable standard variable* (with zero location parameter and unit scale) with index of stability α, $1 < \alpha \leq 2$ and skewness parameter $\beta, -1 \leq \beta \leq 1$. The process in (3.3) is covariance stationary if $(\lambda_{\alpha\beta}\gamma + \delta) < 1$.[1]

If $S(\frac{y_t - \mu_t}{\sigma_t}, \alpha, \beta)$ is the approximate stable density function, the log likelihood function for a series of variable y_t with stability index α, skewness β, scale σ_t and location μ_t is given by

$$\log L = \sum_{t=1}^{T} (\log S(\frac{y_t - \mu_t}{\sigma_t}, \alpha, \beta) - \log \sigma_t) \tag{3.4}$$

The approximate stable density function may be computed based on Zolotarev (1966) and DuMouchel (1973 a, b). The expression in (3.4) is maximized numerically to obtain MLEs of α, β, σ_t, μ_t, δ, and γ.

Mittnik, Rachev, Doganoglu, and Chenyao (1999) describe the maximum likelihood estimation of stable ARMA and stable GARCH models based on the Fourier transform of the stable characteristic function. The estimation of stable ARMA and stable GARCH is nontrivial because of the lack of analytic expression for the probability density function (PDF) of stable distributions. However, stable distributions can be described by characteristic functions (CFs) of the following general form:

$$\varphi(t) = \exp\left\{i\mu y - |ct|^\alpha \left[1 - i\beta\frac{t}{|t|}\omega(|t|, \alpha)\right]\right\}, \tag{3.5}$$

where $\omega(|t|, \alpha) = \begin{cases} \tan\frac{\pi\alpha}{2}, & \text{for } \alpha \neq 1, \\ -\frac{2}{\pi}\log|t|, & \text{for } \alpha = 1. \end{cases}$

The authors show that stable ARMA and GARCH models may be estimated on the basis of approximate stable PDF using the method detailed in Mittnik, Doganoglu, and Chenyao (1999). The authors describe the algorithm for efficient computation of the PDF of stable Paretian distributions. A PDF of a stable distribution $p(x)$ may be expressed as a Fourier transform of its CF given by $\phi(t)$ as follows:

$$p(x; \alpha, \beta, c, \mu) = \frac{1}{2\pi} \int_{-\infty}^{\infty} e^{-ixt}\varphi(t)dt \quad . \tag{3.6}$$

Mittnik, Doganoglu, and Chenyao (1999) show that the integral in (3.6) can be approximated for N equally spaced points with distance h, i.e., for $x(k) = (k - 1 - N/2) * h, k = 1, \ldots, N$ by:

$$\lambda_{\alpha\beta} = \frac{2}{\pi}\Gamma\left(1 - \frac{1}{\alpha}\right)\left(1 + \tau_{\alpha\beta}^2\right)^{\frac{1}{2\alpha}}\cos\left(\frac{1}{\alpha}\arctan\tau_{\alpha\beta}\right), \text{ if } 1 < \alpha < 2,$$

$$= \sqrt{\frac{2}{\pi}}, \text{ if } \alpha = 2 \text{ where } \tau_{\alpha\beta} = \beta\tan(\alpha\pi/2).$$

$$p\left(\left(k - 1 - \frac{N}{2}\right)h\right) \approx s\,(-1)^{k-1-N/2}$$

$$\times \sum_{n=1}^{N} \varphi\left(2\pi s\left(n - 1 - \frac{N}{2}\right)\right)e^{-i2\pi(n-1)(k-1)/N}, \quad (3.7)$$

where $s = (h * N)^{-1}$. The summation in (3.7) can be calculated by applying Fast Fourier Transforms (FFT) to the sequence:

$$(-1)^{n-1}\varphi\left(2\pi s\left(n - 1 - \frac{N}{2}\right)\right), \quad \text{for } n = 1, \ldots, N,$$

where $\phi(\cdot)$ is the CF of a standardized stable distribution, $z = (x - \mu)/c$, such that

$$p(x; \alpha, \beta, c, \mu) = \frac{1}{c}p(z; \alpha, \beta, 1, 0). \quad (3.8)$$

So for $\alpha \neq 1$, $\mu = 0$, and $c = 1$, the CF reduces to the following:

$$\varphi(t) = \exp\left\{-|t|^{\alpha} + i\beta t\,|t|^{\alpha-1} \tan\frac{\pi\alpha}{2}\right\}. \quad (3.9)$$

Standardized PDF values are computed by substituting (3.9) into (3.7) with $t = 2\pi s(n - 1 - N/2)$ and used for inferring any stable PDF parameter values by applying (3.8). Since the obtained PDF values are associated with the equally-spaced x values, at the next step, it is necessary to use interpolation to obtain PDF values for those data points that fall in between the equally spaced x's. In their study, Mittnik, Doganoglu, and Chenyao (1999) consider the issues of accuracy and computational speed of the proposed algorithm. They suggest using powers of 2 for N, i.e., $N = 2^q$. The authors find that setting h=0.01 and q=13 and using linear interpolation is sufficient for financial applications. Having discussed numerical issues of GARCH processes, we next describe applications of models of conditional heteroskedasticity to analyzing energy price dynamics.

3.3   GARCH Processes in Modeling Energy Prices

Examination of the energy return series reveals volatility clustering characterized by periods of high volatility followed by periods of relative tranquility (see Figures 1 and 2). GARCH-type models are suited to describe such conditional heteroskedasticity. There have been several applications of these models to represent the dynamics of energy prices. The existing literature can be classified according to the type of GARCH models used, starting from the simplest one of ARCH and concluding with more complex applications of FIGARCH, switching ARCH, and stable GARCH. Table 1 summarizes the extent of relevant literature and the nature and sources of data used.

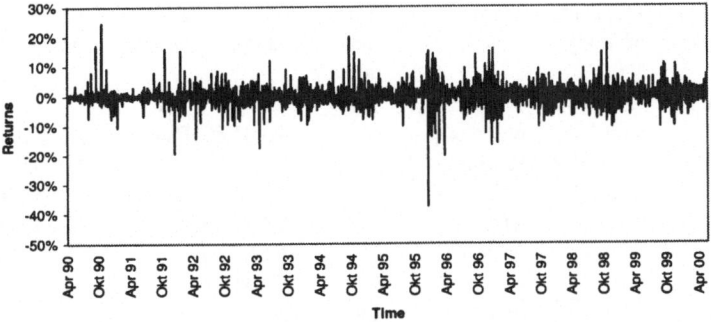

FIGURE 1. Henry Hub One-Month Futures Returns

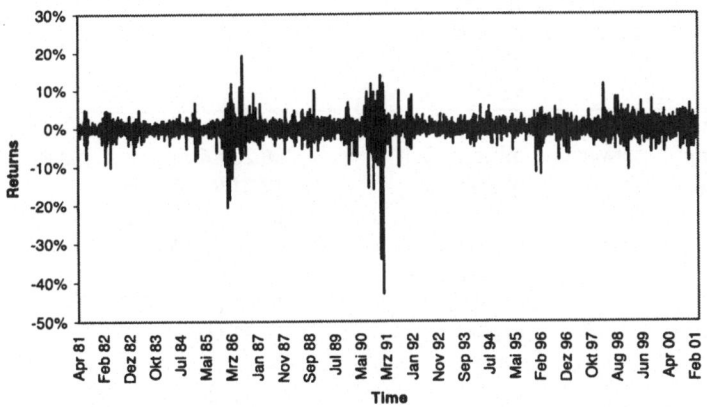

FIGURE 2. Gas Oil One-Month Futures Returns

TABLE 1: SUMMARY OF ENERGY GARCH LITERATURE

Author(s), year	Objective of Study	Series	Sample frequency	Data range	Source of data	Best fitting model
Robinson & Taylor (1998)	Impact of regulatory change on share prices of electric utilities	Share prices of electricity companies	Daily	12.10.90 – 03.11.96	Datastream	ARCH(1)
Deaves & Krinsky (1992)	Risk premium & efficiency in crude oil market.	NYMEX crude oil 1- & 3-mo. futures	Monthly	03.83- 04.90	Techtools	ARCH-M(1)

Author(s), year	Objective of Study	Series	Sample frequency	Data range	Source of data	Best fitting model
Adrangi et al. (2001b)	Sources of non-linearities in energy prices.	NYMEX crude oil, heating oil & unleaded gasoline futures	Daily	10.01.83-03.31.95; 01.02.85-03.31.95; 01.02.85-03.31.95	The Futures Industry Institute	GARCH(1,1), Asymmetric GARCH(1,1), EGARCH(1,1)
Adrangi et al. (2001a)	Link between prices of crude oil & diesel fuel.	Alaskan North Slope crude oil spot; Los Angeles diesel fuel spot	Daily	NA	NA	Bivariate GARCH(1,1)
Ng & Pirrong (1996)	Link between spot & futures prices.	NY heating oil spot, Gulf of Mexico gasoline spot, NYMEX heating oil & gasoline nearby futures	Daily	08.20.84-12.31.90 for heating oil; 12.04.84-12.31.90 for gasoline	Commo-dity Futures Trading Commis-sion (CFTC)	Bivariate GARCH(1,1)
Moosa & Al-Loughani (1994)	Test of un-biasedness & efficiency of futures prices as forecasters of spot prices.	WTI crude: spot 3- & 6-mo. futures	Monthly	01.86-07.90	NA	GARCH-M (1,1)
Antoniou & Foster (1992)	Introduc-tion of futures & spot price volatility.	Brent crude spot	Daily	01.86-07.90	Data-stream	IGARCH (1,1)
Morana (2001)	Forecast-ing.	Brent crude spot	Daily	01.04.82 – 01.21.99	NA	Asymmetric GARCH(1,1)

Author(s), year	Objective of Study	Series	Sample frequency	Data range	Source of data	Best fitting model
Wickham (1996)	Volatility of crude oil prices.	Dated Brent Blend Crude	Monthly	01.80–06.96	NA	AR(4)-GARCH(1,1)
Boyd & Caporale (1996)	Volatility of natural resource prices & rates of economic growth.	Real prices of oil, bituminous coal, nonferrous metals, & iron & steel, PPI	Monthly	01.47 –12.91	Citibase	MA(1or 2)-GARCH(1,1)
Abosedra & Laopodis (1997)	Non-normality of distribution of crude oil prices.	Crude oil spot	Monthly	01.74 –01. 95	EIA Monthly Energy Review	AR(2) – EGARCH -M(1,1)
Mazaheri (1999)	Modeling convenience yield.	Spot & nearby futures: crude oil, unleaded gasoline & heating oil	Daily	1568 obs., 2059 obs., 2422 obs.	Commodity Systems Inc	Asymmetric ARFIMA-GARCH(1,1) & ARFIMA-EGARCH (1,1)
Brunetti & Gilbert (2000)	Link between NYMEX & IPE crude oil prices.	NYMEX & IPE second delivery crude oil futures	Monthly	06.88-03.99	NA	Bivariate FIGARCH (1,1)
Susmel & Thompson (1997)	Modeling regime changes & time-varying volatility.	Natural gas spot	Monthly	01.74 –03.94	USDE Monthly Energy Review	SWARCH (2,2)
Rachev & Mittnik (2000)	Modeling non-normality & time-varying volatility.	AMEX oil index spot	Daily	09.01.88 –07.28.94	NA	AR(4) – stable ARCH(1)

This paper considers modeling of energy series using GARCH type models in the following order: ARCH, GARCH, GARCH-in-mean (GARCH-M), asymmetric GARCH, exponential GARCH (EGARCH), fractionally integrated (FIGARCH), GARCH in combination with the autoregressive moving average process (ARMA-GARCH), GARCH in combination with the autoregressive fractionally integrated moving average process (ARFIMA-GARCH), switching ARCH (SWARCH), Stable GARCH. Most models represent the mean process as $R_t = \mu + \varepsilon_t$, where estimated μ is approximately equal to zero. In such cases, we discuss the specification of the variance process omitting the reference to the mean process.

3.3.1   ARCH Models

ARCH models allow the conditional variance of energy returns to depend on past values of squared residuals. For example, an ARCH(p) process is given by

$$\varepsilon_t = v_t \sigma_t,$$
$$\sigma_t^2 = \omega + \alpha_1 \varepsilon_{t-1}^2 + \alpha_2 \varepsilon_{t-2}^2 + \ldots + \alpha_p \varepsilon_{t-p}^2,$$

where ε_t is the error term of the mean process, σ_t^2 is the conditional variance of ε_t, $v_t \sim$ iid $(0, 1)$, $\omega > 0$, $\alpha_i \geq 0$ for $i = 1, \ldots, p$.

Robinson and Taylor (1998) use this model to study the impact of a regulatory change on share prices of electric utilities in the UK. We consider this work in our present review of GARCH models of energy prices since the changes in share prices may be indirectly related to fluctuations in the value of electricity. Another application of ARCH models is by Deaves and Krinsky (1992) who study risk premia and efficiency of energy markets. They use a slightly modified version of ARCH, ARCH-M.

Robinson and Taylor (1998) use ARCH (1) to study the effect of a regulatory change on share prices R_t of 12 UK regional electricity companies (RECs). Regulation of RECs is based on a system of price caps that are reviewed periodically. The main source of uncertainty for RECs is the nature of such reviews as well as a possibility of regulatory interventions between the price reviews. One such intervention took place on March 7, 1995 when the regulators announced their intention to review price controls prior to the next scheduled price review. To account for the effect of this particular regulatory change, Robinson and Taylor (1998) propose introducing two dummy variables: $D_{It} = 1$ on March 7, 1995 and zero otherwise; and $D_{Lt} = 1$ starting from March 3, 1995 and zero otherwise. Thus their ARCH (1) model is specified as follows:

$$R_t = \beta_0 + \beta_1 D_{It} + \beta_2 D_{Lt} + \varepsilon_t,$$
$$h_t = a_0 + a_1 \varepsilon_{t-1}^2 + c_1 D_{It} + c_2 D_{Lt}$$

where R_t are is the conditional mean of returns, h_t is the conditional variance of returns, and $\varepsilon_t | I_{t-1} \sim N(0, h_t)$ is the disturbance term. Estimated values of β_2 and c_1

are not reported. Robinson and Taylor (1998) find that the estimate of β_1 is significant and negative in all cases. This finding is consistent with the large fall in share prices following the announcement of intervention. The estimate of c_2 is significant and positive in eight out of 12 cases. Thus the regulatory intervention has resulted in a permanent increase in the conditional volatility of share prices of most of the RECs. The implication of this finding is that unexpected regulatory changes may raise the cost of capital for RECs. However, the authors also note that changes in conditional volatility of REC share prices may be associated with large-scale takeover activity in this industry during the period in question.

Deaves and Krinsky (1992) use 84 monthly observations of one- to three-month futures of NYMEX crude oil in order to study the nature of risk premia and efficiency in crude oil markets. Futures returns are calculated as $R_t = F_{T,T} - F_{t,T}$, where $F_{t,T}$ is log price at time t of the futures contract to be delivered on date T. $F_{T,T}$ is the log price of futures on date T at delivery. The authors find that the conditional volatility of R_t is best described by an ARCH (1) process $\sigma_t^2 = \omega + \alpha_1 \varepsilon_{t-1}^2$ and also that the futures returns are positive on average. Then the authors proceed to modeling one-month futures returns using the ARCH-M model:

$$y_t = \beta x_t + \delta \sigma_t + \varepsilon_t \text{ or } y_t = \beta x_t + \delta \sigma_t^2 + \varepsilon_t,$$
$$\sigma_t^2 = \omega + \alpha_1 \varepsilon_{t-1}^2 + \alpha_2 \varepsilon_{t-2}^2 + \ldots + \alpha_p \varepsilon_{t-p}^2.$$

Deaves and Krinsky (1992) find that inclusion of either a variance or a standard deviation term in the mean equation increases the explanatory power of the model. To test the efficiency of futures markets, excess returns are calculated as the difference between returns R_t defined above and returns predicted by the ARCH-M model. If markets are efficient, excess returns have to be zero. Depending on whether the information set (data on past returns, publicly available data or privately available data), market efficiency may hold in its weak, semi-strong or strong form, respectively. Using five lags of excess returns Deaves and Krinsky (1992) fail to reject the efficiency hypothesis in the weak form. However, they do reject it in its semi-strong form, which implies that this hypothesis also does not hold in the strong form. The interpretation of this result is that positive excess returns are possible in the crude oil market if the analysis is based on a superior information set.

Deaves and Krinsky (1992) conclude that the presence of positive returns implies that speculators benefit from futures operations. Such a risk premium is paid by the producers hedging against falling prices. Hedgers pay the speculators for accepting the price risk. The validity of the ARCH-M model implies that this risk premium is dependent on the volatility of oil price changes.

3.3.2   GARCH Models

GARCH models specify conditional variance as a function of both past values of squared residuals and past values of the variance (see (3.1)). Adrangi et al. (2001b) use GARCH to analyze the nature of non-linearities in energy prices. By the non-linearities the authors mean second and higher order dependence between the energy prices. Adrangi et al. (2001a) use bivariate GARCH to investigate the dynamics of crude oil and diesel prices. Bivariate GARCH is also employed by Ng and Pirrong (1996) to study the behavior of refined petroleum spot and futures prices.

Adrangi et al. (2001b) examine the non-linearities in the energy price data. The authors evaluate two alternative approaches to modeling such non-linearities. They conduct tests of chaotic structures and reject them in favor of GARCH-type processes. Specifically, they study deseasonalized daily returns on crude oil, heating oil and unleaded gasoline futures contracts traded at NYMEX over a period of ten years. Adrangi et al. (2001b) find that the non-linear dynamics of energy price changes can be modeled by GARCH (1,1). They propose the following process of conditional volatility of energy prices,

$$h_t = \alpha_0 + \alpha_1 \varepsilon_{t-1}^2 + \beta_1 h_{t-1} + \beta_2 TTM_t ,$$

where h_t is the conditional variance, ε_t is the disturbance term, and TTM is the time to maturity. The time to maturity is introduced in order to test the maturity effect, which implies that the volatility of a futures contract increases as the delivery date approaches. The authors obtain negative and significant estimates of β_2, which they interpret as evidence in support of the maturity effect. The authors also analyze the robustness of their test to structural changes in the data. Given the high level of energy price volatility during 1986 and 1991, the authors introduce dummy variables corresponding to the observations from these two periods. Their results confirm the previous finding that energy price dynamics is not chaotic and that it is consistent with ARCH-type processes. Adrangi et al. (2001a) use GARCH in a bivariate setting in order to analyze the price dynamics of crude oil (Alaska North Slope) and diesel fuel (Los Angeles, CA). The authors test whether the data supports the concept of derived demand, i.e., whether changes in the diesel fuel price are determined by the changes in the crude oil market. Thus they test the proposition that price discovery emanates from the more liquid market, i.e., from the crude oil to the diesel fuel market.[2] Adrangi et al. (2001a) use the vector autoregression (VAR) and bivariate GARCH methodology to show that there is a strong uni-directional causal relationship from the crude oil price to the price of diesel fuel.

The authors initially estimate the following VAR model:

$$\Delta \ln C_t = \mu_0 + \sum_{i=1}^{n} \mu_i \Delta \ln C_{t-i} + \sum_{i=1}^{n} \Delta \theta_i D_{t-i} + \lambda_C (\ln C - \ln D)_{t-1} + \varepsilon_{C,t},$$

$$\Delta \ln D_t = \gamma_0 + \sum_{i=1}^{n} \gamma_i \Delta \ln C_{t-i} + \sum_{i=1}^{n} \Delta \xi_i D_{t-i} + \lambda_D (\ln C - \ln D)_{t-1} + \varepsilon_{D,t},$$

---

[2] Adrangi et al. (2001a) argue that the market for Alaska North Slope crude oil is more liquid relative to the Californian diesel fuel market because the former is related to the international crude oil market while the latter has regional properties.

where C_t and D_t are prices of crude oil and diesel fuel, respectively. $\varepsilon_{C,t} \sim N(0, \sigma_{C,t})$ and $\varepsilon_{D,t} \sim N(0, \sigma_{D,t})$, parameter $\lambda_j, 0 \le |\lambda_j| \le 1$ for $j = C, D$, represents the degree of convergence between the two commodities. Testing for causal relationship corresponds to testing the signs of λ_C and λ_D. If $\lambda_C < 0$ and $\lambda_D = 0$, then the crude oil price is responsible for adjustments that restore the long run equilibrium between the two prices. If $\lambda_C = 0$ and $\lambda_D > 0$, then it is the diesel fuel price that adjusts to short-term fluctuations in the spread between the two prices.

Further on, the authors propose modeling the time varying volatility for both of the series. In addition to modeling the univariate variance persistence, Adrangi et al. (2001a) employ the fact that a linear combination of the two series might have lower persistence. The authors propose a bivariate GARCH (1,1) model, which also allows them to test the direction of volatility spillovers between the two markets as follows:

$$\sigma_{C,t}^2 = \alpha_0 + \alpha_1 \sigma_{C,t-1}^2 + \alpha_2 \varepsilon_{C,t-1}^2 + \alpha_3 \eta \varepsilon_{C,t-1}^2 + \alpha_4 \varepsilon_{D,t-1}^2 + u_{C,t} \,,$$

$$\sigma_{D,t}^2 = \beta_0 + \beta_1 \sigma_{D,t-1}^2 + \beta_2 \varepsilon_{D,t-1}^2 + \beta_3 \eta \varepsilon_{D,t-1}^2 + \beta_4 \varepsilon_{C,t-1}^2 + u_{D,t} \,,$$

$$\sigma_{CD,t} = \pi_0 + \pi_1 \sigma_{CD,t-1} + \pi_2 \varepsilon_{C,t-1} \varepsilon_{D,t-1} + u_{CD,t} \,,$$

where $\sigma_{C,t}^2$ and $\sigma_{D,t}^2$ are the conditional variances of crude oil and diesel fuel, respectively, and $\sigma_{CD,t}$ is their conditional covariance. α_1 and β_1 measure the volatility persistence. α_4 and β_4 measure the volatility spillover between the markets. η is a dummy variable introduced to account for the leverage effect. The estimation results show that there is a strong causality or information flow from the crude oil to the diesel fuel market, i.e., the estimate of λ_D is positive and significant while the estimate of λ_C is insignificant. The estimate of β_4 is significant while α_4 is not, therefore volatility spills over from the crude oil to the diesel fuel market.

Ng and Pirrong (1996) make another application of bivariate GARCH to study the price dynamics of spot and futures prices of two refined petroleum markets: gasoline and heating oil. They propose using a non-linear error-correction model with time-varying volatility in order to account for several observed phenomena in these markets. First of all, according to the theory of storage, there is a long-term equilibrium relationship between the spot and futures prices of physical commodities reflected in the equilibrium level of the convenience yield.[3] This equilibrium relationship tends to be restored after news forces the spot and the futures prices to diverge from their long-term equilibrium spread:

$$z_t = \ln F_t - \ln S_t - r_t(T - t) - w(T - t) + l(T - t) \,,$$

where F_t and S_t are current futures and spot prices, T is the time of delivery of the futures contract, r_t is the current interest rate, w is the proportional warehousing cost, l is the convenience yield, z_t is the deviation of inventory from its long run level. The

---

[3] Convenience yield refers to a stream of benefits to holders of a physical commodity, which do not accrue to the holders of futures contracts on the same commodity.

interdependence between the spot and futures prices may be modeled with the help of an error correction model. It incorporates the adjustment process back to the long-run equilibrium, which can be asymmetric and non-linear. This pattern is especially pronounced in the physical commodities markets. When inventories are low, convenience yield is large and positive. However, it is very costly for producers to increase output significantly and rapidly in order to take advantage of a high spot price. Conversely, when convenience yield is large and negative, producers do not cut output very fast. Thus, this market is characterized by rapid and significant increases in spot prices when the convenience yield is positive and large, as opposed to slow declines in spot prices when the convenience yield is negative and large. Moreover, since the capacity can be increased in large increments, such additions to capacity are possible given sufficiently high spot prices, i.e., the magnitude of the price change affects the adjustment process and causes its non-linearity. Thus, the authors introduce z_{t-i}^+ and z_{t-i}^- to account for the asymmetry of adjustment process[4] to the long-term equilibrium spread. $(z_{t-i}^-)^2$ and $(z_{t-i}^+)^2$ account for non-linearity of this adjustment in the error correction model:

$$\Delta \ln S_t = \alpha_0 + \sum_{i=1}^{n} \alpha_i \Delta \ln S_{t-i} + \sum_{i=1}^{n} \beta_i \Delta \ln F_{t-i} + \sum_{i=1}^{n} \delta_i \Delta y_{t-i}$$
$$+ \theta_1 z_{t-1}^+ + \theta_2 z_{t-1}^- + \theta_3 (z_{t-1}^+)^2 + \theta_4 (z_{t-1}^-)^2 + \varepsilon_t,$$

$$\Delta \ln F_t = \gamma_0 + \sum_{i=1}^{n} \gamma_i \Delta \ln S_{t-i} + \sum_{i=1}^{n} \phi_i \Delta \ln F_{t-i} + \sum_{i=1}^{n} \lambda_i \Delta y_{t-i} + \psi_1 z_{t-1}^+$$
$$+ \psi_2 z_{t-1}^- + \psi_3 (z_{t-1}^+)^2 + \psi_4 (z_{t-1}^-)^2 + \eta_t,$$

where $\Delta y_{t-i} = r_t(T-t) - r_{t-i}(T-t+i)$. The time-varying volatility is modeled as a bivariate GARCH (1,1) including the asymmetry terms:

$$h_{s,t} = \omega + \lambda h_{s,t-1} + \mu_1 \varepsilon_{t-1}^2 + \mu_2 \eta_{t-1}^2 + \mu_3 (z_{t-i}^-)^2 + \mu_4 (z_{t-i}^+)^2,$$
$$h_{f,t} = \sigma + \kappa h_{f,t-1} + \upsilon_1 \varepsilon_{t-1}^2 + \upsilon_2 \eta_{t-1}^2 + \upsilon_3 (z_{t-i}^-)^2 + \upsilon_4 (z_{t-i}^+)^2,$$

where $h_{s,t}$ and $h_{f,t}$ are conditional variances of ε_t and η_t, respectively. The estimation of the above model leads to the following results: spreads between spot and futures prices explain a significant portion of spot volatility; spot prices are more volatile when spot prices are higher than futures prices; shocks to the volatility of futures prices are more persistent than shocks to the volatility of spot prices; the volatility spills over from the futures to the spot market. The implication of these findings is that the fundamentals, i.e., changes in the current inventories, rather than a speculative activity, determine the spot price volatility, and that the futures market is the locus of informed trading in the refined petroleum markets.

---

[4] $z_{t-1}^+ = z_{t-1}$ if $z_{t-1} > 0$ and zero otherwise. Similarly, $z_{t-1}^- = z_{t-1}$ if $z_{t-1} < 0$ and zero otherwise.

3.3.3 GARCH-M

GARCH-M models incorporate either the standard deviation or the variance into the function governing the process of the mean of energy returns.

$$y_t = \beta x_t + \delta \sigma_t + \varepsilon_t$$

or

$$y_t = \beta x_t + \delta \sigma_t^2 + \varepsilon_t \,,$$
$$\sigma_t^2 = \omega + \alpha_1 \varepsilon_{t-1}^2 + \beta_1 \sigma_{t-1}^2 \,,$$

where y_t is the conditional mean of returns, x_t is the exogenous variable, where ε_t is the error term, σ_t^2 is the conditional variance of returns.

Moosa and Al-Loughani (1994) use GARCH-M to test the hypothesis of unbiasedness and efficiency of energy futures prices as forecasters of spot prices. Antoniou and Foster (1992) use GARCH-M to investigate the effect of introduction of futures trading on crude oil spot prices.

Moosa and Al-Loughani (1994) use GARCH-M to model time varying risk premia of energy futures. The authors test the hypothesis of unbiasedness and efficiency of energy futures prices as forecasters of spot prices. They used monthly observations of prices related to WTI crude oil: the spot, three-month and six-month futures. The model specification is as follows:

$$S_t = \beta_0 + \beta_1 F_{i,t-1} + \varepsilon_t \quad \text{for } i = 3, 6 \,,$$

where S_t is the spot price, F_t is the futures price, and ε_t is the error process reflecting the news arriving during the contract period. The test of unbiasedness corresponds to testing if $(\beta_0, \beta_1) = (0, 1)$. The test of efficiency corresponds to testing whether $E(\varepsilon_t, \varepsilon_{t+j}) = 0$ for $j \neq 0$. After checking that all three series are difference stationary of order 1, the authors proceed to specify the following error correction model:

$$\Delta S_t = \alpha_1 \varepsilon_{t-1} + \alpha_2 \Delta F_{i,t-i} + \sum_{j=1}^{m} \theta_j \Delta S_{t-j} + \sum_{j=1}^{k} \phi_j \Delta F_{i,t-i-j} + v_t \text{ for } i = 3, 6 \,.$$

Here m and k are number of lags of the spot and futures prices, included in the error correction model. Equivalently,

$$\Delta S_t = \alpha_0 + \alpha_1 (S_{t-1} - F_{t-i-1}) + \alpha_2 \Delta F_{i,t-i} + \sum_{j=1}^{m} \theta_j \Delta S_{t-j} + \sum_{j=1}^{k} \phi_j \Delta F_{i,t-i-j} + v_t \,.$$

Here $\alpha_0 = -\alpha_1 \beta_0$. Thus, the unbiasedness hypothesis is equivalent to testing $\alpha_0 = 0$, $\alpha_1 = -1$, $\alpha_2 = 1$ and $\theta_j = \phi_j = 0$ for all j. To test the efficiency hypothesis, Moosa and Al-Loughani (1994) propose testing if $\gamma_j = 0$ for all j, where γ_j is obtained from the following regression:

$$S_t - F_{i,t-1} = \gamma_0 + \sum_{j=1}^{m} \gamma_i \Delta S_{t-i} .$$

Moosa and Al-Loughani (1994) reject the unbiasedness and efficiency hypotheses for both the three- and the six-month futures. They argue that rejection of the above hypotheses may have two explanations. One is the lack of rational expectations and the second one is the presence of a time-varying risk premium. The validity of the first reason is impossible to test in the absence of survey data. Therefore, the authors equate failure to reject the unbiasedness and efficiency hypotheses with the presence of time-varying risk premium, which they model using GARCH-M (1,1) defined as follows:

$$\rho_t = S_t - F_{t-1},$$
$$\rho_t = a_0 + a_1 h_t + \varepsilon_t,$$
$$h_t = b_0 + b_1 h_{t-1} + b_2 \varepsilon_{t-1}^2 ,$$

where ρ_t is the risk premium, h_t is the conditional variance of the risk premium, and ε_t is the error term. The coefficients in the GARCH-M model were found to be significant. So the authors conclude that there is a time-varying risk premium in crude oil prices adequately described by the GARCH-M process.

Antoniou and Foster (1992) use GARCH-M to study the effect of the introduction of futures trading on the volatility of Brent crude oil spot prices. The authors estimate a conditional heteroskedasticity model of energy price returns for pre-futures (06/01/86–26/09/88) and post-futures (03/10/88–30/07/90) periods. The GARCH-M (1,1) model was found to have the most appropriate representation of returns y_t for both sample periods:

$$y_t = \mu_t + \delta\sigma_t + \varepsilon_t,$$
$$\sigma_t^2 = t\omega + \alpha_1\varepsilon_{t-1}^2 + \beta_1\sigma_{t-1}^2 .$$

The authors tested the null hypothesis of GARCH parameters being equal across the two sample periods, which was strongly rejected. Ljung–Box test statistics revealed no heteroskedasticity and serial correlation in the GARCH-M residuals. Further analysis of the estimated models revealed a possible presence of integrated variance processes in both pre- and post-futures periods. In other words, the sum of estimated parameters α_1 and β_1 was close to unity: 0.98 and 0.96 for the first and the second period, respectively. This finding is indicative of a possibility of a presence of the non-stationary variance. In this case, a shock to the variance is permanent. Dickey–Fuller tests were conducted for the IGARCH specifications. This test was not rejected for the pre-futures period, but it was rejected for the post-futures period. The interpretation of this result is that the persistence of shocks to the variance has decreased after the introduction of futures trading.

Another interesting finding is that the coefficient of the conditional variance in the conditional mean equation δ is significantly lower in the second period. This means that the possibility of hedging through derivative instruments have reduced the importance of volatility in the mean process for energy returns. The parameters of the conditional variance equation have also changed. The estimates of ω and α_1 are greater for the post-futures period and the estimate of β_1 is smaller. The authors interpret the first observation as an indicator of increased informational efficiency of spot markets, which have become more flexible in terms of their reaction to the arrival of news. The second finding is interpreted as an indication of a lesser importance of the past variance in the conditional variance process due to the greater ability to hedge. Antoniou and Foster (1992) conclude that the introduction of derivative trading in the Brent crude oil market has not led to an increase in the volatility of spot prices. However, the authors discover that the nature of volatility has changed as the spot price volatility implies higher informational efficiency of the post futures market.

3.3.4   Asymmetric GARCH

Asymmetric GARCH models allow the volatility of energy prices to depend on the direction of a price shock:

$$\sigma_t^2 = \omega + (\alpha + \gamma d_{t-1})\varepsilon_{t-1}^2 + \beta \sigma_{t-1}^2 ,$$

where $d_t = 1$ if $\varepsilon_t < 0$ and zero otherwise and γ is the "leverage term". Morana (2001) uses this model to evaluate the performance of one-month ahead forecasts of daily Brent crude oil prices on the basis of their GARCH properties. Asymmetric GARCH has been found to have a superior fit compared to symmetric GARCH. The ex-ante volatility is estimated on the basis of the generalized error distribution. The estimated value of the scaling factor υ of the generalized error distribution is below two, the value associated with the normal distribution, thus indicating thick tails. Morana (2001) compares the performance of one-month ahead forecasts with the random walk model using the decomposition of the mean squared forecast error. Although the former model shows less bias, the variability of its forecasts is higher than that of the random walk model. Both models are able to predict the sign of the oil price change in slightly less than 50% of the time, i.e., they have the same performance as a toss of a fair coin. When re-estimated for the subsample characterized by high volatility, the one-month ahead forecast model actually exhibits higher bias than the random walk model.

Morana (2001) proposes to utilize the long-run unbiasedness of the one-month ahead forecasts to construct confidence intervals of the forecasts of future prices. The author uses the semiparametric approach to oil price forecasting. The procedure is as follows:

A. In-sample parametric analysis:
    1. obtain the demeaned observations of daily returns, $r_1 \ldots r_T$;
    2. estimate an asymmetric GARCH (1,1) model as described above;

3. obtain the standardized residuals $\frac{e_1}{\hat{\sigma}_1}, \ldots, \frac{e_T}{\hat{\sigma}_T}$.

B. Out-of-sample non-parametric analysis:

1. pick randomly one of the observations, e.g. $\frac{e_7}{\hat{\sigma}_7}$;

2. use the last estimate of the variance $\hat{\sigma}_T^2$, to produce the forecast for period T+1:

$$\hat{\sigma}_{T+1}^2 = \hat{\omega} + (\hat{\alpha} + \hat{\gamma} d_T)\hat{\varepsilon}_T^2 + \hat{\beta}\hat{\sigma}_T^2;$$

3. calculate the simulated forecast of innovation at time T+1:

$$\hat{z}_{T=1}^2 = \hat{\sigma}_{T+1}\frac{e_7}{\hat{\sigma}_7};$$

4. For the period T+2, repeat the steps using the simulated variance at T+1 rather than the actual variance.

One may obtain the entire distribution of oil prices by multiple iterations of the above procedure, known as bootstrapping, and thus estimate the quantiles of interest. The simulated density was obtained for the period 11.21.1998 – 01.21.1999. The period 01.04.1982 – 11.20.1998 was used to estimate the asymmetric GARCH (1,1) model. The forecasted density exhibits skewness and kurtosis. Morana (2001) compares the actual price and the one-month ahead forecast and the 80% confidence interval for the forecasted series. His conclusion is that the one-month ahead forecast becomes less reliable when the forecast confidence interval widens. He suggests using both the one-month ahead forecasting and construction of the confidence intervals with changes in the latter serving as a measure of forecast reliability.

3.3.5   EGARCH

Similarly to the asymmetric GARCH, exponential GARCH (EGARCH) may be used to test for the presence of the leverage effect. It is used by Adrangi et al. (2001b) to show that conditional heteroskedasticity is the source of non-linearities in energy price data. In their study of crude oil, heating oil, and unleaded gasoline futures Adrangi et al. (2001b) find that crude oil and unleaded gasoline series may be modeled by EGARCH (1,1) process:

$$\log(h_t) = \alpha_0 + \alpha_1 \frac{\varepsilon_{t-1}}{\sqrt{h_{t-1}}} + \alpha_2 \left| \frac{\varepsilon_{t-1}}{\sqrt{h_{t-1}}} \right| + \beta_1 \log(h_{t-1}) + \beta_2 TTM,$$

where h_t is the conditional variance, ε_t is the disturbance term, and TTM is the time to maturity. The time to maturity is introduced in order to test the maturity effect, which implies that the volatility of a futures contract increases as the delivery date approaches. The parameter estimate α_2 is negative and significant, which provides evidence in support of the leverage effect.

3.3.6 FIGARCH

According to the FIGARCH model, conditional variance is fractionally integrated. In other words, the price shocks have persistent but not permanent effects on the conditional variance. Brunetti and Gilbert (2000) use the bivariate FIGARCH to study the interaction between the NYMEX and the IPE crude oil markets and their volatilities. They find that the volatilities of both series are long memory processes. Also their respective volatility processes are fractionally integrated of the same order. Thus a bivariate FIGARCH and fractional cointegration model may be used by market participants who seek arbitrage opportunities between the two markets.

To define the bivariate GARCH, recall that a univariate GARCH (p,q) process is given by $[1 - \beta(L)]\sigma_t^2 = \omega + \alpha(L)\varepsilon_t^2$, where $\alpha(L)$ and $\beta(L)$ are lag polynomials of orders q and p, respectively. By defining the innovations as $v_t = \varepsilon_t^2 - \sigma_t^2$, GARCH (p,q) process may be expressed in ARMA (m,p) terms:

$$[1 - \alpha(L) - \beta(L)]\varepsilon_t^2 = \omega + [1 - \beta(L)]v_t,$$

where $m = \max\{p, q\}$. Then the univariate FIGARCH can be represented as follows:

$$\phi(L)(1 - L)^d \varepsilon_t^2 = \omega + [1 - \beta(L)]v_t,$$

where $\phi(L) = [1 - \beta(L) - \alpha(L)](1 - L)^{-d}$. The conditional variance process is given by $\sigma_t^2 = \frac{\omega}{1 - \beta(L)} + \lambda(L)\varepsilon_t^2$, where $\lambda(L) = 1 - \{[\phi(L)(1 - L)^d/[1 - \beta(L)]\}$. Then the bivariate constant correlation FIGARCH (1,d,1) may be represented as follows:

$$\sigma_{jj,t}^2 = \frac{\omega_j}{1 - \beta_{jj}(L)} + \lambda_{jj}(L)\varepsilon_{j,t}^2 \text{ and } \sigma_{12} = \rho[\sigma_{11,t}^2, \sigma_{22,t}^2]^{0.5},$$

where $\lambda_{jj}(L) = 1 - \{[\phi_{jj}(L)(1 - L)^{dj}/[1 - \beta_{jj}(L)]\}$ for $j = 1, 2$. The authors test if the two series are cointegrated. Two series of the same order of integration are cointegrated if there is a long-run relationship that tends to be established despite dissimilar behavior of the two series in the short term. In such a case, there is a linear relationship between $y_{1,t} \sim I(d)$ and $y_{2,t} \sim I(d)$, $z_t = y_{1,t} + \gamma_2 y_{2,t} \sim I(d - b)$, where $b > 0$. The variable z_t has an order of integration lower than its components.

The data used in the study by Brunetti and Gilbert (2000) includes monthly returns on second delivery NYMEX and IPE contracts during June 1988 – March 1999 omitting the period of the Gulf war. Both series are found to have fractionally integrated volatility processes. The authors fail to reject the hypothesis of the same order of integration. A cointegrating relationship is found to have a lower level of persistence. Furthermore, Brunetti and Gilbert (2000) discover that the volatility of IPE tends to adjust to the volatility of NYMEX. This finding is consistent with the relatively higher importance of NYMEX due to the greater volume and longer history of operation as compared to the IPE.

3.3.7 ARMA-GARCH

ARMA-GARCH models allow the mean of energy returns to be time-varying and depend on its past values as well as the past values of the error term. Autoregressive moving average process, ARMA (r,s) is given by

$$y_t - \phi_1 y_{t-1} - \phi_2 y_{t-2} - \cdots - \phi_r y_{t-r} = \varepsilon_t + \theta_1 \varepsilon_{t-1} + \theta_2 \varepsilon_{t-2} + \cdots + \theta_s \varepsilon_{t-s}.$$

(3.10)

If $r = 0$, model (3.10) is a moving average (MA) process, if $s = 0$, then model (3.10) is an autoregressive (AR) process. Alternatively, (3.10) can be expressed as

$$\Phi(L)(1 - L)^d y_t = \Theta(L)\varepsilon_t,$$

where $\varepsilon_t | y_{t-1} \sim N(0, \sigma_t^2)$, $\Theta(L)$ is a lag polynomial of order r and lag $\Phi(L)$ is a polynomial of order s. For example, Wickham (1996) uses an AR-ARCH model to study the issue of oil price volatility. In their analysis of the effect of uncertainty in natural resource prices on the economic growth, Boyd and Caporale (1996) find that commodity prices follow MA-GARCH. Abosedra and Laopodis (1997) find that AR-EGARCH-M is best at explaining the asymmetry and heavy tails of the distribution of crude oil prices.

3.3.7.1 AR-GARCH

Wickham (1996) studies the volatility of oil prices on the basis of a sample consisting of 198 observations of average monthly price of Dated Brent Blend Crude, which is the nearest equivalent to a true spot price of North Sea crude oil. The author models the log difference of the price that is characterized by non-normality and excess kurtosis.

As a first step the author models the conditional mean of the series. Although, market efficiency (in weak form) implies that asset prices should follow a random walk, most stock prices and foreign exchange rates are found to exhibit some serial correlation. Since barriers to arbitrage and transaction costs in oil markets are higher than in markets for financial instruments, the presence of serial correlation in oil prices is quite plausible. The author finds that the conditional mean of oil prices r_t is an AR (4) process and the conditional variance σ_t^2 is a GARCH (1,1) process:

$$r_t = \phi_1 r_{t-1} + \phi_2 r_{t-2} + \phi_3 r_{t-3} + \phi_4 r_{t-4} + \varepsilon_t^2,$$
$$\sigma_t^2 = \omega + \alpha \varepsilon_{t-1}^2 + \beta \sigma_{t-1}^2.$$

Standardized residuals exhibit very small excess kurtosis and standardized squared residuals show no serial correlation.

3.3.7.2   MA-GARCH

Boyd and Caporale (1996) examine the impact of natural resource price uncertainty on the aggregate economic growth. In their analysis, they use real prices of oil, bituminous coal, nonferrous metals, iron, and steel, measured by producer price indices for these resources. Having found the evidence of a unit root in log prices, the authors reject the unit root hypothesis for the first difference of log prices of the four series. Prior to modeling the conditional heteroskedasticity, Boyd and Caporale (1996) estimate best-fitting ARIMA processes to the four natural resource series. They find that growth rates of natural resource prices are MA processes of order one or two and that all series follow a GARCH (1,1). To analyze the impact of resource price uncertainty on the rate of economic growth, the authors propose the following vector autoregression (VAR) model:

$$X_t = \Sigma B_s X_{t-s} + u_t ,$$

where $X_t = $ {growth rates of industrial production, prices of oil, coal, nonferrous metals, iron, and steel, as well as the conditional variances of the growth rates of resource prices estimated by GARCH models}, B_s is a vector of parameters and u_t is a vector of heteroskedastic disturbances. The results of VAR estimation show that resource prices and their variances cause the output growth rates to change. The coefficients of the former are all significant and negative, indicating the presence of negative supply shocks.

In order to evaluate the significance of the contribution of resource price volatility to output forecast error, Boyd and Caporale (1996) use the method of variance decomposition. They find that the variances of resource price growth rates play a more important role in output forecast error than the resource price growth rates per se. To test for a possible misspecification of the model, the authors include two other macroeconomic variables, which are often used as exogenous variables in real output growth models. They are the variance of the growth rate of the monetary base and of the 3-month Treasury bill rate. Although these two macroeconomic variables have been indeed found significant, they do not affect the significance of resource price uncertainty. The variances of the growth rates of natural resource prices continue to explain more than 40 per cent of the variance of the output forecast error even after the inclusion of control variables.

3.3.7.3   AR-EGARCH

Abosedra and Laopodis (1997) propose the use of the EGARCH-M model to account for the heteroskedasticity, negative skewness and leptokurtosis of monthly crude oil spot price series. Following Nelson (1991), Abosedra and Laopodis (1997) specified the AR (r) – EGARCH-M (p, q) model as follows:

$$P_t | Q_{t-1} \sim N(\mu_t, \sigma_t)$$

$$\mu_t = a_0 + \sum_{s=1}^{k} a_s P_{t-s} + b \log(\sigma_t)$$

$$\sigma_t^2 = \exp\{\alpha_0 + \sum_{s=1}^{p} \alpha_s \delta(w_{t-s}) + q \sum_{s=1}^{q} \beta_s \log(\sigma_{t-s}^2)\}$$

According to this specification, energy returns follows an AR (r) process while their variance is determined by its own lagged values and the past values of the function $\delta(w_{t-s})$, which allows for the asymmetric effects of shocks of opposite signs. Specifically,

$$\delta(w_t) = |w_t| - E(|w_t|) + \gamma w_t,$$

where $w_t = \varepsilon_t / \sigma_t$ is iid (0,1) and $E(|w_t|) = \sqrt{(2/\pi)}$ (Hamilton, 1994, p. 668-669). The term $\{|w_t| - E(|w_t|)\}$ measures the magnitude of a price shock, while γw_t represents the sign effect. If $\alpha_s > 0$, then the negative value of γ means that a negative shock to prices increases the volatility to a greater extent than the positive shock of the same magnitude, and vice versa. On the basis of numerical maximization of the log likelihood function, Abosdedra and Laopodis (1997) chose AR(2) - EGARCH-M (1,1) specification. Their parameter estimate γ is positive and significant, meaning that positive shocks to prices increase volatility to a greater extent than negative shocks. The authors interpret this finding as an indication that the investors perceive rising crude oil prices as speculative bubbles.

3.3.7.4   ARFIMA-GARCH

ARFIMA models allow the mean process of energy returns to be fractionally integrated. Mazaheri (1999) uses ARFIMA together with GARCH to model the convenience yield in the petroleum markets. Mazaheri (1999) studies the behavior of petroleum markets in the framework of *the theory of storage*. The essence of this theory is that spot prices and futures prices differ due to the presence of the opportunity cost in the form of interest rate foregone, the warehousing cost, and the convenience yield from holding an inventory. "This convenience yield consists of a stream of implicit benefits which the processors or consumers of a commodity receive from holding the inventories of the commodity" (ibid., p. 32). Formally, the theory of storage is expressed through the cost-of-carry relationship:

$$F_t^T = S_t \exp[(r_t + c_t)(T - t)],$$

where F_t^T is the futures price at date t for delivery at date T, S_t is the spot price at date t, r_t is the risk-free rate, c_t is the warehousing cost net of convenience yield. The cost-of-carry relationship implies the absence of arbitrage opportunities. The return from buying a commodity at date t and selling it at date T should be equivalent to the interest rate return from a risk-free asset, plus the warehousing cost minus the convenience yield. In practice, within a relevant range of inventories, marginal warehousing cost is virtually constant, thus it is the interest rate and convenience yield that determine stochastic dynamics of the cost-of-carry relation. In other words, the convenience yield C_t may be obtained from the cost-of-carry relation by adjusting to the interest rate (proxied by Treasury-bill rate) and the warehousing cost (some constant):[5]

$$-C_t = \ln F_t^T - \ln S_t - r_t(T - t) - w(T - t)$$

According to the theory of storage, convenience yield should be mean reverting to ensure that adjustments in demand and supply bring the spot and futures prices back to the long-run equilibrium. However, this reversion may exhibit persistence. To model persistent mean reverting convenience yield, Mazaheri (1999) proposes the following ARFIMA-GARCH model:

$$\Phi(L)(1 - L)^d y_t = \Theta(L)\varepsilon_t,$$
$$\varepsilon_t|y_{t-1} \sim N(0, \sigma_t^2),$$
$$\beta(L)\sigma_t^2 = \lambda + \alpha(L)\varepsilon_t^2,$$

where y_t is the time series in question, $\Phi(L)$ and $\Theta(L)$ are lag polynomial operators from the ARMA process. $\alpha(L)$ and $\beta(L)$ are lag polynomial operators from the GARCH process. 4d4 is the fractional differencing operator. The process is mean reverting for $d < 1$ and it is covariance stationary if $d < 0.5$. If $0.5 \leq d < 1$, then the process is mean reverting but has infinite variance, i.e., shocks to the variance are persistent.

To check if modeling persistence is justified, the author conducts tests of the two null hypotheses: a test of the presence of unit root (Phillips and Perron, 1988) and a test of the absence of unit root (Kwiatkowski et al. 1992). The rejection of both stationarity and non-stationarity for all three series is considered an indication of a long memory process. Next, Mazaheri (1999) considers the fact that the volatility of convenience yield may depend on the direction of the price shock. Negative supply shock or positive demand shock associated with higher spot prices and lower level of inventories cause the convenience yield to be positive. Since negative inventories are impossible and low inventories could not be replenished instantaneously, the convenience yield tends to be more volatile when positive. To model both the persistence of the level of convenience yield and its time-varying volatility characterized by asymmetric response to price shocks, the author proposes ARFIMA-EGARCH and ARFIMA-asymmetric GARCH. As a result of the simultaneous maximum likelihood estimation of the above models, the author finds that the fractional differencing parameter d has estimated values within the range between 0.7 and 0.8. Both

---

[5] The author also accounts for a possible presence of seasonality in the series by introducing seasonal dummy variables.

ARFIMA-EGARCH and ARFIMA-asymmetric GARCH confirmed the premise that positive price shocks lead to a higher volatility of the convenience yield. To test whether interest rates and seasonality are possible sources of mean reversion, the author re-estimated the model without adjusting the convenience yield for the level of interest rates and seasonal dummy variables. His finding is that the interest rate has no effect on the level of persistence in the convenience yield of the three series, while seasonality matters only in the case of heating oil.

3.3.8 SWARCH

Switching ARCH (SWARCH) models are developed for modeling conditional heteroskedasticity if the regimes governing the volatility process change during the time period under consideration. Susmel and Thompson (1997) find that the SWARCH model is best suited to modeling natural gas prices, which were largely deregulated since the end of the 1980s – beginning of the 1990s.

Susmel and Thompson (1997) analyze the recent history of natural gas markets and the volatility of spot prices during a period of 20 years starting from 1974 using SWARCH methodology. During the 1970s, the US natural gas market was strictly regulated. Owners of gas reserves had long-term contracts with regulated utilities. Interstate pipeline companies transported the gas according to minimum volume obligations and acted as regulated monopolists that controlled purchase and resale prices. Some of interstate pipeline companies were allowed to sell gas to large customers at unregulated prices. Storage capacity was mostly under control of the regulated utilities. As a result of oil shocks of the early 1970s, demand for natural gas rose and resulted in rationing. Customers in gas producing areas were willing to pay a higher price than the regulated rate. Since such consumers did not have to use interstate pipelines they were not subject to regulated rates. As a consequence, pipeline companies were allowed to transport gas directly purchased by the customers. There was established a schedule of rates which increased with volume. This schedule was abandoned in 1984. A slump of oil price in early 1980s forced further deregulation of the gas industry. Those customers who had the ability to switch to oil refused to buy gas from the interstate pipelines at regulated rates. Therefore, pipeline companies were allowed to deliver spot gas purchased by large industrial customers. This eliminated the need for regulatory approval of transportation transactions and let the frequency of transactions grow. However, the rates remained regulated for those customers that were not able to switch to alternative fuels. In 1985, several courts classified existing transportation procedures as discrimination between customers. As a result, the regulators required that the interstate pipelines provide equal treatment to all customers. At the same time, obligations from the long-term contracts still had to be met. The regulators had to devise a system of allocating the cost of remaining purchase contracts between the customers. The uncertainty regarding the use of the transportation system was mostly removed by the Natural Gas Decontrol Act of 1989. As of 1995, both transportation and storage services became widely available on a flexible basis. Thus the deregulation of the natural gas industry took place as a sequence of numerous events. These regulatory changes increased the entry and frequency of transactions in the spot market. Susmel and Thompson (1997) study the effect of these structural changes on the volatility of natural gas spot prices.

Because the natural gas market underwent significant structural changes over the previous decade, the authors propose a GARCH model that accounts for changes in the state of the industry. Using deseasonalized monthly spot prices, they found that the Markov Switching ARCH (2,2) model mimics the price and volatility dynamics associated with the regulatory changes that took place in this industry:

$$\frac{\sigma_t^2}{g_{s_t}} = \omega + \sum_{i=1}^{2} \alpha_i \frac{\varepsilon_{t-i}^2}{g_{s_{t-i}}} + \xi \frac{\varepsilon_{t-1}^2}{g_{s_{t-1}}} d_t,$$

where $d_t = 1$ if $\varepsilon_{t-1} < 0$ and zero otherwise, g_{s_t} is the switching parameter corresponding to the state or regime $s_t = 1, 2$ and $\varepsilon_t \sim N(0, \sigma_t^2)$. This model is based on two regimes and two lags in the conditional variance process. According to this specification, a switch to the high volatility regime occurred in February 1990. The estimated smoothed probabilities exhibit an increasing trend from the middle of 1989. These dates are consistent with the anticipation and the actual congressional approval of the Natural Gas Decontrol Act of 1989, which removed the remaining price ceilings and completely deregulated the transportation system. This model outperforms the traditional GARCH (1,1) model and shows that the deregulation of the natural gas market has resulted in a higher level of price volatility.

3.3.9   Stable GARCH

The study by Rachev and Mittnik (2000) incorporates both the time-varying conditional heteroskedasticity of energy returns and the non-normal distribution of the error term. The authors evaluate the performance of a GARCH model that is based on stable Paretian distributions of the error term. Rachev and Mittnik (2000) consider alternative models for the American Exchange (AMEX) oil index. They use the maximized likelihood value (ML) and Kolmogorov distance (KD) to make comparisons across the models. The likelihood function in the case of Pareto stable disturbances is defined as follows:

$$\max L(\theta) = \prod_{t=1}^{T} S_{\alpha,\beta} \left(\frac{r_t - \delta}{c} \right) c^{-1},$$

where α is the index of stability, β is the skewness parameter, c is the scale parameter and δ is the location parameter of a stable Paretian distribution. The ML value is the overall measure of fit that allows one to compare different model specifications. The KD statistic is defined as follows:

$$KD = Sup_{x \in \Re} \left| F_S(x) - \hat{F}(x) \right|,$$

where $F_s(x)$ denotes the empirical sample cumulative distribution and $\hat{F}(x)$ is the estimated cumulative distribution function. The KD statistic compliments the ML value because the former is a robust measure that takes into account only the maximum deviation between the sample and the fitted distributions. The ML value of nested models increases as they become more general. Rachev and Mittnik (2000) first estimate the ARMA models for AMEX oil returns for the Gaussian and stable disturbances:

$$r_t = \mu + \sum_{i=1}^{r} a_i r_{t-i} + \sum_{j=1}^{s} b_j \varepsilon_{t-j} + \varepsilon_t.$$

The authors find that the series follow an AR (4) process under both assumptions regarding the error process. Given the volatility clustering, the authors estimate normal and stable GARCH. The stable GARCH process is given by the following:

$$\varepsilon_t = c_t u_t,$$

$$c_t = \omega + \sum_{i=1}^{p} \alpha_i |\varepsilon_{t-i}| + \sum_{j=1}^{q} \beta_i c_{t-i},$$

where u_t is a stable variate with $\alpha > 0$ and $c_t > 0$. Rachev and Mittnik (2000) have found that the AMEX oil returns are adequately described by GARCH (1,0) or ARCH (1) process for both normal and stable disturbances. By comparing the four alternative models, i.e., AR(4) normal, AR(4) stable, AR(4)-normal ARCH(1) and AR(4)-stable ARCH(1) models, Rachev and Mittnik (2000) find that the highest ML value and the lowest KD value are associated with the AR(4)-stable ARCH(1) model. Thus the authors conclude that the excess kurtosis and asymmetry of energy price changes are best accounted for by a GARCH process that incorporates the stable Paretian error process.

3.3.10   Summary of Existing Literature on Energy GARCH Models

There is an extensive literature on modeling conditional heteroskedasticity of energy prices. These models range from the basic ARCH to complex models of SWARCH, FIGARCH, and stable GARCH. The majority of the reviewed energy GARCH models are of low order, the most common being GARCH (1,1). Many of the applications favor modifications of GARCH, which allow for the asymmetric effects, i.e., asymmetric GARCH and EGARCH. Several authors find that GARCH-M models have best fitting properties. Some researchers employ the bivariate GARCH to study the links between closely related markets. More complex studies involve models based on fractional integration, switching regimes, and stable distributions of energy returns.

Most of the studies use daily or monthly spot and/or futures data on crude oil. GARCH modeling of price changes of petrochemicals is commonly carried out within the framework of error correction models. There are only single cases of studies of natural gas and coal. The only research indirectly related to electricity, deals with share prices of electricity companies. In our research, we use a range of energy commodities, including two crude oil series, two natural gas series, a series of electricity prices, and a series of Gas oil prices.

Applications of energy GARCH models include the following: a study of efficiency of crude oil markets, an investigation of the effect of resource price uncertainty on rates of economic growth, studies of the links between related products, testing the theory of storage and the leverage effect, analysis of the effects of regulatory changes and the introduction of futures trading, and forecasting of energy prices. In our research, we follow the framework of Rachev and Mittnik (2000) and apply stable GARCH to energy risk estimation because this approach will allow us to obtain a closer approximation to the empirical distribution of energy returns and empirical VaR.

3.4   Energy Risk Estimation with Stable GARCH

Following Rachev and Mittnik (2000), we conjecture that distributional properties of energy prices are best accounted for by stable Paretian distributions in the conditional sense. The advantages of using a non-Gaussian distribution as a basis of energy GARCH have also been demonstrated by Mazaheri (1999). To forecast energy prices, he applies ARFIMA-GARCH based on the generalized error distribution that allows for heavy tails of the distribution of the error term. In this section, at first we analyze the performance of traditional GARCH tools in modeling energy price series. We compare the performance of normal GARCH VaR to the VaR based on the unconditionally normal and stable distributions. After that, we present our results of modeling energy prices based on the stable GARCH. This approach allows us to better understand the nature of the volatility of energy prices. Stable GARCH estimates demonstrate that the volatility of several energy prices is non-stationary. In other words, shocks to such volatility are permanent or persistent. Our findings are similar to those of Brunetti and Gilbert (2000) who have successfully applied long memory models to IPE and NYMEX crude oil futures. In those cases, where we find the volatility of energy prices to be stationary, stable GARCH leads to very precise measures of energy risk represented by the VaR.

3.4.1   Energy Risk Estimation on the Basis of Normal GARCH

In this subsection, we estimate the VaR of the one-month energy futures returns (see Table 2) taking into account the time-varying conditional mean and volatility of energy prices. Here we report estimates of energy VaR based on the GARCH models with the conditionally normally distributed error term (hereinafter referred to as the normal GARCH). The next subsection contains results of our estimation of energy GARCH with the stable distributed error term (hereinafter referred to as the stable GARCH).

We conduct energy GARCH modeling in three stages: i) testing for the presence of the unit root in energy return series, ii) fitting an ARMA model, and iii) fitting an

TABLE 2. Energy Price Series

Series	One-month futures		
	Number of observations	Data range	Exchange
Brent	1249	12/6/95–23/05/00	IPE
WTI	4304	31/3/83–23/05/00	NYMEX
Henry Hub	2543	5/4/90–23/05/00	NYMEX
COB	1040	1/4/96–23/05/00	NYMEX
Gas/IPE	1052	3/02/97–30/03/01	IPE
Gas oil	5049	7/04/81–30/03/01	IPE

ARMA-GARCH model.[6] At the first stage of our analysis, the series are tested for the presence of a unit root. The Augmented Dickey-Fuller statistics (Dickey and Fuller, 1979) for all series are considerably greater than McKinnon (McKinnon 1991) critical values, which means that all series are stationary. At stage two, we assess the order of ARMA models by investigating the correlograms of each series:

$$y_t = \phi_0 + \sum_{i=1}^{r} \phi_i y_{t-1} + \sum_{j=1}^{s} \theta_i \varepsilon_{t-i} + \varepsilon_t$$

Alternative ARMA models are evaluated on the basis of the Schwartz Information Criterion (SIC) and the value of the log likelihood function (See Table 3). At stage three, ARMA models determined at the previous stage are further refined through modeling time-varying conditional heteroskedasticity assuming conditionally normally distributed error term. Maximum likelihood parameter estimates of ARMA-normal GARCH models and their corresponding t-statistics are obtained using Eviews 4.0 and are summarized in Table 4. The series for Brent, COB, and Gas oil are represented by moving average processes and asymmetric GARCH, also known as the Threshold GARCH, TGARCH (ARCH in the case of COB and GARCH-M in the case of Gas oil):

$$y_t = \phi_0 + \sum_{i=1}^{s} \theta_i \varepsilon_{t-i} + \varepsilon_t,$$

$$\sigma_t^2 = \omega + (\alpha + \gamma d_{t-1})\varepsilon_{t-1}^2 + \beta \sigma_{t-1}^2,$$

where $d_t = 1$ if $\varepsilon_t < 0$ and zero otherwise. Note that the sums of ARCH and GARCH parameters for Brent and COB are close to one indicating a possibility of non-stationarity of the conditional variance. However, using the Augmented Dickey–Fuller test, we reject the hypothesis of non-stationarity of conditional variances of these series at the 1% level of significance.

---

[6] All test and estimation results of this subsection have been obtained using Eviews 4.0.

TABLE 3. ARMA-GARCH Model Selection

Series	Model	Log Likelihood	SIC
Brent	MA(2)-TGARCH(1,1)[1]	2901.455	-4.606085
COB	MA(1)-TGARCH(1,1)	1759.845	-3.344239
Gas oil	MA(5)[2]-TGARCH(1,0)-M[3]	12651.40	-4.997937
Henry Hub	MA(2)-GARCH(1,0)-M[4]	5132.037	-4.020790
Gas/IPE	AR(2)-GARCH(1,0)-M	2166.430	-4.086783
WTI	MA(3)-EGARCH(1,1)[5]	11233.58	-5.206455

[1] Threshold GARCH is the same as asymmetric GARCH defined in Section Two.
[2] MA(2) and MA(4) terms are omitted.
[3] M corresponds to variance.
[4] MA(1) term is omitted and M corresponds to the standard deviation.
[5] MA(1) term is omitted. Variance process may have a unit root since the sum of GARCH parameter estimates is greater than one

Two natural gas series, Henry Hub and Gas/IPE, are described by a moving average and an autoregression, respectively. For both series, the mean process includes the standard deviation. Conditional variances of both natural gas series are ARCH processes:

$$y_t = \phi + \theta\varepsilon_{t-2} + \delta\sigma_t + \varepsilon_t \text{ (Henry Hub), or}$$
$$y_t = \phi_0 + \phi_1 y_{t-1} + \phi_2 y_{t-2} + \delta\sigma_t + \varepsilon_t \text{ (Gas/IPE)},$$
$$\sigma_t^2 = \omega + \alpha\varepsilon_{t-1}^2.$$

WTI is a moving average process with the asymmetry of the conditional variance process described by EGARCH:

$$\log\sigma_t^2 = \omega + \alpha(\varepsilon_{t-1}/\sigma_{t-1}) + \beta\log\sigma_{t-1}^2 + \gamma|\varepsilon_{t-1}/\sigma_{t-1}|.$$

As follows from these results of normal GARCH estimation, most of the energy series are best modeled with asymmetric GARCH models: TGARCH or EGARCH. It is interesting that the only EGARCH model, that was found to fit the WTI series, exhibits signs of non-stationarity of the variance process. This finding is similar to that of Antoniou and Foster (1992). In their study of the effect of introduction of futures trading on monthly crude oil spot prices volatility, they find that the variance process has a unit root during the period prior to the introduction of futures trading. Although the unit root test for conditional variance of WTI is rejected at a 1% level of significance, the fact that estimates of GARCH parameters are very close to unity may be indicative of a non-stationary variance process in the cases of WTI, Brent, and COB. The choice of GARCH-M models for modeling gas oil and two natural gas series is consistent with the previous study of efficiency and unbiasedness of the crude oil market by Moosa and Al-Loughani (1994) and the study of the impact of the introduction of crude oil futures trading by Antoniou and Foster (1992).

TABLE 4. Normal GARCH Parameter Estimates*

Series	ARMA			GARCH			
	Constant	MA terms	Other	Constant	ARCH term	GARCH term	Other
Brent	0.0005 (0.8798)	MA(1): -0.0630 (-1.9222) MA(2): -0.0609 (-1.9677)	Na	0.0001 (8.1303)	0.2756 (9.5513)	0.6942 (24.5838)	Leverage term: -0.1526 (-4.3953)
COB	0.0017 (1.0655)	MA(1): 0.0911 (2.806)	na	0.0001 (6.3551)	0.0784 (6.2709)	0.9301 (100.9037)	Leverage term: -0.0817 (-6.4170)
Gas oil	0.0029 (11.6573)	MA(1): -0.0178 (-1.7387) MA(3): -0.0579 (-7.5896) MA(5): -0.0260 (-4.1235)	Variance: -7.8389 (-21.6968)	0.0002 (85.9223)	0.53848 (20.2020)	na	Leverage term: -0.3159 (-11.1254)
Henry Hub	0.0033 (2.2558)	MA(2): -0.0342 (-2.9064)	St.Dev.: -0.0923 (-2.2527)	0.0007 (45.9439)	0.5461 (19.0623)	na	na
Gas/IPE	0.0176 (3.2468)	AR(1): 0.2139 (5.8677) AR(2): -0.0913 (-2.7779)	St.Dev.: -0.5281 (-2.9916)	0.0008 (68.1433)	0.3910 (7.6257)	na	na
WTI (log σ_t^2)	0.00005 (-0.3138)	MA(2): -0.0412 (-2.8183) MA(3): -0.0489 (-3.2113)	na	-0.2105 (-14.8334)	-0.0091 (-1.9721)	0.9921 (858.38)	Abs.value (ARCH1): 0.1998 (22.0310)

* t-statistics are reported below the corresponding parameter estimates.

In general, these ARMA-GARCH models are able to explain some of the non-normality in the energy price data. To illustrate, let us compare standardized residual

TABLE 5. Statistics for One-Month Energy Futures Price Change Series

Series	Raw Data		ARMA-GARCH	
	Skewness	Kurtosis	Skewness	Kurtosis
Brent	-0.15	14.69	-0.42	11.69
WTI	-1.38	25.94	-0.26	6.08
Henry Hub	-0.35	10.54	0.00	8.55
COB	-0.01	13.49	0.01	15.25
Gas/IPE	0.34	19.22	0.85	31.49
Gas oil	-2.41	46.30	-0.06	10.97

statistics from the estimated ARMA-GARCH models with the descriptive statistics for the raw data (See Table 5). In the case of four energy series, the value of kurtosis is lower. In three out of these four cases, standardized residuals series exhibit lower asymmetry than the raw data. However, note that the residuals from the ARMA-GARCH models of COB and Gas/IPE have heavier tails than those of the raw data indicated by the higher values of kurtosis.

The implications of GARCH modeling of energy returns to energy risk estimation are evident from a comparison between the VaR estimates based on normal GARCH and the commonly used VaR based on the assumption of the normal distribution of energy returns (hereinafter referred to as the unconditionally normal VaR). The unconditionally normal VaR is discussed in Khindanova and Atakhanova (2002). As follows from our results (See Table 6), energy VaR based on normal GARCH represent marginal improvement over the unconditionally normal VaR. At the 99% confidence level, normal GARCH VaR underestimates losses in all cases but one. (Unconditionally normal VaR leads to the underestimation of losses in all six cases.) At the 97.5% and 95% confidence levels, normal GARCH VaRs are as large as the empirical values with one exception (two exceptions in the case of the unconditionally normal VaR). At all three confidence levels, this exception refers to Gas/IPE. Note from Table 7, that energy VaR based on the unconditional stable distribution are higher than the empirical values of energy VaR at all confidence levels and for all commodities with the same exception of Gas/IPE at the 97.5% and 95% confidence levels. In other words, energy VaR based on the unconditional stable distributions leads to a more conservative estimation of energy risk than conditionally normal VaR.

Our results show that normal GARCH models explain the asymmetry and heavy tails of energy returns series to a limited extent. The standardized residuals from these models are not white noise as they have values of skewness and kurtosis that are not consistent with the normal distribution. In other words, there is some information in the series left after accounting for ARMA-GARCH effects that may be further modeled. The fact that the residuals from ARMA-GARCH models are more

TABLE 6. Energy VaR: Conditionally Normal (GARCH) vs.Unconditionally Normal*

Series	99% VaR			97.5% VaR			95% VaR		
	GARCH	Normal	Empi-rical	GARCH	Normal	Empi-rical	GARCH	Normal	Empi-rical
Brent	0.0659	0.0603	0.0682	0.0556	0.0508	0.0450	0.0466	0.0426	0.0353
WTI	0.0914	0.0547	0.0681	0.0771	0.0461	0.0453	0.0647	0.0387	0.0342
Henry Hub	0.0917	0.0799	0.0943	0.0773	0.0673	0.0691	0.0649	0.0564	0.0507
COB	0.1168	0.1049	0.1306	0.0984	0.0881	0.0833	0.0826	0.0736	0.0612
Gas/IPE	0.0858	0.0737	0.0944	0.0721	0.0620	0.0843	0.0604	0.0520	0.0623
Gas oil	0.0512	0.0517	0.0645	0.0433	0.0435	0.0422	0.0364	0.0366	0.0301

*All VaR estimates have negative values. Here we report their absolute values.

TABLE 7. Energy VaR: Conditionally Normal (GARCH) vs. Unconditionally Stable*

Series	99% VaR			97.5% VaR			95% VaR		
	GARCH	Stable	Empi-rical	GARCH	Stable	Empi-rical	GARCH	Stable	Empi-rical
Brent	0.0659	0.0788	0.0682	0.0556	0.0492	0.0450	0.0466	0.0354	0.0353
WTI	0.0914	0.1052	0.0681	0.0771	0.0566	0.0453	0.0647	0.0361	0.0342
Henry Hub	0.0917	0.1212	0.0943	0.0773	0.0717	0.0691	0.0649	0.0497	0.0507
COB	0.1168	0.1624	0.1306	0.0984	0.0926	0.0833	0.0826	0.0622	0.0612
Gas/IPE	0.0858	0.1418	0.0944	0.0721	0.0697	0.0843	0.0604	0.0395	0.0623
Gas oil	0.0512	0.0839	0.0645	0.0433	0.0471	0.0422	0.0364	0.0312	0.0301

*All VaR estimates have negative values. Here we report their absolute values.

heavy-tailed than the raw data in the cases of COB and Gas/IPE means that a different modeling approach is required. These results support our conjecture that energy returns are not conditionally normally distributed. To account for the skewness and leptokurtosis of the energy series, it is necessary not only to model their time-varying conditional volatility but also to release the assumption of the normal distribution of the error term. In the next subsection, we use Pareto stable distributions that may exhibit asymmetry and heavy tails. We allow the parameters of these distributions to vary with time in order to capture most of the non-normality in the energy data.

Another argument for the stable GARCH is the evidence of persistence of variance processes of WTI, Brent, and COB. In other words, shocks to the variance of these prices take a very long time to fade away. As noted in Bollerslev, Engle and Nelson (1993) with regard to the integrated GARCH " ... it is important to under-

TABLE 8. Estimates of the Index of Stability α under Alternative Specifications[1]

Series	Index of Stability of Unconditionally Stable Distribution[2]	Index of Stability of Stable GARCH[3]
Brent	1.6135±0.0873	1.7024 (0.0056)
WTI	1.3948±0.0468	1.7759 (0.0239)
Henry Hub	1.5317±0.0618	1.6151 (0.0310)
COB	1.4886±0.0964	1.6708 (0.0546)
Gas/IPE	1.1892±0.0880	1.2900 (0.0469)
Gas oil	1.4575±0.0437	1.5928 (0.0227)

[1] Stable GARCH results are obtained using Matlab 6.1.
[2] 95% confidence level bounds given in parentheses.
[3] Numerical standard error given in parentheses.

stand that apparent persistence of shocks may be driven by thick-tailed distributions rather than by inherent non-stationarity." (ibid. p. 35). In other words, possible non-stationarity of the GARCH models of two crude oils and the electricity price series is a sign of the postulated presence of Pareto stable distributed GARCH disturbances.

3.4.2 Energy Risk Estimation on the Basis of Stable GARCH

Following the procedure of Rachev and Mittnik (2000), we estimate energy GARCH with symmetric stable distributed disturbances based on the Fast Fourier Transform of the stable characteristic function, the method proposed by Mittnik, Rachev, Doganoglu, and Chenyao (1999). Parameter estimates from the absolute value GARCH[7] with normal distributed disturbances are used as starting values. We use results from Khindanova and Atakhanova (2002) as initial values for the estimates of the index of stability in the stable GARCH setting. Similarly to the study by Rachev and Mittnik (2000), we find that under the GARCH representation, the estimates of the index of stability have higher values as compared to their unconditional counterparts because allowing the scale parameter to vary over time absorbs some of the leptokurtosis in the standardized residuals (see Table 8). However, the magnitude of the estimated indices of stability is still significantly less than two indicating the presence of heavy tails. Higher values of the log likelihood function of the stable GARCH models of energy returns point to a superior fit to the data as compared to the normal GARCH (see Table 9 and Figures 3 and 4). ARMA — stable GARCH models are given by the following:

---

[7] Taylor (1986) and Schwert (1989) proposed the following absolute value GARCH representation: $\sigma_t = \kappa + \delta\sigma_{t-1} + \gamma|y_{t-1} - \mu_{t-1}|$.

TABLE 9. Values of Log Likelihood Functions of Stable and Normal GARCH

Series	Log Likelihood of Normal GARCH	Log Likelihood of Stable GARCH
Brent	2901.4	3044.9
WTI	11233.6	11387
Henry Hub	5132.0	5328.4
COB	1759.8	1957.1
Gas/IPE	2166.4	2696.6
Gas oil	12651.4	13227

$$y_t = \phi_0 + \phi_1 y_{t-1} + \phi_1 y_{t-2} + \cdots + \theta_1 \varepsilon_{t-1} + \theta_2 \varepsilon_{t-2} + \cdots + \varepsilon_t,$$
$$\sigma_t = \kappa + \delta |y_{t-1} - \mu_{t-1}| + \gamma \sigma_{t-1}.$$

Our estimation results imply that most of the energy returns have an expected mean of zero with the exception of Brent (See Table 10). All other parameter estimates of ARMA-stable GARCH are statistically significant. Note that similarly to the normal GARCH, Henry Hub and Gas/IPE follow ARCH processes.

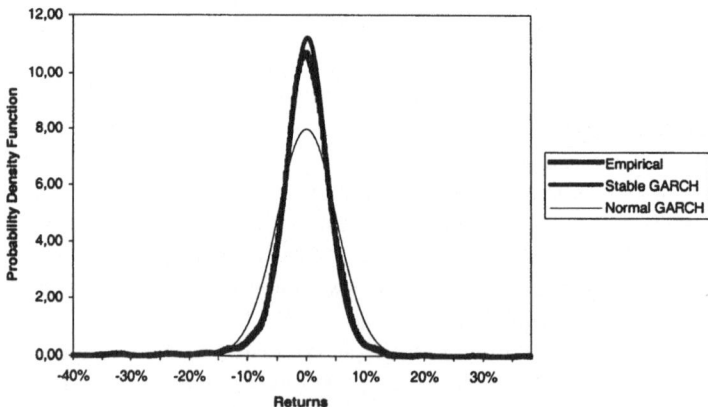

FIGURE 3. COB One-Month Futures. Normal GARCH and Stable GARCH Fit.

Rachev and Mittnik (2000) determine the following condition for the stationarity of the stable GARCH process: $(\lambda_\alpha \gamma + \delta) < 1$. Here $\lambda_\alpha = \frac{2}{\pi} \Gamma \left(1 - \frac{1}{\alpha}\right)$, γ is the estimated ARCH parameter and δ is the estimated GARCH parameter. To evaluate the stationarity of our estimated stable GARCH models of energy returns, we calculate $(\lambda_\alpha \gamma + \delta)$ for each energy price series (See Table 11).

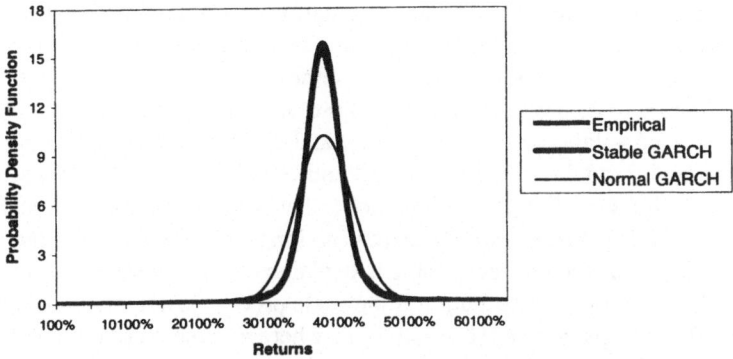

FIGURE 4. One-Month Henry Hub Futures. Normal GARCH and Stable GARCH Fit.

TABLE 10. Parameter Estimates of Stable Energy GARCH[1]

Series	Mean Constant, ϕ_0	ARMA terms, ϕ_i and/or θ_i	Scale Constant, κ	ARCH term, γ	GARCH term, δ	Index of Stability, α
Brent	0.0005 (5.00)	MA(2): -0.0617 (-102.83)	0.0001 (12.50)	0.0222 (14.44)	0.9641 (2410.25)	1.7024 (304.00)
WTI	0.0001 (0.50)	MA(2): -0.0285 (-1.89) MA(3): -0.0284 (-1.76)	0.0001 (2.50)	0.0598 (7.67)	0.9227 (86.23)	1.7759 (74.30)
Henry Hub	0.0003 (0.60)	MA(6): -0.0328 (-2.20)	0.0130 (26.00)	0.2020 (10.52)		1.6151 (52.10)
COB	0.0014 (1.40)	MA(1): 0.0763 (2.96)	0.0013 (2.60)	0.0744 (4.51)	0.8413 (20.82)	1.6708 (30.60)
Gas, IPE	-0.0002 (-0.67)	AR(1): 0.2248 (7.44)	0.0021 (3.50)	0.1980 (6.13)	0.4473 (4.16)	1.29 (27.50)
Gas oil	0.0003 (1.50)	MA(1): 0.0668 (4.67)	0.0082 (41.00)	0.1456 (10.86)		1.5928 (70.17)

[1] t-statistics are reported below the corresponding parameter estimates.

TABLE 11. Stationarity of Stable GARCH

	Brent	WTI	Henry Hub	COB	Gas/IPE	Gas oil
$\lambda_\alpha \gamma + \delta$	0.9945	0.9999	0.3000	0.9578	0.9586	0.2214

As follows from Table 11, WTI is a non-stationary stable GARCH process, while Brent, COB, and Gas/IPE are nearly non-stationary processes. In other words, shocks to the volatility of these energy returns are either permanent or persistent under the stable GARCH hypothesis. Recall, that these are the same series that exhibit either signs of non-stationarity (WTI, Brent, and COB) or greater non-normality of model residuals than the actual data (COB and Gas/IPE) under the normal GARCH approach. Stable GARCH estimation shows that these series have a (nearly) non-stationary volatility. In other words, there is no long-run level, to which the volatility of these prices tends to converge. This is explained by the absence of the unconditional scale of the stable distribution if the scale process is non-stationary. Such unconditional scale, given by $\kappa/(1 - \lambda_\alpha \gamma - \delta)$ where κ is the constant in the GARCH equation, is undefined if $(\lambda_\alpha \gamma + \delta) \approx 1$.

Since two stable GARCH processes satisfy the stationarity requirement, we are able to use our estimation results to obtain the measures of energy price risk. VaR estimates are calculated for Henry Hub and Gas oil. As compared to the normal GARCH estimates of energy VaR, stable GARCH leads to superior measures of energy risk (See Table 12 and Figure 5). Stable VaR estimates are as large as the empirical values at all confidence levels. Note, that normal GARCH underestimates risk at the 99% confidence level. Also note, that compared to the VaR based on the unconditional stable distribution, stable GARCH VaR is more precise, i.e., VaR estimates are closer to the empirical values at the 99% confidence level (See Table 13).

TABLE 12. Energy VaR: Stable GARCH vs. Normal GARCH*

Series	99% VaR			97.5% VaR			95% VaR		
	Stable	Normal	Empirical	Stable	Normal	Empirical	Stable	Normal	Empirical
Henry Hub	0.1133	0.0917	0.0943	0.0713	0.0773	-0.0691	0.0518	0.0649	0.0507
Gas oil	0.0714	0.0512	0.0645	0.0441	0.0435	0.0422	0.0317	0.0366	0.0301

*All VaR estimates have negative values. Here we report their absolute values.

In general, the stable GARCH approach allows us to learn about the nature of the volatility of energy prices. Our results show that the conditional distribution of energy returns has heavy tails. Also we demonstrate that most energy prices have a volatility process that might be covariance non-stationary. In such a case, a shock to the volatility is either permanent or takes a very long time to fade away. A possible approach to modeling such series is the fractionally integrated GARCH that models long-memory processes. Another explanation for a possible non-stationarity of the volatility of energy returns is changes in regimes that may be modeled using switching GARCH. For those series that are covariance stationary, stable GARCH

TABLE 13. Energy VaR: Stable GARCH vs. Unconditionally Stable*

Series	99% VaR			97.5% VaR			95% VaR		
	Stable GARCH	Uncond. Stable	Empirical	Stable GARCH	Stable	Empirical	Stable GARCH	Uncond. Stable	Empirical
Henry Hub	0.1133	0.1212	0.0943	0.0713	0.0717	0.0691	0.0518	0.0497	0.0507
Gas oil	0.0714	0.0839	0.0645	0.0441	0.0471	0.0422	0.0317	0.0312	0.0301

*All VaR estimates have negative values. Here we report their absolute values.

approach leads to superior estimates of the VaR. Unlike the normal GARCH that leads to the underestimation of losses at high confidence levels, the stable GARCH approach yields the VaR estimates that are as large as the empirical ones. Compared to the unconditionally stable VaR, stable GARCH yields more precise results.

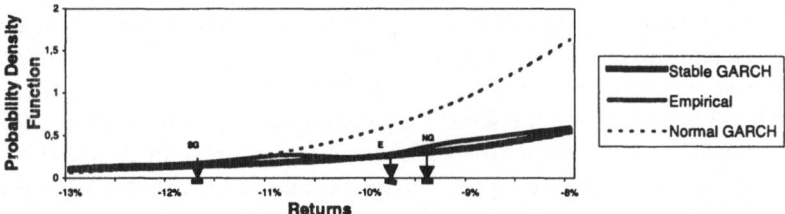

FIGURE 5. One-Month Henry Hub Futures. 99% VAR. Normal GARCH and Stable GARCH Fit

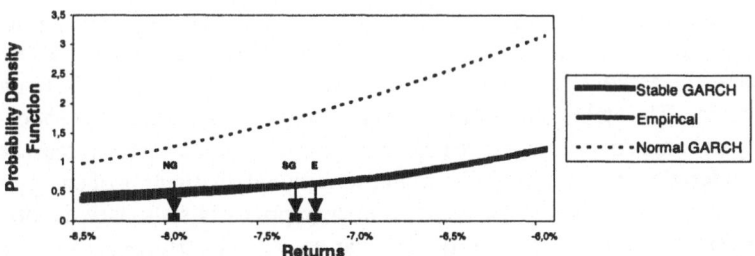

FIGURE 6. One-Month Henry Hub Futures. 97.5% VAR. Normal GARCH and Stable GARCH Fit

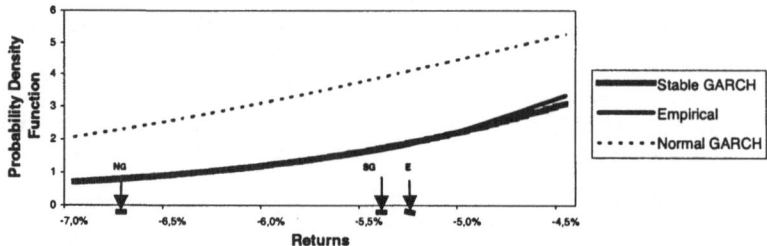

FIGURE 7. One-Month Henry Hub Futures. 95% VAR. Normal GARCH and Stable GARCH Fit

3.5   Conclusions

Models of time-varying conditional heteroskedasticity of energy returns represent an improvement over the models assuming the constancy of the error term variance. Such models are applied to a wide range of economic problems including tests of market efficiency, interrelations between energy markets, regulatory issues, and forecasting. Our research has implications for energy risk management.

We have tested the ability of models of conditional heteroskedasticity to explain the presence of heavy tails and asymmetry of energy returns. Our sample includes a broader set of energy prices as compared to the existing literature. In addition to the crude oil series, we use the series for natural gas, electricity, and gas oil prices. Our conclusion is that normal GARCH models explain some of the non-normality of the distribution of energy prices. When they do, the error term still exhibits skewness and leptokurtosis. At higher confidence levels, normal GARCH based estimates of energy VaR perform marginally better than the ones commonly used by energy companies. To account for non-Gaussian distribution of energy returns and changing volatility, we use the stable GARCH.

We conclude based on the values of the likelihood functions of our estimated models that allowing for the time-varying conditional volatility and Pareto stable conditionally distributed error term represents a better model of energy prices than the normal GARCH. Most importantly, the stable GARCH approach allows us to understand the reasons for the poor performance of the normal GARCH model of energy prices. The stable GARCH approach allows us to conclude that shocks to the volatility of some energy prices are permanent or persistent even if a researcher takes into consideration the non-normality of the conditional distribution of energy returns. In those cases, where stable GARCH of energy prices is stationary, incorporating the time-varying conditional volatility and Pareto stable distributed error term yields very precise estimates of energy risk as measured by the VaR. This approach leads to more conservative conclusions regarding the level of energy price risk at higher confidence levels than the normal GARCH. Stable GARCH VaR estimates are also superior to the unconditionally stable VaR estimates because the latter result in closer approximation to the empirical VaR at higher confidence level. Our future research will focus on the fractionally integrated models for covariance non-stationary models and models of conditional volatility that allow for changes in regimes.

Overall, our research is the first one to evaluate the implications of GARCH modeling for estimation of energy risk. It is the only such research that models energy VaR on the basis of modeling volatility clustering of energy returns. Previous applications of GARCH were concerned with other issues in energy economics and finance. Moreover, stable GARCH is for the first time applied to a wide range of energy commodities. Rachev and Mittnik (2000) showed the applicability of stable GARCH to AMEX oil futures. We have applied their approach to several energy commodity prices and demonstrated gains in precision of energy VaR estimates characterized by stationary volatility processes.

References

[1] Abosedra, S.S. and N.T. Laopodis, 1997, Stochastic Behavior of Crude Oil Prices. A GARCH Investigation, *The Journal of Energy and Development*, 21(2): 283–291.

[2] Adrangi, B., A. Chatrah, K. Raffiee, R.D. Ripple, 2001a, Alaska North Slope Crude Oil Price and the Behavior of Diesel Prices in California, *Energy Economics*, 23:29–42.

[3] Adrangi, B., A. Chatrah, K.K. Dhanda, K. Raffiee, 2001b, Chaos in Oil prices? Evidence from Futures Markets, *Energy Economics*, 23:405–425.

[4] Antoniou, A. and A.J. Foster, 1992, The Effect of Futures Trading on Spot Price Volatility: Evidence for Brent Crude Oil Using GARCH, *Journal of Business Finance and Accounting*, 19(4): 473–484.

[5] Bollerslev, T., 1986, Generalized Autoregressive Conditional Heteroskedasticity, *Journal of Econometrics*, 31:307–327.

[6] Bollerslev, T., 1987, A Conditionally Heteroskedastic Time Series Model for Speculative Prices and Rates of Return, *Review of Economics and Statistics*, 69(3):542–547.

[7] Bollerslev, T. and R.F. Engle, D.B. Nelson, 1993, ARCH Models, UCSD Discussion Paper 93–49, November 1993.

[8] Boyd, R. and T. Caporale, 1996, Scarcity, Resource Price Uncertainty, and Economic Growth, *Land Economics*, 72(3):326–335.

[9] Brunetti, C. and C.L. Gilbert, 2000, Bivariate FIGARCH and Fractional Cointegration, *Journal of Empirical Finance*, 7:509–530.

[10] Deaves, R. and I. Krinsky, 1992, Risk Premiums and Efficiency in the Market for Crude Oil Futures, *The Energy Journal*, 13(2):93–117.

[11] Dickey, D.A. and W.A. Fuller, 1979, Distribution of the Estimators of Autoregressive Time Series with a Unit Root, *Journal of American Statistical Association*, 74:472–431.

[12] DuMouchel, W., 1973a, Stable Distributions in Statistical Inference: 1. Symmetric Stable Distribution Compared to Other Symmetric Long-Tailed Distributions, *Journal of American Statistical Association*, 68:469–477.

[13] DuMouchel, 1973b, On the Asymptotic Normality of the Maximum Likelihood Estimate when Sampling from a Stable Distribution, *Annals of Statistics*, 3:948–957.

[14] Engle, R.F., 1982, Autoregressive Conditional Heteroskedasticity with Estimates of the Variance of the United Kingdom Inflation, *Econometrica*, 50:276–287.

[15] Engle, R. F. and T. Bollerslev, 1986, Modelling Persistence of Conditional Variances, *Econometric Reviews*, 5:1–50.

[16] Hamilton, J.D., 1994, *Time Series Analysis*, Princeton University Press.

[17] Khindanova, I. and Z. Atakhanova, 2002, Stable Modeling in Energy Risk Management, *Mathematical Methods of Operations Research*, 55(2): 225–245.

[18] Kwiatkowski, D., C.B. Phillips, P. Schmidt, Y. Shin, 1992, Testing the Null Hypothesis of Stationarity Against the Alternative of a Unit Root, *Journal of Econometrics*, **54**:159–178.

[19] Liu, S.-M. and B.W. Brorsen, 1995, Maximum Likelihood Estimation of a Garch-Stable Model, *Journal of Applied Econometrics*, **10**(3):273–285.

[20] Mazaheri, A., 1999, Convenience Yield, Mean Reverting Prices, and Long Memory in the Petroleum Market, *Applied Financial Economics*, **9**:31–50.

[21] McKinnon, J., 1991, Critical Values for Cointegration Tests, in: *Long-run Economic Relationships*, R.F. Engle and C.W.J. Granger, eds., Oxford University Press, London, pp. 267–276.

[22] Mittnik, S., T. Doganoglu, D. Chenyao, 1999, Computing the Probability Density Function of the Stable Paretian Distribution, *Mathematical and Computer Modelling*, **29**:235–240.

[23] Mittnik, S., S.T. Rachev, T. Doganoglu, D. Chenyao, 1999, Maximum Likelihood Estimation of Stable Paretian Models, *Mathematical and Computer Modelling*, **29**:275–293.

[24] Moosa, I.A. and N.E. Al-Loughani, 1994, Unbiasedness and Time Varying Risk Premia in the Crude Oil Futures Market, *Energy Economics*, **16**(4):99–105.

[25] Morana, C., 2001, A Semiparametric Approach to Short-term Oil Price Forecasting, *Energy Economics*, **23**:325–338.

[26] Nelson, D.B., 1990, Stationarity and Persistence in the GARCH (1,1) Model, *Econometric Theory*, **6**:318–334.

[27] Nelson, D.B., 1991, Conditional Heteroskedasticity in Asset Returns: A New Approach, *Econometrica*, **59**:229–235.

[28] Ng, V.K. and S.C. Pirrong, 1996, Price Dynamics in Refined Petroleum Spot and Futures Markets, *Journal of Empirical Finance*, **2**:359–388.

[29] Phillips, P.C.B. and P. Perron, 1988, Testing for a Unit-Root in Time Series Regression, *Biometrika*, **75**:335–346.

[30] Rachev, S. and S. Mittnik, 2000, *Stable Paretian Models in Finance*, John Wiley and Sons Ltd., 1–24.

[31] Robinson, T.A. and M.P. Taylor, 1998, Regulatory Uncertainty and the Volatility of Regional Electricity Company Share Prices: The Economic Consequences of Professor Littlechild, *Bulletin of Economic Research*, **50**(1): 37–46.

[32] Schwert, G.W., 1989, Why Does Stock Market Volatility Change Over Time?, *Journal of Finance*, **44**:1115–1153.

[33] Susmel, R. and A. Thompson, 1997, Volatility, Storage and Convenience: Evidence from Natural Gas Markets, *The Journal of Futures Markets*, 1997, **17**(1): 17–43.

[34] Taylor, S., 1986, *Modeling Financial Time Series*, Wiley and Sons, New York, NY.

[35] Wickham, P., 1996, Volatility of Oil Prices, IMF Working Paper WP/96/82.

[36] Zolotarev, V.M., 1966, On Representation of the Stable Laws by Integrals, in *Selected Translations in Mathematical Statistics and Probability*, American Mathematical Society, Providence, RI, 6, 84–88.

4

Malliavin Calculus in Finance

Arturo Kohatsu-Higa

Miquel Montero

ABSTRACT   This article is an introduction to Malliavin Calculus for practitioners. We treat one specific application to the calculation of greeks in Finance. We consider also the kernel density method to compute greeks and an extension of the Vega index called the local vega index.

4.1   Introduction

The purpose of this expository article is to give a basic introduction to Malliavin Calculus and its applications within the area of Monte Carlo simulations in Finance. Instead of giving a general exposition about this theory we will concentrate on one application. That is, the Monte Carlo calculation of financial sensitivity quantities called greeks, where the integration by parts of Malliavin Calculus can be successfully used. Greeks are the general name given to any derivative of a financial quantity with respect to any of its underlying parameters. For example, delta of an option is the derivative of the current option price with respect to the current value of the underlying. Greeks have various uses in Applied Finance such as risk assessment and replication of options between others.

The examples given in this article are derivatives of option prices (mostly European and Asian options) where the payoff function has restricted smoothness. In this case, one is able to carry out the necessary derivatives supposing that the law of the underlying is regular enough. In order to be able to introduce the derivative inside the expectation, one needs to use an integration by parts with respect to the law of the underlying. This is easily done when the law of the underlying is explicitly known (e.g., geometric Brownian motion). But not so easily done if the law is not known (e.g., the integral of a geometric Brownian motion). Here is where Malliavin Calculus has been found to be successful, allowing for explicit expressions of an integration by parts within the expectation although the density is not explicitly known. In this article we stress the applicability rather than the mathematical theory and therefore our discussion will be rather informal as we want to reach also the community of practitioners and we encourage them to try these techniques in their own problems. As with any other technique this one is not the solution to all problems but it could be a helpful tool in various specific problems.

We treat the one-dimensional case for ease of exposition and assume knowledge of basic Itô calculus. Most of the results can be generalized to multi-dimensions. In Section 4.9.2 we briefly describe how to carry out this extension. The article can be divided in the following way.

Index

4.2 Greeks

A greek is the derivative of a financial quantity with respect to any of the parameters in the problem. This quantity could serve to measure the stability of the quantity under study (e.g., vega is the derivative of an option price with respect to the volatility) or to replicate a certain payoff (e.g., delta is the derivative of the option price with respect to the original price of the underlying. This quantity serves to describe the replicating portfolio of an option. For more on this, see Section 4.7.1). As these quantities measure risk, it is important to measure them quickly and with a small order of error. For a careful description of greeks and its uses, see Hull (2000).

One can describe the general problem of greek calculation as follows. Suppose that the financial quantity of interest is described by $E\left(\Phi\left(X(\alpha)\right)Y\right)$ where Φ : $\mathbb{R} \to \mathbb{R}$ is a measurable function and $X(\alpha)$ and Y are two random variables such that the previous expectation exists. α denotes the parameter of the problem. Here we assume that Φ and Y do not depend on α but the general case follows in the same manner as in this simplified case. Now the greek, which we will denote by ϑ, is the derivative of the previous expectation with respect to the parameter α. That is,

$$\vartheta(\alpha) = \frac{\partial}{\partial\alpha}E\left(\Phi\left(X(\alpha)\right)Y\right) = E\left(\frac{\partial}{\partial\alpha}\Phi\left(X(\alpha)\right)Y\right).$$

We will mostly be interested in the case when Φ is a non-smooth function. In our study we will use $\Phi(x) = 1(x \geq K)$ but other functions follow similarly as well. Nevertheless an argument that we will frequently use is to assume that Φ is smooth and then take limits. This argument is usually valid when the densities of $X(\alpha)$ are smooth.

There are essentially three methods to compute the greek ϑ. They are the kernel density estimation method, the integration by parts (ibp) method of Malliavin Calculus and the finite element method.

The finite element method is a numerical method to approach solutions of partial differential equations (pde's). Essentially the method requires to characterize first ϑ as the solution of a pde then one discretizes the differential operators to obtain a system of difference equations that can be solved. This method is entirely deterministic. Although the system of difference equations can be usually solved quickly in low dimensions, the method is not suitable to generate greeks that are not directly related to the derivatives computed in the pde. Cases where it can be applied successfully are the calculation of delta and gamma. In other cases it involves increasing amounts of recalculations which can be cleverly reduced in certain cases. We will not comment further on this method, referring the reader to Wilmott (1998).

A very popular method to compute greeks is the finite difference method. This method only requires to compute the financial quantity of interest at two nearby points and compute the approximative differential. The problem is that the definition of "two nearby points" is not completely clear. An attempt to resolve the issue asymptotically was addressed by L'Ecuyer and Perron (1994) who suggest using $h \sim N^{-1/5}$ where h is the distance between points and N the amount of simulations used in the finite difference method. This method is deeply related with the kernel density estimation method in Statistics. We draw this relationship in the next section.

The third method, usually called the integration by parts of Malliavin Calculus method or the likelihood method, consists of considering

$$\vartheta(\alpha) = E\left(\Phi'(X(\alpha))\frac{\partial X(\alpha)}{\partial \alpha}Y\right)$$

$$= \int \Phi'(x)E\left(\frac{\partial X(\alpha)}{\partial \alpha}Y \,\bigg|\, X(\alpha) = x\right)p(x)dx.$$

Here $p(x)$ is the density of $X(\alpha)$. Then if Φ is irregular one can perform an ibp in order to regularize the integrand. If one can rewrite the integral as an expectation, then one can use the Monte Carlo method in order to approximate ϑ. The success of this method is that the terms within the expectation can be written even in the case that the density p or the conditional expectation are not explicitly known. This will be explained in Section 4.5.

A method that is in between the previous two methods is the likelihood method. This method is an ibp when the density is explicitly known. Otherwise one applies the kernel density estimation method in order to approximate the density.

With these estimations methods one can use various variance reduction techniques to achieve better results. We will not discuss them here. For an exposition on the matter, see Kohatsu–Pettersson (2002) or Bouchard–Ekeland–Touzi (2002).

4.2.1   Examples

Here we briefly describe the examples we will deal with through the article. Consider call (put) *binary* options (for details, see Hull (2000)) which endows an indicator function as payoff. Let us take, for instance[1],

$$\Phi(X) = 1(X > K),$$
$$X(\alpha) = \alpha Z \text{ and}$$
$$Y = e^{-rT}.$$

We will treat two examples with the same underlying asset. That is, we let the underlying asset S be described by a geometric Brownian motion under the risk neutral probability P:

$$S_t = S_0 + \int_0^t rS_s ds + \int_0^t \sigma S_s \, dW_s,$$

where r is the interest rate and σ is the volatility and $\{W_t\}_{t\in[0,T]}$ is the Wiener process. This model is typically used to describe stock prices or stock indices. To simplify the exposition we have assumed that the probability P is the equivalent martingale measure and we compute all expectations with respect to this measure unless stated otherwise. Then we set $Z = S_T/S_0$, $\alpha = S_0$ where

$$S_T = S_0 e^{\{\mu T + \sigma W_T\}},$$

where $\mu = r - \sigma^2/2$. Z follows the lognormal distribution which can be written as

---

[1] Without loss of generality we set the future income equal to one currency unit.

$$p(x) = \frac{1}{x\sqrt{2\pi\sigma^2 T}} \exp\{-[\log(x) - \mu T]^2/2\sigma^2 T\}. \tag{4.1}$$

This example is of educational interest as all greeks have closed formulas. In fact the option price (Π) and delta (Δ) are given by

$$\Pi(S_0) = e^{-rT} E(1(S_T \geq K)) = e^{-rT} \int_{K/S_0}^{+\infty} p(x)dx$$

$$\Delta = \frac{\partial \Pi}{\partial S_0}(S_0) = e^{-rT} \frac{K}{S_0^2} p\left(\frac{K}{S_0}\right).$$

As a second example of real application we use the case of greeks for digital Asian options. That is, we will have

$$Z = \frac{1}{S_0 T} \int_0^T S(s)ds.$$

In this case the density of Z is not explicitly known.

4.3   The kernel density estimation and the finite difference method

In this section we will deduce a generalized finite difference method using ideas taken from kernel density estimation methods. Recall that the goal is to estimate the greek

$$\vartheta = \frac{\partial}{\partial \alpha} E\left(\Phi(X(\alpha))Y\right)\Big|_{\alpha=\alpha_0}$$

where the payoff function Φ is not regular. To solve the problem one convolutes Φ with a regular approximation of the identity (i.e., Dirac's delta function) and then uses methods applicable to regular payoff functions. This argument introduces an approximation parameter as in the finite difference method. Consider first the following alternative expressions for the greek ϑ, if Φ is differentiable $P \circ X(\alpha)^{-1}$ a.s.

$$\vartheta = E\left(\frac{\partial \{\Phi(X(\alpha))\}}{\partial \alpha} Y\right)\Big|_{\alpha=\alpha_0}$$

$$= E\left(\Phi'(X(\alpha))\frac{\partial X(\alpha)}{\partial \alpha} Y\right)\Big|_{\alpha=\alpha_0}.$$

These two formulas help us introduce the estimators

$$\hat{\vartheta} = \frac{1}{Nh} \sum_{i=1}^{N} \int_{\mathbb{R}} \frac{\partial \{\Phi(X(\alpha))\}^i}{\partial \alpha} G\left(\frac{\alpha - \alpha_0}{h}\right) d\alpha Y^i,$$

$$\tilde{\vartheta} = \frac{1}{Nh} \sum_{i=1}^{N} \int_{\mathbb{R}} \Phi'(x) G\left(\frac{x - X(\alpha_0)^i}{h}\right) dx \frac{\partial X(\alpha)^i}{\partial \alpha} Y^i.$$

Both estimators are constructed using an approximation for the derivative using the kernel function $G : \mathbb{R} \to \mathbb{R}_+$ which we assume satisfies that $\int_{\mathbb{R}} G(u)du = 1$ and $\int_{\mathbb{R}} uG(u)du = 0$. The second condition is the parallel to the use of symmetric differences. h is a parameter usually called window size (because it corresponds in a particular case to the interval width used in histograms). These two estimators lead to similar results, therefore we will only consider $\hat{\vartheta}$ because this estimator corresponds to the finite difference estimator when $G(x) = 1(|x| \leq 1/2)$. To simplify our exposition further we consider the case that $\Phi(x) = 1(x \geq K)$ and $X(\alpha) = \alpha Z$ where (Z, Y) has a density $p(z, y)$ which is smooth. The following arguments also follow when p is degenerate with the appropriate modifications. Now, the greek can be written as

$$\vartheta(\alpha) = E(\delta_K(\alpha Z)ZY) = \int_{\mathbb{R}^2} \delta_K(\alpha z)zyp(z, y)dzdy = e^{-rT}\frac{K}{\alpha^2}\int_{\mathbb{R}} p\left(\frac{K}{\alpha}, y\right)dy.$$

Therefore the estimator reduces to

$$\hat{\vartheta} = \frac{1}{Nh}\sum_{i=1}^{N}\int_{\mathbb{R}}\delta_K(\alpha Z^i)G\left(\frac{\alpha - \alpha_0}{h}\right)d\alpha Z^i Y^i.$$

Here δ_x stands for Dirac's delta function. We recall that this generalized functional satisfies that $\int_{\mathbb{R}} \delta_x(u)f(u)du = f(x)$ for f a differentiable function with at most polynomial growth at infinity. First we compute the asymptotic bias of this estimator:

$$E(\hat{\vartheta}) = \frac{1}{h}\int_{\mathbb{R}^3}\delta_K(\alpha z)G\left(\frac{\alpha - \alpha_0}{h}\right)zyp(z, y)d\alpha dzdy$$

$$= K\int_{\mathbb{R}^2}G(u)\frac{y}{(\alpha_0 + uh)^2}p\left(\frac{K}{\alpha_0 + uh}, y\right)dudy.$$

Next in order to expand the bias we use Taylor expansion formulas for $1/x^2$ and $p(K/x, y)$ around $x = \alpha_0$, we assume that $\left|\partial_z^j p(z, y)\right| \leq \mu(y)(1 + |z|)^{-p}$ for some $p > 0$ and $j = 0, \ldots, 3$ where μ is the Radon–Nikodym density of a positive measure with finite expectation. Therefore one obtains that

$$E(\hat{\vartheta}) = K\int_{\mathbb{R}}\frac{y}{\alpha_0^2}p\left(\frac{K}{\alpha_0}, y\right)dy + h^2 f(K, \alpha_0) + O(h^3)$$

$$= \vartheta + h^2 f(K, \alpha_0) + O(h^3),$$

where

$$f(K, \alpha_0) = \frac{K}{2\alpha_0^4}\int_{\mathbb{R}}u^2 G(u)du\int_{\mathbb{R}}\left[\left(\frac{K}{\alpha_0}\right)^2\frac{\partial^2 p}{\partial z^2} + 6\frac{K}{\alpha_0}\frac{\partial p}{\partial z} + 6p\right]\left(\frac{K}{\alpha_0}, y\right)ydy.$$

Similarly, one proves that

$$E(\hat{\vartheta}^2) = \frac{K}{Nh\alpha_0^2} \int_{\mathbb{R}} G(u)^2 du \int_{\mathbb{R}} y^2 p\left(\frac{K}{\alpha_0}, y\right) dy + O(N^{-1})$$

$$+ \left(1 - \frac{1}{N}\right)\left(E(\hat{\vartheta})\right)^2,$$

$$Var(\hat{\vartheta}) = \frac{C_1}{Nh} + O(N^{-1}).$$

Therefore one can minimize the first-order terms in the mean-square error

$$\varepsilon_2 = E\left(\hat{\vartheta} - \vartheta\right)^2 = \frac{C_1}{Nh} + h^4 f(K, \alpha_0)^2 + O(N^{-1}) + O(h^3)$$

to obtain that the optimal value for h is

$$h_0 = \left(\frac{C_1}{4Nf(K, \alpha_0)^2}\right)^{1/5}.$$

Writing explicitly all the terms in the particular case that $p(z, y) = p(z)\delta_{e^{-rT}}(y)$ one obtains the formula

$$\left(\frac{h_0}{\alpha_0}\right)^5 = \frac{C_G p(\frac{K}{\alpha_0})}{N\frac{K}{\alpha_0}\left\{\left(\frac{K}{\alpha_0}\right)^2 \frac{\partial^2 p}{\partial x^2}\left(\frac{K}{\alpha_0}\right) + 6\frac{K}{\alpha_0}\frac{\partial p}{\partial x}\left(\frac{K}{\alpha_0}\right) + 6p\left(\frac{K}{\alpha_0}\right)\right\}^2}, \tag{4.2}$$

with

$$C_G = \int_{\mathbb{R}} G(u)^2 du \left(\int_{\mathbb{R}} u^2 G(u) du\right)^{-2}.$$

Note that the dependence of the error on the kernel G is through the term

$$\left(\int_{\mathbb{R}} G(u)^2 du\right)^2 \int_{\mathbb{R}} u^2 G(u) du.$$

One can therefore carry out the usual variational method to minimize this expression with respect to G subject to $\int_{\mathbb{R}} G(u)du = 1$ and $\int_{\mathbb{R}} uG(u)du = 0$. This leads to the classical Epanechnikov kernel

$$G_0(x) = \frac{3}{4}\mathbb{1}(|x| \le 1)(1 - x^2).$$

Other usual choices are $G_1(u) = \mathbb{1}(|x| \le 1/2)$ (this generates the finite difference method) and $G_2(u) = (2\pi)^{-1/2} \exp\left(-\frac{x^2}{2}\right)$. In each case we have that $C_{G_0} = 15$, $C_{G_1} = 144$, $C_{G_2} = (2\sqrt{\pi})^{-1}$. Nevertheless one should note that in practice the choice of the kernel is not as important as the value of h taken.

In the case of a European call binary option one can obtain explicitly the asymptotically optimal value of h_0. Suppose that $\log(Z) \sim N(\mu T, \sigma^2 T)$, $d = \frac{\ln(x) - \mu T}{\sigma\sqrt{T}}$, then

$$\left(\frac{h_0}{\alpha_0}\right)^5 = \frac{C_G\sqrt{2\pi}\sigma\sqrt{T}\exp\left(\frac{d^2}{2}\right)}{N\left(2 - \frac{1}{\sigma^2 T} - 3\frac{d}{\sigma\sqrt{T}} + \frac{d^2}{\sigma^2 T}\right)^2}. \tag{4.3}$$

In various cases (in particular for options) one usually has that $d \approx 0$ and $\sigma\sqrt{T}$ is small. In such a case an approximate value for h_0 is

$$h_0 \approx \sigma\sqrt{T}\left(\frac{C_G\sqrt{2\pi}}{N}\right)^{1/5}. \tag{4.4}$$

In Figure 1 (a) we have plotted the simulation results for three choices of h for the Delta of a European option. The parameters used are $S_0 = 100$ (in arbitrary cash units), $r = 0.05$, $\sigma = 0.2$ and $T = 0.25$ (in years). The results will be displayed in terms of the present moneyness, S_0/K. The kernel used is G_1 for the European binary and the Epanechnikov kernel G_0 for the binary Asian. In the first case, this kernel generates the finite difference method; in the second the kernel chosen is the asymptotically optimal kernel. We have restricted the plots to a moneyness window which enhances the differences between the three outputs. The results were obtained through direct simulation using Monte Carlo techniques and $N = 10^5$.

The solid line represents the exact value in the case of the European binary, which can be directly computed in the present case. The estimate that uses $h = 3.247$ (empty boxes) is the asymptotically optimal h obtained in (4.4). This is almost indistinguishable from the optimal one (black boxes) which is obtained in (4.3), and both behave better than the heuristic choice of a small h (circles), $h = 0.01S_0 = 1.0$. Therefore one may conclude that choosing h according to (4.4) is not too far from the optimal.

In the case of Asian binary options it is clear that in order to compute the value of h one needs to know the density of (Z, Y) which is not available. In fact this is deeply related with the answer itself, a remark that will be recurrent throughout.

In order to give an estimate of h, one uses qualitative information about the random variables Z. To illustrate this suppose that $Y = e^{-rT}$ and that $Z = T^{-1}\int_0^T S(s)\,ds$ where $S(s) = S(0)\exp\left(\left(r - \frac{\sigma^2}{2}\right)s + \sigma W(s)\right)$ is a geometric Brownian motion. It is well known that the random variable Z has a density that is close to a lognormal density. Therefore we estimate the values related with $p(z)$ using a lognormal density with the same mean and standard deviation as Z. Much more sophisticated approximations can be obtained using results of Dufresne (2000) or Geman and Yor (1993). In our case, this gives that

$$p_1(z) = \frac{1}{\sqrt{2\pi T}\sigma_1 z}\exp\left(-\frac{(\ln(z) - \mu_1)^2 T}{2\sigma_1^2 T}\right),$$

$$r_1 = \frac{1}{T}\ln\left(\frac{e^{rT} - 1}{rT}\right),$$

$$\sigma_1^2 = \frac{1}{T}\ln\left\{\frac{2r^2}{(e^{rT}-1)^2(r+\sigma^2)}\left(\frac{e^{(2r+\sigma^2)T}-1}{2r+\sigma^2} - \frac{e^{rT}-1}{r}\right)\right\}.$$

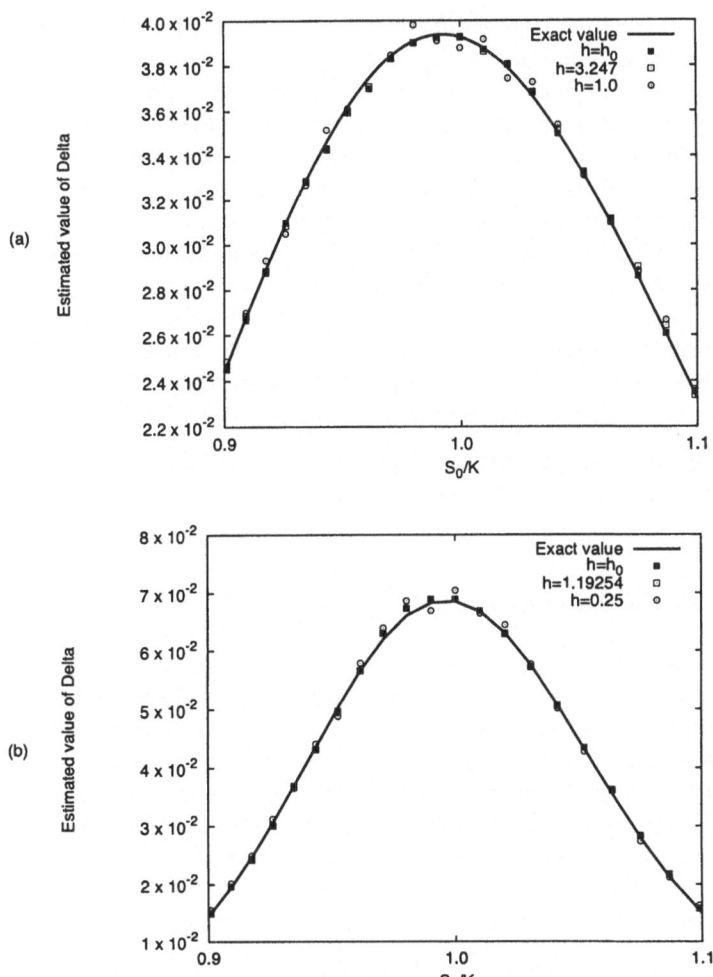

FIGURE 1. Estimated values of Δ for three different choices of h. (a) European binary option (b) Asian binary option

Here $\mu_1 = r_1 - \frac{1}{2}\sigma_1^2$.

In this case the solid line in Figure 1 (b) represents the output of a more precise simulation, since the exact value is not at hand in this case. We have used Malliavin calculus and variance reduction techniques, such as localization (for more on this see the next sections). We have increased the number of simulations to $N = 5 \times 10^5$. In this case the conclusion is the same as in the binary European case: the estimate that uses $h = 1.1937$ (empty boxes) is almost indistinguishable from the (quasi) optimal one (black boxes), and both behave better than the choice of a small h (circles), $h = 2.5 \times 10^{-3} S_0 = 0.25$.

Remark 1. *1. Note that without the condition $\int_{\mathbb{R}} u G(u) du = 0$, then the order of the error in the bias is h instead of h^2, therefore proving the advantage of using symmetric kernels in front of non-symmetric ones. This explains in particular why one has to take symmetric differences when performing the finite difference method.*

2. The bias can also be estimated with similar methods. In general, bias tends to be significantly smaller than the variance of the estimator.

3. If Φ is a differentiable function $P \circ X(\alpha)^{-1}$ a.e., then the optimal value of h is zero. This obviously corresponds to taking the derivative operator inside the expectation. This is the case of the European call or put option.

4. It is well known that the kernel density estimation method increases its bias and increases its variance as the dimension where the random variable leaves Z (here we have considered only dimension 1) becomes higher. This is also expected here and is a drawback in comparison with the methods to follow.

For more comments on practical aspects of the kernel density estimation method and its comparison with the integration by parts method see Section 4.6.

4.4   The likelihood method

The likelihood method, as baptized by Broadie and Glasserman, is one way of calculating the greeks in cases where the joint density of the random variables involved in the problem are explicitly known or can be approximated. The method was proven to be highly effective when applicable. In cases where the density is not known, then a kernel type approximation of the joint density was used. This is obviously related with our previous discussion on the kernel density estimation method. In certain situations it can be considered as a simpler version of the integration by parts of Malliavin Calculus.

Suppose that the vector $(X(\alpha), Y)$ has a joint density $p(x, y, \alpha)$. Then

$$
\vartheta(\alpha) = \frac{\partial}{\partial \alpha} \int_{\mathbb{R}^2} \Phi(x) y p(x, y, \alpha) dx dy \tag{4.5}
$$
$$
= \int_{\mathbb{R}^2} \Phi(x) y \frac{\partial p}{\partial \alpha}(x, y, \alpha) dx dy.
$$

Here, we suppose that one can introduce the derivative in the integrand. Then if the density p is explicitly known one can use numerical integration techniques if the integral can not be computed directly. In the case that p is not explicitly known one can use kernel density estimation techniques to approximate it through Monte Carlo methods to carry out the integration in (4.5). This corresponds to the estimator $\tilde{\vartheta}$ introduced in Section 4.3.

Now we will take this method as a gate to the integration by parts method of Malliavin Calculus. Suppose then that one is interested in using a Monte Carlo method together with the likelihood method. Then we have to rewrite the previous expression (4.5) using a expectation. To do this we divide and multiply by $p(x, y, \alpha)$ to obtain that

$$\vartheta(\alpha) = \int_{\mathbb{R}^2} \Phi(x) y \frac{\partial \log(p)}{\partial \alpha} p(x, y, \alpha) dx dy$$

$$= E(\Phi(X(\alpha)) Y \frac{\partial \log(p)}{\partial \alpha} (X(\alpha), Y, \alpha)).$$

The key point in this argument is that we need to know p in order to carry out the Monte Carlo simulation. The goal of the ibp formula of Malliavin Calculus is to rewrite the previous expression using processes related to $X(\alpha)$ and Y without p. This is related to the alternative expression

$$\vartheta(\alpha) = E\left(\Phi(X(\alpha)) Y \frac{\partial \log E'\left(\delta_{X(\alpha)}(X'(\alpha'))\delta_Y(Y')\right)}{\partial \alpha'}\right).$$

Here $X'(\alpha)$ and Y' are independent copies of $X(\alpha)$ and Y respectively. Making sense of the expression $E'\left(\delta_{X(\alpha)}(X'(\alpha'))\delta_Y(Y')\right)$ is also one of the goals of the ibp formula. We start with the following general definition of the ibp formula. From now on, we will not make any reference to the parameter α as the arguments are valid in general.

Definition 1. *We will say that given two random variables X and Y, the integration by parts (ibp) formula is valid if for any function f in a certain subspace $\mathcal{A}$ of differentiable functions we have that*

$$E(f'(X)Y) = E(f(X)H)$$

for some random variable $H \equiv H_{X,Y}$ (this last notation will be used in the case we want to make clear the dependence of the random variable H upon X and Y).

As it is transparent from this definition, the goal of an integration by parts formula is to convert the derivative f' into its antiderivative f. Note that part of the definition requires the characterization of the subspace $\mathcal{A}$ so that the expectations in the definition are finite. We see that if such a formula is valid, then one could also say that one has an integration by parts formula on $(\Omega, \mathcal{F})$. Sometimes this is also called an infinite dimensional ibp formula in the case Ω is infinite dimensional as is the case when $\Omega = C[0, T]$ (the space of continuous functions) endowed with the Wiener measure.

Most of the application will need to apply this ibp for f measurable. Therefore the ibp formula has to be generalized to include this case. This is generally done via a limit argument.

Let us analyze this definition in more detail with a simple example.

Example 1. *For example, consider $f \in C_p^1 = \{f \in C^1; |f(x)| + |f'(x)| \leq C_f(1 + |x|^{p(f)})$ for some constants C_f and $p(f)\}$ (i.e., the space of continuous differentiable functions with at most polynomial growth at infinity). Let $(\Omega, \mathcal{F}, P)$ be the canonical Wiener space and W denote the Wiener process in this space. Recalling that W_T has an $N(0, T)$ distribution, one can do the following integration by parts:*

$$E(f'(W_T)) = \frac{1}{\sqrt{2\pi T}} \int_{\mathbb{R}} f'(x) \exp(-\frac{x^2}{2T}) dx$$

$$= \frac{1}{\sqrt{2\pi T}} \int_{\mathbb{R}} f(x) \exp(-\frac{x^2}{2T}) \frac{x}{T} dx$$

$$= E(f(W_T) \frac{W_T}{T})$$

$$= E(f(W_T) \int_0^T \frac{1}{T} dW_s).$$

So in this case we can say that we have an integration by parts formula for the random variables $X = W_T$, $Y = 1$ and $f \in C_p^1$. That is,

$$E(f'(X)) = E(f(X)H)$$

for $H = \int_0^T \frac{1}{T} dW_s = \frac{W_T}{T}$.

As noted before, the problem of finding an explicit expression for H is somewhat trivial if the joint density of (X, Y) is known. That is,

$$E(f'(X)Y) = \int_{\mathbb{R}} f'(x) y p(x, y) dx dy$$

$$= -\int_{\mathbb{R}} f(x) y \frac{\partial \log p(x, y)}{\partial x} p(x, y) dx dy$$

$$= -E\left(f(X) Y \frac{\partial \log p(X, Y)}{\partial x} \right).$$

Therefore

$$H = Y \frac{\partial \log p(X, Y)}{\partial x}.$$

The above definition is meaningful when p or some of its properties are known. In particular, suppose the situation where the function p is known but the integral has to be evaluated numerically. Then the above integration by parts formula allows the evaluation through Monte Carlo methods of the expectation.

This argumentation is just for educational purposes as this has a reasonable practical flaw: That is, in most situations $f(x) = 1(x \geq a)$ (almost all other variants follow the same logic) then one obviously has that

$$E(f'(X)Y) = \int_{\mathbb{R}} y p(a, y) dy = p_X(a) \int_{\mathbb{R}} y p_{Y/X=a}(y) dy,$$

therefore an integration by parts is not needed. At most a Monte Carlo simulation to compute the expectation of Y conditioned to $X = a$ solves the problem. Therefore using Monte Carlo simulations in such a situation is not needed.

On the other hand, as we will see in Section 4.5.6, Malliavin Calculus allows writing the above formula in an explicit form, even in situations where the density function is not explicitly known or (X, Y) does not have a joint density, using quantities related to X and Y. This will allow the Monte Carlo simulation of the quantity $E(f'(X)Y)$ using Monte Carlo methodology when f is not smooth.

4.4.1 An application to European options

Here we compare the close formulae for greeks of European options with the likelihood method described previously.

Let Φ denote the payoff function and $\mathcal{P}$, the price of the European option. Then one has

$$\Pi = E(e^{-rT}\Phi(S_T)) = \int_0^\infty e^{-rT}\Phi(x)p(x)dx,$$

where p is given by (4.1). We can now compute the value of Δ (in which case $\alpha = S_0$) in terms of the derivatives of $p(x)$:

$$\Delta = \frac{\partial}{\partial S_0}E(e^{-rT}\Phi(S_T)) = E(e^{-rT}\Phi'(S_T)\frac{S_T}{S_0}).$$

The likelihood method can be applied in this case which gives

$$\Delta = \int_0^\infty e^{-rT}\Phi(x)\frac{\partial p(x)}{\partial S_0}dx = \int_0^\infty e^{-rT}\Phi(x)\frac{\partial \log p(x)}{\partial S_0}p(x)dx$$

$$= e^{-rT}\int_0^\infty \Phi(x)\frac{\log(x/S_0) - \mu T}{S_0\sigma^2 T}p(x)dx.$$

Then we have

$$\Delta = E\left(e^{-rT}\Phi(S_T)\frac{W_T}{S_0\sigma T}\right).$$

This also gives the integration by parts formula for $X = S_T$, $Y = e^{-rT}\frac{S_T}{S_0}$ and $\Phi \in C_p^1$. This gives

$$H = e^{-rT}\frac{W_T}{S_0\sigma T}.$$

A similar procedure applies to the other greeks. We will obtain *Vega* as

$$\mathcal{V} = E\left(e^{-rT}\Phi(S_T)\left(\frac{\partial \log p(x)}{\partial \sigma}\right)_{x=S_T}\right).$$

These formulae coincide with the exact formulas given in Section 4.2.1.

4.5 Malliavin Calculus in finite dimensions

In this setting we have to keep in mind that the density of X is not explicitly known anymore and that we are trying to find a setting where we can write the integration by parts formula in general. One can easily think of examples where this theory can be applied. For example, greeks for Asian options, stochastic volatility models or interest rate models where the densities are not explicitly known.

Here we intend to give a short and basic presentation of Malliavin Calculus. The presentation here is rather informal and does not intend to replace any of the authoritative books on the topic of Malliavin Calculus. See the references for serious mathematical treatment on the topic and for exact statements of the results given here. Similarly, all the results that appear here can be improved as far as hypotheses are concerned. We have preferred to strive for simplicity rather than generality.

In various areas of application one has to use stochastic processes that are generated using one basic stochastic process. This is the case of the abstract Wiener space. To make the presentation simple suppose that on $\Omega = C[0, T]$ (the space of continuous functions) endowed with the sigma field $\mathcal{F}$ generated by the Borel cylindrical sets, one defines the Wiener measure P such that on it the canonical process $W(t)$, $t \in [0, T]$ has Gaussian independent increments with mean zero and variance given by the length of the time interval.

In such a space we can talk of all the random variables generated through various operations of these random variables. Such a space is so rich that it includes all the previous examples mentioned before and in general solutions of stochastic differential equations.

The approach we follow here is through a sequence of extensions of Example 1. Consider the following extension of the ibp formula:

Example 2. *Let $X = F(W(T))$ with $F \in C_p^2$ and $Y \equiv 1$. We want to find an ibp formula. Suppose that there exists a positive constant c such that $\left|F'(x)\right| \geq c > 0$ and let $f \in C_p^1$; then*

$$
\begin{aligned}
E(f'(F(W_T))) &= \frac{1}{\sqrt{2\pi T}} \int_{\mathbb{R}} f'(F(x)) \frac{F'(x)}{F'(x)} \exp(-\frac{x^2}{2T}) dx \\
&= \frac{1}{\sqrt{2\pi T}} \int_{\mathbb{R}} f(F(x)) \exp(-\frac{x^2}{2T}) \left(\frac{x}{T F'(x)} - \frac{F''(x)}{F'(x)^2} \right) dx \\
&= E \left(f(F(W_T)) \left(\frac{W_T}{T F'(W_T)} - \frac{F''(W_T)}{F'(W_T)^2} \right) \right) \\
&= E \left(f(F(W_T)) \left(\frac{1}{T F'(W_T)} \int_0^T 1 dW_s - \frac{F''(W_T)}{F'(W_T)^2} \right) \right).
\end{aligned}
$$

Therefore the integration by parts formula for X is valid if $\left|F'(x)\right| \geq c > 0$ for all $x \in \mathbb{R}$ and

$$
H_X = \frac{1}{T F'(W_T)} \int_0^T 1 dW_s - \frac{F''(W_T)}{F'(W_T)^2}.
$$

The condition $\left|F'(x)\right| \geq c > 0$ is natural as it implies that F is monotone and therefore the density of $F(W_T)$ exists. For example, in the case that F is a constant one can not expect an integration by parts formula as $F(W_T)$ does not have a density.

This example reveals that it is important that some condition relating to the non-degeneracy of $F'(x)$ is needed. This will be later related to the Malliavin derivative of $X = F(W_T)$. In fact, the Malliavin derivative of X will be $F'(W_T)$. The term written as $\int_0^T 1 dW_s = W(T)$ is used to recall the notion of stochastic integral. Another way of writing H is

$$H_X = \left(\frac{1}{F'}\right)(W_T) \int_0^T T^{-1} dW_s + \left(\frac{1}{F'}\right)'(W_T).$$

This formula stresses the fact that H is composed of two terms. The first is the product of a stochastic integral with the inverse of the derivative of F and the second is the derivative of the term $(F')^{-1}$. Later we will see that this structure repeats in other situations.

Example 3. *Let $X = F(W(t_1), ..., W(t_n) - W(t_{n-1}))$ for a partition $\pi : 0 = t_0 < \cdots < t_n = T$ and $F : \mathbb{R}^n \to \mathbb{R}$ so that $F \in C_p^2(\mathbb{R}^n)$; then we can also perform the ibp if for some $i = 1, ..., n$ one has that for all $x \in \mathbb{R}^n$*

$$|\partial_i F(x)| \geq c > 0$$

and the ibp is obtained doing the ibp with respect to the i-th variable. In order to simplify the notation let $W^\pi = (W(t_1), ..., W(t_n) - W(t_{n-1}))$ and $W_i^\pi(x) = (W(t_1), ..., x, ..., W(t_n) - W(t_{n-1}))$ where x is in the i-th component of the vector. Then we have

$$E(f'(F(W^\pi)))$$

$$= \frac{1}{\sqrt{2\pi(t_i - t_{i-1})}} \int_{\mathbb{R}} E(f'(F(W_i^\pi(x_i)))) \exp\left(-\frac{x_i^2}{2(t_i - t_{i-1})}\right) dx_i$$

$$= \frac{1}{\sqrt{2\pi(t_i - t_{i-1})}} \int_{\mathbb{R}} E\left(f(F(W_i^\pi(x_i)))\left(\frac{x_i}{(t_i - t_{i-1})\partial_i F(W_i^\pi(x_i))}\right.\right.$$

$$\left.\left. \times -\frac{\partial_i^2 F(W_i^\pi(x_i))}{\partial_i F(W_i^\pi(x_i))^2}\right)\right) \times \exp\left(-\frac{x_i^2}{2(t_i - t_{i-1})}\right) dx_i$$

$$= E\left(f(F(W^\pi))\left(\frac{W(t_i) - W(t_{i-1})}{(t_i - t_{i-1})\partial_i F(W^\pi)} - \frac{\partial_i^2 F(W^\pi)}{\partial_i F(W^\pi)^2}\right)\right).$$

Therefore in this case we have an ibp formula for $F(W^\pi)$ and

$$H = \frac{W(t_i) - W(t_{i-1})}{(t_i - t_{i-1})\partial_i F(W^\pi)} - \frac{\partial_i^2 F(W^\pi)}{\partial_i F(W^\pi)^2}.$$

Here we have n different ibp formulae. This is natural if one compares with the usual ibp formulae in calculus. The general theory should be obtained by taking limits when the norm of π goes to zero so that $F(W^\pi)$ converges to some random variable in an appropriate topology as to allow taking limits in the ibp formula. For this reason we call the space of all random variables satisfying the conditions in Example 2 the space of smooth random variables and denote it by S. That is,

$$S = \{X \in L^2(\Omega); X = F(W(t_1), ..., W(t_n) - W(t_{n-1})) \text{ with } F \in C_p^2(\mathbb{R}^n)\}.$$

There is one important problem left:

Note that in the previous formula taking the limit is not advisable because in general $\frac{W(t_i) - W(t_{i-1})}{(t_i - t_{i-1})}$ does not converge (see e.g., the law of iterated logarithm for the Wiener process Karatzas–Shreve, Section 2.9.E). Still one may think that this is part of a Riemann sum if one considers instead the sum $\sum_{i=1}^{n}(t_i - t_{i-1})H$. But let us deal with the problems slowly. First we consider a lemma where one obtains the Riemann approximation sums.

Lemma 1. *Let $F \in C_p^2(\mathbb{R}^n)$, $G \in C_p^1(\mathbb{R}^n)$ and let $f \in C_p^1(\mathbb{R})$. Suppose that for all $x \in \mathbb{R}^n$ and all $i = 1, \ldots, n$,*

$$|\partial_i F(x)| \geq c > 0. \tag{4.6}$$

Then we have an ibp formula with $X = F(W^\pi)$, $Y = G(W^\pi)$ and

$$H = \frac{1}{T} G(W^\pi) \left(\sum_{i=1}^{n} \frac{W(t_i) - W(t_{i-1})}{\partial_i F(W^\pi)} - \sum_{i=1}^{n} \frac{\partial_i^2 F(W^\pi)}{\partial_i F(W^\pi)^2}(t_i - t_{i-1}) \right)$$
$$+ \frac{1}{T} \sum_{i=1}^{n} \frac{\partial_i G(W^\pi)}{\partial_i F(W^\pi)}(t_i - t_{i-1}).$$

This formula shows that there is hope in taking limits with respect to n if the random variables X and Y have some stability properties. Here we also see that the second and third sum will converge to Lebesgue integrals of derivatives of the random variables X and Y while the first seems to be an approximation of some kind of extended stochastic integral; $\partial_i F(W^\pi)^{-1}$ is not necessarily $\mathcal{F}_{t_{i-1}}$ measurable as in general it still depends on $W(t_j) - W(t_{j-1})$ for $j \geq i$. This is related with the problem of defining the Itô integral for a general class of integrands (or sometimes called anticipating integrals). On the other hand this result is quite restrictive because it requires the non-degeneracy condition (4.6) in all directions (usually this condition is called strong ellipticity) in comparison with the previous example where only one direction is required but no limit is foreseen.

Proof. First, one generalizes the previous example to obtain that

$$E(f'(F(W^\pi))G(W^\pi))$$
$$= \frac{1}{\sqrt{2\pi(t_i - t_{i-1})}}$$
$$\times \int_{\mathbb{R}} E\left(f'(F(W_i^\pi(x_i)))G(W_i^\pi(x_i))\right) \exp\left(-\frac{x_i^2}{2(t_i - t_{i-1})}\right) dx_i$$

$$= E\left(f(F(W^\pi))G(W^\pi)\left(\frac{W(t_i) - W(t_{i-1})}{(t_i - t_{i-1})\partial_i F(W^\pi)} - \frac{\partial_i^2 F(W^\pi)}{\partial_i F(W^\pi)^2}\right)\right)$$

$$+ E\left(f(F(W^\pi))\frac{\partial_i G(W^\pi)}{\partial_i F(W^\pi)}\right).$$

If we multiply the previous equality by $(t_i - t_{i-1})$ and sum for $i = 1, ..., n$ then we obtain the result.   □

Now we consider the problem of extending this result to allow for a much more general condition for non-degeneracy in the ibp formula and at the same time keeping a formula where one can take limits. It is natural to expect that in some cases the non-degeneration may come from different indexes i in different parts of the whole space $\mathbb{R}^n$; for this reason one needs to develop a theory where one can put all these terms together.

In order to do this we need to practice the previous trick in a reverse way. This may look odd but it does work nicely.

Lemma 2. *Let F, G and f be as in Lemma 1. Suppose the following non-degeneracy condition*

$$|\Delta_1(F)(x)| = \left|\sum_{i=1}^n \partial_i F(x)(t_i - t_{i-1})\right| \geq c > 0,$$

then the ibp formula is valid with

$$H = (\Delta_1(F)(W^\pi))^{-1}\left(G(W^\pi)W(T) + \left(\sum_{i=1}^n \partial_i G(W^\pi)(t_i - t_{i-1})\right)\right)$$

$$- G(W^\pi)(\Delta_1(F)(W^\pi))^{-2}\sum_{j,k=1}^n \partial_{jk}^2 F(W^\pi)(t_j - t_{j-1})(t_k - t_{k-1}).$$

Proof. First consider $I : \mathbb{R}^n \to \mathbb{R}$ to be a $C_p^1(\mathbb{R}^n)$ function

$$E(f(F(W^\pi))I(W^\pi)(W(t_i) - W(t_{i-1})))$$

$$= \frac{1}{\sqrt{2\pi(t_i - t_{i-1})}}$$

$$\times \int_{\mathbb{R}^n} E\left(f(F(W_i^\pi(x_i)))I(W_i^\pi(x_i))\right) x_i \exp\left(-\frac{x_i^2}{2(t_i - t_{i-1})}\right) dx_i.$$

Applying an ibp with respect to the variable x_i one has that

$$E(f(F(W^\pi))I(W^\pi)(W(t_i) - W(t_{i-1})))$$

$$= E\left(f'(F(W^\pi))\partial_i F(W^\pi)I(W^\pi) + f(F(W^\pi))\partial_i I(W^\pi)\right)(t_i - t_{i-1}).$$

Then one moves the last term on the right of the above equation to the left to obtain the ibp formula

$$E\left(f(F(W^{\pi}))\left(I(W^{\pi})(W(t_i) - W(t_{i-1})) + \partial_i I(W^{\pi})(t_i - t_{i-1})\right)\right)$$
$$= E\left(f'(F(W^{\pi}))\partial_i F(W^{\pi})I(W^{\pi})(t_i - t_{i-1})\right). \tag{4.7}$$

Note that this gives exactly the same formula as in Example 3 if one takes $I(x) = (\partial_i F(x)(t_i - t_{i-1}))^{-1}$. Similarly one can also obtain Lemma 1 (exercise). Now sum both sides in (4.7) from $i = 1$ to N to obtain that

$$E\left(f(F(W^{\pi}))\left(I(W^{\pi})W(T) + \sum_{i=1}^{n} \partial_i I(W^{\pi})(t_i - t_{i-1})\right)\right)$$
$$= E\left(f'(F(W^{\pi}))I(W^{\pi})\sum_{i=1}^{n} \partial_i F(W^{\pi})(t_i - t_{i-1})\right).$$

Now we let $I(x) = G(x)\left(\sum_{i=1}^{n} \partial_i F(x)(t_i - t_{i-1})\right)^{-1}$ to obtain the ibp formula. $\square$

This is an ibp formula where one can take limits and the non-degeneracy condition is quite general. In fact, one just needs to define the right concept of derivative and a topology on the space of random variables so that all the partial derivatives above converge. This will be done in the next section. Another way of interpreting the condition (4.6) is that the derivative of F in one particular direction does not cancel. This gives enough room to perform an ibp with respect to that direction. We leave as an exercise to obtain a similar result as in the previous lemma under the condition

$$|\Delta_{\alpha}(F)(x)| = \left|\sum_{i=1}^{n} \alpha_i \partial_i F(x)(t_i - t_{i-1})\right| \geq c > 0.$$

Here $\alpha = (\alpha_1, \dots, \alpha_n) \in \mathbb{R}^n$.

In the next section we will deal in general with any direction. That is, we explain how to obtain an ibp even when the direction wrt which F has an inverse can change according to the value of its argument. This may happen in various diffusion cases.

Lemma 3. *Let F, G and f be as in Lemma 1. Suppose the non-degeneracy condition*

$$\Delta_2(F)(x) = \sum_{i=1}^{n} (\partial_i F(x))^2 (t_i - t_{i-1}) \geq c > 0,$$

then the ibp formula is valid with

$$H = \left(\Delta_2(F)(W^{\pi})\right)^{-1}\left(G(W^{\pi})(W(t_j) - W(t_{j-1}))\right.$$

$$\left. + \sum_{j=1}^{n} \partial_j F(W^{\pi})\partial_j G(W^{\pi})(t_j - t_{j-1})\right) - G(W^{\pi})\left(\Delta_2(F)(W^{\pi})\right)^{-2}$$

$$\times \sum_{j,k=1}^{n} 2\partial_j F(W^{\pi})\partial_{jk}^2 F(W^{\pi})(t_j - t_{j-1})(t_k - t_{k-1}).$$

Proof. We choose $I \equiv I_i$ in (4.7) as

$$I_i(W^\pi) = \frac{\partial_i F(W^\pi) G(W^\pi)}{\sum_{j=1}^n \left(\partial_j F(x)\right)^2 (t_j - t_{j-1})};$$

as before we sum all the equations for $i = 1, ..., n$ to obtain the result. $\square$

This is the formula where one can take limits if the right topology (or norm) is taken on the space S of smooth random variables. In particular it is interesting to look at the approximation to a stochastic integral in the term $\sum_{j=1}^n \partial_j F(W^\pi)(W(t_j) - W(t_{j-1}))$ the problem here is that, as explained before, $\partial_j F(W^\pi)$ is not necessarily adapted as it may depend on $(W(t_k) - W(t_{k-1}))$ for $k \geq j$. This generates a generalization of the stochastic integral. In particular note that one does not have that the expectation of this Riemann sum is zero, as there may be correlations between $\partial_j F(W^\pi)$ and $(W(t_j) - W(t_{j-1}))$ which does not happen in the usual Itô integral. This generalization is usually called the Skorohod integral.

4.5.1 The notion of stochastic derivative

Here we define a derivative with respect to the Wiener process which in the particular case of the previous section will coincide with the partial derivatives. Loosely speaking, we have that for each time t, the r.v. W_t, is the sum of an infinite number of independent increments dW_s for $s \leq t$. In the previous sections we had decomposed $W(T) = \sum_{i=1}^n (W(t_i) - W(t_{i-1}))$ for a partition $0 = t_0 < t_1 < \cdots < t_n = T$. This decomposition defined n derivatives with respect to each component. Therefore in order to take limits, we have to define a derivative in an infinite dimensional space. To explain this better, remember that our purpose here is to do integration by parts for random variables X that have been generated by the Wiener process. Therefore the random variable X is in general a functional of the whole Wiener path W. One way to approach such a functional is to consider that the random variables that we want to consider are limits of functions of increments of the Wiener path. That is, one may suppose that $X = F(W(t_1), ..., W(T) - W(t_{n-1}))$. Therefore if one wants to do an integration by parts for (here $p(t, x)$ stands for the density of a $N(0, t)$ random variable)

$$E(f'(X)) = \int_{\mathbf{R}^{n-1}} f'(F(x_1, \dots, x_{n-1})) p(t_1, x_1) \dots p(T - t_{n-1}, x_{n-1}) dx_1 \dots dx_{n-1},$$

then one can do integration by parts with respect to any of the increment variables x_i as they are independent (this is the case of Example 3. Therefore one needs to have at hand any of the n possible derivatives. In general, as limits are taken one needs an infinite number of derivatives. Therefore stochastic derivatives will be derivatives in infinite dimensional spaces under Gaussian measures. To do this heuristically, note that first we need to decompose the process W in independent pieces. So first we make an independent decomposition of the type

$$W_t = \sum_{s \le t} dW_s.$$

We will denote the derivative of a random variable wrt to dW_s, when it exists, by D_s. In heuristic terms we have

$$D_s = \frac{\partial}{\partial dW_s}.$$

This derivative could be defined using some sort of Fréchet derivative in certain particular directions. Therefore it is only defined in a weak sense. In particular, the definition can be changed at one point for a subset of Ω of null probability without any change in the functional value of the derivative itself.

Definition 2. *Let $X : \Omega \to \mathbb{R}$ be a random variable where $\Omega = C[0, T]$; then we define the stochastic derivative operator (also known as Malliavin derivative), DX, as the Fréchet derivative of X with respect to the subspace $\mathcal{H} = \{h \in C[0, 1]; h' \in L^2[0, T]\}$. That is, DX is defined through the equation*

$$< DX, h >_{L^2[0,T]} = \lim_{\epsilon \to 0} \frac{X(\omega + \epsilon h) - X(\omega)}{\epsilon}.$$

Note that the above definition is local in the sense that it is done for each ω. The reason for defining the directional derivative only with respect to the directions in $\mathcal{H}$ is because most functionals involving stochastic integrals are not continuous in all directions of the space Ω.

Still the idea underneath this stochastic derivative operator D is the limit of the partial differentiation used in the previous section. That is, one starts by considering smooth functionals of $W(t_n) - W(t_{n-1}), \ldots, W(t_2) - W(t_1), W(t_1)$ for a partition $0 < t_1 < \cdots < t_n$ and then takes limits. Instead of taking this long road which can be carried out mathematically with the previous definition, we give some examples that illustrate the intuition behind the operator D. We start with the most simple example of a derivative and the chain rule for $s \le t$:

$$D_s W_t = 1 \text{ (here } X = W_t),$$
$$D_s f(W_t) = f'(W_t) \text{ (here } X = f(W_t)).$$

Note that the derivative $D_s X$ "measures" the change of the random variable X wrt ΔW_s in the sense that X can be written as a functional of the increments of W. This statement can be demonstrated with some examples, let $t' < s < t$ and let $h \in L^2[0, T]$, then

$$D_s W_{t'} = 0,$$
$$D_s (W_t - W_s) = 0,$$
$$D_s W_t = D_s (W_t - W_s) + D_s W_s = 1,$$
$$D_s \left(\int_0^T h(u) dW_u \right) = h(s).$$

This last formula follows because h being deterministic is independent of W_s and furthermore dW_u will be independent of W_s unless $u = s$ and $D_s dW_s = 1$. Finally applying the product formula one obtains that

$$
\begin{aligned}
D_s \left(h(u)dW_u \right) &= D_s h(u)dW_u + h(u)D_s dW_u \\
&= 0 \cdot dW_u + h(u)1(s = u).
\end{aligned}
$$

From here the formula follows. Similarly one also obtains that

$$
\begin{aligned}
D_s \left(\int_0^T f(W_u)dW_u \right) &= \int_0^T D_s f(W_u)dW_u + f(W_u)D_s dW_u \\
&= \int_0^T f'(W_u)D_s W_u dW_u + f(W_u)1(s = u) \\
&= \int_0^T f'(W_u)1(s \le u)dW_u + f(W_s) \\
&= \int_s^T f'(W_u)dW_u + f(W_s).
\end{aligned}
$$

One can also perform high-order differentiation as in the case of

$$
D_s D_t W_u^3 = 6W_u 1(s \vee t \le u).
$$

All the properties we have used so far can be proven using the definition of stochastic derivative. One important aspect to have in mind is that the stochastic derivative is well defined as a random variable in the space $L^2(\Omega, L^2[0, T])$ and therefore will be well defined in the a.s. sense. Therefore, the derivatives D_s are defined only a.s. with respect to the time variable s for almost all $\omega \in \Omega$. Leaving the technicalities aside one can define the derivative as an operator on random variables.

To generate the ibp formula, one way to proceed is to prove that the adjoint operator D^* exists. In order to do this one sufficient condition is to prove that the operator D is closable. In such a case we can define the adjoint operator, denoted by D^* through the formula

$$
E \left(< DZ, u >_{L^2[0,T]} \right) = E(ZD^*(u)).
$$

Here, $D : L^2(\Omega) \to L^2(\Omega, L^2[0, T])$, u is a stochastic process and $D^* : dom(D^*) \subseteq L^2(\Omega, L^2[0, T]) \to L^2(\Omega)$. The above formula is in fact an integration by parts formula! We will show this in Section 4.5.3. The procedure described here is the most classical.

Instead, we will take a different approach. We will use the previous results for random variables depending on only a finite number of increments of W and take limits in the ibp formulas in order to define D^*. At various points we will make reference to the classical approach so that the reader can refer to the textbooks mentioned in the references. To motivate our approach, let us reconsider Example 1:

Let $Z = f(W_T)$ and $u \equiv 1$. Then we have, for $s \le T$,

$$D_s Z = D_s \left(f(W_T) \right) = f'(W_T) D_s W_T$$
$$= f'(W_T).$$

Also

$$< DZ, u > = \int_0^T D_s Z u_s ds = \int_0^T f'(W_T) \cdot 1 ds$$
$$= T f'(W_T).$$

Therefore we have

$$T E \left(f'(W_T) \right) = E \left(< DZ, u > \right) = E(ZD^*(u)) = E(f(W_T)D^*(1)).$$

The conclusion of this small calculation, if one compares with Example 1, is that $D^*(1) = W_T = \int_0^T 1 dW(s)$. In fact, one can easily prove via a density argument that $D^*(h) = \int_0^T h(s) dW_s$ for $h \in L^2[0, T]$. We will be able to say more about this in the next section.

4.5.2   A proof of the duality formula

Here we give a sketch of the proof of the duality principle. This section only gives a mathematical idea of how the duality formula is proved. It is not essentially required to understand the calculations to follow in future sections (except for Remark 2).

We define the norms for $X \in \mathcal{S}$,

$$\|X\|_{1,2} = \left(E \left(|X|^2 + \int_0^T |D_s X|^2 \, ds \right) \right)^{1/2},$$

and let $\mathbb{D}^{1,2} = \overline{\mathcal{S}}$ where the completion in taken in $L^2(\Omega)$ under the norm $\|\cdot\|_{1,2}$. In other words, X is an element of $\mathbb{D}^{1,2}$ if there exists a sequence of smooth random variables X_n such that $E\left(|X_n - X|^2 \right) \to 0$ and there exists a process $Y \in L^2(\Omega \times [0, T])$ such that $E \int_0^T |D_s X_n - Y(s)|^2 \, ds \to 0$. In such a case, X is a differentiable random variable and $DX = Y$.

Similarly, we define the parallel concept for stochastic processes. First, we say that a stochastic process u is a smooth simple process if

$$u(t) = u_{-1} + \sum_{i=1}^N u_{i-1} 1(t_{i-1} < t \le t_i)$$

for some partition $0 = t_0 < t_1 < \cdots < t_N = T$ and where the random variables $u_j \in \mathcal{S}$ for $j = -1, \ldots, N - 1$. We denote the space of smooth simple processes by $\mathcal{S}_p$. Next we define the norm

$$\|u\|_{1,2} = \left(E \left(\int_0^T |u(t)|^2 \, dt + \int_0^T \int_0^T |D_s u(t)|^2 \, ds dt \right) \right)^{1/2}.$$

Here there is a slight abuse of notation as there are two norms $\|\cdot\|_{1,2}$, one for random variables and another one for processes. The nature of the argument will determine the norm we are referring to.

As in the case of random variables we define $\mathbb{L}^{1,2} = \overline{S_p}$ (the closure of S_p with respect to the norm $\|\cdot\|_{1,2}$). u is an element of $\mathbb{L}^{1,2}$ if there exists a sequence of simple smooth processes u_n such that $E\left(\int_0^T |u(t) - u_n(t)|^2\, dt\right) \to 0$ and the sequence Du_n converges in $L^2(\Omega \times [0, T]^2)$. In such a case $u(t) \in \mathbb{D}^{1,2}$ for almost all t.

With these definitions we can state the duality principle.

Theorem 1. *Let $X \in \mathbb{D}^{1,2}$ and $u \in \mathbb{L}^{1,2}$; then there exists a random variable $D^*(u) \in L^2(\Omega)$ such that*

$$E\left(< DX, u >_{L^2[0,T]}\right) = E(X D^*(u)). \tag{4.8}$$

In the particular case that u is an adapted process, then $D^*(u) = \int_0^T u(s) dW(s)$.

In functional analytic terms D^* is the adjoint operator of D. The property $X \in \mathbb{D}^{1,2}$ implies that X is differentiable and that its derivative can be obtained as the limit of the derivatives of the smooth approximating random variables. $u \in \mathbb{L}^{1,2}$ implies that $u \in dom(D^*)$.

In the proof one can also see the properties of D^*. In particular D^* is an extension of the Itô stochastic integral in the sense that, if u is an adapted process, then

$$D^*(u) = \int_0^T u(t) dW(t).$$

Idea of the proof of (4.8):

Step 1: The idea is to prove that (4.8) is true for smooth random variables X_n and simple smooth processes u. Then finish the proof using a limiting procedure. That is, let us assume that $X_n = F(W^\pi)$ and $u(s) = u_{-1} + \sum_{i=1}^n u_{i-1} 1_{(t_{i-1}, t_i]}(s)$. As before, let $W^\pi = (W(t_1), ..., W(t_n) - W(t_{n-1}))$, $W_i^\pi(x) = (W(t_1), ..., x, ..., W(t_n) - W(t_{n-1}))$ and $W_i^\pi = (W(t_1), ..., W(t_i) - W(t_{i-1}))$. Then

$$D_s X_n = \sum_{i=1}^N \partial_i F(W^\pi) 1(t_{i-1} < s \leq t_i),$$

$$< DX_n, u > = \sum_{i,j} \int_0^T \partial_i F(W^\pi) 1(t_{i-1} < s \leq t_i) u_j 1_{(t_{j-1}, t_j]}(s) ds,$$

$$= \sum_{i=1}^N \partial_i F(W^\pi) u_{i-1} (t_{i-1} - t_i).$$

Now we take expectations and integrate by parts to get rid of the partial derivative in the above sum. To do this we also assume that $u_{i-1} = h_i(W^\pi)$ with $h_i : \mathbb{R}^n \to \mathbb{R}$, $h_i \in C_p^2(\mathbb{R}^n)$. One then obtains

$$E\left(\partial_i F(W^\pi)h_i(W^\pi)\right)$$

$$= -E\left(\int_{\mathbb{R}} F(W_i^\pi(x_i))\left(\partial_i h_i(W_i^\pi(x_i))\right.\right.$$

$$\left.\left. -\frac{h_i(W_i^\pi(x_i))x_i}{(t_i - t_{i-1})}\right)\frac{e^{-\frac{x_i^2}{2(t_i - t_{i-1})}}}{\sqrt{2\pi(t_i - t_{i-1})}}dx_i\right).$$

Therefore one finally has

$$E\left(< DX_n, u >\right)$$

$$= E\left(\sum_{i=1}^n \partial_i F(W^\pi)u_{i-1}(t_{i-1} - t_i)\right)$$

$$= \sum_{i=1}^n E\left(F(W^\pi)\left(h_i(W^\pi)(W_{t_i} - W_{t_{i-1}}) - \frac{\partial h_i}{\partial x_i}(W^\pi)(t_i - t_{i-1})\right)\right)$$

$$= E\left(X_n D^*(u)\right)$$

where

$$D^*(u) = \sum_{i=1}^n \left(h_i(W^\pi)(W_{t_i} - W_{t_{i-1}}) - \frac{\partial h_i}{\partial x_i}(W^\pi)(t_i - t_{i-1})\right). \qquad (4.9)$$

The above formula proves our statement in the smooth and simple case. Next we take limits with respect to X_n to obtain that

$$E\left(< DX, u >_{L^2[0,T]}\right) = E\left(X D^*(u)\right),$$

for simple, smooth processes u. To finish we need to take limits in u. For this we use that, if u_n is a sequence of simple smooth processes converging to u in $\mathbb{L}^{1,2}$, then $D^*(u_n)$ converges in $L^2(\Omega)$ to a random variable which we denote by $D^*(u)$. This result is proven in Lemma 4. Therefore we can take limits again in the duality formula to finish the proof.

Next we will prove that $D^*(u)$ coincides with the Itô integral when u is adapted. To prove this it is enough to consider the case when u_{i-1} is $\mathcal{F}_{t_{i-1}}$ adapted in the previous argument. In such a case it is obvious that $h_i(x) = h_i(x_1, ..., x_{i-1})$. Therefore $\frac{\partial h_i}{\partial x_i}(W^\pi) = 0$ and

$$D^*(u) = \sum_{i=1}^n h_i(W^\pi)(W_{t_i} - W_{t_{i-1}})$$

which is the Riemman sum that leads to the Itô integral. This finishes the proof.   □

Some researchers prefer to use the notation $\delta(u)$ instead of $D^*(u)$ to stress the quality of stochastic integral of δ. δ defines what is called the Skorohod integral. When u is not an adapted process, then $\delta(u)$ is not an Itô stochastic integral. Nevertheless in various situations one can find ways to compute such integrals as we will see later.

Lemma 4. *Let u_n be a simple smooth process converging to u in $\mathbb{L}^{1,2}$. Then $D^*(u_n)$ converges in $L^2(\Omega)$ to a random variable which we denote by $D^*(u)$.*

Proof. It is enough to prove that $D^*(u_n)$ is a Cauchy sequence in $L^2(\Omega)$. This will follow immediately if we compute the $L^2(\Omega)$-norm of $D^*(u_n)$. This is done as follows:

$$E\left(D^*(u_n)^2\right) = E\left(\sum_{i=1}^n A_i^2 + 2\sum_{i<j} A_i A_j\right),$$

$$A_i = h_i(W^\pi)(W_{t_i} - W_{t_{i-1}}) - \frac{\partial h_i}{\partial x_i}(W^\pi)(t_i - t_{i-1}).$$

We start by computing

$$E\left(A_i^2\right) = E\left(\frac{\partial h_i}{\partial x_i}(W^\pi)\right)^2 (t_i - t_{i-1})^2$$
$$-2E\left(\frac{\partial h_i}{\partial x_i}(W^\pi)h_i(W^\pi)(W_{t_i} - W_{t_{i-1}})\right)$$
$$\times(t_i - t_{i-1}) + E\left(h_i(W^\pi)^2(W_{t_i} - W_{t_{i-1}})^2\right).$$

Applying again ibp we have that

$$E\left(h_i(W^\pi)^2(W_{t_i} - W_{t_{i-1}})^2\right)$$
$$= 2E\left(\frac{\partial h_i}{\partial x_i}(W^\pi)h_i(W^\pi)(W_{t_i} - W_{t_{i-1}})\right)(t_i - t_{i-1})$$
$$+E\left(h_i(W^\pi)^2\right)(t_i - t_{i-1}).$$

Therefore one has that

$$E\left(A_i^2\right) = E\left(h_i(W^\pi)^2\right)(t_i - t_{i-1}) + B_i(t_i - t_{i-1}),$$

where B_i converges to zero as $n \to \infty$. Similarly one computes $A_i A_j$ for $i < j$ to obtain after some calculations

$$E\left(A_i A_j\right) = E\left(\frac{\partial h_i}{\partial x_j}(W^\pi)\frac{\partial h_j}{\partial x_i}(W^\pi)\right)(t_i - t_{i-1})(t_j - t_{j-1}).$$

Therefore we have that

$$E\left(D^*(u_n)^2\right)$$

$$= \sum_{i=1}^{n} \left(E\left(h_i(W^\pi)^2\right)(t_i - t_{i-1}) + B_i(t_i - t_{i-1}) \right)$$

$$+2\sum_{i<j} E\left(\frac{\partial h_i}{\partial x_j}(W^\pi)\frac{\partial h_j}{\partial x_i}(W^\pi) \right)(t_i - t_{i-1})(t_j - t_{j-1})$$

$$= E\left(\int_0^T u_n(t)^2 dt + \int_0^T \int_0^T D_s u_n(t) D_t u_n(s) ds dt \right) + \sum_{i=1}^{n} B_i(t_i - t_{i-1}).$$

From here the argument is standard. That is, consider the difference between simple smooth processes and use the above equality to prove that their difference goes to zero. Therefore the sequence $D^*(u_n)$ is a Cauchy sequence which should then converge. This finishes the proof. □

Remark 2. *We have various remarks on the proofs we have sketched.*
1. One sees that for $u \in \mathbb{L}^{1,2}$,

$$\left(E\left(D^*(u)^2\right)\right)^{1/2} \le \|u\|_{1,2}. \tag{4.10}$$

Therefore the space $\mathbb{L}^{1,2}$ is smaller than the domain of the operator D^*.
2. In general, if u is not adapted, the classical Riemann sum

$$\sum_{i=1}^{n} u(t_{i-1})(W(t_i) - W(t_{i-1}))$$

does not converge to $D^*(u)$. As it can be seen from (4.9), this converges to the Skorohod integral of u plus a trace term generated by $\sum_i \frac{\partial h_i}{\partial x_i}(W^\pi)(t_i - t_{i-1})$ which is due to the non-adaptedness of the process u and converges to a Lebesgue integral

$$\int_0^T D_s u(s) ds,$$

although this derivative is not well defined. In fact, note that

$$\lim_{v \downarrow s} D_v W_s = 0,$$

$$\lim_{v \uparrow s} D_v W_s = 1.$$

For this reason one needs to define

$$D_{s+} u = \lim_{v \downarrow s} D_v u(s),$$

$$D_{s-} u = \lim_{v \uparrow s} D_v u(s),$$

and therefore the above formula has to be understood as

$$\int_0^T D_{s+} u \, ds$$

so that $D_{s+}u = 0$ if the process is adapted.

3. Maybe for the reader it may feel natural to define the extended stochastic integral as the limit (if it exists) of $\sum_i u(t_{i-1})(W_{t_i} - W_{t_{i-1}})$. First note that the duality formula can obviously be written as

$$E\left(\langle DZ, u \rangle_{L^2[0,T]}\right) = E\left(Z D^*(u)\right).$$

In contrast with this opinion, if the previous limit exists its expectation is not zero in general, while $E(D^*(u)) = 0$ as it can be seen using $Z = 1$ in the duality formula. In terms of approximations we mean that

$$E \sum_i \left(h_i(W^\pi)(W_{t_i} - W_{t_{i-1}}) - \frac{\partial h_i}{\partial x_i}(W^\pi)(t_i - t_{i-1}) \right) = 0$$

while one does not have that

$$E \sum_i h_i(W^\pi)(W_{t_i} - W_{t_{i-1}})$$

is necessarily equal to 0. Obviously, there is no martingale property associated with these integrals as the adaptedness is completely lost. Also there is no L^2-isometry that could help us here. The closest to this property is the inequality (4.10). The continuity property and other related properties can also be studied using this property.

4. To some it may seem that defining stochastic integrals of anticipating processes is just an exercise of generalization. To motivate this issue we will later show the formula

$$F \int_0^T dW(s) = D^*(F) + \int_0^T D_s F \, ds.$$

Here $F \in \mathbb{D}^{1,2}$ is a random variable. The problem is the natural extension of the linearity property of integrals extended to random variables.

*   This problem was first studied by K. Itô. Note that on the right one needs to use an anticipating type of integral in order to make sense of the integral as F is not adapted to the filtration (except in the trivial case that F is a constant). This formula also helps to compute integrals of non-adapted processes using adapted ones. For more on this, see Section 7 in Kohatsu–Pettersson (2002).*

5. Another approach to the definition of the stochastic derivative and the adjoint operator is through chaos decompositions of functionals. This approach, which is morally equivalent to the one presented here, is based in some kind of approximations for functionals. Nevertheless its applications have been limited to very specific cases such as calculations regarding local times.

4.5.3   Obtaining the ibp formula from the duality formula

Now to obtain an ibp formula, we consider the random variable $Z = f(X)$ with $X \in \mathbb{D}^{1,2}$, $f \in C_b^1$, $Y \in L^2(\Omega)$. Then $Z \in \mathbb{D}^{1,2}$ and

$$D_s Z = f'(X) D_s X.$$

From here we multiply the above by $Y D_s X$. Then we obtain that

$$D_s Z Y D_s X = f'(X) D_s X Y D_s X;$$

integrating this for $s \in [0, T]$, we have that

$$\int_0^T D_s Z Y D_s X \, ds = \int_0^T f'(X) (D_s X)^2 Y \, ds = f'(X) Y \int_0^T (D_s X)^2 \, ds$$

so that finally we have that

$$\int_0^T \frac{Y D_s Z D_s X}{\int_0^T (D_v X)^2 \, dv} ds = f'(X) Y,$$

$$E\left(<DZ, u>_{L^2[0,T]} \right) = E\left(f'(X) Y \right)$$

with

$$u_s = \frac{Y D_s X}{\int_0^T (D_v X)^2 \, dv}.$$

Finally, we have the following result:

Theorem 2. *Assume that $f \in C_b^1$, $X \in \mathbb{D}^{1,2}$, $Y \in L^2$ and $u \in \mathbb{L}^{1,2}$, then we have that*

$$E(f(X) D^*(u)) = E\left(f'(X) Y \right),$$

$$E\left(f(X) D^* \left(\frac{Y D.X}{\int_0^T (D_v X)^2 \, dv} \right) \right) = E\left(f'(X) Y \right), \tag{4.11}$$

In other words, the ibp formula is valid with

$$H \equiv H_{XY} = D^* \left(\frac{Y D.X}{\int_0^T (D_v X)^2 \, dv} \right).$$

As we have seen in Remark 2, this is again another situation where one finds naturally the stochastic integral of an anticipating process u. In fact in the above integral if $Y \in \mathcal{F}_T$, then the integral is in fact an anticipative integral. Even if $Y \equiv 1$, then $A = \int_0^T (D_v X)^2 \, dv \in \mathcal{F}_T$ in general.

Note that for the above formulas to hold one needs that the variable A (the so-called Malliavin variance) has to be different from zero. This is the non-degeneracy condition that we have required through Example 2 and Lemma 3. In fact in the case of Example 2, we have that $X = F(W_T)$ and $A = F'(W_T)^2 T \geq c^2 T > 0$. Therefore the condition $u \in \mathbb{L}^{1,2}$ contains in itself the non-degeneracy condition.

It is known that in the case that X is a diffusion with sufficiently smooth coefficients evaluated at a positive time, the Hörmander condition implies that A is well defined and that the anticipating stochastic integral of u is well defined.

The above formulas are useful because they give a general explicit expression for integration by parts of smooth variables without using explicitly the density of (X, Y). That is, we are giving an explicit formula for the ibp formula which was not generally possible with the likelihood method. Furthermore it has enough flexibility as to give different versions of the integration by parts. Let us discuss one of the many different possibilities available. As before let's start with

$$D_s Z = f'(X) D_s X.$$

Now we integrate both sides without multiplication by $D_s X$ as we did before. Supposing that the random variable $\int_0^T D_s X ds \neq 0$ a.s., besides other smooth properties we have that

$$E\left(f'(X)Y\right) = E\left(\int_0^T \frac{Y D_u Z}{\int_0^T D_s X ds} du\right) \tag{4.12}$$

$$= E\left(f(X)D^*\left(\frac{Y}{\int_0^T D_s X ds}\right)\right).$$

Remark 3. *There are other possible variations that can be applied according to the situation at hand. These include the following:*
1. One can do various combinations of components in the case that the driving process is multidimensional as well as obtaining integration by parts formulae for partial derivatives. Here is where the so-called Malliavin covariance matrix appears. In fact, a more general formula of integration by parts is given by

$$E\left(f(X)D^*\left(\frac{Yh(\cdot)}{\int_0^T h(v)D_v X dv}\right)\right) = E\left(f'(X)Y\right). \tag{4.13}$$

As before, this formula makes sense if $\int_0^T h(v)D_v X dv$ is different from 0 and has the necessary properties so that all terms make complete sense. Sometimes this is called the non-degeneracy condition. Previously we had taken in Theorem 2 $h(v) = D_v X$ and in the previous discussion $h \equiv 1$.

2. Perform various time changes so that one obtains a variation of the above formula. That is, using an interval [a, b] instead of the standard [0, T].

3. One can do various localizations before the stochastic integration by parts is done so that one does not need to integrate in a big portion of the state space.

4. Perform a change of measure so that the integration by parts formula could be weighted as desired.

5. Change the random variables in the problem by others which have the same law but can be differentiated easily or where the non-degeneracy is easier to obtain.

6. Changing the function f by f + c for a constant c so that certain optimal property is achieved (e.g., variance reduction).

4.5.4   Extracting r.v.'s from anticipating stochastic integrals

Before tackling the problem of greek estimation we will prove a formula to extract random variables out of anticipating integrals. This is another interesting application of the integration by parts formula and in particular the interpretation of D^*. This is a non-trivial generalization of the formula

$$\int_0^T X u_s ds = X \int_0^T u_s ds$$

to the case when Lebesgue integrals are replaced by stochastic integrals. The following formula, for u an adapted process, is not true in general:

$$\int_0^T X u_s dW_s = X \int_0^T u_s dW_s$$

unless X is a constant. First, the integral on the left has to be re-interpreted as a Skorohod integral because $X u_s$ is not necessarily adapted to $\mathcal{F}_s$ unless X is constant. The integral on the right has the usual meaning of stochastic integral for adapted process as u is adapted. This problem has been studied by many authors and it seems to go back to K. Itô.

The formula we will prove is

$$D^*(Xu) = X D^*(u) - < DX, u > .$$

This formula will be applied many times in order to carry out simulations of H. In other words, this is equivalent to saying

$$\int_0^T X u_s dW_s = X \int_0^T u_s dW_s - \int_0^T D_s X u_s ds.$$

Therefore the random variables can be taken out of the stochastic integrals but an extra term appears. This extra terms disappears if X is constant as $DX \equiv 0$. Obviously there are other situations when the extra term $\int_0^t D_s X u_s ds = 0$. (Exercise for the reader: Find some examples!)

Theorem 3. *Let* Xu, $u \in \mathbb{L}^{1,2}$ *and* $X \in \mathbb{D}^{1,2}$; *then* $D^*(Xu) = XD^*(u) - < DX, u >$.

Proof. To prove this formula one proceeds as follows: Let Y be any smooth random variable; then using the duality formula we have

$$
\begin{aligned}
E\left(YD^*(Xu)\right) &= E < DY, Xu >_{L^2[0,T]} \\
&= E < XDY, u >_{L^2[0,T]} \\
&= E < D(XY), u >_{L^2[0,T]} - EY < DX, u >_{L^2[0,T]} \\
&= E\left(Y\left(XD^*(u) - < DX, u >_{L^2[0,T]}\right)\right).
\end{aligned}
$$

As the above formula is satisfied for any Y, then the formula follows. $\square$

With this formula and under appropriate conditions we have that (4.12) can be written as

$$
E\left(f'(X)Y\right) = E\left(f(X)\left(\frac{YW(T)}{\int_0^T D_s X ds} - \int_0^T D_t\left(\frac{Y}{\int_0^T D_s X ds}\right) dt\right)\right). \quad (4.14)
$$

Therefore if one has explicit expressions for $D_s X$, $D_t Y$ and $D_t D_s X$ one can hope to be able to simulate H in this case.

4.5.5   Ibp formula for irregular functions

So far we have considered functions $f \in C_b^1$. Nevertheless, in applications one is interested in functions f that are irregular. Therefore we need a density argument to obtain the ibp formula in such a case. This is the following result:

Theorem 4. *Assume the same conditions as in Theorem 2. Then X has a density and furthermore*

$$
E\left(\delta_a(X)Y\right) = E(Y|X = a)p_X(a) = E\left(1(X \geq a)D^*\left(\frac{YD.X}{\int_0^T (D_v X)^2 \, dv}\right)\right).
$$

This theorem also shows that one can give mathematical meaning to expectations of generalized functions such as Dirac delta functions multiplied by smooth random variables. In the rest of the article we use this notation with the understanding that the expectation operator has been generalized to include this situation.

Proof. To prove the existence of the density one has to prove that the law of X is absolutely continuous with respect to the Lebesgue measure. For this we have used the ibp formula

$$
E\left(1(a \leq X \leq b)\right) = E\left(((X - a) \wedge b) H\right).
$$

The right side converges to zero if $b - a$ converges to zero. Therefore the law of X is absolutely continuous and has a density. Let $\phi_h(x) = (2\pi h)^{-1/2} \exp\left(-\frac{x^2}{2h}\right)$. Then applying Theorem 2 for $\Phi_h(x) = \int_{-\infty}^{x} \phi_h(y)dy$ we have that

$$\int_{\mathbb{R}} \phi_h(x - a)E(Y/X = x)p_X(x)dx = E\left(\phi_h(X - a)Y\right) = E\left(\Phi_h(X - a)H\right).$$

Taking limits the result follows.   □

The result in this theorem can obviously be stated for generalized functions with a similar argument. Similarly, one can also prove that the density function is bounded and smooth with bounded derivatives under the appropriate hypotheses. In order to apply this theorem we need to check the conditions stated in Theorem 2. If X and Y can be differentiated a sufficient number of times with their derivatives in $L^p(\Omega)$ for p big enough and importantly $\left(\int_0^T (D_s X)^2 ds\right)^{-1} \in L^p(\Omega)$ for p big enough, then the conditions in Theorem 2 are satisfied. We briefly sketch this in the next lemma. The generalization of the spaces $\mathbb{D}^{n,p}$ and $\mathbb{L}^{n,p}$ are defined as the natural extension of spaces previously defined. For example,

$$\|X\|_{n,p} = \left(E\left(|X|^p + \sum_{j=1}^{n} \left\|D^j X\right\|_{L^2[0,T]^j}^p\right)\right)^{1/p}.$$

Lemma 5. *Assume that* $X \in \mathbb{D}^{2,16}$ *and* $\left(\int_0^T h(v)D_v X dv\right)^{-1} \in L^{16}(\Omega)$, $Y \in \mathbb{D}^{1,16}$ *and* $h \in \mathbb{L}^{1,16}$ *Let*

$$u(t) = \frac{Yh(t)}{\int_0^T h(v)D_v X dv}.$$

Then $u \in \mathbb{L}^{1,2}$.

Proof. First, we need to compute the derivative of u which gives

$$D_s u(t) = \frac{D_s Y h(t) + Y D_s h(t)}{\int_0^T h(v)D_v X dv} - \frac{Yh(t)\int_0^T (D_s h(v)D_v X + h(v)D_s D_v X)dv}{\left(\int_0^T h(v)D_v X dv\right)^2}.$$

Then is a matter of using Holder's inequality to obtain the result.   □

Obviously the above result is not optimal.

4.5.6   Greeks for options using the ibp formula

As an application of the previous integration by parts formula we will obtain the same formulas for the greeks of European options as the one obtained through the finite difference or the likelihood method. In the case of digitals of asians we provide formulas that are not available using other methods.

As before $X(\alpha) = S_T$, $\alpha = S_0$. Here, the payoff Φ is differentiable a.e. such as $(x - K)_+$ or $1(x \geq K)$. Therefore when applying the ibp formula we need to use the results in the previous section.

Let us start computing *Delta* for a European digital option.

$$\Delta = \frac{\partial}{\partial S_0} E\left(e^{-rT}\Phi(S_T)\right) = \frac{e^{-rT}}{S_0} E\left(\frac{\partial S_T}{\partial S_0}\Phi'(S_T)\right) = \frac{e^{-rT}}{S_0} E\left(\Phi'(S_T)S_T\right).$$

We intend to apply the ibp formula given in (4.12) with $X = S_T$ and $Y = S_T$. Therefore we need to check the hypotheses which require that enough derivatives exist with a right amount of moments as in Lemma 5. In fact, if one differentiates $S(T)$ one has

$$D_u S_T = \sigma S_T D_u W_T = \sigma S_T 1(u \leq T).$$

Therefore it is clear that $D_u D_v S(T) = \sigma S_T 1(u \vee v \leq T)$ and that $S(T) \in \mathbb{D}^{2,16}$. Furthermore, $\int_0^T D_v S_T dv = \sigma T S_T$ and $E(S(T)^{-16}) < \infty$. Therefore we can apply the ibp formula to obtain that

$$\Delta = \frac{e^{-rT}}{S_0} E\left(\Phi(S_T)D^*\left(\frac{S_T}{\int_0^T D_v S_T dv}\right)\right). \tag{4.15}$$

Then we are able to perform the stochastic integral in (4.15),

$$D^*\left(\frac{S_T}{\int_0^T D_v S_T dv}\right) = D^*\left(\frac{S_T}{\int_0^T \sigma S_T dv}\right) = D^*\left(\frac{1}{\sigma T}\right) = \frac{W_T}{\sigma T}.$$

Then the expression for Δ reads,

$$\Delta = E\left(e^{-rT}\Phi(S_T)\frac{W_T}{S_0\sigma T}\right).$$

One can also compute gamma to obtain that (we leave the details of the calculation to the reader)

$$\Gamma = \frac{e^{-rT}}{S_0^2} E\left(\Phi'(S_T)D^*\left(\frac{S_T^2}{\int_0^T D_v S_T dv}\right)\right)$$

$$= E\left(\frac{e^{-rT}}{S_0^2\sigma T}\left\{\frac{W_T}{\sigma T} - W_T - \frac{1}{\sigma}\right\}\Phi(S_T)\right).$$

Simulations that show their performance in Monte Carlo simulations can be seen in Fournié et. al. (1999).

Now we consider greeks for options written on the average of the stock price $\frac{1}{T}\int_0^T S_s ds$. Note that in this particular case the density function of the random variable does not have a known closed formula. *Delta* in this case is given by

$$\Delta = \frac{\partial}{\partial S_0} E\left(e^{-rT}\Phi\left(\frac{1}{T}\int_0^T S_s ds\right)\right)$$
$$= \frac{e^{-rT}}{S_0} E\left(\Phi'\left(\frac{1}{T}\int_0^T S_s ds\right)\frac{1}{T}\int_0^T S_u du\right).$$

In this example we will show the versatility of the integration by parts formula (see Remark 3 1.), obtaining different expressions for Δ. First of all, we find in Fournié et. al. (1999) the expression

$$\Delta = \frac{2e^{-rT}}{S_0\sigma^2} E\left(\Phi\left(\frac{1}{T}\int_0^T S_s ds\right)\left(\frac{S_T - S_0}{\int_0^T S_t dt} - \mu\right)\right). \qquad (4.16)$$

Proof. To obtain this expression one uses (4.13) with $X = \frac{1}{T}\int_0^T S_s ds$, $Y = \frac{1}{T}\int_0^T S_u du$, $h_t = S_t$ so that one has that

$$E\left(\Phi'\left(\frac{1}{T}\int_0^T S_s ds\right)\frac{1}{T}\int_0^T S_u du\right)$$
$$= E\left(\Phi(X)D^*\left(\frac{YS.}{\sigma\int_0^T S(v)\int_v^T S(u)du}\right)\right)$$
$$= E\left(\Phi(X)\frac{2}{\sigma}\int_0^T S_t dW_t\right).$$

$\square$

Now one can deduce a different expression using the ibp formula (4.11). In such a case one obtains

$$\frac{e^{-rT}}{S_0} E\left(\Phi\left(\frac{1}{T}\int_0^T S_s ds\right)\left(\frac{1}{<S>}\left\{\frac{W_T}{\sigma} + \frac{<S^2>}{<S>}\right\} - 1\right)\right) \qquad (4.17)$$

where $<S> = \left(\int_0^T tS_t dt\right)\left(\int_0^T S_s ds\right)^{-1}$ and $<S^2> = \left(\int_0^T t^2 S_t dt\right)\left(\int_0^T S_s ds\right)^{-1}$.

In the next section, we will show some simulations of these ibp methods. Other simulation results can be seen in Fournié et. al. (1999), (2001).

4.6   Comparison and efficiency of the estimation methods

Now that we have introduced both methods of estimating a greek, kernel density estimation and the integration by parts of Malliavin Calculus, we can carry out a comparison between them to discern when to use a particular method. This also implies that we have to discuss some practical aspects of each method. Let us start with some comments about the kernel density method. We illustrate the case of estimating the value of *Delta* for a European and an Asian binary option, within the following scenario: $S_0 = 100$ (in arbitrary cash units), $r = 0.05$, $\sigma = 0.2$ and $T = 0.25$ (in years). In Figure 2 (a), we plot the value of the absolute bias and the root of the variance of the finite-difference estimate for the European binary Δ, in the case that we choose $h = h_0$. Let us observe that in spite of the fact that variance carries the main contribution to the total error, the effect of the bias is not negligible. The variance error can be evaluated through the *sampling variance* of the Δ estimate, but the value of the bias is not directly measurable from the estimate itself. Thus, in general, we should consider a new estimate for the bias in order to compute the magnitude of the total error. In this example, however, one is able to compute exactly both bias and variance contribution. In the case of Asians one can see that the estimate of the bias is not reliable. Therefore it is necessary to study this problem further.

The criteria that one can choose to do the comparison between the kernel density method and the ibp method may be varied. Here we narrow this discussion to the bias and the variance. The bias of the kernel density estimation method can already be seen in Figure 1. As we have said previously the integration by parts method does not create any bias (at least theoretically). Therefore our comparison can now be restricted to the variance. In the case of the kernel density estimation method the mean square error can be measured and the result is in Figure 3.

We consider the same example as in Figure 1. The curves depict the root of the mean squared error, ε_2 in Section 4.3, corresponding to $N = 10^5$ and the respective selection for h.

In the case of the European binary option the asymptotically optimal value of h_0 (4.2) and $h = 3.247$ (4.3), differ significantly only in the values of the moneyness for which the bias is close to zero. In any case, both of them are significantly lower than the error level that we achieve considering a smaller choice for h, $h = 1.0$. This result contradicts the naive rule that dictates that h should be chosen as small as possible so that the simulations lead to a stable result.

The results for the digital Asian are similar. Again h_0 and its at-the-money approximation $h = 1.1937$ differ only in the low bias regions. In any case, both of them show a better performance than the third proposal, $h = 2.5 \times 10^{-3} S_0 = 0.25$, which represents a very small value for the parameter.

An issue related with the choice of h is that of the three possible choices proposed in Figure 1 (a), the asymptotically best one is not constant and sometimes may have a non-smooth behavior. We have computed the behavior of these choices of h in Figure 4. We consider different contract specifications K, ranging from $K = 2S_0/3$ to $K = 2S_0$, and the results will be displayed in terms of the so-called present moneyness, S_0/K. We use in this case the kernel that conduces to the classical finite differences method, G_1. The curve labelled h_0 corresponds to the optimal value for h (4.2). It is notorious that in the vicinity of $S_0/K = 0.893$ and $S_0/K = 1.087$ the value of h_0 grows dramatically. The presence of these two critical points is a consequence of the existence of two particular values of the moneyness that make

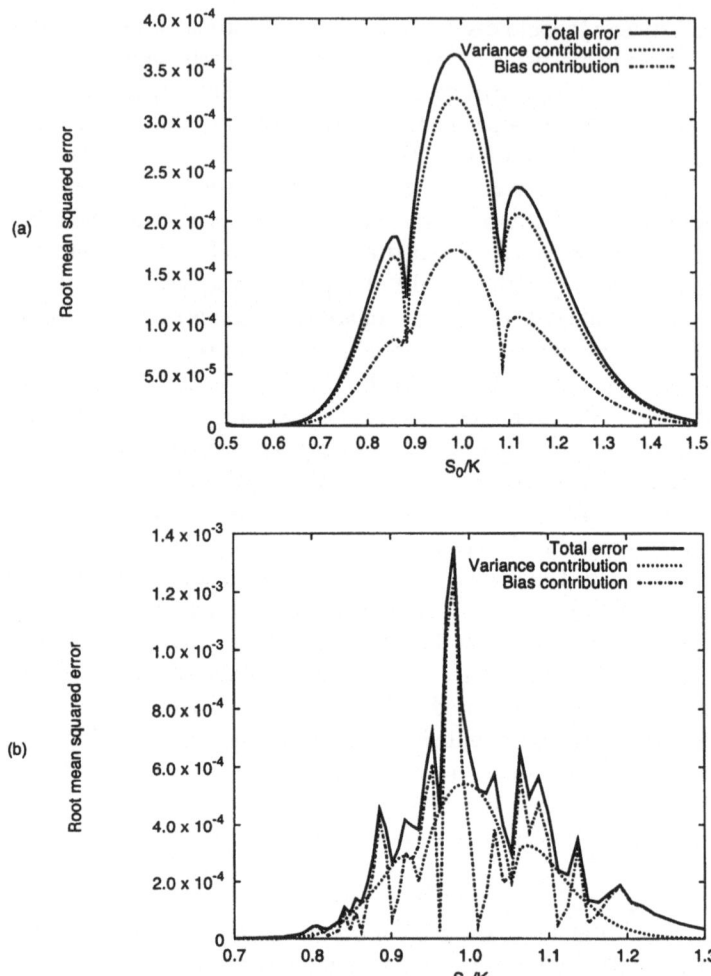

FIGURE 2. Relative weight of the bias and the variance in the total error. The value of parameters are —$S_0 = 100$ (in arbitrary cash units), $r = 0.05$, $\sigma = 0.2$, $T = 0.25$ (in years), and $N = 10^5$. (a) European binary option (b) Asian binary option.

unbiased the estimate. For binary options, the leading term of the bias in computing Δ is proportional to $\partial_z^2(z^3 p(z))$, where $p(z)$ is the probability density function of Z. Whenever $p(z)$ is a bell-shaped function and $\lim_{z\to\pm\infty} \partial_z(z^3 p(z)) = 0$, there will always be two and only two values $z_{1,2}$ such that $\partial_z^2(z^3 p(z))|_{z=z_{1,2}} = 0$. The straight line $h = 3.247$ corresponds to the asymptotically optimal value of h_0, mostly valid when d and $\sigma\sqrt{T}$ are small. The third proposal, marked as $h = 1.0$, will highlight the fact that reducing the value of h, $h = 0.01S_0$ within our setup, is not a good procedure in this case.

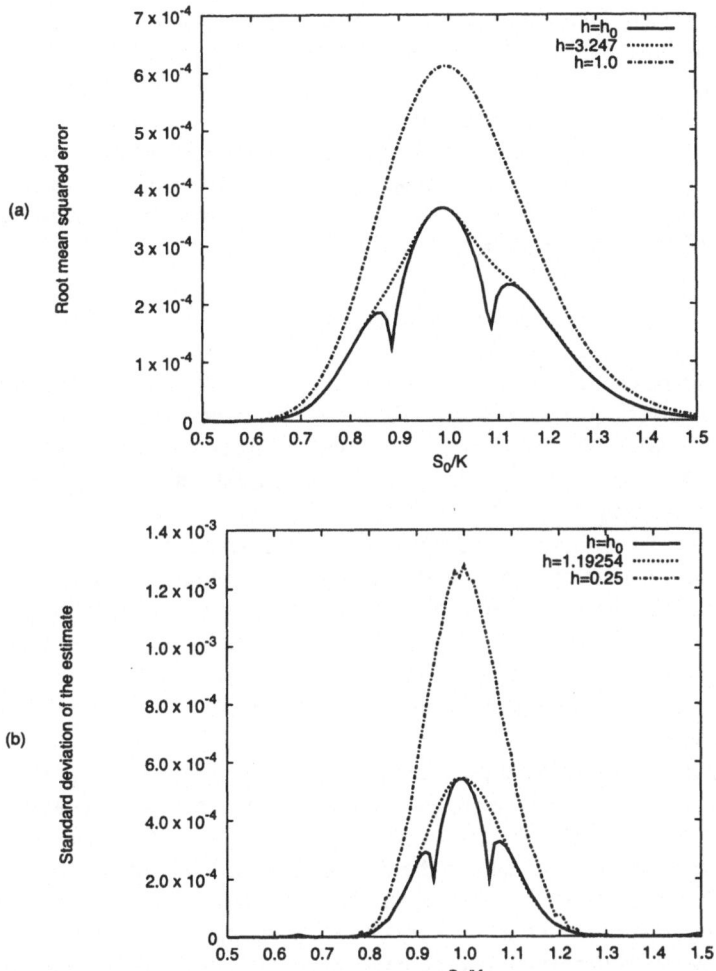

FIGURE 3. Statistical errors associated to the Δ estimate for three different choices of h. (a) European binary option (b) Asian binary option.

The results for Asians are similar, although the involved probability function has no closed expression. In fact, when dealing with binary options, the Δ itself is proportional to this unknown density function. Of course we could proceed in a recursive way: we can estimate a probability density function, then we can use it to compute h, and with this value we can start the whole process anew.

Now we compare the results and variances of the various kernel density estimation methods together with Malliavin type estimators. Before that we recall that the payoff function being $\Phi(x) = 1(x > K)$ one can do the integration by parts to recover the same function. We call this integration by parts, the Malliavin method or plain/non-symmetrical integration by parts method. In the case that the integration by parts is done in such a way as to recover the function $1(x > K) - 0.5$ (see Remark 3.7) then the method is called the symmetric Malliavin/integration by parts method. This method, as it will shown shortly, introduces some variance reduction and is the parallel of the control variate method.

Another interesting method of variance reduction is localization which can be briefly introduced as follows. Suppose that we want to perform an integration by parts for $E(\delta_K(X)Y)$. Then one can use a smooth even function φ such that $\varphi(0) = 1$, $\varphi, \varphi' \in L^2(\mathbb{R})$ as a localization to obtain

$$E(\delta_K(X)Y) = E(\delta_K(X)Y\varphi(\frac{X-K}{r}))).$$

One can then compute through variational calculus the optimal values for φ and r as to obtain an effective reduction of variance. These give $\varphi(x) = \exp(-|x|)$ and an explicit expression for r. For details, see either Kohatsu–Pettersson (2002) or Bermin et. al. (2003). This method has been shown to be quite effective and we call it the localization method.

In Figure 5 we observe the superposition of several estimates for the European (a) and Asian (b) binary option, and different values of the moneyness. In this plot we put together the output of five different estimates of *Delta*: finite differences, Gaussian kernel, Epanechnikov kernel (all of them use the corresponding optimal value of parameter h_0), Malliavin and symmetrical Malliavin. All the estimations are good enough to be indistinguishable from the exact value, depicted in the graph with empty boxes.

In Figure 5 (b), we have used the corresponding approximation to the optimal value of parameter h_0, Malliavin, Symmetrical Malliavin. In general all the estimations give similar results, except in two regions well apart from the at-the-money range. The discrepancy appears in the three kernel-related estimates, and it is originated in the choice of h_0. We must remember that in the Asian framework the exact value of h_0 is as unknown as the greek itself. The approximation that we have introduced works better when dealing with values of K near S_0. If we disregard this disfunction, all the methods closely reproduce the *exact value*, depicted again in the graph with empty boxes. Obviously, since no closed expression for the Asian *Delta* exists, we have simply used a better estimate (the localization method for the Malliavin integration by parts method with $N = 5 \times 10^5$) in order to simulate it.

We have introduced the localization method as a way to obtain an accurate answer. Also as means to show that if the integration by parts method is used appropriately it gives very accurate answers, as we will also see shortly.

In the next Figure 6 we show the statistical errors associated to the previous estimates. In the case of the European binary Δ estimates, it is clear that the kernel-based methods give worse estimations for the greek if the moneyness is near the at-the-money value. Even if we are interested in values of the moneyness that are either in the in-the-money or in the out-of-the-money range, we can pick the appropriate non-symmetrical Malliavin estimate which show a similar degree of accuracy.

In the case of binary Asians, we must remember that all the kernel-based estimates are biased estimates, and that the exact amount of bias they present is also

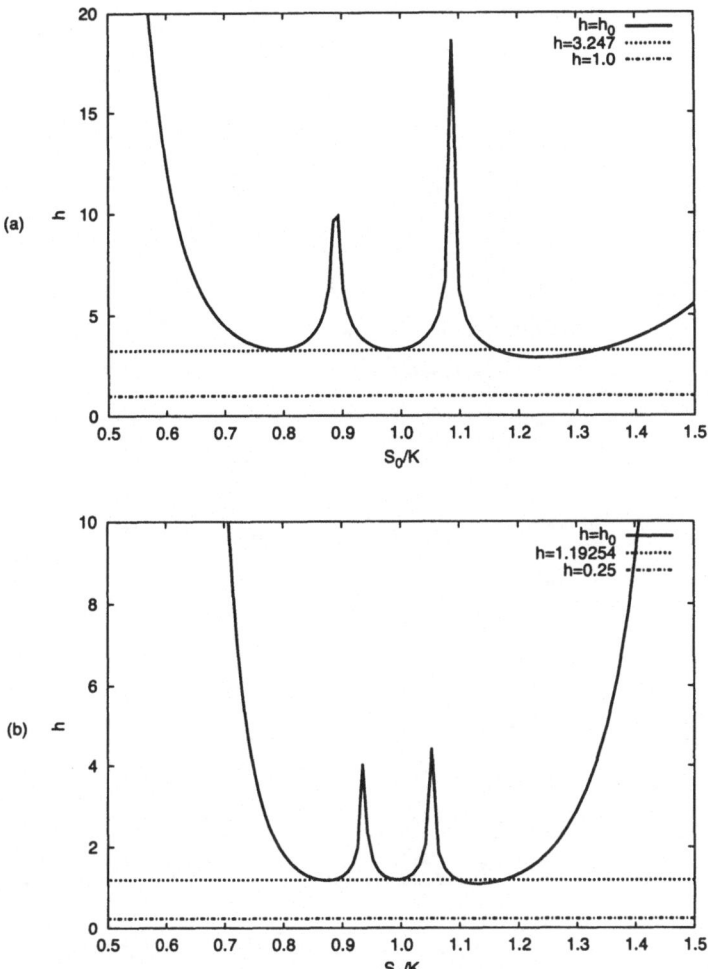

FIGURE 4. Three criteria for choosing the value of parameter h in the kernel density estimation framework. (a) European binary option (b) Asian binary option.

unknown. This means that these three estimates will show an even higher level of error than the one depicted here. With this feature in mind, it is clear that the kernel-based methods give definitively worse estimations for the greek if the moneyness is near the at-the-money value. Again, even in the case we are interested in values of the moneyness that are either in the in-the-money or in the out-of-the-money range, we can pick the appropriate non-symmetrical Malliavin estimate, or even better the localized Malliavin estimate, in order to achieve a similar degree of accuracy.

Next we show a table that describes times of computation. These experiments were carried out on a desktop PC with a Pentium III-500 MHz, runing under Win-

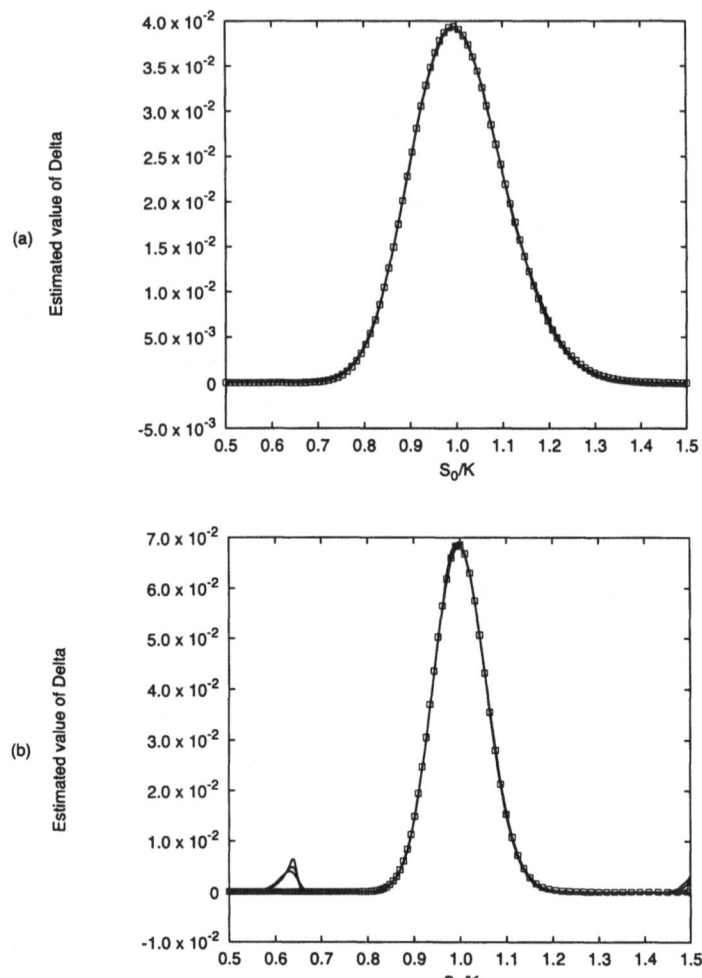

FIGURE 5. Comparison between the kernel density estimates and the Malliavin estimates for the (a) European binary option (b) Asian binary option.

dows 2000 Professional. The programs were written in ANSI C, and compiled with Microsoft Visual Studio C++ 6.0, in "Release" mode.

We present the mean time associated to each numerical method which we have previously used when obtaining the value of *Delta*. As there is no seeming variation in the times of computation as the moneyness changes, we present here an average of computation times ratios for different degrees of moneyness.

Let us start analyzing the results for the European option. As it can be seen, among the kernel-related methods, the Gaussian one, is the most time-consuming, whereas the other two are very similar. This is certainly due to the more intricate form of the kernel itself. It is also evident that the use of Malliavin techniques does not lead

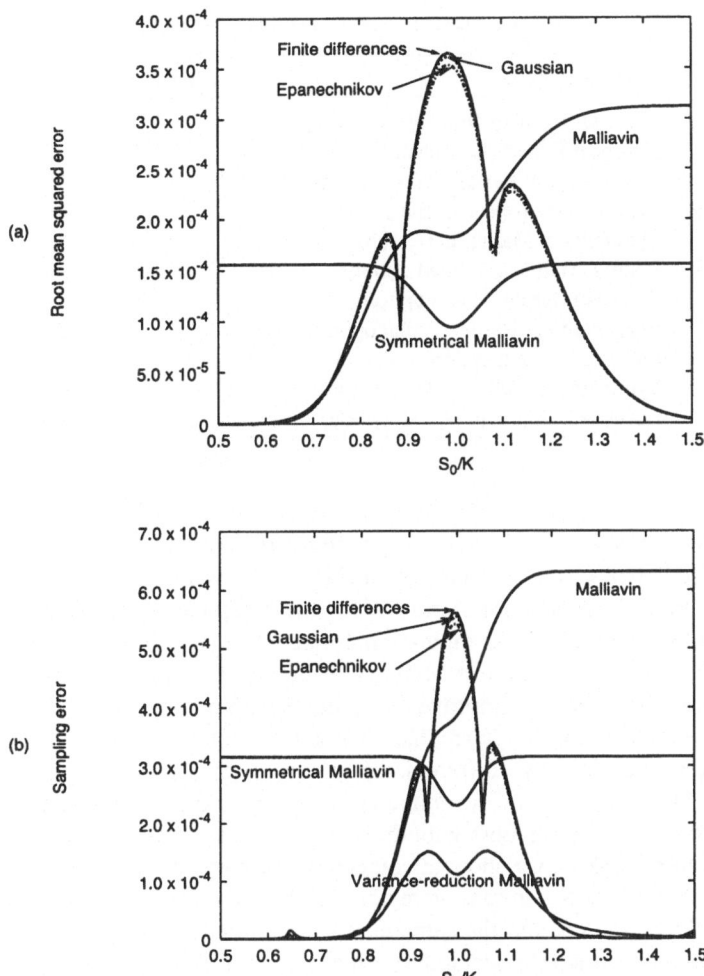

FIGURE 6. Statistical errors of the kernel density estimates and the Malliavin estimates for the (a) European binary option (b) Asian binary option

in this case to slower estimates of the greek. Both Malliavin-based proposals defeat the previous algorithms. The fastest is also the simplest estimate, the plain Malliavin estimate, although the more elaborate "Symmetrical Malliavin" is better than any of the kernel-based procedures. These results go against the typical argument that the use of Malliavin calculus leads to cumbersome estimates. Time units have been chosen in such a way that the time corresponding to finite differences is set to one.

In the case of the digital Asian, the higher complexity in the simulation of the involved random variable, $Z = \frac{1}{T} \int_0^T S(t)dt$, virtually eliminates any differential behaviour arising from the functional form of the kernel. In this case, these kernel-oriented algorithms work faster than the Malliavin ones. We must point out, however,

TABLE 1. Comparison between computational times.

Numerical method	Computational time
Finite Differences (European option)	1.00
Gaussian Kernel (European option)	1.45
Epanechnikov Kernel (European option)	1.11
Malliavin (European option)	0.87
Symmetrical Malliavin (European option)	0.95
Finite Differences (Asian option)	24.2
Gaussian Kernel (Asian option)	24.5
Epanechnikov Kernel (Asian option)	23.2
Malliavin (Asian option)	31.3
Symmetrical Malliavin (Asian option)	30.4
Localized Malliavin (Asian option)	23.9

that both Malliavin estimates, the non-symmetric and the symmetrical one, are based upon the Malliavin formula (4.17) which showed a lower variance than (4.16) (see discussion below). This expression involves the computation of several additional integrals which increase the computational time. We can follow an alternative path instead. We can use the simplest Malliavin estimate, and upgrade it using localization. With this approach we improve the estimate with no time penalty.

Now we discuss the two formulae obtained in Section 4.5.6 for Asian options. The fact that there are two different ibp formulas for the same greek may seem strange at first but these two formulae are different and therefore their simulation will lead to different estimators with different variances. We can observe these features in Figure 7, where we show the outcome of the Monte Carlo simulation using these estimators. The two graphs were obtained with the two different estimates presented in Section 4.5.6, and the same ensemble of random variables. The thin one corresponds to the first estimate, and the thick one to the second, and more complex, estimate. It is clear that they numerically differ, and that the second one displays a lower level of variance. This fact is in contradiction with what is claimed in Fournié et al. (2001).

A general rule of thumb is that if the estimator used invokes a higher number of statistics, then the estimator will have smaller variance. An open problem is how to obtain the most significant statistics in order to optimize the integration by parts formula. Therefore one can not expect that all ibp formulae will lead to the same estimator. The main reason being that this is equivalent to knowing the probability density of the random variable in question. To expose the main ideas that also appear in Fournié et. al. (2001) one can note first that there is an integration by parts that is the "most" straightforward but highly unrealistic. For this, consider the generalized problem

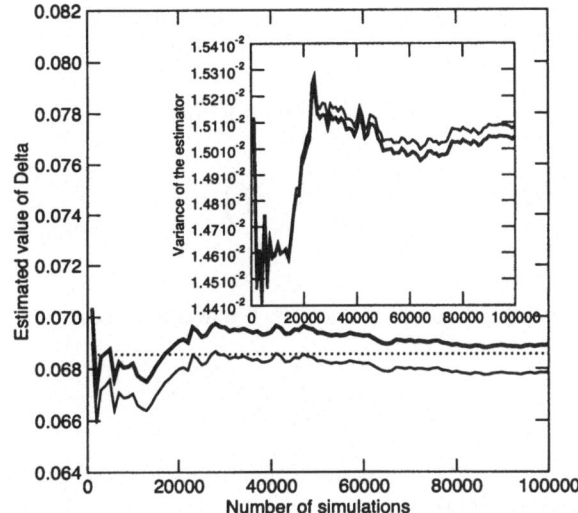

FIGURE 7. Computed value of *Delta* for an Asian call (the parameters are the same as in Figure 2), using Monte Carlo techniques (N is the number of simulations of the integral), for the estimators presented in Section 4.5.6. We have broken the interval of integration in 60 pieces, representing the approximate number of trading days in three months ($T = 0.25$). The exact result is represented by the dotted line.

$$E\left(\Phi'\left(\int_0^T S_s ds\right)\int_0^T S_s ds\right) = \int_{\mathbb{R}} \Phi'(x)xp(x)dx.$$

Here p denotes the density of $\int_0^T S_s ds$ which exists and is smooth (it is an interesting exercise of Malliavin Calculus). Therefore one can perform the integration by parts directly in the above formula, thus obtaining that

$$E\left(\Phi'\left(\int_0^T S_s ds\right)\int_0^T S_s ds\right) = \int_{\mathbb{R}} \Phi(x)(p(x) + xp'(x))dx$$

$$= E\left(\Phi(\int_0^T S_s ds)\left(1 + \frac{\int_0^T S_s ds\, p'(\int_0^T S_s ds)}{p(\int_0^T S_s ds)}\right)\right).$$

Now we proceed to prove that the above gives the minimal integration by parts in the sense of variance. Obviously it is not possible to carry out the simulations unless p'/p is known. Let us construct the set of all possible integration by parts. Suppose that Y is a random variable such that

$$E\left(\Phi'\left(\int_0^T S_s ds\right)\int_0^T S_s ds\right) = E\left(\Phi\left(\int_0^T S_s ds\right)Y\right),$$

for any function $\Phi \in C_p^{+\infty}$, then it is not difficult to deduce that

$$E\left(Y\Big/\sigma\left(\int_0^T S_s ds\right)\right) = 1 + \frac{\int_0^T S_s ds p'(\int_0^T S_s ds)}{p(\int_0^T S_s ds)}.$$

Therefore the set of all possible integration by parts can be characterized as

$$\mathcal{M} = \left\{ Y \in L^2(\Omega);\ E\left(Y\Big/\sigma\left(\int_0^T S_s ds\right)\right) = 1 + \frac{\int_0^T S_s ds p'(\int_0^T S_s ds)}{p(\int_0^T S_s ds)} \right\}.$$

Next in order we want to find the element in Y that minimizes

$$\inf_{Y \in \mathcal{M}} E\left(\Phi\left(\int_0^T S_s ds\right)^2 Y^2\right).$$

As in Fournié et. al. (2001) it is not difficult to see which Y achieves the minimum. This is done as follows:

$$E\left(\Phi(\int_0^T S_s ds)^2 Y^2\right)$$

$$= E\left(\Phi(\int_0^T S_s ds)^2 \left(Y - 1 - \frac{\int_0^T S_s ds p'(\int_0^T S_s ds)}{p(\int_0^T S_s ds)}\right)^2\right)$$

$$+ E\left(\Phi(\int_0^T S_s ds)^2 \left(1 + \frac{\int_0^T S_s ds p'(\int_0^T S_s ds)}{p(\int_0^T S_s ds)}\right)^2\right),$$

since the cross product is 0, due to the property of the set $\mathcal{M}$. Therefore the minimum is achieved at $Y = \left(1 + \frac{\int_0^T S_s ds p'(\int_0^T S_s ds)}{p(\int_0^T S_s ds)}\right)$. This is clearly impossible to write explicitly as p is unknown in the case of Asian options. Therefore it is still an open problem to devise good ways to perform an efficient integration by parts so that the variance is made small rapidly and efficiently.

4.7   Other examples of applications

4.7.1   The Clark–Ocone formula

As another application of stochastic derivatives we discuss the Clark–Ocone formula that can be used to obtain replicating hedging strategies for options. As before let $X \in \mathbb{D}^{1,2}$, then the problem consists in finding an adapted process u such that

$$X = E(X) + \int_0^T u_s dW_s.$$

To find u, differentiate the above equation to obtain

$$D_t X = u_t + \int_t^T D_t u_s dW_s.$$

Next take the conditional expectation with respect to $\mathcal{F}_t$, which gives

$$E(D_t X / \mathcal{F}_t) = u_t.$$

In Finance one actually has that X is a contingent claim and one wants to find u in the expression

$$e^{-rT} X = E(e^{-rT} X) + \int_0^T u_s d\widehat{S}_s$$

where $\widehat{S}(t) = S_0 \exp\left(-\frac{\sigma^2}{2}s + \sigma W(s)\right)$ represents the discounted stock and r is the interest rate. In this case, we have that

$$e^{-rT} X = E(e^{-rT} X) + \int_0^T E\left(e^{-rT} D_t X / \mathcal{F}_t\right) dW(t)$$

$$= E(e^{-rT} X) + \int_0^T (\sigma \widehat{S}(t))^{-1} E\left(e^{-rT} D_t X / \mathcal{F}_t\right) d\widehat{S}(t).$$

Then it is not difficult to prove that the integrand corresponds to the greek Δ (we leave this as an exercise for the reader).

4.7.2   Ibp for the maximum process

In this section we are interested in an application of the ibp formula where the properties of differentiability of the processes at hand are limited. This is the case of the maximum process. In this section we consider the integration by parts formula of Malliavin Calculus associated to the maximum of the solution of a one-dimensional stochastic differential equation. The problem of obtaining such an integration by parts formula has already been considered by Nualart and Vives (1988) where the absolute continuity of the maximum of a differentiable process is proven. Later in Nualart (1995), the smoothness of the density of the Wiener sheet was obtained.

The ideas presented here have appeared in Gobet–Kohatsu (2001) and Bernis et. al. (2003). In the following example we consider the delta of an up in & down in Call option. That is, let $0 < t_1 < \cdots < t_N = T$ be monitoring times for the underlying S. Then the payoff of the up in & down out Call option is

$$\Phi = 1(\min_{i=1,\ldots,N} S_{t_i} \leq D)1(\max_{i=1,\ldots,N} S_{t_i} \geq U)1(S_T < K).$$

The payoff in this case is path-dependent as in the case of Asian options. Nevertheless the maximum function is not as smooth (in path space) as the integral function.

Therefore this example lies in the boundaries of application of Malliavin Calculus. Interestingly, the law of the minimum and maximum processes are smooth enough therefore the calculations are still possible (this is related with our Remark 3). In this case one could also apply the likelihood method although the problem involves a cumbersome expression. First $\Delta = e^{-rT} \lim_{n \to \infty} E(\Phi_n)$ where

$$S_0 \Phi_n = -\phi_n \left(\min_{i=1,...,N} S_{t_i} - D \right) \min_{i=1,...,N} S_{t_i} 1(\max_{i=1,...,N} S_{t_i} \geq U) 1(S_T < K)$$

$$+1(\min_{i=1,...,N} S_{t_i} > D) \phi_n \left(\max_{i=1,...,N} S_{t_i} - U \right) \max_{i=1,...,N} S_{t_i} 1(S_T < K)$$

$$-1(\min_{i=1,...,N} S_{t_i} > D) 1(\max_{i=1,...,N} S_{t_i} \geq U) \phi_n(S_T - K) S_T$$

where $\phi_n(x) = n^{-1} \phi(nx)$ with ϕ a positive smooth function with support in $[-1, 1]$ satisfying that $\int_{\mathbb{R}} \phi(x) dx = 1$. In other words ϕ_n is an approximation of the Dirac delta function at zero (previously we had used the density of a normal random variable). We can therefore apply the ibp (4.12) with $X = \Phi$, $Y = 1$ and $T = t_1$ (see Remark 3). Then we have that

$$\Delta = e^{-rT} E \left(\Phi \frac{W_{t_1}}{\sigma S_0 t_1} \right). \tag{4.18}$$

To obtain this formula we have used that for $t < t_1$ (the formula is not valid for $t > t_1$),

$$D_t \min_{i=1,...,N} S_{t_i} = \sigma \min_{i=1,...,N} S_{t_i},$$

$$D_t \max_{i=1,...,N} S_{t_i} = \sigma \max_{i=1,...,N} S_{t_i}.$$

This ibp avoids the non-smoothness of X but the problem with the simulation of this expression is the instability of $\frac{W_{t_1}}{t_1}$ when t_1 is close to zero. Therefore the ideas exposed up to this point have to be revised to try to improve this formula.

Instead, we will use a localization process h (see Remark 3.) in order to integrate by parts the processes involved in the whole time interval $[0, T]$ therefore avoiding the instability mentioned previously.

In order to do this, we first compute in general the formula for the stochastic derivative of Φ. Using the local property of the derivative we have that

$$D_t \min_{i=1,...,N} S_{t_i} = D_t \sum_{j=1}^{N} S_{t_j} 1 \left(\min_{i=1,...,N} S_{t_i} = S_{t_j} \right)$$

$$= \sum_{j=1}^{N} D_t S_{t_j} 1 \left(\min_{i=1,...,N} S_{t_i} = S_{t_j} \right)$$

$$= \sigma \sum_{j=1}^{N} S_{t_j} 1(t \leq t_j) 1 \left(\min_{i=1,...,N} S_{t_i} = S_{t_j} \right)$$

$$= \sigma S_{\tau} 1(t \leq \tau),$$

where $\tau = \inf\{t_i; S_{t_i} = \min_{i=1,\dots,N} S_{t_i}\}$. Similarly,

$$D_t \max_{i=1,\dots,N} S_{t_i} = \sigma S_{\tau'}$$

with $\tau = \inf\{t_i; S_{t_i} = \max_{i=1,\dots,N} S_{t_i}\}$. Now we can see the non-smoothness of the maximum or minimum process. A second derivative of the maximum will involve the differentiation of $1(t \leq \tau')$ which is not a differentiable random variable (exercise for the reader). Now, to do the integration by parts we perform the integration by parts using what we call a dominating process.

Let Y be defined as

$$Y_t = \sqrt{N \sum_{\substack{1 \leq i \leq N \\ t_i \leq t}} (S_{t_i} - S_{t_{i-1}})^2}.$$

Lemma 6. *Suppose that $U > S_0 > D$. Then one has:*

i) *For any $t \in \{t_i : 0 \leq i \leq N\}$, one has $|S_t - S_0| \leq Y_t$.*
ii) *There exists a positive function $\alpha : \mathbb{N} \mapsto \mathbb{R}_+$, with $\lim_{q \to \infty} \alpha(q) = \infty$, such that, for any $q \geq 1$, one has:*

$$\forall t \in [0, T] \quad E(Y_t^q) \leq C_q t^{\alpha(q)}.$$

iii) *For any $q \geq 1$, choose a C_b^∞ function $\Psi : [0, \infty) \mapsto [0, 1]$, with*

$$\Psi(x) = \begin{cases} 1 & \text{if } x \leq a/2, \\ 0 & \text{if } x \geq a, \end{cases}$$

with $U > S_0 + \frac{a}{2} > S_0 - \frac{a}{2} > D$. The random variable $\Psi(Y_t)$ belongs to $\mathbb{D}^{q,\infty} = \cap_{p>1} \mathbb{D}^{q,p}$ for each t. Moreover, for $j = 1, \dots, q$, one has

$$\forall p \geq 1 \quad \sup_{r_1,\dots,r_j \in [0,T]} E\left(\sup_{r_1 \vee \cdots \vee r_j \leq t \leq T} \|D^j_{r_1,\dots,r_j} \Psi(Y_t)\|^p \right) \leq C_p.$$

Proof. For $t = t_j$, one has $|S_t - S_0| \leq \sum_{i=1}^j |S_{t_i} - S_{t_{j-1}}| \leq Y_t$, using Jensen's inequality: this proves Assertion i). The other assertions are also easy to justify, we omit details. □

Now we are ready to perform the integration by parts. That is, we compute the stochastic derivative of Φ in general to obtain

$$\sigma^{-1} D_t \Phi_n$$

$$= -1(t < \tau) S_\tau \phi_n \left(\min_{i=1,\dots,N} S_{t_i} - D \right) 1(\max_{i=1,\dots,N} S_{t_i} \geq U) 1(S_T < K)$$

$$+ 1(t < \tau') S_{\tau'} 1(\min_{i=1,\dots,N} S_{t_i} \leq D) \phi_n \left(\max_{i=1,\dots,N} S_{t_i} - U \right) 1(S_T < K)$$

$$- 1(\min_{i=1,\dots,N} S_{t_i} \leq D) 1(\max_{i=1,\dots,N} S_{t_i} \geq U) \phi_n (S_T - K) S_T.$$

We multiply this expression by $\Psi(Y_t)$ to obtain that

$$\sigma^{-1}D_t\Phi_n\Psi(Y_t)$$
$$= -\Psi(Y_t)S_\tau\phi_n\left(\min_{i=1,\dots,N}S_{t_i} - D\right)1(\max_{i=1,\dots,N}S_{t_i} \geq U)1(S_T < K)$$
$$+\Psi(Y_t)S_{\tau'}1(\min_{i=1,\dots,N}S_{t_i} \leq D)\phi_n\left(\max_{i=1,\dots,N}S_{t_i} - U\right)1(S_T < K)$$
$$-1(\min_{i=1,\dots,N}S_{t_i} \leq D)1(\max_{i=1,\dots,N}S_{t_i} \geq U)\phi_n(S_T - K)S_T\Psi(Y_t).$$

Note that in this expression we have deleted the indicator functions. The reason for this is that if $\Psi(Y_t) \neq 0$, then $Y_t \leq a/2$ and therefore $S_{t_i} \geq S_0 - a/2$ for all $t_i \leq t$ and if we also assume that $\min_{i=1,\dots,N} S_{t_i} \leq D + \frac{1}{n}$, then it means that $t < \tau$ for n big enough. In all other cases this term is zero.

Similarly for the second term we have that if $\Psi(Y_t) \neq 0$, then if $\max_{i=1,\dots,N} S_{t_i} > U - \frac{1}{n}$, then $t < \tau'$. Now we can perform the integration by parts to obtain that

$$\Delta = \frac{S_0}{\sigma}E\left[\Phi D^*\left(\frac{\Psi(Y_.)}{\int_0^T \Psi(Y_t)dt}\right)\right].$$

Here it should be clear that the integration by parts is carried out in the whole time interval therefore allowing for stable simulations. The simulations results which appear in Gobet–Kohatsu (2001) show that this last methodology gives better results than the finite difference and the previous integration by parts formula (4.18).

4.8   The local Vega index

In this section we introduce a generalization of the Vega index which we call the local Vega index (lvi) which measures the stability of option prices in complex models. In other words, the lvi weights the local effect of changes in the volatility structure of a stochastic volatility model. This index comes naturally under the general framework introduced in Section 3.3 in Fournié et al. (1999).

The first natural interpretation of the lvi measure is to consider them as the Fréchet derivatives of option prices with respect to changes in the volatility structure, therefore naturally generalizing the concept of greek. The Vega index measures the perturbations of the option prices under perturbations of volatility structure. In the particular case that this volatility is constant, then this sensibility index is characterized by a classical derivative. If instead one wants to consider general volatility models, then one has to consider Fréchet derivatives. These derivatives therefore become also functions which are parametrized as the perturbations themselves.

Here let ε denote the perturbation parameter and $\hat{\sigma}(t, x)$ is the direction of perturbation. Then the goal is to obtain the corresponding sensibility weight that corresponds to this direction. For this, let's suppose that we want to test the robustness of our original model for S and consider S^ε to be a positive diffusion process, defined as the solution to

$$\begin{cases} dS^\varepsilon(t) = r(t, S^\varepsilon(t))dt + \sigma^\varepsilon(t, S^\varepsilon(t))dW(t), \\ S^\varepsilon(0) = S_0, \end{cases}$$

where $\sigma^{\varepsilon,r} : \mathbb{R}_+ \times \mathbb{R} \to \mathbb{R}$ are smooth functions with bounded derivatives. W is a one-dimensional Brownian motion. Here we assume that the equivalent martingale measure is independent of ε. Furthermore, suppose that σ^ε is of the form $\sigma^\varepsilon(t, x) = \sigma(t, x) + \varepsilon \hat{\sigma}(t, x)$. S^0 is the basic model which we are perturbing, we will use S^0 or just S to denote our base model.

Definition 3. *Given a financial quantity Π^ε based on S^ε, we say that it has a local Vega index if*

$$\left. \frac{\partial \Pi^\varepsilon}{\partial \varepsilon} \right|_{\varepsilon=0} = \int_0^T E\left[\mu(s, T, S^0(s))\hat{\sigma}(s, S^0(s)) \right] ds.$$

In most applications $\Pi^\varepsilon = E(\Phi(F(S^\varepsilon)))$ where $\Phi : \mathbb{R} \to \mathbb{R}$ is the payoff function and F is a functional. For general results on the existence of the lvi for option prices see Bermin et. al. (2003).

The kernel $\mu(s, T, x)$ measures the importance or the effect of the local changes in volatility of the underlying and of the noise at time s and value of the underlying x standardized in perturbation units. If such a weight is comparatively big, then small changes in volatility will be important. The most important point is the fact that this formula gives quantitative meaning to various expected qualitative behavior of option prices.

A way to define a global Vega index rather than a local one is to choose a uniform deformation of volatility. That is, $\hat{\sigma} \equiv 1$. Suppose we denote this global index by $\frac{\partial \Pi}{\partial \sigma}$, then we have the relationship

$$\left. \frac{\partial \Pi^\varepsilon}{\partial \varepsilon} \right|_{\varepsilon=0} = \int_0^T E\left[\hat{\mu}(s, T, S^0(s))\hat{\sigma}(s, S^0(s)) \right] ds \frac{\partial \Pi}{\partial \sigma}.$$

The only formal difference with respect to the expression in the previous theorem is that now the weights $\hat{\mu}$ integrate to 1 and therefore one can interpret the comparative values of these indices easily.

The ibp formula plays a role in the construction of the lvi index. In fact, in most situations μ involves the conditional expectation of the second derivative of the payoff function and therefore to give meaning and to compute such a term one needs to use the integration by parts formula.

Example 4. *Let us consider a standard call option with payoff $G^\varepsilon = \max(S^\varepsilon(T) - K, 0)$ for some constant strike price K and assume $\sigma^\varepsilon(t, x) = \sigma x + \varepsilon \hat{\sigma}(t)x$ and $r(t, x) = rx$. It is easily verified that at time 0 the price is given by*

$$\Pi^\varepsilon = S_0 N\left(d_1^\varepsilon \right) - e^{-rT} K N\left(d_1^\varepsilon - \sqrt{\Sigma^\varepsilon} \right),$$

where $N(\cdot)$ denotes the cumulative distribution function of a standard normal random variable, and d_1^ε is defined by

$$d_1^\varepsilon = \frac{\ln(S_0/K) + rT + \frac{1}{2}\Sigma^\varepsilon}{\sqrt{\Sigma^\varepsilon}} \quad ; \quad \Sigma^\varepsilon = \int_0^T (\sigma + \varepsilon\hat{\sigma}(t))^2 dt.$$

Straightforward calculations then give, denoting $\varphi(\cdot) = \frac{dN}{dx}(\cdot)$, that

$$\left.\frac{\partial\Pi^\varepsilon}{\partial\varepsilon}\right|_{\varepsilon=0} = S_0\varphi\left(d_1^0\right)\frac{1}{\sqrt{T}}\int_0^T \hat{\sigma}(t)\,dt.$$

Finally, denoting $\frac{\partial\Pi}{\partial\sigma}(0) = \frac{\partial\Pi^0}{\partial\sigma}(0)$, we get the relationship

$$\left.\frac{\partial\Pi^\varepsilon}{\partial\varepsilon}\right|_{\varepsilon=0} = \frac{1}{T}\int_0^T \hat{\sigma}(t)\,dt\,\frac{\partial\Pi}{\partial\sigma}(0).$$

Therefore we see here that $\frac{\partial\Pi^\varepsilon}{\partial\varepsilon}(0)$ is the measure of robustness of the quantity Π^0 as long as volatility perturbations are concerned.

In order to explain how the ibp formula of Malliavin Calculus can help generalize this calculation we repeat this deduction using a different argument. That is,

$$\frac{\partial\Pi^\varepsilon}{\partial\varepsilon} = e^{-rT}E\left[1(S^\varepsilon(T) \geq K)\frac{dS^\varepsilon(T)}{d\varepsilon}\right].$$

In our case we have that

$$S^\varepsilon(T) = S_0\exp\left(rT - \frac{1}{2}\Sigma^\varepsilon + \int_0^T (\sigma + \varepsilon\hat{\sigma}(t))dW(t)\right),$$

therefore

$$\frac{dS^\varepsilon(T)}{d\varepsilon} = S^\varepsilon(T)\left(-\int_0^T (\sigma + \varepsilon\hat{\sigma}(t))\hat{\sigma}(t)\,dt + \int_0^T \hat{\sigma}(t)\,dW(t)\right).$$

We can then rewrite using the duality formula (4.8) together with a density argument as in Section 4.5.5,

$$\left.\frac{\partial\Pi^\varepsilon}{\partial\varepsilon}\right|_{\varepsilon=0} = -e^{-rT}E\left[1(S(T) \geq K)S(T)\int_0^T \sigma\hat{\sigma}(t)\,dt\right]$$

$$+ e^{-rT}E\left[1(S(T) \geq K)S(T)\int_0^T \hat{\sigma}(t)\,dW(t)\right]$$

$$= e^{-rT}E\left[\delta_K(S(T))S(T)^2\right]\sigma\int_0^T \hat{\sigma}(t)\,dt$$

$$= \sigma e^{-rT}\int_0^T E\left[E\left[\delta_K(S(T))S(T)^2 S(t)^{-1}\Big/ S(t)\right]\hat{\sigma}(t)S(t)\right]dt.$$

Therefore $\mu(s, T, x) = \sigma e^{-rT}E\left[\delta_K(S(T))S(T)^2 S(t)^{-1}\Big/ S(t) = x\right]$ and obviously $E(\mu(s, T, S(s))S(s)) = S_0\varphi\left(d_1^0\right)\frac{1}{\sqrt{T}}$.

One can generalize the previous discussion to obtain a European contingent claim with payoff $\Phi(S^\varepsilon(T))$ where Φ is differentiable once a.e. and option price $\Pi^\varepsilon = e^{-rT} E(\Phi(S^\varepsilon(T)))$, $r > 0$. Then if the coefficients of S^ε do not depend on the time variable, we assume that the Hörmander condition for the diffusion $S(T)$ is satisfied, otherwise we assume the restricted Hörmander condition (see Cattiaux and Mesnager (2002)). Under any of these two assumptions one has that the Malliavin covariance matrix of $S(T)$ is non-degenerate and therefore one can integrate by parts and we have that

$$\mu(u, T, x) = E\left[\Phi''(S(T))\left(U(T)U(u)^{-1}\right)^2 \sigma(u, x) \bigg/ S(u) = x\right].$$

Here U denotes the stochastic exponential associated with the derivative of S with respect to its initial value S_0 (for a specific definition, see Section 4.9.1). From this formula we can conclude that $\mu(\cdot, T, x)$ is a positive kernel if Φ is a convex function that is independent of $\hat{\sigma}(\cdot, \cdot)$ but depends on $S(\cdot)$, $\Phi(\cdot)$ and its derivatives.

For example in the case of a plain vanilla call option with strike price $K > 0$ one has

$$\mu(s, T, x) = e^{-rT} K^2 x^{-1} p_T(K \mid S(s) = x).$$

The above result is also true for digital call options although the kernel μ is no longer positive. That is, $\Phi(x) = 1\{x \geq K\}$, one has

$$\mu(s, T, x) = -e^{-rT} K x^{-1}\left(K p_T(K \mid S(s) = x) + 2p'_T(K \mid S(s) = x)\right)$$

where $p_T(\cdot \mid S(s) = x)$ denotes the conditional density of $S(T)$ given $S(s)$, and $p'_T(\cdot \mid S(s) = x)$ its derivative. Now we consider another example related to Asian options.

Example 5. *For Asian options, i.e., contingent claims with payoff*
$\Phi(\int_0^T w(s)S^d(s)dv(s))$, *where $w \in L^2[0, T]$, then*

$$\mu(s, T, x) = E\left[\Phi''\left(\int_0^T w(t)S(t)\,dv(t)\right)\int_s^T w(t)U(t)U(s)^{-1}\sigma(s, x)dv(t)\right.$$
$$\left.\int_s^T w(u)U(u)U(s)^{-1}dv(u)\bigg/ S(s) = x\right].$$

Here we assume that the integration by parts can be performed. Now if $d = 1$ and Φ is convex, we have that

$$\frac{\partial \Pi^\varepsilon}{\partial \varepsilon}\bigg|_{\varepsilon=0} = \int_0^T E\left[\mu(s, T, S(s))\hat{\sigma}(s, S(s))\right] ds,$$

where $\mu(\cdot, T, x)$ is a positive kernel where

$$\mu(s, T, x)$$
$$= e^{-rT}\sigma(s, x)E\left[\Phi''\left(\int_0^T S(t)\,dv(t)\right)\left(\int_s^T \frac{U(t)}{U(s)}dv(t)\right)^2\bigg/ S(s) = x\right].$$

This example therefore includes basket options as well as continuous Asian options. Sometimes for practical purposes it is better to condition not only on the current value of the underlying but also on the current value of the integral. We will do so in the examples to follow.

Although it is not included in the above theorem, the principle given here is far more general in various senses. In the case of lookback options we have the following result:

Example 6. *Under the Black–Scholes set-up and for lookback options, i.e., contingent claims with payoff* $\Phi\left(\sup_{0\leq t\leq T} S^{\varepsilon}(t)\right)$, *we have that*

$$\left.\frac{\partial \Pi^{\varepsilon}}{\partial \varepsilon}(0)\right|_{\varepsilon=0} = \int_0^T \mu(s,T)\hat{\sigma}(s)\,ds$$

where the density function $\mu(\cdot, T)$ *is given by*

$$\mu(s,T) = e^{-rT} E\left[\Phi'\left(\sup_{0\leq t\leq T} S(t)\right) \sup_{0\leq t\leq T} S(t)\left(\frac{2r}{\sigma}1_{s\leq\tau} + X\right)\right].$$

The random time τ *is implicitly defined by the relation* $\sup_{0\leq t\leq T} S(t) = S(\tau)$ *and* X *is an appropriate random variable that belongs to* $L^p(\Omega, \mathcal{F}, P)$ *for any p. Furthermore, if* $\Phi(\cdot)$ *is monotone, then* $\mu(\cdot, T)$ *is decreasing and if* $\Phi'(0) \geq 0$, *then* $\lim_{s\to T} \mu(s,T) \geq 0$.

Now we study another possible interpretation of the lvi provided by quantile or VaR type problems.

4.8.1 Asymptotic behavior of quantile hedging problems

Suppose that we have sold an option considering some volatility structure that later is discovered to have been underestimated. In such a case, we ran into the danger of not being able to replicate the option. Another similar situation is when we are not willing to invest the full price of the option and for a lower price we are willing to take some risk of not being covered. These problems fall in the general category of quantile type problems.

As an example let us start with a simple goal problem in the Black–Scholes setup. Suppose that we have incurred a misspecification of volatility which therefore implies that hedging is not possible. We want to take the decision of either selling the option at loss or keeping it under the risk of not being able to hedge it. We want therefore to compute the probability of perfect hedging. We show that the local Vega index determines how close this probability is to 1. We refer to Karatzas (1996) for details and further references. Let us recall that the discounted value process of a self-financing portfolio is given by the expression

$$e^{-rt} X^{x_0,\xi}(t) = x_0 + \int_0^t e^{-rs}\xi(s)\sigma(s)\,dW(s) \; ; \; \sigma(t) := \sigma + \varepsilon\hat{\sigma}(t).$$

Here x_0 is the initial wealth in our portfolio and $\xi(\cdot)$ is the portfolio or strategy which represents the amount of money that is invested in the stock at each point in time. Suppose that G^ε is the payoff of the option, hence starting with the initial wealth $u_0 := \Pi^\varepsilon = e^{-rT} E_{P*,\varepsilon} [G^\varepsilon]$ there exists a strategy $\bar{\pi}(\cdot)$ which achieves a perfect hedge and $P^{*,\varepsilon}$ is the equivalent martingale measure associated with the problem.

Now, suppose that our initial wealth x_0 is less than the money required to obtain a perfect hedge, i.e., we assume $0 < x_0 \le u_0$, then as we can no longer obtain a perfect hedge we will instead try to maximize the probability of a perfect hedge:

$$p(\varepsilon) := \sup_{\substack{\xi(\cdot) \text{ tame} \\ X^{x_0+u_0,\xi}(T) \ge G^\varepsilon \text{ a.s.}}} P\left(X^{x_0,\xi}(T) \ge G^\varepsilon\right).$$

That is, the above is the probability that, if given a loan of extra u_0 monetary units, one can cover for the option considering that the loan has to be returned at the end of the expiration time. Obviously as $\varepsilon \to 0$, then $p(\varepsilon) \to 1$. The following proposition gives the rate at which this quantity converges.

Proposition 1. *Assume that the perturbed price Π^ε has a Taylor expansion of order 2 around $\varepsilon = 0$, in the sense that*

$$\Pi^\varepsilon = \Pi + \left.\frac{\partial \Pi^\varepsilon}{\partial \varepsilon}\right|_{\varepsilon=0} \varepsilon + G(\varepsilon) \varepsilon^2,$$

where $G(\cdot)$ is differentiable around 0, and $|G(\varepsilon)| \le C_1$ for $\varepsilon \le 1$.
Then the maximal probability of obtaining a perfect hedge $p(\varepsilon)$, has the property

$$\lim_{\varepsilon \to 0} \frac{1 - p(\varepsilon)}{\varepsilon \exp\left(-cN^{-1}(1 - \varepsilon)\right)} = \exp\left(-c^2/2\right) \left.\frac{\partial \Pi^\varepsilon}{\partial \varepsilon}\right|_{\varepsilon=0} /\Pi,$$

where $c = \frac{|\alpha-r|}{\sigma}\sqrt{T}$ and α is the stock appreciation rate under P.

Note that $N^{-1}(1 - \varepsilon) \approx \sqrt{-\ln \varepsilon}$, hence $\exp\left(-cN^{-1}(1 - \varepsilon)\right)$ goes to zero slower than any polynomial.

The proof is done through an asymptotic study of the probability of perfect hedging which can be obtained explicitly. In fact,

$$p(\varepsilon) = N\left(N^{-1}\left(\frac{x_0}{u_0}\right) + |\alpha - r|\sqrt{\int_0^T [\sigma + \varepsilon\hat{\sigma}(t)]^{-2} dt}\right).$$

The main issue in the above proposition is that the lvi determines the speed at which the probability of perfect hedging goes to 1. This principle is a generalization of the interpretation of greeks. In fact in other similar setups the same result seems to hold. For example, let us consider the quantile hedging problem of Föllmer and Leukert (1999) with $x_0 = E[G^0]$ where we assume that $r \equiv 0$ without loss of generality. Then define the probability of perfect hedging as

$$p(\varepsilon) = \sup_{\xi(\cdot) \text{ self-financing}} P\left(X^{x_0,\xi}(T) \geq G^\varepsilon\right).$$

Then it is known that the solution of the above problem supposing the existence of a unique equivalent martingale measure $P^{*,\varepsilon}$, is to replicate $G^\varepsilon 1_{A^\varepsilon}$ where

$$A^\varepsilon = \{G^\varepsilon \frac{dP^{*,\varepsilon}}{dP} < a\}$$

and $a \equiv a(\varepsilon)$ is such that $E[G^\varepsilon 1_{A^\varepsilon}] = x_0$ and $p(\varepsilon) = P(A^\varepsilon)$. The main constant in the asymptotic behavior of $1 - p(\varepsilon)$ is $C \left.\frac{\partial \Pi^\varepsilon}{\partial \varepsilon}\right|_{\varepsilon=0}$ for a positive constant C independent of the lvi. In order to avoid long arguments and conditions we give a brief heuristic argument of the idea of the proof.

First we consider the first order term of $1 - p(\varepsilon)$ which is characterized by the derivative of $P(A^\varepsilon)$ wrt ε. Then we have that

$$1 - p(\varepsilon) \approx \varepsilon E\left[\delta_a\left(G^\varepsilon \frac{dP^{*,\varepsilon}}{dP}\right)\left(\frac{da}{d\varepsilon} - \frac{d}{d\varepsilon}\left(G^\varepsilon \frac{dP^{*,\varepsilon}}{dP}\right)\bigg|_{\varepsilon=0}\right)\right]. \qquad (4.19)$$

Here δ stands for the Dirac delta function. To compute $\frac{da}{d\varepsilon}$, one differentiates implicitly the equation $E[G^\varepsilon 1_{A^\varepsilon}] = x_0$, obtaining that

$$\frac{da}{d\varepsilon} \approx \frac{E\left[\frac{d}{d\varepsilon}\left(G^\varepsilon \frac{dP^{*,\varepsilon}}{dP}\right)\bigg|_{\varepsilon=0}\right]}{E\left[G\delta_a\left(G^\varepsilon \frac{dP^{*,\varepsilon}}{dP}\right)\right]},$$

which replaced in (4.19) gives that the main error term is

$$1 - p(\varepsilon) \approx \varepsilon \frac{E\left[\frac{d}{d\varepsilon}\left(G^\varepsilon \frac{dP^{*,\varepsilon}}{dP}\right)\bigg|_{\varepsilon=0}\right] E\left[\delta_a\left(G \frac{dP^*}{dP}\right)\right]}{E\left[G\delta_a\left(G \frac{dP^*}{dP}\right)\right]}$$

$$= \varepsilon \frac{\left.\frac{\partial \Pi^\varepsilon}{\partial \varepsilon}\right|_{\varepsilon=0}}{E\left[G/G \frac{dP^*}{dP} = a(\varepsilon)\right]}.$$

Similar considerations can be used to analyze the shortfall risk (see Föllmer and Leukert (2000)) if enough conditions on the loss function between other assumptions are made. The same remark is also true for other quantile hedging problems. For instance, it is easily shown that our results still hold in the setting of Spivak and Cvitanić (1999).

4.8.2   Computation of the local Vega index

So far we have discussed the issue of the theoretical properties and uses of the lvi index. Here we will show some simulations of these quantities using various techniques of Monte Carlo simulation. In particular we consider the Black–Scholes setup and then a stochastic volatility model for Asian options. We will also show that these calculations can be performed in various ways. We use the integration by parts formula as proposed by Fournié et. al. (1999) (2001) which does not always give reliable results unless some variance reduction methods are performed.

4.8.2.1   Asian options within the Black–Scholes setup

As stated before we plan to start the presentation of our numerical work focusing the case of Asian options in the frame of the classical Black–Scholes scenario. We also consider that the payoff function is associated with a call option with strike K. In terms of the notation we have:

$$r(t, x) = xr,$$
$$\sigma(t, x) = x\sigma,$$
$$dv(t) = \frac{1}{T}dt,$$
$$\Phi(x) = e^{-rT}(x - K)_+,$$

and S is geometric Brownian motion which can be written as $S(t) = S_0 U(t)$. Then $\mu(s, T, x)$ can be rewritten as

$$\mu(s, T, x)$$
$$= e^{-rT}\sigma x E\left[\delta\left(\frac{1}{T}\int_0^T S(t)\,dt - K\right)\left(\frac{1}{T}\int_s^T \frac{U(t)}{U(s)}dt\right)^2 \bigg/ S(s) = x\right].$$

In Figure 8 we have simulated the conditional expectation as above to obtain $\mu(s, T, x; S_0, K)$. We have used $g(x) = 1\{x \geq 0\} - 1/2$ in this case as this generates smaller variance (see Remark 3.7 and section 3.3 in Bermin et. al. (2003)) .

The picture presented so far may be misleading for a practitioner, since we are not including all the information we have at hand by the time s, in the calculation of the weight. In particular we know the average of the underlying up to that moment. Therefore we may define a new weight $\mu(s, T, x, y)$,

$$\mu(s, T, x, y) = e^{-rT}\sigma x E\left[\delta\left(\frac{1}{T}\int_0^T S(t)\,dt - K\right)\right.$$
$$\left.\times\left(\frac{1}{T}\int_s^T \frac{U(t)}{U(s)}dt\right)^2 \bigg/ S(s) = x, \frac{1}{T}\int_0^s S(t)\,dt = y\right],$$

where the above information has been added. We show an lvi index in Figure 9 that shows an interesting pattern.

It is not difficult to see that there is a line that defines the maximum of the index and in fact one can show that the lvi index depends on a single parameter, which we denote by α, $\alpha = (K - y)/x$,

$$\mu(s, T; \alpha) = e^{-rT}\sigma E\left[\delta\left(\frac{1}{T}\int_s^T \frac{U(t)}{U(s)}dt - \alpha\right)\left(\frac{1}{T}\int_s^T \frac{U(t)}{U(s)}dt\right)^2\right]. \quad (4.20)$$

α represents the *effective* remaining fraction of the integral we must fulfill in order to obtain some gross profit with the option. Its inverse would also be linked to an

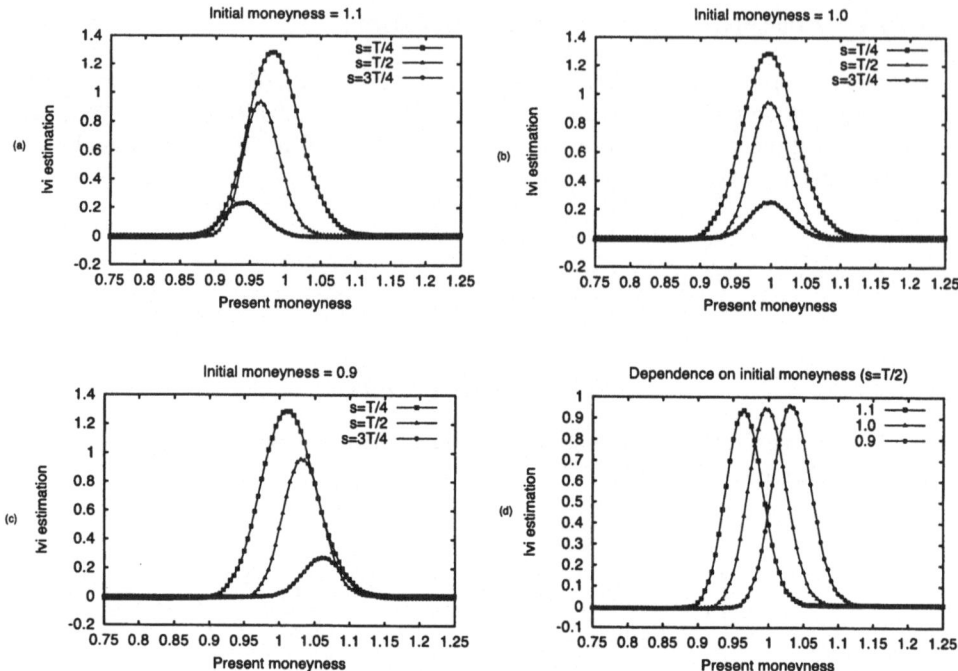

FIGURE 8. Estimated value of weight $\mu(s, T, x; S_0, K)$ of an Asian call with global parameters $r=0.05$, $\sigma=0.2$, $T=0.2$ (in years), for different values of the present time and the initial moneyness (in, at and out of the money): (a) $S_0/K=1.1$, (b) $S_0/K=1.0$, (c) $S_0/K=0.9$. We also present in graph (d) some plots showing the dependence of the results on the initial moneyness for a given fixed time $s=0.1$. The numerical simulations have been performed using Malliavin Calculus and Monte Carlo techniques. We have broken the whole time interval into 200 discrete time steps, and we have computed 10 000 paths at each point.

effective quantity, the effective present moneyness, since our contract is equivalent to another with maturity time equal to $T-s$, and strike price $K-y$.

We can also analyze heuristically the value $\alpha = \hat{\alpha}$ which maximizes the result of $\mu(s, T; \alpha)$, that is, the value of effective moneyness which is the most sensible to changes in pricing. If we compute the first derivative of equation (4.20), with respect to α, we arrive at the following condition that $\hat{\alpha}$ must fulfill:

$$2\alpha P_{s,T}(\alpha) + \alpha^2 P'_{s,T}(\alpha)\Big|_{\alpha=\hat{\alpha}} = 0, \qquad (4.21)$$

where $P_{s,T}(\cdot)$ is the probability density function of $T^{-1}\int_s^T U(t)/U(s)dt$, which does not depend either on x or on y. The mean and variance of this random variable are given by

$$a \equiv E\left[\frac{1}{T}\int_s^T \frac{U(t)}{U(s)}dt\right] = \frac{1}{rT}\left(e^{r(T-s)} - 1\right),$$

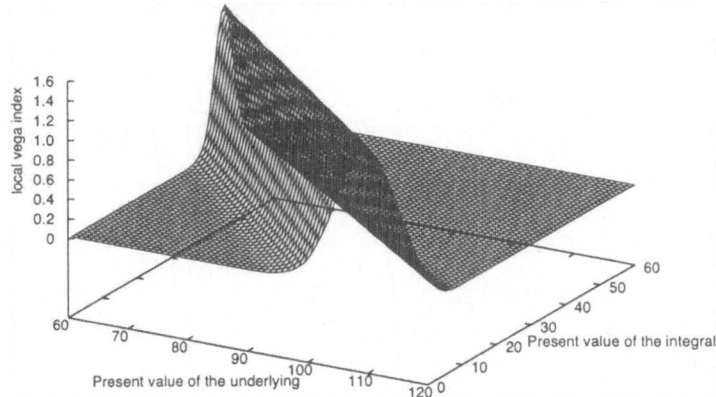

FIGURE 9. Estimated value of weight $\mu(s, T; x, y)$ for an Asian call with parameters $r = 0.05$, $\sigma = 0.2$, $T = 0.2$ (in years), for $s = T/4$. The numerical simulations have been performed using Malliavin Calculus and Monte Carlo techniques as described in Figure 1.

and its variance, b^2,

$$b^2 \equiv \frac{2}{rT^2} \left\{ \frac{re^{r(T-s)} \left(e^{(r+\sigma^2)(T-s)} - 1 \right)}{(2r + \sigma^2)(r + \sigma^2)} - \frac{(e^{r(T-s)} - 1)}{2r + \sigma^2} \right\} - a^2.$$

The asymptotic behavior of a and b when rT and $\sigma^2 T$ are small is

$$a \approx 1 - \frac{s}{T}, \text{ and}$$

$$b^2 \approx \frac{\sigma^2(T - s)^3}{3T^2}.$$

Note in particular that $b^2 \ll a^2$. In this case, we assume that we can take a Gaussian approximation for the probability density function, at least in the vicinity of a as we did for the kernel density estimation method in Section 4.3.

$$P_{s,T}(\alpha) \sim \frac{1}{\sqrt{2\pi b^2}} e^{-\frac{(\alpha-a)^2}{2b^2}}.$$

Now we can solve (4.21), and find that $\hat{\alpha} \approx a$, and, since $\mu(s, T; \hat{\alpha}) = e^{-rT} \sigma \hat{\alpha}^2 P_{s,T}$ ($\hat{\alpha}$), the maximum sensibility follows:

$$\mu(s, T; \hat{\alpha}) \approx e^{-rT} \sqrt{\frac{3(T - s)}{2\pi T^2}}.$$

The simulations to calculate the lvi are shown in Figure 10, for a variety of values for the parameters. Note that the results show a very good agreement with the outcome of our previous discussion.

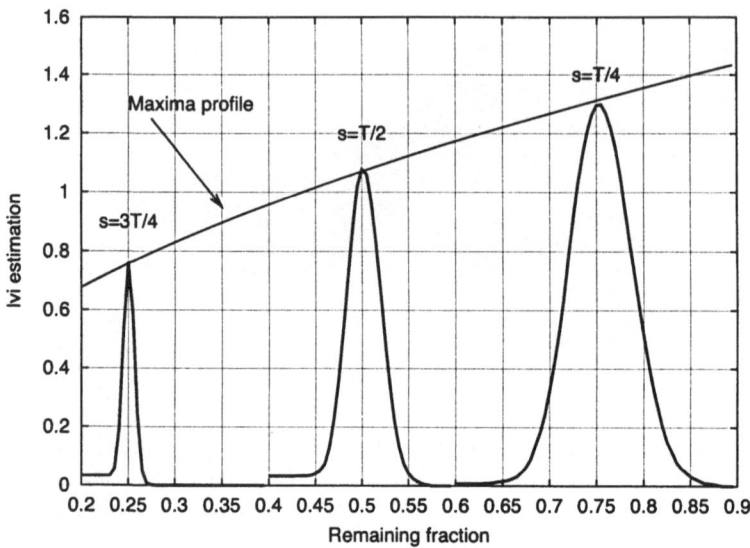

FIGURE 10. Estimated value of weight $\mu(s, T; \alpha)$ for an Asian call with parameters $r = 0.05$, $\sigma = 0.2$, $T = 0.2$ (in years), for different values of the present time and the parameter α. We also display the approximate values of the maxima of the weight, obtained in Section 4.8.2.1. The numerical simulations have been performed using Malliavin Calculus and Monte Carlo techniques as described in Figure 1.

4.8.2.2   Asian options within a stochastic volatility model

The purpose of this subsection is just to show that the lvi can be computed in more complex financial models than the Black–Scholes model and that some of the conclusions reached in previous sections seem to be also valid here. Let us consider the Asian option with the same payoff function as before but where the underlying process has a stochastic volatility driven by the noise driving the stock and an independent noise. That is,

$$S(t) = S_0 + r \int_0^t S(u)du + \int_0^t \sigma(u)S(u)dW(u),$$

$$\sigma(t) = \sigma_0 + a \int_0^t (b - \sigma(u))du + \rho_1 \int_0^t \sigma(u)dW(u) + \rho_2 \int_0^t \sigma(u)dW'(u).$$

Here, W and W' are two independent Wiener processes. Once more the weight $\mu(s, T, x, y, z)$ can be constructed:

$$\mu(s, T, x, y, z) = e^{-rT} z x \rho_2^2$$
$$\times E\left[A(s, T, \alpha, z) \,\Big/\, S(s) = x, \frac{1}{T} \int_0^s S(t)dt = y, \sigma(s) = z \right],$$

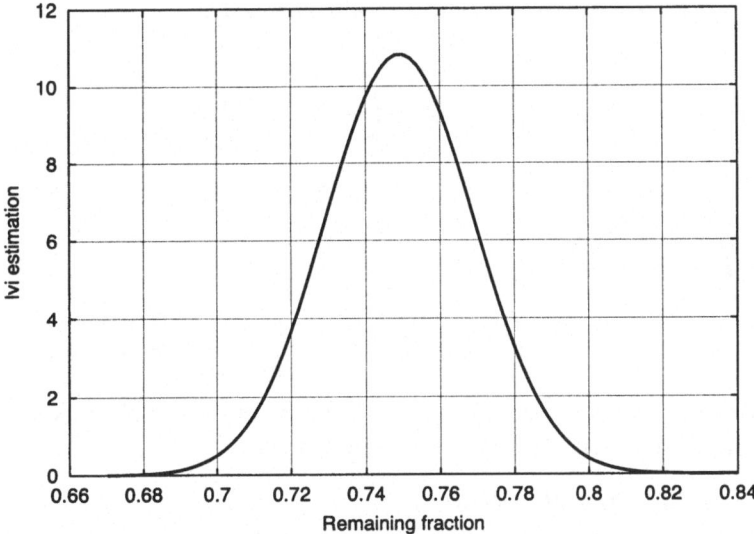

FIGURE 11. Estimated value of the weight $\mu(s, T, x, z, \alpha)$ for an Asian within a stochastic volatility framework, in terms of α. The selected values for the parameters were $r = 0.05$, $a = 0.695$, $b = 0.1$, $\rho_1 = 0.21$, $\rho_2 = 0.9777$, $x = 100.0$, $z = 0.2$, $T = 0.2$ and $s = 0.05$ (in years). A Gaussian kernel with parameter $h = 0.02$ was chosen when computing A. We have broken the whole time interval into 20 discrete time steps, and we have simulated 10 000 paths at each point.

where A is a stochastic process with a long explicit expression, that we will not detail here. Prior to presenting the output of the simulation, let us note that the variable $\alpha = (K - y)/x$ plays again an important role. We have indeed that

$$\mu(s, T, x, y, z) = \mu(s, T, x, z, \alpha) = e^{-rT} zx\rho_2^2 f(\alpha, z).$$

Unfortunately, the dependence in z cannot be fully factorized.

We conclude by introducing Figure 11, where we plot $\mu(s, T, x, z, \alpha)$ in terms of α.

4.9 Appendix

4.9.1 Stochastic derivative of a diffusion

Differentiating a diffusion is not so complicated from the heuristic point of view. The ideas involved are the same as when differentiating the solution of an ordinary differential equation with respect to a parameter. That is, let X be the solution of the stochastic differential equation

$$X(t) = x + \int_0^t b(X_s)ds + \int_0^t \sigma(X_s)dW_s,$$

where $b, \sigma : \mathbb{R} \to \mathbb{R}$ are smooth functions with bounded derivatives. Then

$$
\begin{aligned}
D_u X(t) &= D_u \left\{ \int_0^t b(X_s) ds \right\} + D_u \left\{ \int_0^t \sigma(X_s) dW_s \right\} \\
&= \int_0^t b'(X_s) D_u X_s ds + \int_0^t \sigma'(X_s) D_u X_s dW_s + \int_0^t \sigma(X_s) D_u(dW_s) \\
&= \int_0^t b'(X_s) D_u X_s 1(u \leq s) ds + \int_0^t \sigma'(X_s) D_u X_s 1(u \leq s) dW_s + \sigma(X_u) \\
&= \int_u^t b'(X_s) D_u X_s ds + \int_u^t \sigma'(X_s) D_u X_s dW_s + \sigma(X_u).
\end{aligned}
$$

If one regards the previous equation as a linear equation on $D_u X_s$ with u fixed and $s \in [u, T]$, one obtains as an explicit solution that

$$
\begin{aligned}
D_u X_t &= \sigma(X_u) U(t) U(u)^{-1}, \\
U(t) &= \exp \left(\int_0^t b'(X_s) - \frac{1}{2} (\sigma'(X_s))^2 ds + \int_0^t \sigma'(X_s) dW_s \right). \quad (4.22)
\end{aligned}
$$

Obviously the above argument is just heuristic. The mathematical proof is much longer because one needs to prove that the process X is differentiable. This is done through an approximation procedure as in Theorem 1.

4.9.2   The multidimensional case

Here we deal with the task of repeating the previous steps in many dimensions. In particular, we will show as before that there are many different ways of performing the ibp. The most common one generates the Malliavin covariance matrix.

Let us suppose that $W = (W^1, W^2, \dots, W^k)$ is a k-dimensional Wiener process. Then suppose that we want to find an integration by parts formula for $f_i(W_T^1, W_T^2, \dots, W_T^k)$ where f_i denotes the partial derivative with respect to the i-th coordinate of the smooth function f. Then as before we have

$$
\begin{aligned}
&E(f_i(W_T^1, W_T^2, \dots, W_T^k)) \\
&= \frac{1}{(2\pi T)^{k/2}} \int_{\mathbb{R}^k} f_i(x_1, \dots, x_k) \exp\left(-\frac{1}{2T} \sum_{i=1}^k x_i^2\right) dx_1 \dots dx_k \\
&= \frac{1}{\sqrt{2\pi T}} \int_{\mathbb{R}^k} f(x_1, \dots, x_k) \exp\left(-\frac{1}{2T} \sum_{i=1}^k x_i^2\right) \frac{x_i}{T} dx_1 \dots dx_k \\
&= E(f(W_T^1, W_T^2, \dots, W_T^k) \frac{W_T^i}{T}) \\
&= E(f(W_T^1, W_T^2, \dots, W_T^k) \int_0^T \frac{1}{T} dW_s^i).
\end{aligned}
$$

Of course one could continue with this calculation and other similar examples as before. Instead of repeating all the argument in the previous section, we just show informally how to deal with the ibp formula in multidimensional cases. We will consider an integration by parts formula for ∇f for $f : \mathbb{R}^d \to \mathbb{R}$, a smooth function and X a smooth random variable in the Malliavin sense. First, let's start by denoting D^i, the derivative with respect to the i-th component of the Wiener process W^i. $D = (D^1, ..., D^k)$ is the vector of derivatives. Then as before we have that

$$D_s Z = D_s X \nabla f(X),$$

$$D_s^j Z = \sum_{i=1}^{d} \partial_i f(X) D_s^j X^i,$$

where in this case

$$D_s X = (D_s^j X^i)$$

so that if we multiply the equation for DZ by a smooth $d \times d$-dimensional matrix process u and integrate we have

$$\langle DZ, u \rangle_{L^2[0,T]} = \langle (DX) \nabla f(X), u \rangle_{L^2[0,T]}$$

$$= \sum_{i=1}^{d} \sum_{j=1}^{k} \int_0^T f_i(X) D_s^j X^i u_s^{jl} ds$$

$$= A \nabla f(X)$$

where $A_{il} = \sum_{j=1}^{k} \int_0^T D_s^j X^i u_s^{jl} ds$. Suppose that there exists a $d \times d$ matrix B so that $BA = I$. Then one has for a d-dimensional random variable Y,

$$E\left[(B \langle DZ, u \rangle_{L^2[0,T]})Y\right] = E[\nabla f(X)Y]$$

which after using an extension of the duality principle gives as a result

$$E[\nabla f(X)Y] = \sum_{l,m=1}^{d} \sum_{j=1}^{k} E\left[ZD^{*j}(B_{ml}Y_m u^{jl})\right]$$

$$= \sum_{l,m=1}^{d} \sum_{j=1}^{k} E\left[f(X)D^{*j}(B_{ml}Y_m u^{jl})\right].$$

Here D^{*j} stands for the adjoint of D^j which is the extension of the stochastic integral with respect to W^j. That is, $D^{*j}(1) = W_T^j$ and $D^{*j}(u) = \int_0^T u_s dW_s^j$ if u is an $\mathcal{F}_t^j$-adapted process. As in Remark 3.1 the problem is to find the right process B. In the particular case that $u_s^{jl} = D_s^j X^l$, then one obtains that B is the inverse of the Malliavin covariance matrix which should belong to $L^p(\Omega)$ for p big enough in order for the integration by parts formula to be valid. For details, see Malliavin (1997), Ikeda–Watanabe (1989) or Nualart (1995).

4.10   Conclusion and Comments

The goal of this article is to introduce Malliavin Calculus for practitioners and to show one of its applications compared with classical techniques. Other applications of Malliavin Calculus in Finance and other areas appear frequently in specialized journals. Such are the cases of models for asymmetric information (see Imkeller et al (2001) and the references therein). The Clark–Ocone formula has led to various applications in financial economics where it has become a natural tool. There are also applications in asymptotic statistics, see Gobet (2001).

Other extensions of Malliavin Calculus for random variables generated by Lévy processes are still under study with partial results available in Bichteller, Gravereaux and Jacod (1987), Picard (1996) and Privault (1998).

The extension of the Itô stochastic integral, which in our exposition leads to the Skorohod integral, is an independent area of study which has seen various extensions defined (most of these extensions are related) through the last 30 years.

In this article we have dealt with the simulation of greeks for binary type options. In the original article of Fournié et.al. (1999), the ibp formula is applied to European type options after a proper localization argument. Localizations for greeks of binary type options appear in Kohatsu–Pettersson (2002) where it is proven that these lead to effective variance reduction.

Acknowledgements

AK-H has been supported by grants BFM 2000-807 and BFM 2000-0598 of the Spanish Ministerio de Ciencia y Tecnologia. MM has been supported in part by Dirección General de Proyectos de Investigación under contract No.BFM2000-0795, and by Generalitat de Catalunya under contract 2001 SGR-00061.

References

[1] Bally V., Talay D.: The law of the Euler scheme for stochastic differential equations (I): convergence rate of the distribution function, *Probab. Rel. Fields* **104** (1996), 43–60.

[2] Bally V., Talay D.: The law of the Euler scheme for stochastic differential equations (II): convergence rate of the distribution function, *Monte Carlo Methods Methods Appl* **2** (1996), 93–128.

[3] Bermin H.-P., Kohatsu-Higa A., Montero M.: Local Vega index and variance reduction methods. *Math. Finance* **13** (2003), 85–97.

[4] Bernis G., Gobet E., Kohatsu-Higa A.: Monte Carlo evaluation of Greeks for multidimensional barrier and lookback options. *Math. Finance* **13** (2003), 99–113.

[5] Bichteler K., Gravereaux J. and Jacod J.: *Malliavin Calculus for Processes with Jumps.* Gordon and Breach Science Publishers, 1987.

[6] Bouchard B., Ekeland I., Touzi N.: On the Malliavin approach to Monte Carlo approximations to conditional expectations. To appear in *Finance and Stochastics*, 2004.

[7] Broadie, M., Glasserman P.: Estimating security price derivatives using simulation. *Management Science* **42** (1996), 269–285 .

[8] Cattiaux, P., Mesnager L.: Hypoelliptic non homogeneous diffusions. *Probability Theory and Related Fields* **123** (2002), 453–483.

[9] Cvitanic, J., Spivak G., Maximizing the probability of a perfect hedge. *Annals of Applied Probability*, **9**(4) (1999), 1303–1328.

[10] Dufresne, D.: Laguerre series for Asian and other options. *Math. Finance* **10** (2000), 407–428.

[11] Föllmer H., Leukert P.: Quantile Hedging. Finance and *Stochastics* **3** (1999), 251–273.

[12] Föllmer H., Leukert P.: Efficient hedging: Cost versus shortfall risk. *Finance and Stochastics* **4** (2000), 117–146.

[13] Fournié, E., Lasry, J.-M., Lebuchoux, J., Lions, P.-L., Touzi, N.: An application of Malliavin calculus to Monte Carlo methods in finance. *Finance and Stochastics* **3** (1999), 391–412.

[14] Fournié E., Lasry J.M., Lebuchoux J., Lions P.L. : Applications of Malliavin calculus to Monte Carlo methods in finance II. *Finance Stochast.* **5** (2001), 201–236.

[15] Geman H., Yor M.: Bessel processes, Asian options and perpetuities. *Math. Finance*, **3** (1993), 349–375.

[16] Glynn P.W.: Likelihood ratio gradient estimation: an overview. In: *Proceedings of the 1987 Winter Simulation Conference*, A. Thesen, H. Grant and W.D. Kelton, eds, 366–375, 1987.

[17] Gobet, E.: LAMN property for elliptic diffusion: a Malliavin Calculus approach. *Bernoulli* **7** (2001), 899–912.

[18] Gobet E., Kohatsu-Higa A.: Computation of Greeks for barrier and lookback options using Malliavin calculus. *Electronic Communications in Probability* **8** (2003), 51–62.

[19] Gobet E., Munos R.: Sensitivity analysis using Itô–Malliavin Calculus and Martingales. Application to stochastic optimal control. Preprint (2002).

[20] Hull J.C.: *Options, Futures and Other Derivatives*, Fourth Ed., Prentice-Hall, Upper Saddle River, NJ, 2000.

[21] Ikeda N. and Watanabe S.: *Stochastic Differential Equations and Diffusion Processes*, North-Holland, Amsterdam, 1989.

[22] Imkeller, P., Pontier, M., Weisz, F.: Free lunch and arbitrage possibilities in a financial market with an insider. *Stochastic Proc. Appl.* **92** (2001), 103–130.

[23] Karatzas, I.: *Lectures on the Mathematics of Finance*. CRM Monographs 8, American Mathematical Society, 1996 .

[24] Karatzas, I., Shreve S.: *Brownian Motion and Stochastic Calculus*. Springer-Verlag, 1991.

[25] Kohatsu-Higa, A.: High order Itô–Taylor approximations to heat kernels. *Journal of Mathematics of Kyoto University* **37** (1997), 129–151.

[26] Kohatsu-Higa, A., Pettersson R.: Variance reduction methods for simulation of densities on Wiener space. *SIAM Journal of Numerical Analysis* **40** (2002), 431–450.

[27] Kulldorff, M.: Optimal control of favorable games with a time limit. *SIAM Journal of Control & Optimization* **31** (1993), 52–69.

[28] Kunitomo, N., Takahashi A.: The asymptotic expansion approach to the valuation of interest rate contingent claims. *Math. Finance* **11** (2001), 117–151.

[29] Kusuoka, S., Stroock, D.W.: Application of the Malliavin calculus I. In: *Stochastic Analysis*, Itô, K., ed. Proceedings of the Taniguchi International Symposium on Stochastic Analysis, Katata and Kyoto, 1982. Tokyo: Kinokuniya/North-Holland , 271–306, 1988.

[30] L'Ecuyer P., Perron G.: On the convergence rates of IPA and FDC derivative estimators. *Operations Research* **42** (1994), 643–656.

[31] Nualart D.: *The Malliavin Calculus and Related Topics*, Springer-Verlag, Berlin (1995).

[32] Nualart D., Vives J.: Continuité absolue de la loi du maximum d'un processus continu. C.R. Acad. sci. Paris **307** (1988), 349–354.

[33] Malliavin P.: *Stochastic Analysis*. Springer-Verlag, Berlin, Heidelberg, New York, 1997.

[34] Oksendal B.: An Introduction to Malliavin Calculus with Applications to Economics. Working Paper 3, Norwegian School of Economics and Business Administration (1996).

[35] Picard, J.: On the existence of smooth densities for jump processes. *Probab. Theory Related Fields* **105** (1996), 481–511.

[36] Privault N.: Absolute continuity in infinite dimensions and anticipating stochastic calculus. *Potential Analysis* **8** (1998), 325–343.

[37] Silverman W.: *Density Estimation*, Chapman Hall, London, 1986.

[38] Üstünel, A.S.: *An introduction to stochastic analysis on Wiener space*. Springer-Verlag, Lecture Notes in Mathematics 1610, 1996.

[39] Wilmott, P., *Derivatives*, Wiley, Chichester 1998.

5

Bootstrap Unit Root Tests for Heavy-Tailed Time Series

Piotr Kokoszka
Andrejus Parfionovas

ABSTRACT   We explore the applicability of bootstrap unit root tests to time series with heavy-tailed errors. The size and power of the tests are investigated using simulated data. Applications to financial time series are also presented. Two different bootstrap methods and the subsampling approach are compared. Conclusions on the optimal bootstrap parameters, the range of applicability, and the performance of the tests are presented.

5.1   Introduction

In econometric theory and practice it is very important to be able to tell if the observations $y_1, \ldots, y_n$ follow a stationary time series model or if their differences $\Delta y_t = y_t - y_{t-1}$ are stationary. If the Δy_t form a stationary process, we say that the original series $\{y_t\}$ has a unit root or is integrated. In unit root tests, under the null hypothesis, the y_t have a unit root and under the alternative they form a stationary sequence. Unit root tests in various settings have been extensively studied in econometric and statistical literature in the last two decades, we refer to Chapter 17 of [10] for an introduction.

Essentially all work in this area assumed that the noise sequence has at least finite variance. The goal is then to establish the null distribution of a test statistic. Typically the limiting distribution is a complicated functional of a Brownian motion (or a related process with stationary increments) and the parameters describing the noise process $\{u_t\}$, see e.g., [24], [15] and [16] for recent contributions in this direction.

The author of [20], Chapter 12, and [19] used, respectively, subsampling and residual block bootstrap to develop unit root tests for which the exact form of the limiting null distribution is not required. Other work on bootstrap procedures includes [3], [8], [11] and [23]. All these papers assume that the noise process has at least finite variance, usually finite fourth moment.

Some economic and financial time series are believed not to have finite variance, see e.g., [9] and [21]. The unit root problem in the setting of infinite variance errors was first considered by [4] whose theory was extended in Chapter 15 of [21] to more complex models and to t-tests. The approach of [4] and [21] relies on finding the asymptotic distribution of a test statistic which is shown to depend on the tail index parameter α, as explained in Section 5.3. The parameter α is known to be difficult to estimate, especially if the data generating process is unknown, see e.g., Section 6.4 of [7].

Assuming that a unit root null hypothesis is true, [25] and [12] established the validity of certain bootstrap approximations to the distribution of the unit root statistic when the errors are heavy-tailed with infinite variance. In order to ensure the convergence of the bootstrap distribution in that case, it is necessary to assume that the bootstrap sample size m satisfies $m \to \infty$ and $m/n \to 0$ as the sample size $n \to \infty$. These theoretical papers did not address the question of the practical choice of m.

The present paper explores the applicability of residual bootstrap to testing the unit root hypothesis when the model errors are heavy-tailed with infinite variance. The method we study is similar to the one developed by [19] who carefully explain that in order to derive a useful test, it must be ensured that the null distribution of a test statistic can be approximated no matter whether the data actually follow the null or the alternative. As in [19], we construct centered residuals and then many pseudo-processes which satisfy the null hypothesis. These pseudo-processes allow us to approximate the null distribution of test statistics.

We show by means of simulations and applications to financial data that the residual bootstrap can be used with confidence when testing the unit root hypothesis even when the model errors have very heavy tails. The bootstrap method is simple and computationally efficient. It does not assume any prior knowledge of the tail index α and does not rely on any estimates of α.

The paper is organized as follows. In Section 5.2 we briefly review modeling of heavy-tailed time series and define the tail index α. Section 5.3 focuses on the heavy-tailed unit root models we study in the present paper and presents the asymptotic theory of unit root tests for these models. In Sections 5.4 and 5.5, the test procedures are described in detail and their properties explored by means of simulations. In particular, in Section 5.4.4 we address the problem of choosing an appropriate bootstrap sample size m. Section 5.6 compares the subsampling tests of [13] with the bootstrap tests explored in the present paper. After applying in Section 5.7 the bootstrap tests to several financial time series, we present the main conclusions of our study in Section 5.8.

5.2   Modeling heavy-tailed observations

Before introducing in Section 5.3 the unit root models we study in the present paper, we describe in this section the random variables we use to model the errors.

The iid errors ε_t are said to have a heavy-tailed distribution if $P(|\varepsilon_t| > x) \sim x^{-\alpha}$, as $x \to \infty$, for some $\alpha > 0$. This means that the tails of the distribution decay much slower than for normal random variables. Also, unlike for the normal distribution, there is a large probability that a heavy-tailed random variable takes a value far away from its center of distribution.

In this paper we focus on the *tail index* α in the range $1 < \alpha < 2$. This is because for $\alpha > 2$ the standard asymptotic theory of unit root testing applies, and if $\alpha \leq 1$, the ε_t do not have a finite expected value and exhibit very high variability not encountered in econometric applications.

Heavy-tailed observations with $1 < \alpha < 2$ are often modeled by symmetric α-stable distributions. These distributions are bell-shaped but have much longer tails than the normal distribution. There is no closed-form formula for their density function, but they have a simple characteristic function: $E \exp[i\theta\varepsilon_t] = \exp[-\sigma^\alpha |\theta|^\alpha]$, $\sigma > 0$. These distributions have been extensively studied and applied in many contexts. The monographs of [22] and [17] study their mathematical theory, whereas [21] focus on applications in finance and econometrics.

As can be seen from the form of the characteristic function, the case $\alpha = 2$ corresponds to a normal distribution which is not heavy-tailed. We however do include this case in our paper because we want to see if the procedures we study are also applicable in the case of normal errors. Figure 1 shows realizations of iid symmetric α-stable random variables for several values of α. It shows that as α decreases from 2 to 1, the tails become heavier.

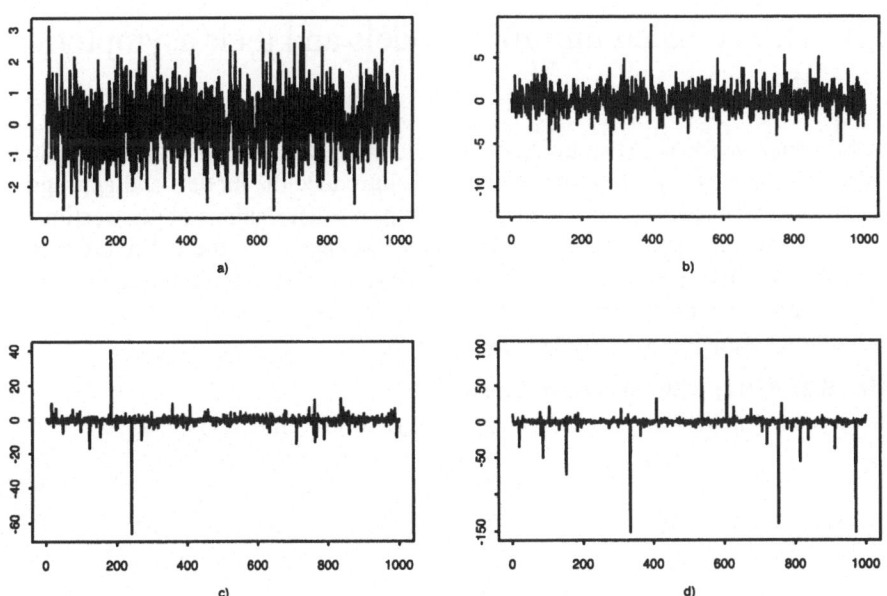

FIGURE 1. Realizations of 1000 iid random random variables following a symmetric stable distribution with (a) $\alpha = 2$ (normal), (b) $\alpha = 1.9$, (c) $\alpha = 1.5$, (d) $\alpha = 1.2$.

The iid symmetric α-stable random variables, and more generally random variables from their domain of attraction, obey a functional central limit theorem similar to Donsker's theorem. The limit process is a symmetric Lévy process $L_\alpha(t)$, $0 \leq t \leq 1$. The process L_α is right continuous with left limits and has stationary increments whose distribution is symmetric α-stable. If $\alpha = 2$, L_α is proportional to the Brownian motion.

More specifically,

$$\frac{1}{n^{1/\alpha}l(n)} \sum_{t=1}^{[nt]} \varepsilon_t \to L_\alpha(t), \quad \text{in } D[0, 1], \tag{5.1}$$

where $l(n)$ is a slowly varying function. It can also be shown that

$$\left(\frac{1}{n^{1/\alpha}l(n)} \sum_{t=1}^{[nt]} \varepsilon_t, \frac{1}{n^{2/\alpha}l^2(n)} \sum_{t=1}^{[nt]} \varepsilon_t^2 \right) \to \left(L_\alpha(t), [L_\alpha](t) \right), \text{in} D([0, 1] \times [0, 1]), \tag{5.2}$$

where $[L_\alpha](t) = L_\alpha^2(t) - 2 \int_0^t L_\alpha(s-)dL_\alpha(s)$, $t \geq 0$. Relations (5.1) and (5.2) are used to derive the asymptotic distribution of unit root test statistics for heavy-tailed observations with $\alpha < 2$. We refer to [4] and Chapter 15 of [21] for further details.

5.3   Heavy-tailed unit root models and their asymptotic theory

In this paper we focus on two unit root models, which are referred to in [10] as Model 1 and Model 2. Our goal in this section is to introduce the models and appropriate estimators and to show explicitly that the asymptotic distributions of these estimators and unit root test statistics are very complex and depend on the tail index α in an intricate way. These results motivate the use of simple bootstrap procedures.

We assume that the errors ε_t are symmetric α-stable with $1 < \alpha \leq 2$. The exposition below follows closely Chapter 15 of [21].

Model 1: AR(1) with no constant term:
We assume that the observations $y_1, y_2, \ldots, y_n$ follow an AR(1) model

$$y_t = \rho y_{t-1} + \varepsilon_t. \tag{5.3}$$

We wish to test

$$H_0 : \rho = 1 \quad \text{versus} \quad H_A : |\rho| < 1. \tag{5.4}$$

Using (5.1) and (5.2), it is easy to verify that under H_0 the OLS estimator

$$\hat{\rho}_n := \frac{\sum_{t=1}^n y_t y_{t-1}}{\sum_{t=1}^n y_{t-1}^2} \tag{5.5}$$

satisfies

$$n\left(\hat{\rho}_n - 1\right) \overset{L}{\to} \frac{\int_0^1 L_\alpha(s-)dL_\alpha(s)}{\int_0^1 L_\alpha^2(s)ds}.$$

The t-statistic $\hat{t}_\rho = \left(\hat{\rho}_n - 1\right)/s_{\hat{\rho}}$, where

$$s_{\hat{\rho}}^2 = \frac{\sum_{t=1}^n (y_t - \hat{\rho})y_{t-1}}{n \sum_{t=1}^n y_{t-1}^2}$$

also converges weakly:

$$\hat{t}_\rho \overset{L}{\to} \frac{\int_0^1 L_\alpha(s-)dL_\alpha(s)}{\sqrt{[L_\alpha](1) \int_0^1 L_\alpha^2(s)ds}}.$$

Model 2: AR(1) with a constant term:
We assume that the observations $y_1, y_2, \ldots, y_n$ follow the specification

$$y_t = \tau + \rho y_{t-1} + \varepsilon_t. \tag{5.6}$$

We wish to test

$$H_0 : \rho = 1 \text{ and } \tau = 0 \quad \text{versus} \quad H_A : |\rho| < 1 \text{ or } \tau \neq 0. \tag{5.7}$$

In this case the OLS estimators are given by

$$\begin{bmatrix} \hat{\tau}_n \\ \hat{\rho}_n \end{bmatrix} = \begin{bmatrix} n & \sum y_{j-1} \\ \sum y_{j-1} & \sum y_{j-1}^2 \end{bmatrix}^{-1} \begin{bmatrix} \sum y_j \\ \sum y_{j-1}y_j \end{bmatrix} \tag{5.8}$$

and satisfy

$$n\left(\hat{\rho}_n - 1\right) \overset{L}{\to} \frac{\int_0^1 L_\alpha(s-)dL_\alpha(s) - L_\alpha(1)\int_0^1 L_\alpha(s)ds}{\int_0^1 L_\alpha^2(s)ds - \left(\int_0^1 L_\alpha(s)ds\right)^2},$$

and

$$\frac{\hat{\tau}_n n^{1-1/\alpha}}{l(n)} \overset{L}{\to} \frac{L_\alpha(1)\int_0^1 L_\alpha^2(s)ds - \int_0^1 L_\alpha(s)ds \int_0^1 L_\alpha(s-dL_\alpha(s))}{\int_0^1 L_\alpha^2(s)ds - \left(\int_0^1 L_\alpha(s)ds\right)^2}.$$

The t-statistic in this case has the form

$$\hat{t}_{\rho,\tau} = \frac{\left(\hat{\rho}_n - 1\right)}{s_{\hat{\rho},\hat{\tau}}},$$

where

$$s_{\hat{\rho},\hat{\tau}}^2 = \frac{\sum_{t=1}^n \left(y_t - \hat{\tau}_n - \hat{\rho}_n y_{t-1}\right)^2}{n \sum_{t=1}^n y_{t-1}^2 - \left(\frac{1}{n} \sum_{t=1}^n y_{t-1}\right)^2}.$$

The asymptotic distribution of this statistic is also more complicated:

$$\frac{(\hat{\rho}_n - 1)}{s_{\hat{\rho},\hat{\tau}}} \xrightarrow{L} \frac{\int_0^1 L_\alpha^- dL_\alpha - L_\alpha(1) \int_0^1 L_\alpha ds}{\int_0^1 L_\alpha^2 ds - \left(\int_0^1 L_\alpha ds\right)^2} \sqrt{\frac{\int_0^1 L_\alpha^2 ds}{[L_\alpha](1)}}.$$

5.4   Bootstrap unit root tests for Model 1

In this section we assume that the observations $y_1, y_2, \ldots, y_n$ follow the AR(1) model (5.3) and we consider the testing problem (5.4). In Sections 5.4.1–5.4.4 we study a residual bootstrap test which relies on approximating the null distribution of the test statistic. In Section 5.4.5, we discuss an alternative approach based on constructing a confidence interval for ρ.

5.4.1   Residual bootstrap

In this section we describe the residual bootstrap test. Using the estimator $\hat{\rho}_n$ (5.5), we construct the residuals:

$$\hat{\varepsilon}_j = y_j - \hat{\rho}_n y_{j-1}, \quad j = 2, 3, \ldots, n. \tag{5.9}$$

The residual $\hat{\varepsilon}_1$ remains undefined, but it is not needed in our algorithm. The residuals are centered as follows:

$$\bar{\hat{\varepsilon}}_j = \hat{\varepsilon}_j - \frac{1}{n-1} \sum_{j=2}^n \hat{\varepsilon}_j. \tag{5.10}$$

Since we are interested in estimating the density of $\hat{\rho}_n$ under the null hypothesis $H_0 : \rho = 1$, the following formula is used to generate each bootstrap sample:

$$y_j^* = y_{j-1}^* + \varepsilon_j^*, \tag{5.11}$$

for $j = 2, 3, \ldots, n$, and we set for convenience $y_1^* = y_1$. The ε_t^* are randomly sampled with replacement from the set of the centered residuals $\bar{\hat{\varepsilon}}_j$, $j = 2, 3, \ldots, n$. Each bootstrap sample $\{y_j^*, \ j = 1, 2, \ldots, n\}$ gives the bootstrap estimator:

$$\hat{\rho}_n^* = \frac{\sum_{j=1}^n y_{j-1}^* y_j^*}{\sum_{j=1}^n y_{j-1}^{*2}}, \quad i = 1, 2, \ldots, B,$$

where B is the number of the bootstrap samples.

The sampling distribution of the test statistic $n(\hat{\rho}_n - 1)$ is approximated by the bootstrap empirical distribution function

$$F_n^*(x) = \frac{1}{B} \sum_{i=1}^{B} \mathbf{1} \left\{ n(\hat{\rho}_i^* - 1) \leq x \right\}.$$

Denote by $q_n^*(\gamma)$ the γth quantile of F_n^*. The unit root null hypothesis is rejected at level γ if

$$n(\hat{\rho}_n - 1) < q_n^*(\gamma).$$

5.4.2   Size and power of the test

To estimate the size and power of the residual bootstrap test, we applied it to the simulated series of process (5.3) with different values of $\rho = 1, 0.99, 0.95, 0.9, 0.8, 0.5$, $\alpha = 2, 1.75, 1.5, 1.25, 1.1$ and lengths $n = 500, 375, 250, 125, 60$. The power was estimated as the ratio of the number of rejections to the total number of experiments $N = 10000$. Decision about rejecting the null hypothesis $H_0 : \rho = 1$ every time was made based on the distribution F_n^* computed from $B = 2000$ bootstrap samples.

To give a rough idea about the results we have obtained, we present in Table 1 the empirical rejection probabilities for $\alpha = 1.75$ and the nominal size $\gamma = 5\%$.

TABLE 1. Empirical rejection probabilities for $\alpha = 1.75$ and the nominal size $\gamma = 5\%$ for the residual bootstrap test for the problem (5.4).

Length of the series	ρ					
	1	0.99	0.95	0.9	0.8	0.5
60	5.11	10.38	22.54	48.28	92.56	100
125	4.50	12.11	50.26	93.38	99.99	100
250	4.64	17.21	93.49	99.99	100	100
375	4.72	24.98	99.84	100	100	100
500	4.74	35.46	99.99	99.99	100	100

It is seen that the empirical size is very close to the nominal size for all series lengths considered in our study and the power of the test is also satisfactory. In Section 5.4.3 we study in greater detail the performance of the test for various choices of α whereas in Section 5.4.4 we study a variant of the residual bootstrap in which the bootstrap sample size is allowed to be different from the size of the original sample.

5.4.3   Performance of the test for different stability indexes α

To investigate that dependence on α, we compared type-I errors (percentage of the rejections of the true null-hypothesis) for the series generated using stable distribution with different α. As always, we used $N = 10000$ replications and $B = 2000$ bootstrap samples. The results of the experiments are presented in Figure 2.

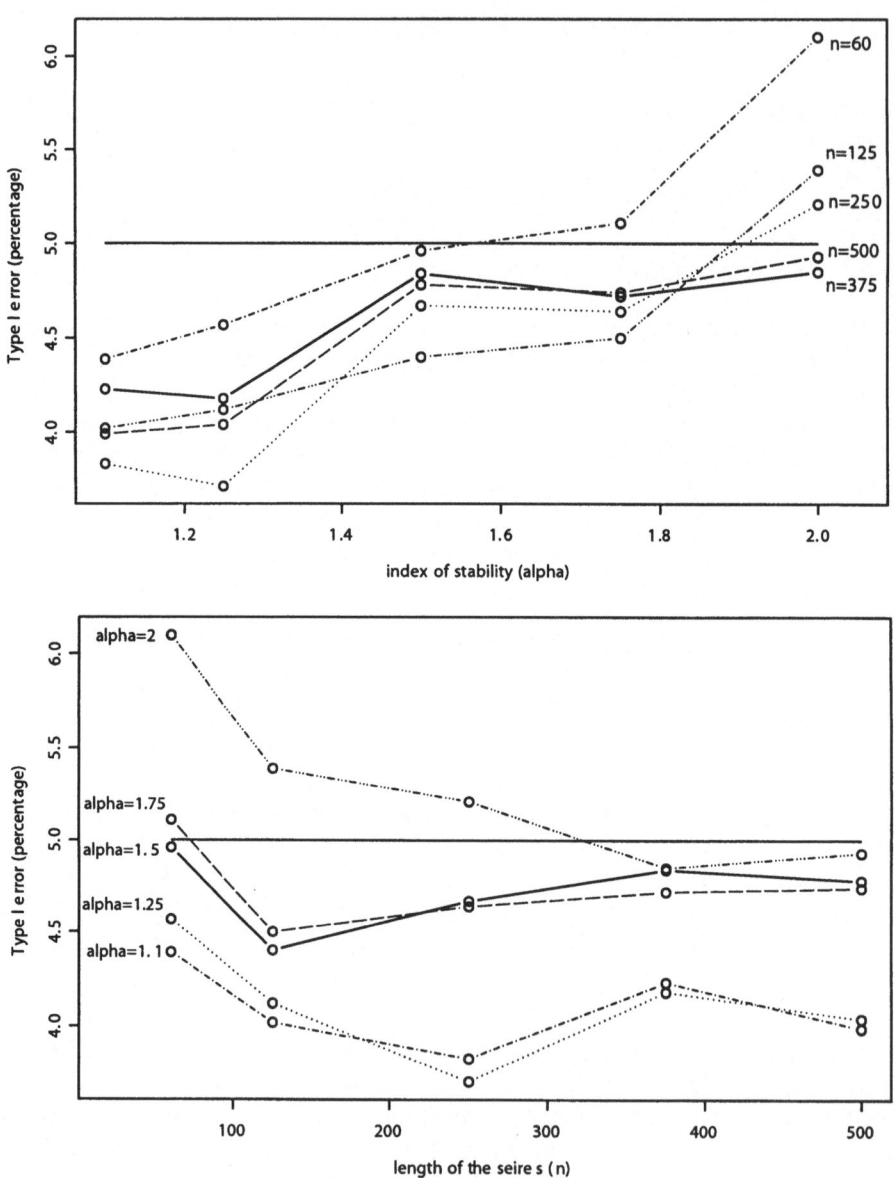

FIGURE 2. Type-I errors for the series of different lengths ($n = 500, 375, 250, 125, 60$) and different values of the stability index ($\alpha = 2, 1.75, 1.5, 1.25, 1.1$).

The top panel of Figure 2 shows that, at least for the longer series, the type I error approaches the nominal size as the index α approaches 2. This is probably due to the fact that for $\alpha = 2$ the stable distribution becomes Gaussian, while for $\alpha = 1$ it is the Cauchy distribution, which is much less "regular" than the normal distribution. Such a dependence on α makes the test somewhat non-robust, because in practice the index α is not known. On the other hand, for most combinations of n and α, the empirical sizes are between 4% and 6%. This means that despite this shortcoming, the test is very precise.

The bottom panel of Figure 2 shows that for $\alpha \geq 1.5$ the empirical size is close to the nominal size for sample sizes $n \geq 125$. For $\alpha < 1.5$, the empirical size is about 4%, 1% less than the nominal size.

5.4.4   Changing the bootstrap sample size

It is well known, see e.g., [2], [1], [14] and [12], that if the observations are heavy tailed with $\alpha < 2$, then in order to ensure a consistency of a bootstrap procedure, it is often necessary to reduce the bootstrap sample size. In our context, the appropriate modification would be to reject H_0 if $n(\hat{\rho}_n - 1) < q_m^*(\gamma)$, where

$$F_m^*(x) = \frac{1}{B} \sum_{i=1}^{B} 1 \left\{ m(\hat{\rho}_{i,m}^* - 1) \leq x \right\},$$

and where each $\hat{\rho}_{i,m}^*$ is now computed from the bootstrap sample $y_1^*, \ldots y_m^*$ with m not necessarily equal to n, the length of the original series.

We performed a number of experiments to estimate type-I error when the bootstrap sample size is different from the length of the observed series. We are not aware of any corresponding theoretical results for the residual bootstrap considered in this paper, so it was not *a priori* clear if m should be taken smaller or larger than n.

The results of our experiments for $\alpha = 1.75$ are shown in Figure 3. For $n \geq 125$, the values of m equal to 85%–90% of n give empirical sizes very close to the nominal size of $\gamma = 5\%$. The graphs for other values of α are similar. Other experiments showed that for $n \geq 125$ and $1.5 \leq \alpha \leq 1.75$, taking $m = 0.9n$ gives empirical sizes within a quarter of a percent of the nominal size of 5%.

The results described in this section strongly suggest that for $\alpha < 2$, in order to establish the consistency of the residual bootstrap test, it must be assumed that $m = m(n) \leq n$. A theoretical result of this type remains an open challenge. On the other hand, for the sample sizes $100 \leq n \leq 500$, the bottom panel of Figure 2 shows that taking $m = n$ still leads to a very precise test.

5.4.5   Confidence interval based bootstrap test

In this section we investigate an approach motivated by the work of [18]. The idea of the method is to find a level $(1 - \gamma)$ confidence interval (CI) of the form $(-1, \hat{u}]$ for the parameter ρ. A size γ test then rejects H_0 if $\hat{u} < 1$. To find the above confidence interval, we consider the differenced series

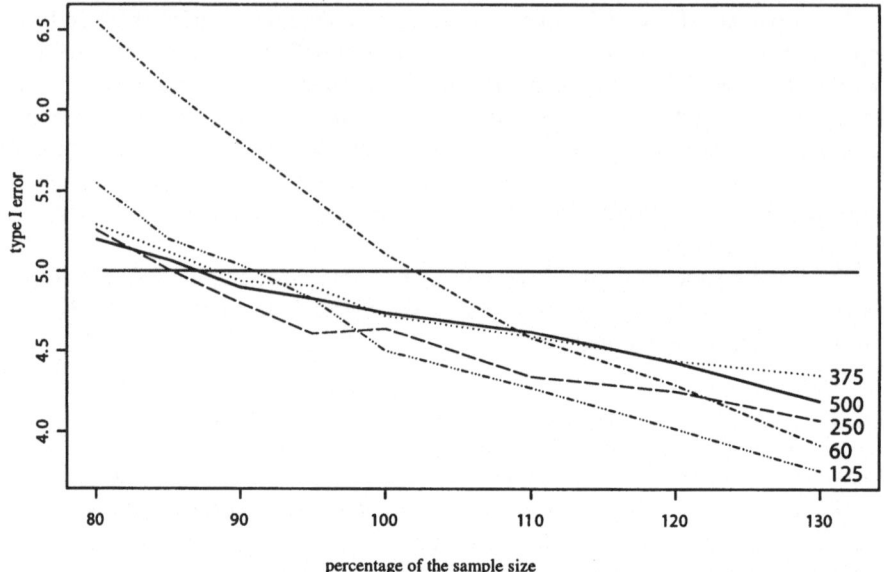

FIGURE 3. Type-I errors for series of different lengths versus the bootstrap sample size expressed as its percentage of the original series length. ($\alpha = 1.75, n = 500, 375, 250, 125, 60$. Nominal size $\gamma = 5\%$.)

$$Y_t = y_t - y_{t-1}.$$

Observe that

$$Y_t = \sum_{j=0}^{\infty} c_j \varepsilon_{t-j}, \text{ where}$$

$$c_j = \begin{cases} 1, \text{ if } j = 0 \\ 0, \text{ if } j > 0 \end{cases}, \text{ if } \rho = 1$$

and

$$c_j = \begin{cases} 1, \text{ if } j = 0 \\ (\rho - 1)\rho^{j-1}, \text{ if } j > 0 \end{cases}, \text{ if } |\rho| < 1.$$

Theorem 4.2 of [5] implies that

$$\hat{\rho}_Y := \frac{\sum_{j=2}^{n-1} Y_j Y_{j+1}}{\sum_{j=2}^{n} Y_j^2} \overset{P}{\to} \frac{\sum_{j=0}^{\infty} c_j c_{j+1}}{\sum_{j=0}^{\infty} c_j^2} = \frac{\rho - 1}{2} =: \rho_Y$$

under both H_0 and H_A. If we can construct a level $(1 - \gamma)$ confidence interval $(-1, \hat{u}_Y]$ for ρ_Y, then $(-1, 2\hat{u}_Y + 1]$ is the corresponding interval for ρ.

The interval $(-1, \hat{u}_Y]$ can be constructed as follows. Using $\hat{\rho} = 2\hat{\rho}_Y + 1$, we get bootstrap residuals as in Section 5.4.1. The bootstrap observations are however generated according to

$$y_i^* = \hat{\rho}y_{i-1}^* + \varepsilon_i^*, \ i = 2, 3, \dots, n, \quad y_1^* = y_1.$$

This is because now we want to construct many "bootstrap copies" of the original series rather than many series satisfying H_0.

After generating B bootstrap series $\{y_1^*, y_2^*, \dots, y_n^*\}_j, \ = 1, 2, \dots, B$, we compute B estimates $\hat{\rho}_Y^*$, which are used to construct the confidence interval $(-1, \hat{u}_Y]$, with $\hat{u}_Y$ being the $(1 - \gamma)$th empirical quantile of the $\hat{\rho}_Y^*$.

Table 2 shows that the above procedure gives empirical sizes comparable to those of the test described in Section 5.4.1 only for n close to 500. For smaller n, H_0 is accepted too often. Reducing the bootstrap sample size does not lead to any noticeable improvement. (The standard errors in Table 2 are about 0.2%.)

TABLE 2. Type-I errors for the test based on confidence intervals

The original length of the series	Length of the bootstrap samples (percents of the original length)			
	100%	95%	90%	85%
60	1.39	1.29	1.58	1.45
125	2.13	2.24	2.12	2.22
250	3.20	2.28	3.19	2.99
375	4.12	4.30	4.50	4.27
500	5.18	5.51	5.61	4.98

5.5   Bootstrap unit root test for Model 2

In this section we assume that the observations follow specification (5.6) and we consider the testing problem (5.7).

Following [6] we use the statistic

$$\Phi_n = \frac{(n - 3)(RSS_0 - RSS_1)}{2RSS_1}. \tag{5.12}$$

In (5.12), RSS_1 is the unconstrained residual sum of squares

$$RSS_1 = \sum_{k=2}^{n}(y_k - \hat{\rho}_n y_{k-1} - \hat{\tau}_n)^2 \tag{5.13}$$

and

$$RSS_0 = \sum_{k=2}^{n}(y_k - y_{k-1})^2 \tag{5.14}$$

is the residual sum of squares under the constraints $\rho = 1, \tau = 0$. The OLS estimators $\hat{\rho}_n$ and $\hat{\tau}_n$ are defined in Section 5.3.

Notice that $\Phi_n \geq 0$ and large values of Φ_n lead to the rejection of H_0. [6] derived the asymptotic null distribution of Φ_n under the assumption that the ε_t are normal and obtained the critical values by Monte Carlo simulations. The statistic Φ_n also has a limiting null distribution if the ε_t are heavy-tailed with infinite variance, but the limit is then different from that obtained by [6] and depends on α. Its precise form is quite complex and can be found in Appendix B of [13]. The bootstrap procedure described in Section 5.5.1 approximates the null distribution of Φ_n for finite n.

5.5.1   Residual bootstrap algorithm

In this section we explain how the residual bootstrap test introduced in Section 5.4.1 should be modified to apply to model (5.6) and the testing problem (5.7).

Using the estimators $\hat{\rho}_n$ and $\hat{\tau}_n$ computed according to (5.8), we construct the residuals

$$\hat{\varepsilon}_j = y_j - \hat{\rho}_n y_{j-1} - \hat{\tau}_n, \quad j = 2, 3, \dots, n. \tag{5.15}$$

After centering the residuals according to (5.10), we construct the bootstrap sequence $\{y_j^*, j = 1, 2, \dots, n\}$ using (5.11) and setting $y_1^* = y_1$.

Thus, the only modification that is required to construct a bootstrap process satisfying the null hypothesis consists in using the residuals (5.15) rather than the residuals (5.9).

Each bootstrap series $\{y_j^*, j = 1, 2, \dots, n\}$ gives the bootstrap estimates $\hat{\rho}_n^*$ and $\hat{\tau}_n^*$ that are used to construct the $\hat{\Phi}_n^*$, according to (5.12).

The set of bootstrap estimates $\hat{\Phi}_i^*$, $i = 1, 2, \dots B$, where B is the number of the bootstrap series, is then used to construct the empirical distribution function

$$F_{\Phi,n}^*(x) = \frac{1}{B}\sum_{i=1}^{B}\mathbf{1}\{\hat{\Phi}_i^* \leq x\}. \tag{5.16}$$

Denoting by $q_{\Phi,n}^*(1 - \gamma)$ the $(1 - \gamma)$th quantile of $F_{\Phi,n}^*$, we reject H_0 if $\hat{\Phi}_n > q_{\Phi,n}^*(1 - \gamma)$.

As in Section 5.4, we used in our experiments $N = 10000$ replications and $B = 2000$ bootstrap samples.

5.5.2   Size and power of the test

Figure 4 shows the empirical size of the test described in Section 5.5.1 for several series lengths n and indexes of stability α. The top panel shows that the empirical size

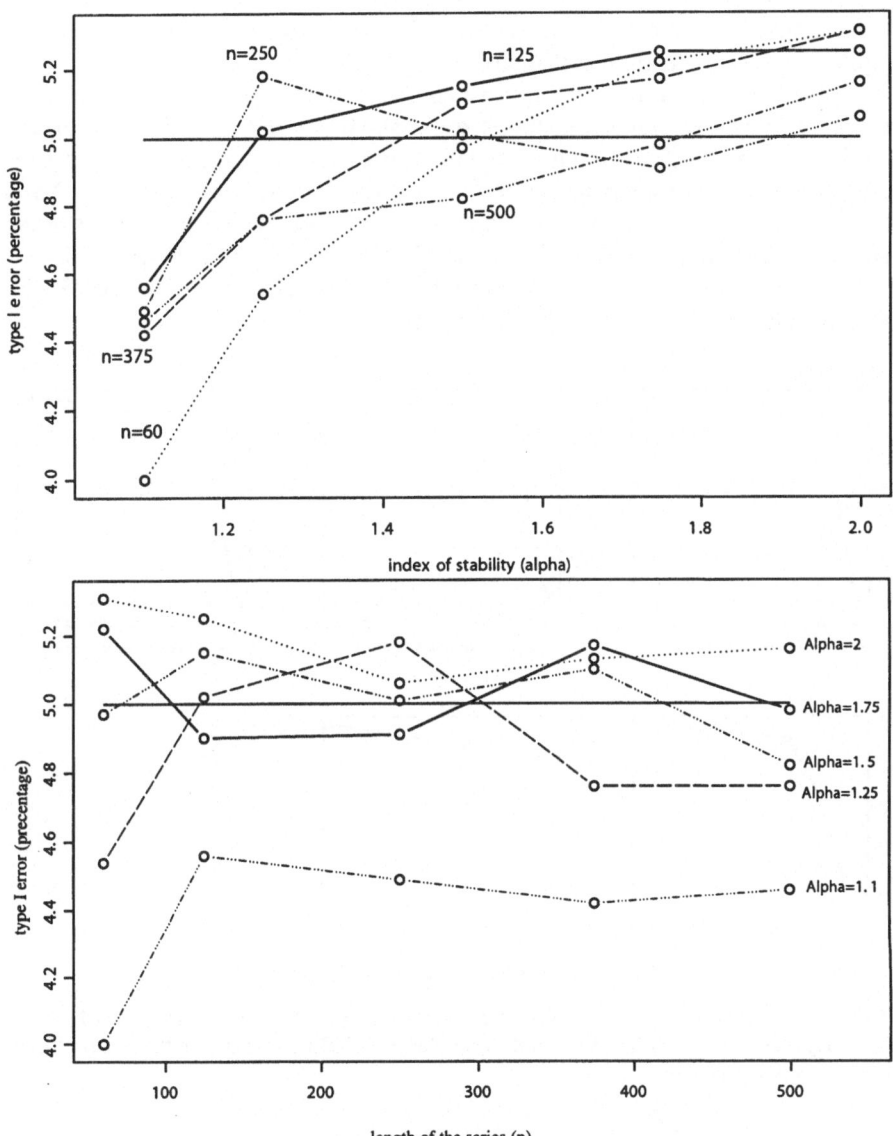

FIGURE 4. Type-I error (percentage of the rejections of the true null-hypothesis $H_0 : \rho = 1$ and $\tau = 0$) for series of length $n = 500, 375, 250, 125, 60$ with different values of the index of stability $\alpha = 2, 1.75, 1.5, 1.25, 1.1$.

is essentially equal to the nominal size for $\alpha \geq 1.5$ and all series lengths considered in our experiments (the standard error for the size estimates is about 0.2%). The type I error rates are in the interval $(4.5, 5.3)$ for all $\alpha \in [1.25, 2]$, regardless the length of the series. The bottom panel shows that for the index of stability $\alpha = 1.1$ the empirical size is about 1% less than the nominal test level.

To evaluate the power of the test we applied it to simulated realizations of the process (5.6) with several values of $0 \le \rho \le 1$ and $0 \le \tau \le 5$ and selected values of α. Due to the multitude of resulting simple alternatives, we present here the results only for two types of alternatives: $H_A : \rho < 1, \tau = 0$ (Figure 7, $\alpha = 1.5$) and $H_A : \rho = 1, \tau > 0$ (Figure 5, $\alpha = 1.75$)). The results for other combinations of the parameters are similar and show that the test has very good power for $n \ge 125$.

For $\alpha = 2$ (the normal case) the results of our simulations are in accord with those of [23] who considered local alternatives.

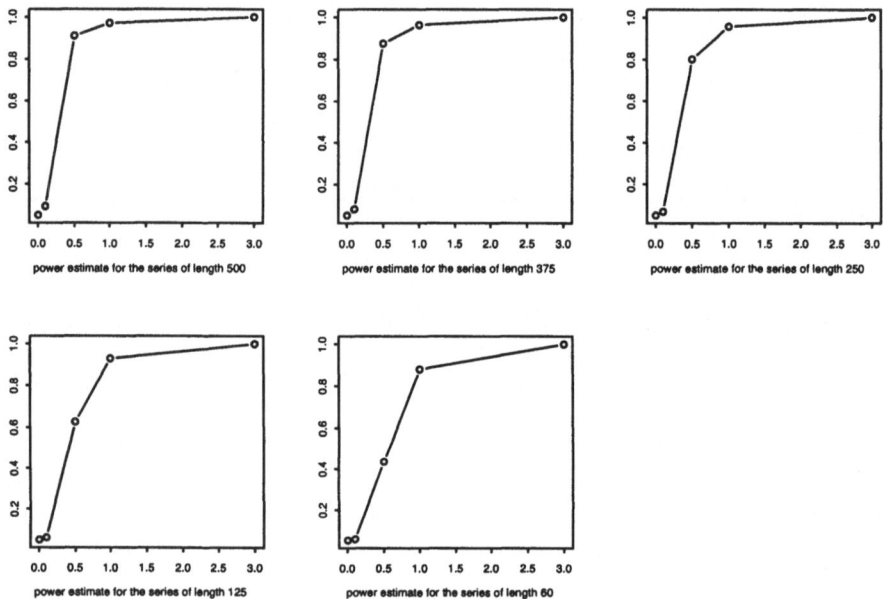

FIGURE 5. Empirical power for $\alpha = 1.75$ of the residual bootstrap test of $H_0 : \rho = 1, \tau = 0$ versus $H_A : \rho = 1, \tau > 0$ for series of different lengths. Values of τ are on the X-axis, power estimates are on the Y-axis.

5.6 Comparison with subsampling unit root tests

In this section we compare the residual bootstrap tests of Sections 5.4.1 and 5.5.1 with their subsampling counterparts introduced and studied in [13]. An excellent reference on subsampling is the monograph of [20]. Here we merely outline the idea of the subsampling test for Model 1 and refer the reader to [13] for the details and for the description of the subsampling test for Model 2.

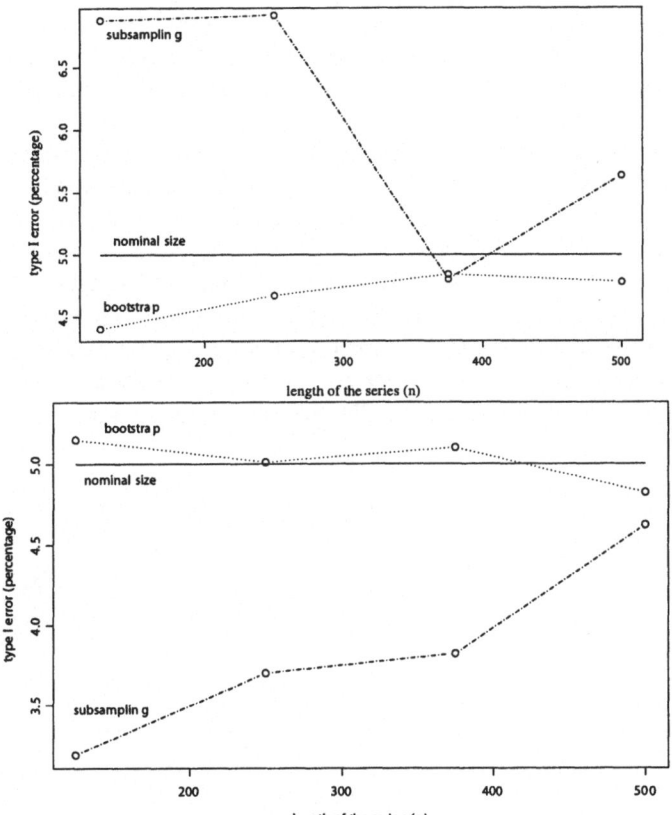

FIGURE 6. Empirical sizes of the residual bootstrap and subsampling tests for Model 1 (top panel) and Model 2 (bottom panel), $\alpha = 1.5$. Bootstrap sizes are based on 10000 replications and 2000 bootstrap resamples. Subsampling sizes are based on $R = 5000$ replications.

Suppose then that the observations follow model (5.3) and consider the testing problem (5.4). After computing the centered residuals $\bar{\hat{\varepsilon}}_j$, $j = 2, \dots, n$ exactly as in Section 5.4.1, we construct $n - b$ series of length $b < n$ which follow the null-hypothesis. For $k = 2, \dots, n - b + 1$, the kth series is defined by

$$y_0(k) = 0, \quad y_1(k) = \bar{\hat{\varepsilon}}_k, \quad y_2(k) = \bar{\hat{\varepsilon}}_k + \bar{\hat{\varepsilon}}_{k+1}, \dots, \tag{5.17}$$

$$y_b(k) = \bar{\hat{\varepsilon}}_k + \bar{\hat{\varepsilon}}_{k+1}, \dots, \bar{\hat{\varepsilon}}_{k+b-1}.$$

For every subsample $\{y_j(k), j = 0, \dots, b - 1\}$, the OLS estimator $\hat{\rho}_b(k)$ is computed ($k = 2, \dots, n - b + 1$). The empirical distributions of the $n - b$ values $T_{b,k} = b(\hat{\rho}_b(k) - 1)$ approximates the null distribution of the test statistic $T_n = n(\hat{\rho}_n - 1)$. The null hypothesis is rejected if $n(\hat{\rho}_n - 1) < q_{T,b}(\gamma)$, where $q_{T,b}(\gamma)$ is the γth quantile of the empirical distribution of the $T_{b,k}$.

Subsampling tests depend on the choice of the parameter b, the length of the subseries (5.11). Theory implies merely that b must be chosen so that $b \to \infty$ and $b/n \to 0$, as $n \to \infty$. In the comparisons below, we use the "optimal" b found by [13] by means of simulations.

Figure 6 compares the empirical sizes of the bootstrap and subsampling tests for Models 1 and 2 and $\alpha = 1.5$. The bootstrap test is seen to have type I error closer to the nominal size of the test than the subsampling test. This conclusion also holds for other choices of α.

Figure 7 compares the power of the two tests for Model 2, again with $\alpha = 1.5$, for the alternatives $\rho < 1, \tau = 0$. For Model 1, the results are similar, but in that case the power of the subsampling test is somewhat closer to the power of the bootstrap test and in some cases the two power curves overlap. In no case, however, is the power of the bootstrap test visibly smaller than that of the subsampling case.

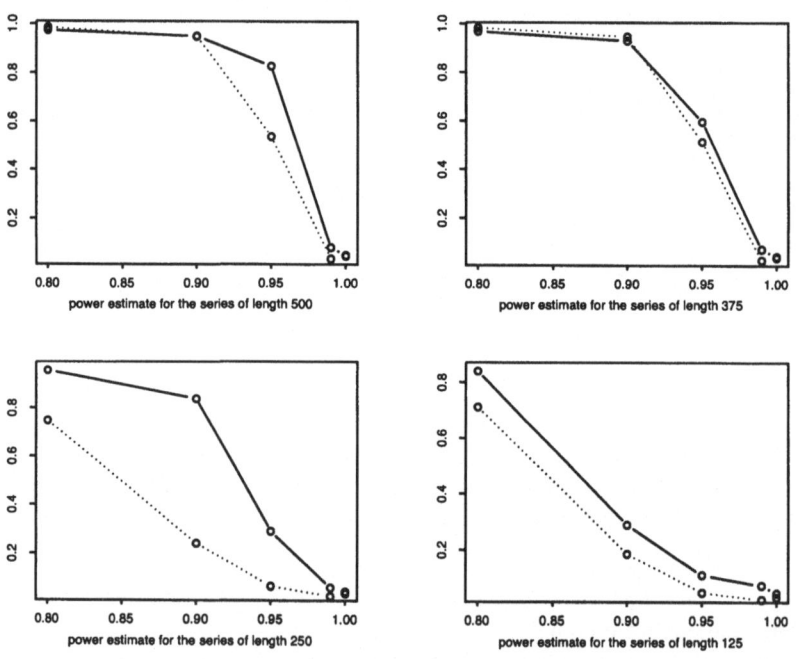

FIGURE 7. Empirical power of the residual bootstrap (continuous line) and subsampling (dotted line) tests of the null hypothesis H_0; $\rho = 1, \tau = 0$ against the alternative H_A : $\rho < 1, \tau = 0$ in Model 2, $\alpha = 1.5$. Parameter ρ is on the X-axis, power on the Y-axis.

5.7 Application to yield curves on corporate bonds

In this section we apply the residual bootstrap tests of Sections 5.4.1 and 5.5.1 to yield curves on corporate bonds. The data have been kindly made available by Pro-

TABLE 3. The OLS estimates $\hat{\rho}_{500}$ for the yield curves on corporate bonds.

Least Square	Yield curve					
Estimate	A	B	C	D	E	F
$\hat{\rho}_{500}$	0.99628	1.00204	0.98646	0.94641	0.99859	0.98591

fessor Svetlozar Rachev of the University of Karlsruhe and UC Santa Barbara. The data consist of six financial time series, each of length $n = 500$, which we labeled A, B, C, D, E and F, so as not to disclose the corporations issuing the bonds. Four of the six series are shown in Figure 8 together with their first differences (returns) $y_t - y_{t-1}$. The differences, especially A and C, are seen to have highly non-normal distribution with several spikes which are suggestive of heavy tails.

Consider first fitting Model 1. Since Model 1 assumes that the data have mean zero, we subtracted the non-zero sample average from every data set before applying the tests. The Least Square estimates of ρ for the six series are given in Table 3. All estimates, except series D, are quite close to 1 and strongly suggest the presence of the unit root. In fact, all of them, except series D, are well above the 5% critical values for $\alpha = 1.5$ and $\alpha = 2$ which are slightly larger than 0.97, see Table 1 of [4]. The bootstrap test allows us to calculate the P-values which are shown in Table 4. We used bootstrap sample size equal to 100% of the length of the original series and to 85% of that length, see Section 5.4.4. For each series the two P-values are very close, indicating (see also the discussion at the end of Section 5.4.4) that for series lengths considered in this paper it is not critical to reduce the bootstrap sample size. In each case, the P-values were computed using $B = 1000$ bootstrap replications.

The unit root null hypothesis can be rejected only for series D. The rejection is very strong and indicates that the yield curve D can possibly be viewed as a realization of a stationary process. This is confirmed by a visual examination of the graph of the yield curve D which appears roughly stationary. For the remaining five processes, the model $(y_t - \mu) = (y_{t-1} - \mu) + \varepsilon_t$, where μ is the mean, appears to be a reasonable approximation to the underlying data generating mechanism.

Now let us fit Model 2 to the data. In this case we have to estimate two parameters: ρ and τ. Their Least Square estimates for the six series are given in Table 5.

TABLE 4. The P-values for the test of $H_0 : \rho = 1$ applied to the yield curves.

Bootstrap	Yield curve					
sample size	A	B	C	D	E	F
100%	0.361	0.906	0.075	0	0.552	0.052
85%	0.353	0.911	0.077	0	0.537	0.066

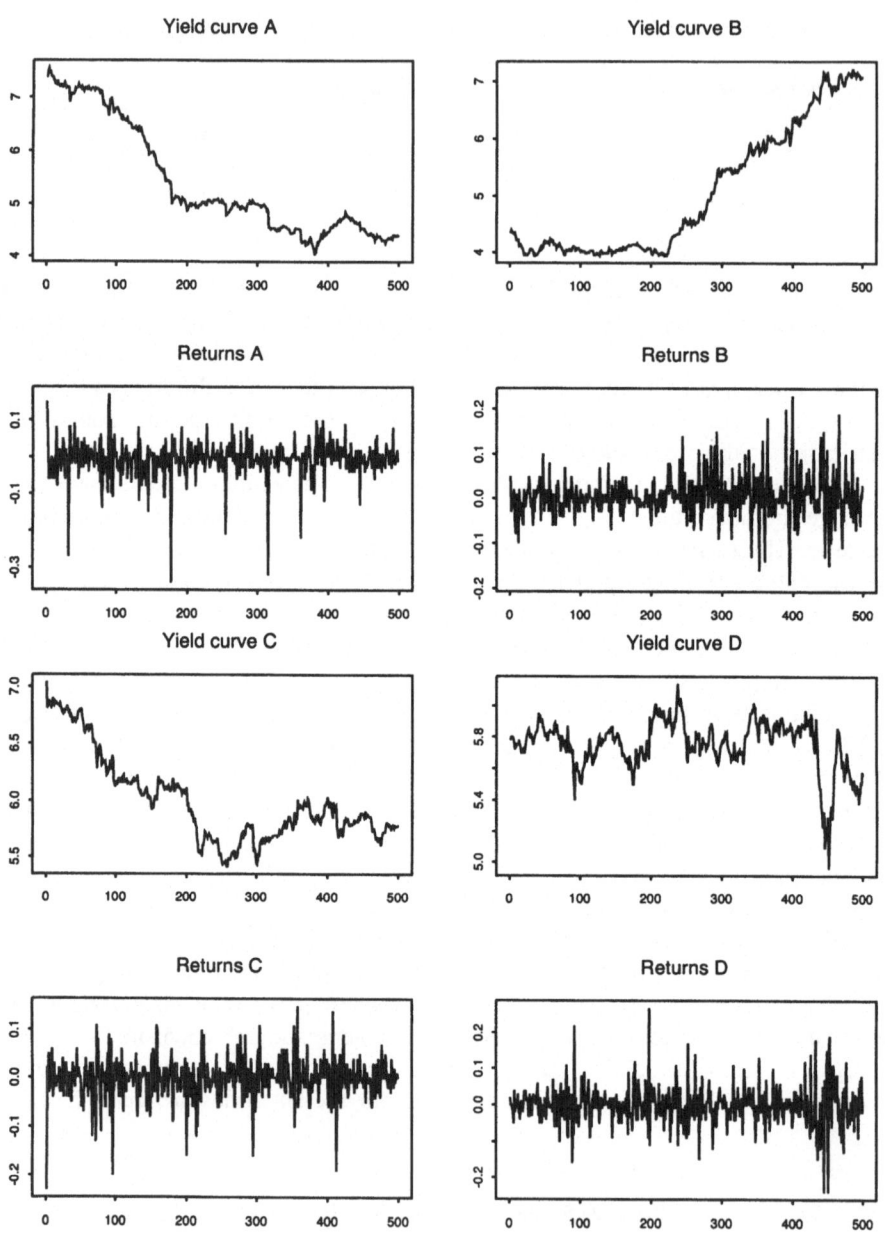

FIGURE 8. Yield curves on corporate bonds and the corresponding returns (first differences).

TABLE 5. Summary of the bootstrap test for Model 2 applied to the six yield curves.

Statistic	Yield curve					
	A	B	C	D	E	F
$\hat{\rho}_{500}$	0.99642	1.00202	0.99089	0.98504	0.99884	0.99612
$\hat{\tau}_{500}$	0.01327	-0.00468	0.05197	0.08545	0.01097	-0.00352
$\hat{\Phi}_{500}$	5.57901	4.04260	4.34621	3.17112	1.46973	3.39928
P-value	0.021	0.085	0.06	0.178	0.621	0.225

We see that by estimating a more flexible model with a possibly nonzero intercept, we did not obtain a highly significant evidence against the unit root null hypothesis for any of the six series. For series A, the P-value is smaller than 5% which is usually interpreted as a significant, but not highly significant, rejection. Note however, that the parameter estimates for series A are closer to the null hypothesis $\rho = 1$, $\tau = 0$ than the parameter estimates for series C and D, for which we cannot reject H_0 at the 5% level. It thus appears that the rejection for series A stems from the fact that we use the statistic $\hat{\Phi}_n$ rather than a two-dimensional rejection region. Even though the test does not allow us to reject the null hypothesis $\rho = 1$, $\tau = 0$ for series D, in light of the strong rejection of Model 1 with unit root, it appears that a suitable model D might be Model 2 with parameters ρ and τ close to their estimates reported in Table 5, i.e., with ρ slightly smaller than unity and τ a small positive value.

We may thus conclude that for all yield curves, possibly except series D, a unit root model is appropriate.

5.8   Conclusions

The goal of this paper was to explore and implement bootstrap methods for unit root testing for heavy-tailed infinite variance time series.

Among various conceivable bootstrap tests, we focused on the residual bootstrap because it is simple and has been extensively studied and applied in the context of time series with finite high moments. We also considered a confidence interval (CI) based test, but our results clearly showed the superiority of the residual bootstrap over the CI-based bootstrap when applied to samples of size smaller than 500. Our conclusions below apply to the residual bootstrap unless specified otherwise.

We focused on two data generating processes, (5.3) and (5.6). In our experiments we used errors that followed the stable distribution with the index of stability $1 < \alpha \leq 2$. The type I error rate was found to be within 1% of the nominal size of 5% when the errors have the index of stability $\alpha \geq 1.5$ and slightly worse in some cases for $1 < \alpha < 1.5$. Since the values of $\alpha < 1.5$ are unlikely to be encountered in econometric applications, see panel d) of Figure 1, we conclude that the tests can be used with confidence, especially for series of length $n \geq 250$.

To improve the performance of the test we considered the reduction of the size of bootstrap samples. The results of the experiments when fitting the Model 1 (5.3) demonstrated an improvement in the accuracy of the residual bootstrap test, but not the CI-based test. If Model 1 is believed to hold almost exactly, the optimal bootstrap sample size should be around $0.85n$ for series of length $250 \leq n \leq 500$. In practice, however, for the series of that length, using the bootstrap sample size equal to the length of the original series gives P-values that lead to the same decision.

We also compared the performance of the residual bootstrap with the subsampling method. The type I error of the bootstrap method is closer to the nominal size than the error of the subsampling method. The bootstrap test rejects the null-hypothesis more often when it is false. For series of lengths $n = 500$ bootstrap and subsampling tests are comparable.

The present paper explored bootstrap unit root tests for infinite variance observations only in two archetypal models. The residual bootstrap can however be easily extended to more complicated models. Roughly speaking, all that is required, is to compute the centered residuals and construct many bootstrap series obeying the null hypothesis. The null distribution of a test statistic can then be readily approximated. Detailed exploration of the method in the multitude of specifications used in practice would go beyond the intended scope of the paper. Our results do however indicate that the method has a good potential for unit root testing in the presence of heavy-tailed errors.

References

[1] Arcones, M. and Giné, E. (1989) The bootstrap of the mean with arbitrary bootstrap sample size. *Annals of the Institute Henri Poincaré*, **22**, 457–481.

[2] Athreya, K. B. (1987) Bootstrap of the mean in the infinite variance case. *The Annals of Statistics*, **15**:2, 724–731.

[3] Basawa, I. V., Mallik, A. K., McCormick, W. P., and Taylor, R. L. (1991) Bootstrap test of significance and sequential bootstrap estimation for unstable first order autoregressive processes. *Communications in Statististics – Theory and Methods*, **20**, 1015–1026.

[4] Chan, N. H. and Tran, L. T. (1989) On the first order autoregressive process with infinite variance. *Econometric Theory*, **5**, 354–362.

[5] Davis, R. A. and Resnick, S. I. (1985) Limit theory for moving averages of random variables with regularly varying tail probabilities. *The Annals of Probability*, **13**:1, 179–195.

[6] Dickey, D. A. and Fuller, W. A. (1981) Likelihood ratio statistics for autoregressive time series with unit root. *Econometrica*, **49**, 1057–1074.

[7] Embrechts, P., Klüppelberg, C. and Mikosch, T. (1997) *Modelling Extremal Events for Insurance and Finance*. Springer–Verlag, Berlin.

[8] Ferretti, N. and Romo, J. (1996) Unit root bootstrap tests for ar(1) models. *Biometrika*, **83**, 849–860.

[9] Guillaume, D. M., Dacorogna, M. M., Dave, R. D., Müller, U. A., Olsen, R. B. and Pictet, O. V. (1997) From the bird's eye to the microscope: a survey of new stylized facts of the intra-daily foreign exchange markets. *Finance and Stochastics*, **1**, 95–129.

[10] Hamilton, J. D. (1994) *Time Series Analysis*. Princeton University Press, Princeton, NJ.

[11] Heimann, G. and Kreiss, J-P. (1996) Bootstrapping general first order autoregression. *Statistics and Probability Letters*, **30**, 87–98.

[12] Horváth, L. and Kokoszka, P. S. (2003) A bootstrap approximation to a unit root test statistic for heavy-tailed observations. *Statistics and Probability Letters*, **62**, 163–173.

[13] Jach, A. and Kokoszka, P. (2003) Subsampling unit root tests for heavy-tailed observations. *Methodology and Computing in Applied Probability*, **6**, 73–97.

[14] Knight, K. (1989) On the bootstrap of the sample mean in the infinite variance case. *The Annals of Statistics*, **17**, 1168–1175.

[15] Ling, S. and Li, W. K. (2003) Asymptotic inference for unit root processes with GARCH(1,1) errors. *Econometric Theory*, **19**, 541–564.

[16] Ling, S. and McAleer, M. (2003) On adaptive estimation in nonstationary ARMA models with GARCH errors. *Annals of Statistics*, **31**, 642–674.

[17] Meerschaert, M. M. and Scheffler, H. P. (2001) *Limit Theorems for Sums of Independent Random Vectors*. Wiley, New York.

[18] Paparoditis, E. and Politis, D. N. (2001) Large sample inference in the general AR(1) case. *Test*, **10**:1, 487–589.

[19] Paparoditis, E. and Politis, D. N. (2003) Residual based block bootstrap for unit root testing. *Econometrica*, **71**:3, 813–855.

[20] Politis, D. N., Romano, J. P. and Wolf, M. (1999) *Subsampling*. Springer-Verlag.

[21] Rachev, S. and Mittnik, S. (2000) *Stable Paretian Models in Finance*. John Wiley & Sons Ltd.

[22] Samorodnitsky, G. and Taqqu, M. S. (1994) *Stable Non-Gaussian Random Processes: Stochastic Models with Infinite Variance*. Chapman & Hall.

[23] Swensen, A. R. (2003) A note on the power of bootstrap unit root tests. *Econometric Theory*, **19**, 32–48.

[24] Wang, Q., Lin, Y-X. and Gulati, C. M. (2003) Asymptotics for general fractionally integrated processes with applications to unit root tests. *Econometric Theory*, **19**, 143–164.

[25] Zarepour, M. and Knight, K. (1999) Bootstrapping unstable first order autoregressive process with errors in the domain of attraction of stable law. *Communications in Statistics -Stochastic Models*, **15(1)**, 11–27.

6

Optimal Portfolio Selection and Risk Management: A Comparison between the Stable Paretian Approach and the Gaussian One

Sergio Ortobelli

Svetlozar Rachev

Isabella Huber

Almira Biglova

ABSTRACT This paper analyzes stable Paretian models in portfolio theory, risk management and option pricing theory. Firstly, we examine investor's optimal choices when we assume respectively either Gaussian or stable non-Gaussian distributed index returns. Thus, we approximate discrete time optimal allocations assuming different distributional assumptions and considering several term structure scenarios. Secondly, we compare some stable approaches to compute VaR for heavy-tailed return series. These models are subject to backtesting on out-of-sample data in order to assess their forecasting power. Finally, when asset prices are log-stable distributed, we propose a numerical valuation of option prices and we describe and compare delta hedging strategies when asset prices are either log-stable distributed or log-normal distributed.

S. Rachev's research was supported by grants from Division of Mathematical, Life and Physical Sciences, College of Letters and Science, Univeristy of California, Santa Barbara and the Deutschen Forschungsgmeinschaft.

S. Ortobelli's research has been partially supported under Murst 40%, 60%, 2002, 2003, Vigoni project and CNR-MIUR-Legge 95/95.

A. Biglova's research has been partially supported by DAAD.

6.1   Introduction

This paper fulfils a two-fold objective. Firstly, it proposes several stable models, that can be used either for the optimal portfolio selection and value of the risk of a given portfolio or for pricing contingent claims. Secondly it compares models in portfolio theory, risk management and option pricing theory considering normal distributed returns and stable non-Gaussian distributed returns alike.

It is well known that asset returns are not normally distributed, but many of the concepts in theoretical and empirical finance developed over the past decades rest upon the assumption that asset returns follow a normal distribution. The fundamental work of Mandelbrot (1963a-b, 1967) and Fama (1963,1965a-b) has sparked considerable interest in studying the empirical distribution of financial assets. The excess kurtosis found in Mandelbrot's and Fama's investigations led them to reject the normal assumption and to propose the stable Paretian distribution as a statistical model for asset returns. The Fama and Mandelbrot's conjecture was supported by numerous empirical investigations in the subsequent years, (see Mittnik, Rachev and Paolella (1996) and Rachev and Mittnik (2000)).

The practical and theoretical appeal of the stable non-Gaussian approach is given by its attractive properties that are almost the same as the normal ones. A relevant desirable property of the stable distributional assumption is that stable distributions have a domain of attraction. The Central Limit Theorem for normalized sums of i.i.d. random variables determines the domain of attraction of each stable law. Therefore, any distribution in the domain of attraction of a specified stable distribution will have properties close to those of the stable distribution. Another attractive aspect of the stable Paretian assumption is the stability property, i.e., stable distributions are stable with respect to the summation of i.i.d. random stable variables. Hence, the stability governs the main properties of the underlying distribution. Detailed accounts of theoretical aspects of stable distributed random variables can be found in Samorodnitsky and Taqqu (1994) and Janicki and Weron (1994).

First, the paper presents a unifying framework for understanding the stable Paretian approach to portfolio theory. In particular, we recall some symmetric and asymmetric stable models to study the multivariate portfolio selection (see Rachev, Ortobelli and Schwartz (2002)). We develop three alternative stable models under the assumption that investors allocate their wealth across the available assets in order to maximize their expected utility of final wealth. We first consider the portfolio allocation between n sub-Gaussian symmetrical α-stable distributed risky assets (with $\alpha > 1$), and the riskless one. The joint sub-Gaussian α-stable family is an elliptical family. Hence, as argued by Owen and Rabinovitch (1983), we can extend the classic mean variance analysis to a mean dispersion one. The resulting efficient frontier is formally the same as Markowitz–Tobin's mean-variance one, but, instead of the variance as a risk parameter, we have to consider the scale parameter of the stable distributions. Unlike Owen and Rabinovitch, we propose a method based on the moments to estimate all stable parameters. Calculation of the efficient frontier exhibits the two-fund separation property and, in equilibrium, we obtain an α-stable version of Sharpe–Lintner–Mossin's CAPM. Using Ross' necessary and sufficient conditions of the two-fund separation model (see Ross (1978)), we can link this stable version of asset pricing to that in Gamrowski and Rachev (1999). In order to consider the possible asymmetry of asset returns, we describe a three-fund separation model for returns in the domain of attraction of a stable law. In case of asymmetry, the model results from a new stable version of Simaan's model (see Simaan (1993)). In case of symmetry of returns, we obtain a version of a model recently studied by Götzenberger, Rachev and Schwartz (2001), that can also be viewed as a particular version of the two-fund separation of Fama's (1965b) model. However, this model distinguishes itself from Götzenberger, Rachev and Schwartz's, as well as from the Simaan and Fama models because it takes into consideration both the heavy tails and the asymmetry of the empirical return distributions. Using the stochastic dominance rules, (see Ortobelli (2001), Levy (1992) and the references therein), we show how to determine the investor's efficient choices. Similarly to the sub-Gaussian approach, it is possible to estimate all parameters with an OLS method. One of the most severe restrictions in performance measurement and asset pricing in the stable case is the assumption of a common index of stability for all assets. Hence, the last model we propose deals with the case of optimal allocation between stable distributed portfolios with different indexes of stability. In order to overcome the difficulties in the most general case of the stable law, we introduce a $k + 1$ fund separation model. Then we show how to express the model's multi-parameter efficient frontier. The evidence on temporal dependence for daily stock returns generally is not strong (see, among other, Fama and French (1988)), and the magnitudes of the autocorrelations are clearly too small to form profitable trading rules. As a matter of fact, these unconditional models were applied successfully in some recent empirical analyses which consider daily data (see Ortobelli, Huber and Schwartz (2002), Ortobelli, Huber, Rachev and Schwartz (2002), Ortobelli, Huber, Höchstötter, and Rachev (2001)).

The literature in the multi-period portfolio selection has been dominated by the results of maximizing expected utility functions of terminal wealth and/or multi-period consumption. Differently from classic multi-period approaches, we generalize Li and Ng's (2000) mean-variance analysis, by giving a mean-dispersion and a mean-dispersion-skewness formulation of the optimal dynamic portfolio selection. These alternative multi-period approaches analyze portfolio selection, taking into consideration all the admissible risk averse investors' optimal portfolio choices. In particular, analytical optimal portfolio policies are derived for the multi-period mean-dispersion and mean-dispersion-skewness formulation. In order to compare the Gaussian and the stable sub-Gaussian dynamic models, we analyze two investment allocation problems. The primary contribution of this empirical comparison is the analysis of the impact of distributional assumptions and different term structures on the multi-period asset allocation decisions. Thus, we propose a performance comparison between the multiperiod mean-variance approach and the stable sub-Gaussian one taking into consideration eleven risky indexes and five different implicit term structures. For this purpose we analyze two allocation problems for investors with different risk aversion coefficients. We determine the multiperiod efficient frontiers given by the minimization of the dispersion measures for different levels of expected value of final wealth. Each investor, characterized by his/her utility function, will prefer the mean-dispersion model which maximizes his/her expected utility on the efficient frontier. The portfolio policies obtained with this methodology represent the optimal investors' choices of the different approaches. Therefore, we examine the differences in optimal investor's strategies for a given implicit term structure under the Gaussian or the stable distributional hypothesis.

Secondly, we study stable autoregressive models in portfolio theory and risk management. In particular, we describe dynamic portfolio choices when returns follow an ARMA(1,1) model and we present a comparison among different conditional Value at Risk models. We recall the portfolio choice empirical analysis proposed by Ortobelli, Biglova, Huber, Racheva, Stoyanov (2003). Thus, in the multistage portfolio allocation problem we analyze the investor's choices under the assumption that we generate future scenarios considering an ARMA(1,1) model for the portfolio returns. Then, we compare investor's optimal allocations obtained when the residuals are either stable distributed or Gaussian distributed. In addition, considering the conditional extension of the previous mean-dispersion and mean-dispersion-skewness models proposed by Lamantia, Ortobelli and Rachev (2003), we evaluate and compare the performance of symmetric and asymmetric stable conditional VaR models.

Finally, we analyze the dynamic structure of return processes using stable laws and we show how inter-temporal stable models can be used to price contingent claims. Thus, we examine the subordinated stable model in option pricing and we propose an empirical comparison between dynamic delta hedging strategies when we adopt either the classic Black and Scholes option pricing model (B&S) or the stable subordinated one.

In Section 2 we introduce several static and dynamic portfolio choice models when returns are unconditionally stable distributed. In Section 3, we introduce a comparison between stable and Gaussian portfolio choice strategies. Section 4 proposes a conditional extension of previous models and a comparison among multi-stage conditional portfolio choice models. In Section 5 we backtest different stable Value at Risk models. Section 6 reviews recent stable option pricing models and compares different delta hedging strategies. Finally, we briefly summarize the results.

6.2   Portfolio selection with stable distributed returns

In this section, we consider several multivariate estimable stable portfolio selection models consistent with the maximization of the expected utility developed by Rachev, Ortobelli and Schwartz (2003). For each portfolio choice model we first present a static version remarking on the differences among the most known models. Secondly, we extend the static analysis by studying the problem of the optimal discrete time allocations.

Basically, we analyze the problem of optimal allocation among $n + 1$ assets: n of those assets are stable distributed risky assets with returns $z = [z_1, ..., z_n]'$, and the $(n + 1)th$ asset is risk-free with return z_0. Therefore, we examine the admissible optimal allocations for the following stable models of asset returns.

6.2.1   The sub-Gaussian α- stable model

Assume the vector of returns $z = [z_1, ..., z_n]'$ is sub-Gaussian α-stable distributed with $1 < \alpha < 2$. Then, the characteristic function of z has the form

$$\Phi_z(t) = E(\exp(it'z)) = \exp\left(-\left(t'Qt\right)^{\frac{\alpha}{2}} + it'\mu\right), \qquad (6.1)$$

where $Q = \left[\frac{R_{i,j}}{2}\right]$ is a positive definite $(n \times n)$-matrix, $\mu = E(z)$ is the mean vector, and $\gamma(ds)$ is the spectral measure with support concentrated on $S_n = \{s \in R^n / \|s\| = 1\}$. The term $R_{i,j}$ is defined by

$$\frac{R_{i,j}}{2} = [\tilde{z}_i, \tilde{z}_j]_\alpha \|\tilde{z}_j\|_\alpha^{2-\alpha}, \qquad (6.2)$$

where $\tilde{z}_j = z_j - \mu_j$ are the centralized return, the covariation $[\tilde{z}_i, \tilde{z}_j]_\alpha$ between two jointly symmetric stable random variables $\tilde{z}_i$ and $\tilde{z}_j$ is given by

$$[\tilde{z}_i, \tilde{z}_j]_\alpha = \int_{S_2} s_i \, |s_j|^{\alpha-1} \, sgn(s_j)\gamma(ds),$$

in particular, $\|\tilde{z}_j\|_\alpha = \left(\int_{S_2} |s_j|^\alpha \gamma(ds)\right)^{\frac{1}{\alpha}} = ([\tilde{z}_j, \tilde{z}_j]_\alpha)^{\frac{1}{\alpha}}$. Here the spectral measure $\gamma(ds)$ has support on the unit circle S_2.

This model can be considered as a special case of Owen–Rabinovitch's elliptical model (see Owen and Rabinovitch (1983)). However, no estimate procedure of the model parameters is given in the elliptical models with infinite variance. In our approach we use (6.1) and (6.2) to provide a statistical estimator of the stable efficient frontier. To estimate the efficient frontier for returns given by (6.1), we need to consider an estimator for the mean vector μ and an estimator for the dispersion matrix Q. The estimator of μ is given by the vector $\hat{\mu}$ of sample averages. Using lemma 2.7.16 in Samorodnitsky, Taqqu (1994) we can write for every $p \in (1, \alpha)$,

$$\frac{[\tilde{z}_i, \tilde{z}_j]_\alpha}{\|\tilde{z}_j\|_\alpha^\alpha} = \frac{E\left(\tilde{z}_i \tilde{z}_j^{\langle p-1\rangle}\right)}{E\left(|\tilde{z}_j|^p\right)}, \qquad (6.3)$$

where $\tilde{z}_j^{\langle p-1\rangle} = sgn\left(\tilde{z}_j\right)|\tilde{z}_j|^{p-1}$, and the scale parameter can be written $\sigma_j = \|\tilde{z}_j\|_\alpha$. Then, σ_j can be approximated by the moment method suggested by Samorodnitsky, Taqqu (1994) Property 1.2.17 in the case $\beta = 0$,

$$\sigma_j^p = \|\tilde{z}_j\|_\alpha^p = \frac{p \int_0^{+\infty} u^{-p-1} \sin^2 u \, du}{2^{p-1}\Gamma\left(1 - \frac{p}{\alpha}\right)} E\left(|\tilde{z}_j|^p\right). \qquad (6.4)$$

Moreover the following lemma holds (see Ortobelli, Biglova, Huber, Racheva, Stoyanov (2003)).

Lemma 1. *For any* $p \in (0, 2)$ *we get* $\int_0^{+\infty} u^{-p-1} \sin^2 u\, du = \frac{\Gamma(1-\frac{p}{2})\sqrt{\pi}}{2p\Gamma(\frac{p+1}{2})}$.

It follows that

$$\sigma_j^p = \frac{\Gamma\left(1 - \frac{p}{2}\right)\sqrt{\pi}}{2^p \Gamma\left(1 - \frac{p}{\alpha}\right)\Gamma\left(\frac{p+1}{2}\right)} E\left(|\tilde{z}_j|^p\right),$$

$$\frac{R_{i,j}}{2} = \sigma_j^2 \frac{E\left(\tilde{z}_i \tilde{z}_j^{\langle p-1 \rangle}\right)}{E\left(|\tilde{z}_j|^p\right)} = \sigma_j^{2-p} \frac{\Gamma\left(1 - \frac{p}{2}\right)\sqrt{\pi}}{2^p \Gamma\left(1 - \frac{p}{\alpha}\right)\Gamma\left(\frac{p+1}{2}\right)} E\left(\tilde{z}_i \tilde{z}_j^{\langle p-1 \rangle}\right).$$

The above suggests the following estimator $\widehat{Q} = \left[\frac{\widehat{R}_{i,j}}{2}\right]$ for the entries of the unknown covariation matrix Q,

$$\frac{\widehat{R}_{i,j}}{2} = \widehat{\sigma}_j^{2-p} \frac{\Gamma\left(1 - \frac{p}{2}\right)\sqrt{\pi}}{2^p \Gamma\left(1 - \frac{p}{\alpha}\right)\Gamma\left(\frac{p+1}{2}\right)} \frac{1}{N} \sum_{k=1}^{N} \tilde{z}_i^{(k)} \left(\tilde{z}_j^{(k)}\right)^{\langle p-1 \rangle}, \tag{6.5}$$

where the σ_j is estimated as

$$\widehat{\sigma}_j^2 = \frac{\widehat{R}_{j,j}}{2} = \left(\frac{\Gamma\left(1 - \frac{p}{2}\right)\sqrt{\pi}}{2^p \Gamma\left(1 - \frac{p}{\alpha}\right)\Gamma\left(\frac{p+1}{2}\right)} \frac{1}{N} \sum_{k=1}^{N} \left|\tilde{z}_j^{(k)}\right|^p\right)^{\frac{2}{p}}. \tag{6.6}$$

The rate of convergence of the empirical matrix $\widehat{Q} = \left[\frac{\widehat{R}_{i,j}}{2}\right]$ to the unknown matrix Q (to be estimated), will be faster, if $p = 1$, see Rachev (1991)[1] and Lamantia et al. (2003).

In order to describe the optimal choices, we have to know the stochastic dominance rules existing among different portfolios (see, among others, Ortobelli (2001)). Thus, consider two α stable distributed random variables $W_i = S_\alpha(\sigma_{W_i}, \beta_{W_i}, m_{W_i})$, $i = 1, 2$ with $\alpha > 1$, which have:

1. the same skewness parameters $\beta_{W_1} = \beta_{W_2}$,
2. the same mean $m_{W_1} = m_{W_2} = E(W_i)$,
3. $\sigma_{W_1} > \sigma_{W_2}$.

Then W_2 dominates W_1 in the sense of Rothschild–Stiglitz and every risk averse investor prefers W_2 to W_1. As a consequence of this stochastic dominance rule, when the returns $z = [z_1, ..., z_n]'$ are jointly α-stable distributed (not necessarily

---

[1] We also observe that as $p \to \alpha \to 2$, then $Q \to \frac{V}{2}$, where $V = [v_{i,j}]$ is the variance-covariance matrix.

sub-Gaussian distributed), every risk averse investor will choose a solution of the following optimization problem for some m_W and β^*:

$$\min_{x} \sigma_{x'z} \text{ subject to}$$
$$x'E(z) + (1 - x'e)z_0 = m_W$$
$$\beta_{x'z} = \beta^*, \quad 0 \le x'e \le 1 \qquad (6.7)$$
$$\text{and } x_i \ge 0, \quad i = 1, ..., n,$$

where $e = [1, ..., 1]$ and $W = x'z + (1 - x'e)z_0$. In order to determine estimates of the scale parameter and of the skewness parameter, we can consider the tail estimator for the index of stability α and the estimator for the spectral measure $\gamma(ds)$ proposed by Rachev and Xin (1993) and Cheng and Rachev (1995). However, even if the estimates of the scale parameter and the skewness parameter are computationally feasible, they require numerical calculations. Considering a sub-Gaussian α-stable vector of returns z, we simplify the portfolio choice problem (6.7) because

$$W = x'z + (1 - x'e)z_0 \stackrel{d}{=} S_\alpha(\sigma_{x'z}, \beta_{x'z}, E(W))$$
$$\text{and } W = z_0 \text{ when } x = 0,$$

where α is the index of stability, $\sigma_{x'z} = \sqrt{x'Qx}$ is the scale (dispersion) parameter, $\beta_{x'z} = 0$ is the skewness parameter and $E(W) = x'E(z) + (1 - x'e)z_0$. Thus, when the returns $z = [z_1, ..., z_n]'$ are jointly sub-Gaussian α-stable distributed and unlimited short sales are allowed, every risk averse investor will choose an optimal portfolio among the portfolio solutions of the following optimization problem:

$$\min_{x} x'Qx \text{ subject to} \qquad (6.8)$$
$$x'\mu + (1 - x'e)z_0 = m_W,$$

for some given mean m_W. Therefore, every optimal portfolio that maximizes a given concave utility function u, belongs to the mean-dispersion frontier

$$\sigma = \begin{cases} \dfrac{m - z_0}{\sqrt{(\mu - ez_0)'Q^{-1}(\mu - ez_0)}} & \text{if } m \ge z_0, \\[3mm] \dfrac{z_0 - m}{\sqrt{(\mu - ez_0)'Q^{-1}(\mu - ez_0)}} & \text{if } m < z_0, \end{cases} \qquad (6.9)$$

where $\mu = E(z)$; $m = x'\mu + (1 - x'e)z_0$; $e = [1, ..., 1]'$; and $\sigma^2 = x'Qx$. Besides, the optimal portfolio weights x satisfy the following relation:

$$x = Q^{-1}(\mu - z_0e)\frac{m - z_0}{(\mu - ez_0)'Q^{-1}(\mu - ez_0)}. \qquad (6.10)$$

Note that (6.9) and (6.10) have the same forms as the mean-variance frontier. However, even if Q is a symmetric matrix (it is positive definite), the estimator proposed in the sub-Gaussian cases (see formulas (6.5) and (6.6)) generally is not symmetric.

Therefore, we could obtain the inconsistent situation of stable distributions associated to portfolios $x'z$ whose square scale parameter estimator is lower than zero. However, most of the times $x'\widehat{Q}x > 0$ for every vector $x \in R^n$, because generally $\frac{\widehat{Q}+(\widehat{Q})'}{2}$ is a positive definite matrix[2]. This is the first reason for considering and studying the convergence properties of the estimator (see Rachev (1991)) and the suitability of the model. Moreover, (6.10) exhibits the two-fund separation property for both the stable and the normal case, but the matrix Q and the parameter σ have different meaning. In the normal case, Q is the variance-covariance matrix and σ is the standard deviation, while in the stable case Q is a dispersion matrix and $\sigma = \sqrt{x'Qx}$ is the scale (dispersion) parameter. According to the two-fund separation property of the sub-Gaussian α-stable approach, we can assume that the market portfolio is equal to the risky tangent portfolio under the equilibrium conditions (as in the classic mean-variance Capital Asset Pricing Model (CAPM)). Therefore, every optimal portfolio can be seen as the linear combination between the market portfolio

$$\overline{x}'z = \frac{z'Q^{-1}(\mu - z_0 e)}{e'Q^{-1}\mu - e'Q^{-1}ez_0}, \tag{6.11}$$

and the riskless asset return z_0. Following the same arguments as in Sharpe, Lintner, Mossin's mean-variance equilibrium model, the return of asset i is given by:

$$E(z_i) = z_0 + \beta_{i,m}(E(\overline{x}'z) - z_0), \tag{6.12}$$

where $\beta_{i,m} = \frac{\overline{x}'Qe^i}{\overline{x}'Q\overline{x}}$, with $e^i = [0, ..., 0, 1, 0, ..., 0]'$ the vector with 1 in the i-th component and zero in all the other components. As a consequence of the two fund separation property, we can apply Ross' necessary and sufficient conditions of the two-fund separation model (see Ross (1978)). Therefore, the above model admits the form

$$z_i = \mu_i + b_i Y + \varepsilon_i, \qquad i = 1, ..., n, \tag{6.13}$$

where $\mu_i = E(z_i)$, $E(\varepsilon/Y) = 0$, $\varepsilon = [\varepsilon_1, ..., \varepsilon_n]'$, $b = [b_1, ..., b_n]'$ and the vector $bY + \varepsilon$ is sub-Gaussian α-stable distributed with zero mean.

Hence, the above sub-Gaussian α-stable version of CAPM is not much different from Gamrowski and Rachev's (1999) version of the two-fund separation α-stable model. As a matter of fact, Gamrowski and Rachev (1999) propose a generalization of Fama's α-stable model (1965b) assuming $z_i = \mu_i + b_i Y + \varepsilon_i$, for every $i = 1, ..., n$, where ε_i and Y are α-stable distributed and $E(\varepsilon/Y) = 0$. In view of their assumptions,

---

[2] Observe that for every $x \in R^n$, we get $x'\widehat{Q}x > 0$ if and only if $\frac{(\widehat{Q}+(\widehat{Q})')}{2}$ is a positive definite matrix. Thus, we can verify that $\frac{(\widehat{Q}+(\widehat{Q})')}{2}$ is positive definite in order to avoid stable portfolios $x'z$ with negative scale parameter estimators. Moreover, we observe that the symmetric matrix $\frac{(\widehat{Q}+(\widehat{Q})')}{2}$ is an alternative estimator of the dispersion matrix Q whose statistical properties have to be proved.

$$E(z_i) = z_0 + \widetilde{\beta}_{i,m}(E(\overline{x}'z) - z_0),$$

where $\widetilde{\beta}_{i,m} = \dfrac{1}{\alpha\|\overline{x}'z\|_{\alpha}^{\alpha}}\dfrac{\partial\|\overline{x}'z\|_{\alpha}^{\alpha}}{\partial\overline{x}_i} = \dfrac{[\widetilde{z}_i, \overline{x}'z]_{\alpha}}{\|\overline{x}'z\|_{\alpha}^{\alpha}}$. Furthermore, the coefficient $\dfrac{[\widetilde{z}_i, \overline{x}'z]_{\alpha}}{\|\overline{x}'z\|_{\alpha}^{\alpha}}$ can be estimated using formula (6.3).

Differently from Gamrowski and Rachev we impose conditions on the joint distribution of asset returns, we observe that the sub-Gaussian model verifies the two-fund separation property and we apply Ross' necessary and sufficient conditions in order to obtain the linear pricing relation (6.13). In addition, observe that, in the above sub-Gaussian α-stable model, $\overline{x}'Q\overline{x} = \|\overline{x}'z\|_{\alpha}^{2}$ and $\overline{x}'Qe^{i} = \frac{1}{2}\dfrac{\partial\|\overline{x}'z\|_{\alpha}^{2}}{\partial\overline{x}_i}$. Thus, the coefficient $\beta_{i,m}$ of model (6.12) is equal to coefficient $\widetilde{\beta}_{i,m}$ of Gamrowski and Rachev's model, i.e.,

$$\beta_{i,m} = \frac{\overline{x}'Qe^{i}}{\overline{x}'Q\overline{x}} = \frac{1}{\sigma_{\overline{x}'z}}\frac{\partial\sigma_{\overline{x}'z}}{\partial\overline{x}_i} = \frac{[\widetilde{z}_i, \overline{x}'z]_{\alpha}}{\|\overline{x}'z\|_{\alpha}^{\alpha}} = \widetilde{\beta}_{i,m} ,$$

where $\sigma_{\overline{x}'z}$ is the scale parameter of the market portfolio.

Next, we can consider a discrete time extension of the two-parameter mean-dispersion analysis. Let W_{t_j} be the wealth of the investor at the beginning of the period $[t_j, t_{j+1})$, and let x_{i,t_j} $i = 1, ..., n$; $t_j = t_0, t_1, ..., t_{T-1}$ (with $t_0 = 0$ and $t_i < t_{i+1}$) be the amount invested in the i-th risky asset at the beginning of the time period $[t_j, t_{j+1})$. x_{0,t_j}; $t_j = 0, t_1, ..., t_{T-1}$ is the amount invested in the riskless asset at the beginning of the time period $[t_j, t_{j+1})$. We suppose that the wealth process is uniquely determined by two parameters: the mean and a dispersion parameter. Besides, we assume that the initial wealth $W_0 = \sum_{i=0}^{n}x_{i,0}$ is known and we point the vector of risky returns on the time period $[t_j, t_{j+1})$ as $z_{t_j} = [z_{1,t_j}, ..., z_{n,t_j}]'$. Therefore, a risk averse investor is seeking a better investment strategy $x_{0,t} = W_t - \sum_{i=1}^{n}x_{i,t}$ and $x_t = [x_{1,t}, ..., x_{n,t}]'$ for $t = t_0, t_1, ..., t_{T-1}$, so that the dispersion of the terminal wealth, $\sigma(W_{t_T})$, is minimized if the expected terminal wealth $E(W_{t_T})$ is not smaller than a pre-selected level. Mathematically, a mean-dispersion formulation for multiperiod portfolio selection can be posed as the following form:

$$\min_{\{x_t\}_{t=0,1,...,T-1}} \sigma(W_{t_T})$$
$$\text{s. t. } E(W_{t_T}) \geq m$$
$$W_{t_{k+1}} = \sum_{i=0}^{n}x_{i,t_k}(1 + z_{i,t_k})$$
$$= (1 + z_{0,t_k})W_{t_k} + x_{t_k}'p_{t_k} \quad k = 0, 1, 2, ..., T - 1, \tag{6.14}$$

where $p_t = [p_{1,t}, ..., p_{n,t}]'$ is the vector of excess returns $p_{i,t} = z_{i,t} - z_{0,t}$ for $t = t_0, t_1, ..., t_{T-1}$. Therefore, any risk averse investor will choose a strategy $\{x_{t_j}\}_{0 \leq j \leq T-1}$ among the solutions of problem (6.14) which maximizes his/her expected utility. Generally speaking, we can solve the above problem simulating a defined joint distribution of the return vector. However, in the case where the vectors of risky returns $z_t = [z_{1,t}, ..., z_{n,t}]'$ $t = 0, t_1, ..., t_{T-1}$ are statistically independent and we consider as a dispersion measure the variance of the final wealth W_{t_T}, Li and Ng

(2000) have proposed an analytical solution to problem (6.14) for any given m and W_{t_0}. As previously observed, the choice of the dispersion measure in problem (6.14) could be strategic either in order to consider the characteristics of the returns (such as heavy tails and asymmetry) or in order to reduce the complexity of the problem. For example, we can consider dispersion measures such as MAD (mean absolute deviation $\sigma(W_{t_T}) = E\left(\left|W_{t_T} - E\left(W_{t_T}\right)\right|\right)$) which do not imply the existence of the second moment finite[3]. Moreover, using the mean absolute deviation with respect to the variance, we can reduce the complexity of the problem (6.14). Suppose the vectors of returns $z_{t_j} = [z_{1,t_j}, ..., z_{n,t_j}]'$ $t_j = t_0, t_1, ..., t_{T-1}$ are statistically independent and α-stable sub-Gaussian distributed with $\alpha > 1$. That is z_t admits the characteristic function

$$\Phi_{z_t}(u) = \exp(-\left(u'Q_t u\right)^{\alpha/2} + iu'\mu_t)$$

where $\mu_t = E(z_t)$ and Q_t is the positive definite dispersion matrix associated to return z_t at time t for $t = t_0, t_1, ..., t_{T-1}$. As in the analysis proposed by Li and Ng (2000), we want to determine the optimal strategies $\{x_{t_j}\}_{0 \le j \le T-1}$ that minimize the dispersion of final wealth

$$W_{t_T} = W_0 \prod_{k=0}^{T-1}(1 + z_{0,t_k}) + \sum_{i=0}^{T-2} x'_{t_i} p_{t_i} \prod_{k=i+1}^{T-1}(1 + z_{0,t_k}) + x'_{t_{T-1}} p_{t_{T-1}}, \qquad (6.15)$$

for any fixed mean and initial wealth W_0. The multiperiod portfolio policies in the risky assets $x_{t_j} = [x_{1,t_j}, ..., x_{n,t_j}]'$ for any j are the deterministic variable of the problem, while the wealth invested in the riskless return at time t_j is given by $W_{t_j} - x'_{t_j} e$ where $W_{t_j} = \sum_{i=0}^{n} x_{i,t_{j-1}}(1 + z_{i,t_{j-1}})$ and $e = [1, ..., 1]'$. Thus, differently from Li and Ng (2000), we do not need to constrain the optimization problem at any intermediate time $t = t_1, ..., t_{T-1}$ with the wealth process relations $W_{t_j} = \sum_{i=0}^{n} x_{i,t_{j-1}}(1 + z_{i,t_{j-1}})$. Therefore, considering that the final wealth is determined by relation (6.15) and the statistically independent vectors of returns are α-stable sub-Gaussian distributed, for any given strategy $\{x_{t_i}\}_{0 \le i \le T-1}$, W_{t_T} is α-stable sub-Gaussian distributed with mean

$$\begin{aligned} E(W_{t_T}) = W_0 \prod_{k=0}^{T-1}(1 + z_{0,t_k}) \\ + \sum_{i=0}^{T-2} E(x'_{t_i} p_{t_i}) \prod_{k=i+1}^{T-1}(1 + z_{0,t_k}) + E(x'_{t_{T-1}} p_{t_{T-1}}), \end{aligned} \qquad (6.16)$$

and scale parameter $\sigma(W_{t_T})$ defined by

---

[3] The two parametric distribution families (for example, elliptical families with finite mean) could admit more than one parameterization (see Ortobelli (2001)). For example, for any fixed $\alpha > 1$ the stable sub-Gaussian distributions are uniquely determined either by the mean and the classical scale parameter or by the mean and the mean absolute deviation (MAD). Then we can choose the most opportune parameterization of the distribution family in order to improve and simplify the optimization model.

$$\sigma^{\alpha}(W_{tT}) = \sum_{i=0}^{T-2} \left(x_{t_i}' Q_{t_i} x_{t_i}\right)^{\alpha/2} \left(\prod_{k=i+1}^{T-1}(1+z_{0,t_k})\right)^{\alpha}$$
$$+ \left(x_{tT-1}' Q_{tT-1} x_{tT-1}\right)^{\alpha/2}. \tag{6.17}$$

Hence, in a market where unlimited short selling is allowed, any risk averse investor will choose a strategy $\{x_{t_j}\}_{0 \le j \le T-1}$ among the solutions of the following problem:

$$\min_{\{x_{t_j}\}_{j=0,1,\dots,T-1}} \quad \tfrac{1}{2}\sigma^{\alpha}(W_{tT}),$$
$$\text{s. t. } E(W_{tT}) = m. \tag{6.18}$$

Imposing the first order conditions on the Lagrangian $L(x_{t_j}, \lambda) = \sigma^{\alpha}(W_{tT}) - \lambda(E(W_{tT}) - m)$ and considering that:

1. $\left(x_{t_j}' Q_{t_j} x_{t_j}\right)^{\alpha-1} = \frac{4}{\alpha^2}\left(B_{j+1}\right)^{2-2\alpha}\lambda^2 C_{t_j}$ for $j = 0, 1, \dots, T-2$,

2. $\left(x_{tT-1}' Q_{t_j} x_{tT-1}\right)^{\alpha-1} = \frac{4}{\alpha^2}\lambda^2 C_{tT-1}$ and

3. $0 = \sum_{j=0}^{T-1} x_{t_j}' \frac{\partial L(x_{t_j}, \lambda)}{\partial x_{t_j}} = m - W_0 B_0 - \left(\frac{2}{\alpha}\lambda\right)^{\frac{1}{\alpha-1}} \sum_{j=0}^{T-1} \left(C_{t_j}\right)^{\frac{\alpha}{2(\alpha-1)}},$

where $B_i = \prod_{k=i}^{T-1}(1+z_{0,t_k})$ and $C_{t_j} = E(p_{t_j})' Q_{t_j}^{-1} E(p_{t_j})$, then we obtain the optimal multiperiod portfolio policy x_{t_j}:

$$x_{t_j} = \left(C_{t_j}\right)^{\frac{2-\alpha}{2(\alpha-1)}} \frac{m - W_0 B_0}{B_{j+1} \sum_{j=0}^{T-1}\left(C_{t_j}\right)^{\frac{\alpha}{2(\alpha-1)}}} Q_{t_j}^{-1} E(p_{t_j}),$$
$$\forall j = 0, 1, \dots, T-2, \tag{6.19}$$
$$x_{tT-1} = \left(C_{tT-1}\right)^{\frac{2-\alpha}{2(\alpha-1)}} \frac{m - W_0 B_0}{\sum_{j=0}^{T-1}\left(C_{t_j}\right)^{\frac{\alpha}{2(\alpha-1)}}} Q_{tT-1}^{-1} E(p_{tT-1}).$$

The wealth invested in the riskless asset at time t_0 is the deterministic quantity $W_0 - x_0'e$, while at time t_j it is given by the random variable $W_{t_j} - x_{t_j}'e$, where $W_{t_1} = (1+z_{0,0})W_0 + x_0' p_0$ and for any $j \ge 2$,

$$W_{t_j} = W_0 \prod_{k=0}^{j-1}(1+z_{0,t_k}) + \sum_{i=0}^{j-2} x_{t_i}' p_{t_i} \prod_{k=i+1}^{j-1}(1+z_{0,t_k}) + x_{t_{j-1}}' p_{t_{j-1}}, \tag{6.20}$$

Observe that in this case the results are much more useful than those obtained by Li and Ng's model because the investor can address future wealth in a more precise way.

Thus, as we could expect, the two-fund separation property holds at any time t_j. Moreover, when $\alpha = 2$ (i.e., we implicitly suppose that the vectors of returns are Gaussian distributed), the relation (6.17) gives the standard deviation of W_t and we obtain optimal portfolio policy x_{t_j} (6.19) for the mean-variance case. Thus, the main difference between the Stable and Gaussian multiperiod portfolio policies is represented by the risk factor $(C_{t_j})^{\frac{2-\alpha}{2(\alpha-1)}}$ which is equal to 1 in the Gaussian case.

6.2.2. A three fund separation model in the domain of attraction of a stable law

It is difficult to believe that the distributions of all admissible portfolios in the market are uniquely determined by two parameters. Next, suppose the vector of returns z is α-stable distributed i.e.,

$$W = x'z + (1 - x'e)z_0 \overset{d}{=} S_\alpha(\sigma_{x'z}, \beta_{x'z}, E(W))$$
$$\text{and } W = z_0 \text{ when } x = 0,$$

where $\beta_{x'z} = \dfrac{\int_{S_n} |x's|^{\langle\alpha\rangle} \gamma(ds)}{\sigma_{x'z}^\alpha}$ is the skewness parameter, α is the index of stability,

$\sigma_{x'z} = \left(\int_{S_n} |x's|^\alpha \gamma(ds)\right)^{1/\alpha}$ is the scale parameter, $\gamma(ds)$ is the spectral measure concentrated on S_n, and $E(W) = x'E(z) + (1 - x'e)z_0$. Under these assumptions we can only state that every solution of the risk averse investor's allocation problem is one of the solutions of the optimization problem (6.7). In order to find an analytical version of a three-parameter stable efficient frontier, we can consider the following three-fund separation model of security returns:

$$z_i = \mu_i + b_i Y + \varepsilon_i, \qquad i = 1, ..., n, \tag{6.21}$$

where the random vector $\varepsilon = (\varepsilon_1, \varepsilon_2, ..., \varepsilon_n)'$ is independent of Y and follows a joint sub-Gaussian α_1-stable distribution ($1 < \alpha_1 < 2$), with zero mean and characteristic function

$$\Phi_\varepsilon(t) = \exp\left(-|t'Qt|^{\frac{\alpha_1}{2}}\right),$$

where Q is the positive definite dispersion matrix. On the other hand,

$$Y \overset{d}{=} S_{\alpha_2}(\sigma_Y, \beta_Y, 0)$$

is an α_2-stable distributed random variable, independent of ε, with $1 < \alpha_2 < 2$ and zero mean. Under these assumptions, the portfolios are in the domain of attraction of an α-stable law with $\alpha = \min(\alpha_1, \alpha_2)$. A testable case in which Y is α_2-stable symmetric distributed (i.e., $\beta_Y = 0$), was recently studied by Götzenberger, Rachev and Schwartz (2001). When $\beta_Y = 0$ and $\alpha_1 = \alpha_2$, this model can lead to Fama's two-fund separation model. The characteristic function of the vector of returns $z = [z_1, z_2, ..., z_n]'$ is given by:

$$\begin{aligned}
\Phi_z(t) &= \Phi_\varepsilon(t)\Phi_Y(t'b)e^{it'\mu} = \exp\left(-|t'Qt|^{\frac{\alpha_1}{2}}\right. \\
&\left. - |t'b\sigma_Y|^{\alpha_2}\left(1 - i\beta_Y sgn(t'b)\tan\left(\tfrac{\pi\alpha_2}{2}\right)\right) + it'\mu\right),
\end{aligned} \tag{6.22}$$

where $b = [b_1, ..., b_n]'$ is the coefficient vector and $\mu = [\mu_1, ..., \mu_n]'$ is the mean vector.

Next, we shall estimate the parameter in model (6.21), (6.22). The estimator of μ is given by the vector $\widehat{\mu}$ of sample average. Then, we consider as factor Y a centralized index return (for example the centralized market portfolio (6.11) given by the above sub-Gaussian model). Regressing the centered returns $\widetilde{z}_j = z_j - \widehat{\mu}_j$ on Y we write the following OLS estimators for $b = [b_1, ..., b_n]'$ and Q:

$$\widehat{b}_i = \frac{\sum_{k=1}^{N} Y^{(k)} \widetilde{z}_i^{(k)}}{\sum_{k=1}^{N} (Y^{(k)})^2}; \quad i = 1,...,n, \tag{6.23}$$

$$\text{and } \widehat{Q} = \left[\frac{\widehat{q}_{i,j}}{2} \right]$$

where

$$\frac{\widehat{q}_{i,j}}{2} = \widehat{\sigma}_j^{2-p} \frac{\Gamma\left(1 - \frac{p}{2}\right)\sqrt{\pi}}{2^p \Gamma\left(1 - \frac{p}{\alpha}\right)\Gamma\left(\frac{p+1}{2}\right)} \frac{1}{N} \sum_{k=1}^{N} \widetilde{\varepsilon}_i^{(k)} \left(\widetilde{\varepsilon}_j^{(k)}\right)^{(p-1)}$$

$$\widehat{\sigma}_j^2 = \frac{\widehat{q}_{j,j}}{2} = \left(\frac{\Gamma\left(1 - \frac{p}{2}\right)\sqrt{\pi}}{2^p \Gamma\left(1 - \frac{p}{\alpha}\right)\Gamma\left(\frac{p+1}{2}\right)} \frac{1}{N} \sum_{k=1}^{N} \left|\widetilde{\varepsilon}_j^{(k)}\right|^p \right)^{\frac{2}{p}},$$

$p \in (1, \alpha_1)$ and $\widetilde{\varepsilon}^{(k)} = \widetilde{z}^{(k)} - \widehat{b} Y^{(k)}$ are the sample residuals. The asymptotic properties of the above estimator can be derived arguing similarly with Rachev (1991), Götzenberger, Rachev and Schwartz (2000), Tokat, Rachev and Schwartz (2003).

In order to determine portfolios that are R-S non-dominated when unlimited short selling is allowed, we have to minimize the scale parameter $\sigma_W = \sqrt{x'Qx}$ for some fixed mean $m_W = x'E(z) + (1 - x'e)z_0$ and $\widehat{b} = \frac{x'b}{\sqrt{x'Qx}}$ (see Ortobelli 2001). Alternatively, consider two portfolios $x'z$ and $y'z$ with the same mean, the same parameter $x'b = y'b$, so that $x'Qx > y'Qy$. Then, for any real t,

$$X_{/Y=t} = \frac{x'z - x'E(z) - x'bt}{\sqrt{x'Qx}} \overset{d}{=} \frac{x'\varepsilon}{\sqrt{x'Qx}} \overset{d}{=} \frac{y'\varepsilon}{\sqrt{y'Qy}} \overset{d}{=} S_{\alpha_1}(1, 0, 0).$$

Thus, $y'z$ dominates $x'z$ in the sense of Rothschild–Stiglitz because for every real v :

$$\int_{-\infty}^{v} \left(\Pr\left(y'z \leq s\right) - \Pr\left(x'z \leq s\right) \right) ds$$

$$= \int_{-\infty}^{v} \int_{R} \left(\Pr\left(X \leq \frac{s - y'E(z) - y'bt}{\sqrt{y'Qy}} \,\middle|\, Y = t \right) \right.$$

$$\left. - \Pr\left(X \leq \frac{s - x'E(z) - x'bt}{\sqrt{x'Qx}} \,\middle|\, Y = t \right) \right) f_Y(t) dt ds \tag{6.24}$$

$$= \int_{R} \int_{-\infty}^{v} \left(\Pr\left(X \leq \frac{s - y'E(z) - y'bt}{\sqrt{y'Qy}} \,\middle|\, Y = t \right) \right.$$

$$\left. - \Pr\left(X \leq \frac{s - x'E(z) - x'bt}{\sqrt{x'Qx}} \,\middle|\, Y = t \right) \right) ds f_Y(t) dt \leq 0$$

where f_Y is the density of Y. Therefore, when unlimited short selling is allowed, the efficient frontier of non-dominated portfolios is given by the solution of the following quadratic programming problem:

$$\min_x x'Qx \text{ subject to}$$
$$x'\mu + (1 - x'e)z_0 = m_W,$$
$$x'b = b^*$$

(6.25)

for some m_W and b^*. We know that generally the efficient set does not present an analytical form. In particular, Dybvig (1985), Markowitz (1959), and Bawa (1976–1978) show that the mean variance efficient set for the risk averse investors and for the non-satiable investors with restrictions on short sales, consists of segments which are parabolic or horizontal line segments. In addition, kinks in the efficient sets are the rule rather than the exception. We cannot expect the multi-parameter efficient set to take a simpler form. However, under our assumptions, every portfolio that maximizes the expected value of a given concave utility function u, belongs to the frontier

$$(1 - \lambda_2 - \lambda_3)\, z_0 + \lambda_2 \frac{z'Q^{-1}(\mu - z_0 e)}{e'Q^{-1}(\mu - z_0 e)} + \lambda_3 \frac{z'Q^{-1}b}{e'Q^{-1}b}$$

(6.26)

spanned by the riskless return z_0, and the two risky portfolios

$$u^{(1)} = \frac{z'Q^{-1}(\mu - z_0 e)}{e'Q^{-1}(\mu - z_0 e)} \quad \text{and} \quad u^{(2)} = \frac{z'Q^{-1}b}{e'Q^{-1}b}.$$

Similarly to other three-fund separation models (see, among others, Kraus and Litzenberger (1976), Ingersoll (1987), Simaan (1993)), we get a pricing linear relation for asset returns:

$$E(z_i) = \mu_i = z_0 + \tilde{b}_{i,2}\delta_2 + \tilde{b}_{i,3}\delta_3$$

where δ_p for $p = 1, 2$ are the risk premiums relative to a market factor and a skewness factor.

When $\alpha = \alpha_1 = \alpha_2 > 1$ in (6.22), every portfolio $x'z$ is an α-stable distribution and satisfies the relation

$$W = (1 - x'e)z_0 + x'z \overset{d}{=} S_\alpha(\sigma_{x'z}, \beta_{x'z}, (1 - x'e)z_0 + m_{x'z})$$

and $W = z_0$ when $x = 0$, where

$$\sigma_{x'z}^\alpha = (x'Qx)^{\frac{\alpha}{2}} + |x'b\sigma_Y|^\alpha, \quad \beta_{x'z} = \frac{|x'b\sigma_Y|^\alpha \, \text{sgn}(x'b)\beta_Y}{\sigma_{x'z}^\alpha}, \quad m_{x'z} = x'E(z)$$

(6.27)

and β_Y and σ_Y are respectively the skewness and the scale parameter of Y. Hence, this jointly α-stable model is a fund separation model whose optimal solutions are given by the optimization problem (6.7) or equivalently, by the quadratic programming problem (6.25).

As for the mean-dispersion model, we can consider a discrete time extension of the three-parameter analysis. In fact we get analogous results if we extend these considerations to a multiperiod analysis. Suppose the vectors of returns $z_t = [z_{1,t}, ..., z_{n,t}]'$ $t = t_0, t_1, ..., t_{T-1}$ are statistically independent of any t and follow the stable law

$$z_{i,t} = \mu_{i,t} + b_{i,t} Y + \varepsilon_{i,t}$$

where Y is an α_2-stable distributed asymmetric index return independent of vectors of residuals $\varepsilon_t = [\varepsilon_{1,t}, ..., \varepsilon_{n,t}]'$ which are statistically independent of any t and sub-Gaussian α_1-stable distributed.

Thus, the vector of returns z_t admits the following characteristic function:

$$\exp\left(- \left(u' Q_t u\right)^{\alpha_1/2} - |u' b_t \sigma_Y|^{\alpha_2} \left(1 - i \left(u' b_t \sigma_Y\right)^{\langle \alpha_2 \rangle} \beta_Y \tan \frac{\pi \alpha_2}{2}\right) + i u' \mu_t\right)$$

where $b_t = [b_{1,t}, ..., b_{n,t}]'$, $\mu_t = E(z_t)$, Q_t is the positive definite dispersion matrix associated to vector $\varepsilon_t = [\varepsilon_{1,t}, ..., \varepsilon_{n,t}]'$ at time t, σ_Y and β_Y are respectively the scale and the skewness parameter of the centred index return Y (independent of ε_t). In this case the final wealth is given by:

$$
\begin{aligned}
W_{t_T} &= W_0 \prod_{k=0}^{T-1}(1 + z_{0,t_k}) + \sum_{i=0}^{T-2} x'_{t_i} p_{t_i} \prod_{k=i+1}^{T-1}(1 + z_{0,t_k}) + x'_{t_{T-1}} p_{t_{T-1}} \\
&= W_0 \prod_{k=0}^{T-1}(1 + z_{0,t_k}) + \sum_{i=0}^{T-2} x'_{t_i} E(p_{t_i}) \prod_{k=i+1}^{T-1}(1 + z_{0,t_k}) \\
&\quad + x'_{t_{T-1}} E(p_{t_{T-1}}) + Y \left(\sum_{i=0}^{T-2} x'_{t_i} b_{t_i} \prod_{k=i+1}^{T-1}(1 + z_{0,t_k}) + x'_{t_{T-1}} b_{t_{T-1}}\right) \\
&\quad + \sum_{i=0}^{T-2} x'_{t_i} \varepsilon_{t_i} \prod_{k=i+1}^{T-1}(1 + z_{0,t_k}) + x'_{t_{T-1}} \varepsilon_{t_{T-1}},
\end{aligned}
$$

where $p_{t_i} = [p_{1,t_i}, ..., p_{n,t_i}]'$ is the vector of excess of returns $p_{k,t_i} = z_{k,t_i} - z_{0,t_i}$. Therefore, the mean of final wealth W_{t_T} is given by relation (6.16). While $\sigma_{\left(x'_{t_i} \varepsilon_{t_i}\right)}$ points out the scale parameter of the α_1-stable sub-Gaussian distribution $\sum_{i=0}^{T-2} x'_{t_i} \varepsilon_{t_i} \prod_{k=i+1}^{T-1}(1 + z_{0,t_k}) + x'_{t_{T-1}} \varepsilon_{t_{T-1}}$ and it is defined by

$$
\sigma^{\alpha_1}_{\left(x'_{t_i} \varepsilon_{t_i}\right)} = \sum_{i=0}^{T-2} \left(x'_{t_i} Q_{t_i} x_{t_i}\right)^{\alpha_1/2} \left(\prod_{k=i+1}^{T-1}(1 + z_{0,t_k})\right)^{\alpha_1}
$$
$$
+ \left(x'_{t_{T-1}} Q_{t_{T-1}} x_{t_{T-1}}\right)^{\alpha_1/2}.
$$

As for relation (6.24), when the random variable $W_{t_T}^{(1)}$ has:

1. the same asymmetry coefficient $\sum_{i=0}^{T-2} x'^{(1)}_{t_i} b_{t_i} \prod_{k=i+1}^{T-1}(1 + z_{0,t_k}) + x'^{(1)}_{t_{T-1}} b_{t_{T-1}}$ as the wealth $W_{t_T}^{(2)}$,

2. the same mean as the wealth $W_{t_T}^{(2)}$, and

3. $\sigma^{(1)}_{\left(x'^{(1)}_{t_i}\varepsilon_{t_i}\right)} < \sigma^{(2)}_{\left(x'^{(2)}_{t_i}\varepsilon_{t_i}\right)}$,

then every risk averse investor prefers $W^{(1)}_{t_T}$ to $W^{(2)}_{t_T}$. In particular, any risk averse investor will choose one of the multi-portfolio policies solutions of the following optimization problem for some m, v and W_0:

$$
\begin{aligned}
&\min_{\{x_{t_j}\}_{j=0,1,\ldots,T-1}} \tfrac{1}{2}\sigma^{\alpha_1}_{\left(x'_{t_i}\varepsilon_{t_i}\right)} \\
&\text{s. t. } E(W_{t_T}) = m; \\
&\textstyle\sum_{i=0}^{T-2} x'_{t_i} b_{t_i} \prod_{k=i+1}^{T-1}(1+z_{0,t_k}) + x'_{t_{T-1}} b_{t_{T-1}} = v.
\end{aligned}
\tag{6.28}
$$

Thus, all the multi-portfolio policies solutions of problem (6.28) are:

$$
x_{t_j} = \left(\frac{2}{\alpha_1}\right)^{\frac{1}{(\alpha_1-1)}} \frac{\left(\left(\lambda_1 E(p_{t_j})+\lambda_2 b_{t_j}\right)' Q_{t_j}^{-1}\left(\lambda_1 E(p_{t_j})+\lambda_2 b_{t_j}\right)\right)^{\frac{2-\alpha_1}{(\alpha_1-1)2}}}{B_{j+1}}
$$
$$
\times Q_{t_j}^{-1}\left(\lambda_1 E(p_{t_j})+\lambda_2 b_{t_j}\right),
$$
$$
\forall j = 0, 1, \ldots, T-2,
$$

$$
x_{t_{T-1}} = \left(\left(\lambda_1 E(p_{t_{T-1}})+\lambda_2 b_{t_{T-1}}\right)' Q_{t_{T-1}}^{-1}\left(\lambda_1 E(p_{t_{T-1}})+\lambda_2 b_{t_{T-1}}\right)\right)^{\frac{2-\alpha_1}{(\alpha_1-1)2}}
$$
$$
\times \left(\frac{2}{\alpha_1}\right)^{\frac{1}{(\alpha_1-1)}} Q_{t_{T-1}}^{-1}\left(\lambda_1 E(p_{t_{T-1}})+\lambda_2 b_{t_{T-1}}\right),
$$

where $B_i = \prod_{k=i}^{T-1}(1+z_{0,t_k})$ and λ_1, λ_2 are uniquely determined by the relations

$$
\textstyle\sum_{i=0}^{T-2} x'_{t_i} b_{t_i} B_{i+1} + x'_{t_{T-1}} b_{t_{T-1}} = v,
$$
$$
\textstyle\sum_{i=0}^{T-2} x'_{t_i} E(p_{t_i}) B_{i+1} + x'_{t_{T-1}} E(p_{t_{T-1}}) = m - W_0 B_0.
$$

Besides, the wealth invested in the riskless asset at the beginning of the period $[t_k, t_{k+1})$ is the deterministic wealth $W_0 - x'_0 e$ in t_0, while, for any $k \geq 1$, it is given by the random variable $W_{t_k} - x'_{t_k} e$, where W_{t_k} is determined by the relation (6.20). In particular, when the vector $\varepsilon_t = [\varepsilon_{1,t}, \ldots, \varepsilon_{n,t}]'$ is Gaussian distributed (i.e., $\alpha_1 = 2$), we obtain the following analytical solution to the optimization problem (6.28)

$$
x_{t_j} = \frac{(m-W_0 B_0)A - vD}{B_{j+1}(AC-D^2)} Q_{t_j}^{-1} E(p_{t_j}) + \frac{vC-(m-W_0 B_0)D}{B_{j+1}(AC-D^2)} Q_{t_j}^{-1} b_{t_j},
$$
$$
\forall j = 0, 1, \ldots, T-2,
$$
$$
x_{t_{T-1}} = \frac{(m-W_0 B_0)A - vD}{AC-D^2} Q_{t_{T-1}}^{-1} E(p_{t_{T-1}}) + \frac{vC-(m-W_0 B_0)D}{AC-D^2} Q_{t_{T-1}}^{-1} b_{t_{T-1}},
$$

where

$$
A = \textstyle\sum_{i=0}^{T-1} b'_{t_i} Q_{t_i}^{-1} b_{t_i}, \quad B_i = \prod_{k=i}^{T-1}(1+z_{0,t_k}),
$$
$$
C = \textstyle\sum_{i=0}^{T-1} E(p_{t_i})' Q_{t_i}^{-1} E(p_{t_i})
$$
$$
\text{and } D = \textstyle\sum_{i=0}^{T-1} E(p_{t_i})' Q_{t_i}^{-1} b_{t_i}.
$$

In both cases (Gaussian and stable non-Gaussian) the multi-portfolio policies in the risky assets x_{t_j} are spanned by vectors $Q_{t_j}^{-1} b_{t_j}$, $Q_{t_j}^{-1} E(p_{t_j})$, and at any time t_j the three-fund separation property holds.

We draw our conclusion on the three-fund separation model, underlining that in the three-moment model (see for example Simaan (1993)), the solution of any allocation problem depends on the choice of the asymmetric random variable Y. Clearly, one should expect that the optimal allocation will be different assuming that asset returns are in the domain of attraction of a stable law, or depend on the three moments.

6.2.2   A $k + 1$ fund separation model in the domain of attraction of a stable law

As empirical studies show, in the stable case one of the most severe restrictions of performance measurement and asset pricing is the use of a common index of stability for all assets (individual securities and portfolio alike). It is well understood that the asset returns are not normally distributed. We also know that the return distributions do not have the same index of stability. However, under the assumption that returns have different indexes of stability, it is not generally possible to find a closed form to the efficient frontier. Generalizing the above model instead, we get the following $k + 1$-fund separation model, (for details on k-fund separation models see Ross (1978)):

$$z_i = \mu_i + b_{i,1} Y_1 + \cdots + b_{i,k-1} Y_{k-1} + \varepsilon_i, \qquad i = 1, ..., n. \qquad (6.29)$$

Here, $n \geq k \geq 2$, the vector $\varepsilon = (\varepsilon_1, \varepsilon_2, ..., \varepsilon_n)'$ is independent of $Y_1, ..., Y_{k-1}$ and follows a joint sub-Gaussian symmetric α_k-stable distribution with $1 < \alpha_k < 2$, zero mean and characteristic function $\Phi_\varepsilon(t) = \exp\left(-\left|t' Q t\right|^{\frac{\alpha_k}{2}}\right)$, and the random variables $Y_j \overset{d}{=} S_{\alpha_j}\left(\sigma_{Y_j}, \beta_{Y_j}, 0\right)$, $j = 1, \ldots, k - 1$ are mutually independent[4] α_j-stable distributed with $1 < \alpha_j < 2$ and zero mean. If we need to insure the separation obtained in situations where the above model degenerates into a p-fund separation model with $p < k + 1$, we require the rank condition (see Ross (1978)). Using

---

[4] In order to estimate the parameters, we need to know the joint law of the vector $(Y_1, ..., Y_{k-1})$. Therefore, we assume independent random variables Y_j, $j = 1, \ldots, k-1$. Then the characteristic function of the vector of returns $z = [z_1, ..., z_n]'$ is given by

$$\Phi_z(t) = \Phi_\varepsilon(t) \prod_{j=1}^{k-1} \Phi_{Y_j}(t' b_{.,j}) e^{it'\mu}.$$

Under this additional assumption, we can approximate all parameters of any optimal portfolio using a similar procedure of the previous three-fund separation model. However, if we assume a given joint $(\alpha_1, ..., \alpha_{k-1})$ stable law for the vector $(Y_1, ..., Y_{k-1})$, we can generally determine estimators of the parameters by studying the characteristics of the multivariate stable law.

the same arguments of (6.24) we can prove that $y'z$ dominates $x'z$ in the sense of Rothschild–Stiglitz when the portfolios $y'z$ and $x'z$ have the same mean, the same parameters $x'b_{.,j} = y'b_{.,j} = c_j$ and $x'Qx > y'Qy$. Thus, when unlimited short selling is allowed every risk averse investor will choose the solution of the following quadratic programming problem:

$$
\min_x x'Qx \quad \text{subject to}
$$
$$
x'\mu + (1 - x'e)z_0 = m_W \quad , \tag{6.30}
$$
$$
x'b_{.,j} = c_j \quad j = 1, ..., k - 1
$$

for some m_W and c_j ($j = 1, ..., k - 1$). By solving the optimization problem (6.30), we obtain that the riskless portfolio and other k-risky portfolios span the efficient frontier which is given by

$$
\left(1 - \sum_{j=1}^{k} \lambda_j\right) z_0 + \lambda_1 \frac{z'Q^{-1}(\mu - z_0 e)}{e'Q^{-1}(\mu - z_0 e)} + \sum_{j=1}^{k-1} \lambda_{j+1} \frac{z'Q^{-1}b_{.,j}}{e'Q^{-1}b_{.,j}}.
$$

The above multivariate models are motivated by arbitrage considerations as in the Arbitrage Pricing Theory (APT) (see Ross (1976)). Without going into details, it should be noted that there are two versions of the APT for α-stable distributed returns, a so-called equilibrium (see Chen and Ingersoll (1983), Dybvig (1983), Grinblatt and Titman (1983)) and an asymptotic version (see Huberman (1982)). Connor (1984) and Milne (1988) introduced a general theory which encompassed the equilibrium APT as well as the mutual fund separation theory for returns belonging to any normed vector space (hence also α-stable distributed returns). While Gamrowski and Rachev (1999) provide the proof for the asymptotic version of α-stable distributed returns. Hence, it follows from Connor and Milne's theory that the above random law in the domain of attraction of a stable law of the return is coherent with the classic arbitrage pricing theory and the mean returns can be approximated by the linear pricing relation

$$
\mu_i \sim z_0 + \widetilde{b}_{i,1}\delta_1 + \cdots + \widetilde{b}_{i,k}\delta_k,
$$

where δ_p, for $p = 1, ..., k - 1$, are the risk premiums relative to the different factors.

6.3   A first comparison between Gaussian and stable dynamic strategies

In this section we examine the performances of Gaussian and stable non-Gaussian unconditional approaches and we compare the Gaussian and stable sub-Gaussian dynamic portfolio choice strategies when short sales are allowed. As a matter of fact, in the previous sections we have underlined and discussed the theoretical differences among portfolio choice models. Now, we evaluate and compare their real performances.

First, assuming that unlimited short sales are allowed, we analyze the optimal dynamic strategies during a period of five months, among the riskless return and eleven index-monthly returns from 01/01/88 until 01/30/98 (CAC 40,Coffee Brazilian, Corn No2 Yellow, DAX 100, DAX 30, Dow Jones Commodities, Dow Jones Industrials, Fuil Oil No2, Nikkei 225, Reuters Commodities, S&P 500). We start with a riskless asset with a 1.5% annual rate and we examine the different allocations considering five different implicit term structures. Table 1 describes the implicit term structures that we will use in this comparison.

TABLE 1. Implicit term structures

	$z_{0,0}$	$z_{0,1}$	$z_{0,2}$	$z_{0,3}$	$z_{0,4}$
Term 1	0.00125	0.00119	0.00113	0.00107	0.00101
Term 2	0.00125	0.00131	0.00137	0.00143	0.00149
Term 3	0.00125	0.00125	0.00125	0.00125	0.00125
Term 4	0.00125	0.00119	0.00113	0.00119	0.00125
Term 5	0.00125	0.00131	0.00137	0.00131	0.00125

In this analysis we approximate optimal solutions to the utility functional:

$$\max_{\{x_{t_j}\}_{j=0,1,\dots,T-1}} E(W_T) - cE\left(|W_T - E(W_T)|^{1.4}\right) \tag{6.31}$$

where c is an indicator of the aversion to the risk and W_T is defined by formula (6.15).

Secondly, we consider the negative exponential utility function

$$u(x) = -\exp(-\gamma x)$$

with risk aversion coefficient $\gamma > 0$. In this case, the absolute risk aversion function $\frac{-u''(x)}{u'(x)} = \gamma$ is constant. Hence, for every distributional model considered we are interested in finding optimal solutions to the functional

$$\max_{\{x_{t_j}\}_{j=0,1,\dots,T-1}} -E\left(\exp(-\gamma W_T)\right). \tag{6.32}$$

Observe that in the case of α-stable distributed returns with $1 < \alpha < 2$, the expected utility of formula (6.32) is infinite. However, assuming that the returns are truncated far enough, formula (6.32) is formally justified by pre-limit theorems (see Klebanov, Rachev, Szekely (1999) and Klebanov, Rachev, Safarian (2000)), which provide the theoretical basis for modeling heavy-tailed bounded random variables with stable distributions. On the other hand, it is obvious that the incomes are always bounded random variables. Typically, the investor works with a finite number of data so he/she can always approximate his/her expected utility. Therefore, we use diverse utility functions which differ in their absolute risk aversion functions and depend on a risk aversion coefficient. The presence of a parameter enables us to study the investor optimal portfolio selection for different degrees of risk aversion. Practically, we distinguish three separate steps in the decision process:

1. Choose the distributional model.
2. Calculate the optimal strategies of the multiperiod efficient frontier.
3. Express a preference among efficient portfolios. (In particular, we assume that the investor's distributional belief is not correlated to his/her expected utility. Therefore, the investor finds efficient frontiers assuming stable or Gaussian distributed returns, but his/her utility function can be any increasing concave utility function. This hypothesis is realistic enough because investors try to approximate their maximum expected utility among the efficient portfolios previously selected.)

We suppose the vectors of returns $z_{t_j} = [z_{1,t_j}, ..., z_{11,t_j}]'$ $t_j = 0, 1, ..., 4$ are statistically independent and α-stable sub-Gaussian distributed with $\alpha = 2$ (the Gaussian case) or $\alpha = \frac{1}{11} \sum_{i=1}^{11} \alpha_i = 1.7374$ and α_i the maximum likelihood estimates of indexes of stability which are reported in Table 2.

TABLE 2. Stable distributional parameters for unconditional monthly returns

	α	β	γ	δ
CAC 40	1.8981	0.2456	0.0337	0.0074
Coffee Brazilian	1.4142	-0.0964	0.0512	-0.0024
Corn No 2 Yellow	1.6804	-0.3997	0.0255	-0.0004
DAX 100	1.513	-0.2485	0.0259	0.0072
DAX 30	1.4812	-0.2375	0.0392	0.0072
Dow Jones Com.	1.9772	-1	0.0183	-0.0004
Dow Jones Industrials	1.8275	-0.5275	0.0222	0.0108
Fuil Oil No2	1.6997	0.1677	0.0556	0.0014
Nikkei 225	1.8903	-0.2546	0.0441	-0.0033
Reuters Com.	1.956	-0.9999	0.0208	-0.0012
S&P 500	1.7739	-0.3021	0.02	0.0098

Thus, we compare the performance of Gaussian and sub-Gaussian approaches for each optimal allocation proposed. In view of these comparisons, we discuss and study the differences in maximum expected utility for each allocation problem ((6.31) and (6.32)) for every portfolio choice model (Gaussian or sub-Gaussian) proposed.

Note that every model, Gaussian or sub-Gaussian (6.8), is based on a different risk perception. In order to compare the different models, we use the same algorithm proposed by Giacometti and Ortobelli (2003). Thus, first we consider the optimal strategies (6.19) for different levels of the mean. The efficient frontiers have been obtained for each model, discretizing properly the expected optimal final wealth. Second, we select the portfolio strategies on the efficient frontiers that maximize some parametric expected utility functions for different risk aversion coefficients.

Thus, we need to select portfolios belonging to the efficient frontiers such that:

$$x_t^* = \arg \left(\max_{x_t \in \text{efficient frontier}} E(u(W_T)) \right), \quad t = 0, 1, 2, 3, 4$$

where u is a given utility function. Finally, in Tables 3 and 4 we compare the maximum expected utility obtained with the stable or normal model for different risk aversion coefficients.

Therefore, considering N i.i.d. observations. $z^{(i)}$ ($i = 1, \dots, N$) of the vector $z_t = [z_{1,t}, z_{2,t}, \dots, z_{11,t}]'$, the main steps of our comparison are the following:

Step 1 Fit the five multiperiod efficient frontiers corresponding to the different distributional hypotheses: Gaussian and sub-Gaussian. Therefore, for every term structure we estimate 150 optimal portfolio weights varying the monthly mean m in the formulas of optimal portfolio policies:

$$x_t = (C_t)^{\frac{2-\alpha}{2(\alpha-1)}} \frac{m - W_0 B_0}{B_{t+1} \sum_{j=0}^{T-1} (C_j)^{\frac{\alpha}{2(\alpha-1)}}} Q_t^{-1} E(p_t),$$
$$\forall t = 0, 1, 2, 3, \tag{6.33}$$
$$x_4 = (C_4)^{\frac{2-\alpha}{2(\alpha-1)}} \frac{m - W_0 B_0}{\sum_{j=0}^{4} (C_j)^{\frac{\alpha}{2(\alpha-1)}}} Q_4^{-1} E(p_4)$$

where $W_0 = 1$. We assume constant over the time t the vector mean $E(z_t)$ and the dispersion matrix Q_t that is one of the estimated dispersion matrices (either the covariance matrix or α-stable sub-Gaussian dispersion matrix).

Step 2 Choose a utility function u with a given coefficient of aversion to risk.

Step 3 Calculate for every multiperiod efficient frontier (6.33)

$$\max_{\{x_j\}_{j=0,1,\dots,4}} \sum_{i=1}^{N} u\left(W_5^{(i)}\right)$$
$$\text{subject to}$$
$$\{x_j\}_{j=0,1,\dots,4} \text{ are portfolio strategies}$$
$$\text{that generate the efficient frontier,}$$

where $W_5^{(i)} = \prod_{k=0}^{4}(1 + z_{0,k}) + \sum_{j=0}^{3} x_j' p_j^{(i)} \prod_{k=j+1}^{4}(1 + z_{0,k}) + x_4' p_4^{(i)}$ is the i-th observation of the final wealth and $p_t^{(i)} = [p_{1,t}^{(i)}, \dots, p_{n,t}^{(i)}]'$ is the i-th observation of the vector of excess returns $p_{k,t}^{(i)} = z_{k,t}^{(i)} - z_{0,t}$ relative to the t-th period.

Step 4 Repeat steps 2 and 3 for every utility function and for every risk aversion coefficient.

Finally, we obtain Tables 3 and 4 with the approximated maximum expected utility. In fact, we implicitly assume the approximation

$$\frac{1}{N} \sum_{i=1}^{N} u\left(W_5^{(i)}\right) \approx E\left(u\left(W_5^{(i)}\right)\right).$$

Moreover in order to obtain significant results, we calibrate the risk aversion coefficients so that the optimal portfolio strategies which maximize the expected utility are optimal portfolios in the segment of the efficient frontier considered.

As we can observe from Tables 3 and 4 it follows that the sub-Gaussian models present a superior performance with respect to the mean-variance model. In addition, the differences obtained comparing the expected utility among the five term structures with the same return distributions are lower than those observed between the two distributional approaches. This ex-ante comparison confirms that the stable risk measure, the scale parameter σ, capture the data distributional behavior (typically the component of risk due to heavy tails) better than the Gaussian model. Moreover, this issue implicitly supports that stable distributions fit real data better than Gaussian distributions.

Even if in Tables 3 and 4 the stable sub-Gaussian approach does not seem to diverge significantly from the mean-variance approach, we could ascertain that optimal portfolio weights which maximize the expected utility in the different distributional frameworks are quite divergent as shown by Tables 5, 6, 7 and 8.

In Tables 5 and 6 we consider optimal choices of an investor with utility functional $E(W_5) - 2.5E\left(|W_5 - E(W_5)|^{1.4}\right)$ for five scenarios of term structures and with stable or Gaussian return distributions. Analogously, in Tables 7 and 8 we consider optimal choices of an investor with utility functional $E\left(-\exp(-30W_5)\right)$. The term structure determines the biggest differences in the portfolio weights of the same strategy and different periods. When the interest rates of the implicit term structure are growing (decreasing) we obtain that the investors are more (less) attracted to invest in the riskless asset in the sequent periods. Generally speaking, a common multiplicative factor does not exist between portfolio weights of different periods and of the same strategy. However, when we consider the flat term structure (3-rd term structure), the portfolio weights change over the time with the capitalization factor 1.00125. Moreover, the portfolio weights do not change excessively changing the implicit term structure, whilst we obtain greater differences considering different distributional hypotheses. The investor who approximates the returns with stable distributions invest less in the riskless asset than the investor who approximates the returns with Gaussian distributions. This behavior is compensated by a greater expected utility of the stable Paretian approach observed in the Tables 3 and 4.

TABLE 3. Expected utility of final wealth $E(W_5) - cE\left(|W_5 - E(W_5)|^{1.4}\right)$ assuming either Gaussian or stable sub-Gaussian distributed returns

	Term 1		Term 2	
Parameter c	Stable	Gaussian	Stable	Gaussian
2.5	1.0193353	1.01577256	1.0199845	1.016614189
3	1.0143304	1.01207174	1.01518337	1.013046658
3.5	1.0115582	1.01002216	1.01252421	1.011071059
4	1.0098849	1.00878464	1.01091906	1.009878245
4.6	1.0086397	1.00786409	1.00972471	1.008990315
	Term 3		Term 4	
Parameter c	Stable	Gaussian	Stable	Gaussian
2.5	1.0196455	1.01618152	1.01958453	1.016052814
3	1.0147477	1.01255174	1.01462064	1.01238181
3.5	1.0120351	1.01054142	1.01187158	1.010348667
4	1.0103975	1.00932735	1.01021203	1.009121343
4.6	1.009179	1.00842475	1.00897673	1.008207863
	Term 5			
Parameter c	Stable	Gaussian		
2.5	1.0197143	1.01631572		
3	1.0148794	1.01272506		
3.5	1.0122019	1.0107363		
4	1.0105852	1.00953597		
4.6	1.0093827	1.00864236		

TABLE 4. Expected utility of final wealth $E\left(-\exp(-\gamma W_5)\right)$ assuming either Gaussian or stable sub-Gaussian distributed returns

	Term 1		Term 2	
Parameter γ	Stable	Gaussian	Stable	Gaussian
18	-8.8305E-09	-9.6330E-09	-8.7187E-09	-9.4640E-09
22	-1.5811E-10	-1.7250E-10	-1.5536E-10	-1.6864E-10
26	-2.8311E-12	-3.0880E-12	-2.7684E-12	-3.0051E-12
30	-5.0692E-14	-5.5300E-14	-4.9332E-14	-5.3549E-14
34	-9.0766E-16	-9.9010E-16	-8.7905E-16	-9.5419E-16
	Term 3		Term 4	
Parameter γ	Stable	Gaussian	Stable	Gaussian
18	-8.7759E-09	-9.5497E-09	-8.7903E-09	-9.5766E-09
22	-1.5676E-10	-1.7058E-10	-1.5717E-10	-1.7123E-10
26	-2.8001E-12	-3.0470E-12	-2.8101E-12	-3.0614E-12
30	-5.0016E-14	-5.4425E-14	-5.0243E-14	-5.4737E-14
34	-8.9339E-16	-9.7219E-16	-8.9832E-16	-9.7867E-16
	Term 5			
Parameter γ	Stable	Gaussian		
18	-8.7607E-09	-9.5220E-09		
22	-1.5633E-10	-1.6992E-10		
26	-2.7898E-12	-3.0323E-12		
30	-4.9784E-14	-5.4111E-14		
34	-8.8841E-16	-9.6562E-16		

TABLE 5. Optimal portfolio strategies with expected utility $E(W_5) - cE\left(|W_5 - E(W_5)|^{1.4}\right)$ assuming Gaussian distributed returns and considering implicit term structures 1, 2 and 3

	Gaussian		Term 1		
	x_0	x_1	x_2	x_3	x_4
CAC 40	-0.0792	-0.0792	-0.07923	-0.07923	-0.0792
Coffee Bra.	0.04795	0.04765	0.04735	0.04704	0.04673
Corn No2	0.0189	0.01894	0.01897	0.01901	0.01904
DAX 100	-0.3988	-0.4001	-0.40127	-0.40245	-0.4036
DAX 30	0.54008	0.54141	0.5427	0.54396	0.5452
DowJonesCom.	-0.2154	-0.2111	-0.20669	-0.20228	-0.1979
DowJonesInd.	0.47216	0.47097	0.46974	0.46848	0.46719
Fuil Oil	0.0118	0.01213	0.01246	0.01279	0.01312
Nikkei 225	-0.1011	-0.1013	-0.10154	-0.10174	-0.1019
Reuters	-0.1228	-0.1208	-0.1188	-0.1168	-0.1148
S&P 500	0.0327	0.0375	0.0423	0.0471	0.0519
Riskless	0.7939	W1-0.216	W2-0.226	W3-0.236	W4-0.246

	Gaussian		Term 2		
	x_0	x_1	x_2	x_3	x_4
CAC 40	-0.0768	-0.0769	-0.07712	-0.07732	-0.0775
Coffee Bra.	0.04646	0.04686	0.04727	0.04768	0.0481
Corn No2	0.01832	0.01833	0.01834	0.01836	0.01837
DAX 100	-0.3865	-0.3863	-0.38606	-0.38588	-0.3857
DAX 30	0.52335	0.52337	0.52343	0.52352	0.52363
DowJonesCom.	-0.2088	-0.2135	-0.2183	-0.2231	-0.2279
DowJonesInd.	0.45753	0.45984	0.46218	0.46455	0.46695
Fuil Oil	0.01143	0.01114	0.01085	0.01056	0.01026
Nikkei 225	-0.098	-0.098	-0.09807	-0.09812	-0.0982
Reuters	-0.11901	-0.12124	-0.12348	-0.12573	-0.128
S&P 500	0.03164	0.02707	0.022486	0.01789	0.01328
Riskless	0.80027	W1-0.191	W2-0.182	W3-0.172	W4-0.163

	Gaussian		Term 3		
	x_0	x_1	x_2	x_3	x_4
CAC 40	-0.0776	-0.0777	-0.07784	-0.07794	-0.078
Coffee Bra.	0.047	0.04706	0.04712	0.04718	0.04724
Corn No2	0.01853	0.01856	0.01858	0.0186	0.01863
DAX 100	-0.391	-0.3915	-0.39196	-0.39245	-0.3929
DAX 30	0.52944	0.5301	0.53076	0.53143	0.53209
DowJonesCom.	-0.2112	-0.2115	-0.21173	-0.21199	-0.2123
DowJonesInd.	0.46286	0.46344	0.46402	0.4646	0.46518
Fuil Oil	0.01157	0.01158	0.0116	0.01161	0.01163
Nikkei 225	-0.0991	-0.0993	-0.09938	-0.0995	-0.0996
Reuters	-0.1204	-0.12055	-0.1207	-0.12085	-0.121
S&P 500	0.03201	0.03205	0.03209	0.03213	0.03217
Riskless	0.79795	W1-0.202	W2-0.2026	W3-0.2028	W4-0.203

TABLE 6. Optimal portfolio strategies with expected utility $E(W_5) - cE\left(|W_5 - E(W_5)|^{1.4}\right)$ assuming stable sub-Gaussian distributed returns and considering implicit term structures 1, 2 and 3

	Stable	Term 1		
	x_0	x_1	x_2	x_3	x_4
CAC 40	-0.1059	-0.10628	-0.10666	-0.107	-0.10741
Coffee Bra.	0.05768	0.05771	0.05773	0.05775	0.057771
Corn No2	0.03291	0.03347	0.03404	0.03461	0.035174
DAX 100	0.79483	0.79679	0.79873	0.80064	0.80252
DAX 30	-0.5637	-0.56542	-0.56708	-0.5687	-0.57036
DowJonesCom.	-0.2892	-0.28581	-0.28237	-0.2789	-0.27541
DowJonesInd.	0.63635	0.6354	0.63443	0.63342	0.632379
Fuil Oil	0.03441	0.03522	0.03603	0.03684	0.037651
Nikkei 225	-0.1386	-0.13914	-0.13965	-0.1402	-0.14065
Reuters	-0.1775	-0.1751	-0.1727	-0.1702	-0.1678
S&P 500	-0.0008	0.006	0.0128	0.0197	0.0265
Riskless	0.7196	W1-0.293	W2-0.305	W3-0.318	W4-0.330

	Stable	Term 2		
	x_0	x_1	x_2	x_3	x_4
CAC 40	-0.1037	-0.1036	-0.10349	-0.1034	-0.10332
Coffee Bra.	0.05648	0.0566	0.05672	0.05685	0.056982
Corn No2	0.03222	0.03175	0.03128	0.03081	0.030349
DAX 100	0.77835	0.77839	0.77851	0.77868	0.778927
DAX 30	-0.552	-0.55179	-0.55159	-0.5514	-0.55131
DowJonesCom.	-0.2832	-0.28728	-0.29136	-0.2955	-0.29962
DowJonesInd.	0.62315	0.62565	0.62821	0.63082	0.633485
Fuil Oil	0.0337	0.033	0.03229	0.03159	0.030896
Nikkei 225	-0.1358	-0.1356	-0.13546	-0.1353	-0.1352
Reuters	-0.1738	-0.177	-0.1793	-0.1821	-0.185
S&P 500	-0.0008	-0.007	-0.0141	-0.0208	-0.0275
Riskless	0.7254	W1-0.263	W2-0.252	W3-0.24	W4-0.229

	Stable	Term 3		
	x_0	x_1	x_2	x_3	x_4
CAC 40	-0.1049	-0.105	-0.10513	-0.1053	-0.1054
Coffee Bra.	0.05712	0.05719	0.05726	0.05733	0.057406
Corn No2	0.03259	0.03263	0.03267	0.03271	0.03275
DAX 100	0.7871	0.78808	0.78907	0.79006	0.791043
DAX 30	-0.5583	-0.55895	-0.55965	-0.5603	-0.56105
DowJonesCom.	-0.2864	-0.28677	-0.28713	-0.2875	-0.28784
DowJonesInd.	0.63016	0.63095	0.63174	0.63253	0.633318
Fuil Oil	0.03408	0.03412	0.03416	0.03421	0.03425
Nikkei 225	-0.1373	-0.13746	-0.13763	-0.1378	-0.13797
Reuters	-0.17573	-0.17595	-0.17617	-0.17639	-0.17661
S&P 500	-0.0008	-0.0008	-0.0008	-0.00081	-0.00081
Riskless	0.7223	W1-0.278	W2-0.2784	W3-0.279	W4-0.2791

TABLE 7. Optimal portfolio strategies with expected utility $E\left(-\exp(-\gamma W_5)\right)$ assuming either Gaussian or stable sub-Gaussian distributed returns and considering implicit term structures 3, 4 and 5

	Gaussian		Term 3		
	x_0	x_1	x_2	x_3	x_4
CAC 40	-0.0533	-0.0534	-0.05346	-0.05353	-0.0536
Coffee Bra.	0.03228	0.03232	0.03236	0.0324	0.03244
Corn No2	0.01273	0.01274	0.01276	0.01278	0.01279
DAX 100	-0.2685	-0.2688	-0.26918	-0.26952	-0.2699
DAX 30	0.3636	0.36406	0.36451	0.36497	0.36542
DowJonesCom.	-0.145	-0.1452	-0.14541	-0.14559	-0.1458
DowJonesInd.	0.31788	0.31827	0.31867	0.31907	0.31947
Fuil Oil	0.00794	0.00795	0.00796	0.00797	0.00798
Nikkei 225	-0.068	-0.068	-0.068	-0.068	-0.068
Reuters	-0.083	-0.083	-0.083	-0.083	-0.083
S&P 500	0.022	0.022	0.022	0.022	0.022
Riskless	0.861	W1-0.1389	W2-0.1391	W3-0.1393	W4-0.1395

	Gaussian		Term 4		
	x_0	x_1	x_2	x_3	x_4
CAC 40	-0.053	-0.053	-0.05298	-0.05309	-0.0532
Coffee Bra.	0.03206	0.03186	0.03166	0.03193	0.03221
Corn No2	0.01264	0.01266	0.01269	0.01269	0.0127
DAX 100	-0.2667	-0.2675	-0.26831	-0.26813	-0.268
DAX 30	0.36113	0.36202	0.36288	0.36286	0.36285
DowJonesCom.	-0.1441	-0.1411	-0.13821	-0.14147	-0.1447
DowJonesInd.	0.31572	0.31492	0.3141	0.31565	0.31722
Fuil Oil	0.00789	0.00811	0.00833	0.00813	0.00793
Nikkei 225	-0.0676	-0.0678	-0.0679	-0.06792	-0.0679
Reuters	-0.08212	-0.08079	-0.07945	-0.08098	-0.08252
S&P 500	0.02183	0.02504	0.02826	0.0251	0.02194
Riskless	0.86218	W1-0.144	W2-0.151	W3-0.145	W4-0.138

	Gaussian		Term 5		
	x_0	x_1	x_2	x_3	x_4
CAC 40	-0.0529	-0.053	-0.05317	-0.05319	-0.0532
Coffee Bra.	0.03203	0.03231	0.03259	0.0324	0.0322
Corn No2	0.01263	0.01264	0.01265	0.01267	0.0127
DAX 100	-0.2665	-0.2663	-0.26616	-0.26702	-0.26785
DAX 30	0.36081	0.36083	0.36087	0.3618	0.36271
DowJonesCom.	-0.1439	-0.1472	-0.1505	-0.1476	-0.14469
DowJonesInd.	0.31544	0.31703	0.31864	0.31788	0.3171
Fuil Oil	0.00788	0.00768	0.00748	0.0077	0.00792
Nikkei 225	-0.0676	-0.0676	-0.06761	-0.06777	-0.06791
Reuters	-0.08205	-0.08359	-0.08513	-0.08381	-0.08248
S&P 500	0.02182	0.01866	0.0155	0.01871	0.02193
Riskless	0.8623	W1-0.131	W2-0.125	W3-0.132	W4-0.138

TABLE 8. Optimal portfolio strategies with expected utility $E\left(-\exp(-\gamma W_5)\right)$ assuming either Gaussian or stable sub-Gaussian distributed returns and considering implicit term structures 3, 4 and 5

	Stable			Term 3	
	x_0	x_1	x_2	x_3	x_4
CAC 40	-0.0674	-0.06749	-0.06758	-0.0677	-0.06774
Coffee Bra.	0.03671	0.03676	0.03681	0.03685	0.036898
Corn No2	0.02095	0.02097	0.021	0.02102	0.02105
DAX 100	0.50591	0.50654	0.50718	0.50781	0.508445
DAX 30	-0.3588	-0.35927	-0.35972	-0.3602	-0.36062
DowJonesCom.	-0.1841	-0.18432	-0.18455	-0.1848	-0.18501
DowJonesInd.	0.40504	0.40554	0.40605	0.40656	0.407067
Fuil Oil	0.0219	0.02193	0.02196	0.02199	0.022014
Nikkei 225	-0.088	-0.088	-0.088	-0.089	-0.089
Reuters	-0.113	-0.113	-0.113	-0.113	-0.114
S&P 500	-0.001	-0.001	-0.001	-0.001	-0.001
Riskless	0.822	W1-0.1787	W2-0.1789	W3-0.1792	W4-0.1794
	Stable			Term 4	
	x_0	x_1	x_2	x_3	x_4
CAC 40	-0.0677	-0.06792	-0.06816	-0.0681	-0.068
Coffee Bra.	0.03686	0.03688	0.03689	0.03696	0.037037
Corn No2	0.02103	0.02139	0.02175	0.02144	0.02113
DAX 100	0.50795	0.5092	0.51043	0.51038	0.510368
DAX 30	-0.3603	-0.36134	-0.3624	-0.3622	-0.36198
DowJonesCom.	-0.1848	-0.18265	-0.18045	-0.1831	-0.18571
DowJonesInd.	0.40667	0.40606	0.40543	0.407	0.408606
Fuil Oil	0.02199	0.02251	0.02302	0.02256	0.022098
Nikkei 225	-0.0886	-0.08892	-0.08924	-0.0891	-0.08902
Reuters	-0.1134	-0.11188	-0.11034	-0.11214	-0.11394
S&P 500	-0.00052	0.00383	0.00819	0.00384	-0.00052
Riskless	0.82079	W1-0.187	W2-0.195	W3-0.188	W4-0.180
	Stable			Term 5	
	x_0	x_1	x_2	x_3	x_4
CAC 40	-0.0679	-0.06781	-0.06775	-0.068	-0.06824
Coffee Bra.	0.03698	0.03705	0.03713	0.03715	0.037169
Corn No2	0.02109	0.02079	0.02048	0.02084	0.021205
DAX 100	0.50951	0.50954	0.50961	0.51091	0.512184
DAX 30	-0.3614	-0.36121	-0.36107	-0.36217	-0.36327
DowJonesCom.	-0.1854	-0.18805	-0.19073	-0.18856	-0.18637
DowJonesInd.	0.40792	0.40955	0.41123	0.41065	0.41006
Fuil Oil	0.02206	0.0216	0.02114	0.02166	0.022176
Nikkei 225	-0.0889	-0.08877	-0.08867	-0.089	-0.08933
Reuters	-0.11375	-0.11557	-0.11739	-0.11588	-0.11435
S&P 500	-0.00052	-0.00487	-0.00923	-0.00489	-0.00052
Riskless	0.82024	W1-0.172	W2-0.165	W3-0.173	W4-0.181

6.4 Dynamic autoregressive models

The study and analysis of empirical behavior of data (see, among others, Akigiray (1989) Campbell (1987), Fama and French (1988) French, Schwert and Stambaugh (1987)) have reported evidence that conditional first and second moments of stock returns are time varying and potentially persistent especially when returns are measured over long horizons. However, the evidence observed on temporal dependence for daily stock returns is not strong (see among others Fama and French (1988)) and it is still a source of controversy whether stock market price models should include long memory or not (see among others Lo (1991)). In this sense the unconditional models of previous sections represent valid and indicative portfolio choice models when returns are measured over short horizons.

If financial modeling involves information on past market movements, it is not the unconditional return distribution which is of interest but the conditional distribution which is conditioned on information contained in past return data, or a more general information set. The class of auto-regressive moving average (ARMA) models is a natural candidate for conditioning on the past of a return series. These models have the property of the conditional distribution being homoskedastic (i.e., constant conditional volatility). If we model asset returns with an ARMA(p, q) model of autoregressive order p and a moving average order q, returns assume the following form:

$$z_t = a_0 + \sum_{i=1}^{p} a_i z_{t-i} + \varepsilon_t + \sum_{j=1}^{q} b_j \varepsilon_{t-j},$$

where $\{\varepsilon_t\}$ is a white noise process. To specify the orders p and q, we can follow the standard Box–Jenkins identification techniques (see among others, Box and Jenkins (1976), Brokwell and Davis (1991)) and we inspect the sample autocorrelation functions (SACFs) and sample partial autocorrelation functions (SPACFs) of the return series. In particular, we can assume that the sequence of innovations $\{\varepsilon_t\}$ is an infinite variance process consisting of i.i.d. symmetric random variables in the domain of normal attraction of a symmetric stable distribution with index of stability $\alpha \in (0, 2)$ and scale parameter $\sigma_0 > 0$. That is,

$$N^{-1/\alpha} \sum_{t=1}^{N} \varepsilon_t \xrightarrow{d} Y \text{ as } N \longrightarrow \infty,$$

where Y has characteristic function $\Phi_Y(u) = E(e^{iuY}) = e^{-|\sigma_0 u|^\alpha}$. Under these assumptions, Mikosh, Gadrich, Kluppelberg and Adler (1995) determined estimators for this process based on the sample periodogramm of z_t, and studied their asymptotic properties. Other estimators (a Gaussian Newton type and an M-estimator) for the parameters of the ARMA process with infinite variance innovations were proposed and studied by Davis (1996). It is interesting to observe that differently from ARMA processes with finite variance, in the stable case we generally obtain an estimator whose rate of convergence is considerably faster.

Next we consider a portfolio choice ARMA(1,1) model and we want to compare the impact of stable residuals and Gaussian ones. In particular we recall the recent Ortobelli, Biglova, Huber, Racheva, Stoyanov (2003) analysis. So we propose a dynamic portfolio choice among three risky indexes that Dow Jones Industrial, DAX 100 and FTSE all share. We consider 5000 portfolios of these assets and we assume that portfolio returns admit the form $x'z_t = a_0 + a_1 x' z_{t-1} + \varepsilon_t + b_1 \varepsilon_{t-1}$ where we suppose the sequence of innovations $\{\varepsilon_t\}$ are either stable or Gaussian distributed. We consider 803 daily observations of index returns from 01/04/1995 till 01/30/98 and for each portfolio we verify the stationarity and we estimate the parameters of the model. By a first comparison between the Gaussian and the stable non-Gaussian hypothesis it appears clear that stable distributions approximate much better the residuals than the Gaussian one. As a matter of fact, with the Kolmogorov–Smirnov test we could compare the empirical cumulative distribution $F_E(x)$ of several portfolio residuals with either a simulated Guassian or a simulated stable distribution. Therefore by this first analysis we can generally reject at a 5% level of confidence the hypothesis of normality of residuals because we obtain that probability, that the empirical distribution of the residuals is Gaussian, is on average (among different portfolios) equal to $1.2 \times 10^{-6} \ll 5\%$. In addition we could observe that in average (among different portfolios) $k = \sup_x |F_E(x) - F(x)| = 0.1875$ when $F(x)$ is the cumulative Gaussian distribution. These differences are also shown in Figure 1.

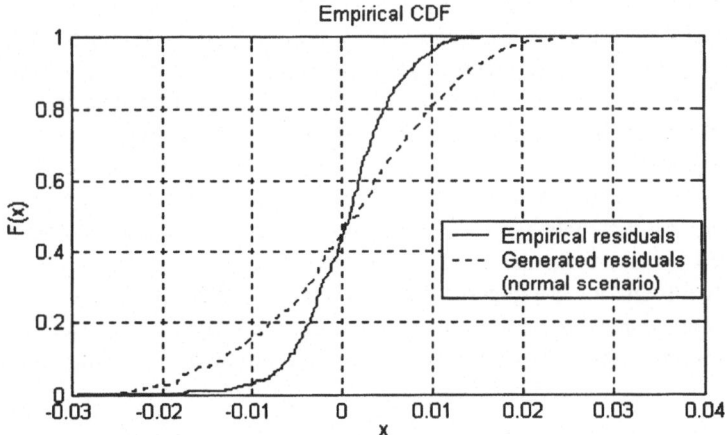

FIGURE 1. Kolmogorov–Smirnov test for residuals Gaussian distributed

Analogously, we cannot generally reject at a 5% level of confidence the hypothesis that residuals are stable distributed because we obtain this probability, that the empirical distribution of the residuals is stable distributed, is on average (among different portfolios) equal to 20.22%. In addition we could observe that on average (among different portfolios) $k = \sup_x |F_E(x) - F(x)| = 0.075$ when $F(x)$ is a cumulative stable distribution. These differences are also shown in Figure 2.

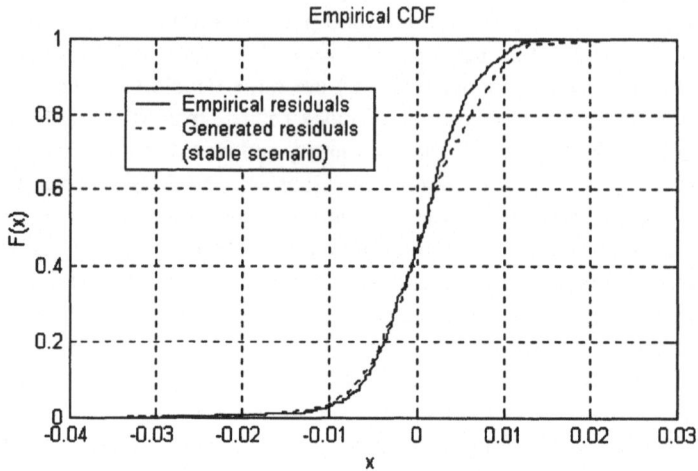

FIGURE 2. Kolmogorov–Smirnov test for residuals stable distributed

In order to value the impact of different distributional approximations in the investor's portfolio choices, we consider a dynamic asset allocation approach very similar to those proposed by Boender (1997) and Tokat, Rachev and Schwartz (2003). So we generate about 2500 initial asset allocations. These allocations are then simulated into the future by using the economic scenarios, which are generated under the Gaussian and stable assumptions for the innovations of the time series models. Future economic scenarios are simulated at daily intervals. One set of scenarios is generated by assuming that residuals of the variables are iid normal and another set of scenarios is generated by assuming that residuals are iid stable. The horizon of interest is 10 days and two scenarios are generated for each day, so 1024 possible economic scenarios are considered for each initial portfolio. The 10-day scenario tree is repeated 10000 times. We assume that investors wish to maximize the following functional of final wealth:

$$U(W_T) = E(W_T) - cE\left(|W_T - E(W_T)|^{1.5}\right)$$

where c is a coefficient of investor's risk aversion. Therefore, the investor will choose among the initial portfolios the portfolio weight vector $x = [x_1, x_2, x_3]'$ ($x_i \geq 0$, $x_1 + x_2 + x_3 = 1$) which maximizes the utility functional associated $U(W_T^x)$ considering that $E(W_T^x) = \frac{1}{S}\sum_{s=1}^{S} W_{s,T}^x$ is the mean of final wealth that we obtain with portfolio x; $E\left(|W_T^x - E(W_T^x)|\right) = \frac{1}{S}\sum_{s=1}^{S} \left|W_{s,T}^x - E(W_T^x)\right|$ is the measure of risk associated to x ; $W^x_{s,T} = \Pi^T_{t=1} R^x_{s,t}$ is the final wealth that we obtain with portfolio x under scenario $s \in \{1, 2, ..., S\}$; $R^x_{s,t} = x_1 z_{1,s,t} + x_2 z_{2,s,t} + x_3 z_{3,s,t}$ is the return of portfolio x under scenario $s \in \{1, 2, ..., S\}$ in time period t and $z_{i,s,t}$ is the rate of return of i-th asset under scenario $s \in \{1, 2, ..., S\}$ in time period t. In this empirical analysis we generally program with Delphi and Matlab.

Table 9 summarizes the results of this comparison. In particular we observe that for each risk aversion coefficient we obtain a greater expected utility using stable distributed residuals. This implicitly confirms that a better distributional approximation is permitted to implement the investor's choices. As for the previous empirical comparison we observe that there exist important differences between the portfolio allocations under different distributional approaches.

TABLE 9. Maximum $E(W_{10}) - cE(|W_{10} - E(W_{10})|)$ and portfolio composition considering return scenario generated with an ARMA(1,1) model under the Gaussian and stable assumptions for the innovations of the time series models

Risk aversion parameter c	Expected Utility	Stable Innovations		
		Optimal portfolio composition		
		Dow Jones Industrial	DAX 100	FTSE all share
0.1	0.00318	0.01643	0.96489	0.01866
0.3	-0.00286	0.01643	0.96489	0.01866
0.5	-0.006	0.06585	0.33784	0.5963
0.7	-0.00853	0.06585	0.33784	0.5963
0.8	-0.0097	0.05988	0.31699	0.62312
2.2	-0.0247	0.0813	0.21418	0.70451
Risk aversion parameter c	Expected Utility	Gaussian Innovations		
		Optimal portfolio composition		
		Dow Jones Industrial	DAX 100	FTSE all share
0.1	0.00106	0.004192	0.520012	0.475796
0.3	-0.00295	0.04557	0.305768	0.648662
0.5	-0.0061	0.047021	0.268504	0.684475
0.7	-0.00902	0.05344	0.288484	0.658076
0.8	-0.01013	0.053643	0.239928	0.706429
2.2	-0.02575	0.053643	0.239928	0.706429

A peculiar feature of ARMA models is that the conditional volatility is independent of past realizations. The assumption of conditional homoskedasticity is often violated in financial data where we could observe volatility clusters implying that a large absolute return is often followed by larger absolute returns, which more or less slowly decay. Such behavior can be captured by conditional heteroskedastic models, such as Engle's (Engle (1982)) auto-regressive conditional heteroskedastic (ARCH) models and Bollerslev's generalization (GARCH models, see Bollerslev (1986)). If we model asset returns with a GARCH (p, q) model which admits Gaussian innovations, then we assume that:

$$z_t = \sigma_t u_t \, ,$$

where $u_t \simeq N(0, 1)$ and

$$\sigma_t^2 = a_0 + \sum_{i=1}^{p} a_i z_{t-i}^2 + \sum_{j=1}^{q} b_j \sigma_{t-j}^2 \quad \sigma_t > 0.$$

When $q = 0$ we obtain an ARCH (p) process. A standard approach to detect GARCH dependencies in a time series y_t, is to compute the SACF of the squared series y_t^2. It turns out that GARCH-type models imply unconditional distributions which themselves possess heavier tails. Besides, there exists a relationship between unconditional stable and ARCH models with Gaussian innovations. As a matter of fact, in the limit an ARCH process has a Pareto tail and in many cases it is in the domain of attraction of stable or geometric stable laws see Kesten (1973), de Haan, Resnik, Rootzen and Vries (1989). However, as argued by McCulloch (1996), the GARCH models and α-stable distributions cannot be viewed as competing hypotheses. In fact, if conditional heteroskedasticity is present in the infinite variance stable case as well as in the Gaussian case, it is advisable to eliminate it. Recent studies have shown that GARCH-filtered residuals are themselves heavy tailed, so that α-stable distributed innovations would be a reasonable assumption. Davis and Resnik (1986, 1996), Davis, Mikosch and Basrak (1999), Mikosch and Starica (1998) studied the asymptotic behavior of sample auto-correlation functions of linear and non-linear stationary sequences with heavy-tailed distributions (*regularly varying finite-dimensional distributions*). Panorska, Mittnik and Rachev (1995) and Mittnik, Rachev and Paolella (1996) proposed the stable GARCH model and derived necessary and sufficient conditions for existence and uniqueness of stationary stable GARCH processes. In particular the model takes the form $z_t = \sigma_t u_t$ where u_t is symmetric α stable distributed with $\alpha > 1$ and

$$\sigma_t = a_0 + \sum_{i=1}^{p} a_i |z_{t-i}| + \sum_{j=1}^{q} b_j \sigma_{t-j} \quad \sigma_t > 0.$$

Many other autoregressive models were proposed in literature such as Integrated GARCH (IGARCH), Exponential GARCH (EGARCH), regime ARCH (SWARCH), just to name a few. A particular GARCH type model is the EWMA model that has been set out below in order to value the risk of a portfolio and to propose alternative portfolio choice models.

6.4.1   Value at Risk and portfolio choice with other conditional stable distributed returns

In this subsection, we recall an intertemporal extension of the above unconditional approaches proposed and tested by Lamantia, Ortobelli and Rachev (2003). First, we consider an exponentially weighted moving average model with α-stable distributions ($\alpha > 1$) (SEWMA) that generalize the classic EWMA model (see Longestaey and Zangari 1996). In particular, we assume that the conditional distribution of the centered continuously compounded returns vector $\tilde{z}_t = [\tilde{z}_{1,t}, ..., \tilde{z}_{n,t}]'$ is α-stable sub-Gaussian (in the period $[t-1, t]$) with characteristic function

$$\Phi_{\tilde{z}_t}(u) = \exp\left(-\left(u' Q_{t/t-1} u\right)^{\alpha/2}\right).$$

Thus, for any time t, the centered continuously compounded returns $\tilde{z}_{i,t}$ and the elements of matrix $Q_{t/t-1} = \left[\sigma_{ij,t/t-1}^2\right]$ are generated as follows:

$$\tilde{z}_{i,t} = \sigma_{ii,t/t-1}\varepsilon_{i,t}, \tag{6.34}$$

$$\sigma_{ii,t/t-1}^p = (1-\lambda)\left|\tilde{z}_{i,t-1}\right|^p A(p) + \lambda\sigma_{ii,t-1/t-2}^p,$$

$$B_{ij,t/t-1}(p) = (1-\lambda)\tilde{z}_{i,t-1}\left(\tilde{z}_{j,t-1}\right)^{\langle p-1 \rangle} A(p) + \lambda B_{ij,t-1/t-2}(p),$$

$$\sigma_{ij,t/t-1}^2 = B_{ij,t/t-1}(p)\sigma_{jj,t/t-1}^{2-p},$$

where $A(p) = \dfrac{\Gamma(1-\frac{p}{2})\sqrt{\pi}}{2^p\Gamma(1-\frac{p}{\alpha})\Gamma(\frac{p+1}{2})}$, λ is a parameter (decay factor) that regulates the weighting on past covariation parameters and $\varepsilon_{i,t} \sim S_\alpha(1,0,0)$. The time t stable scale parameter of i-th return $z_{i,t}$ is given by

$$\sigma_{ii,t/t-1} = \left(E_{t-1}(\left|\tilde{z}_{i,t}\right|^p)A(p)\right)^{1/p}$$

$$\simeq \left(A(p)(1-\lambda)\sum_{k=0}^{K}\lambda^{K-k}\left|\tilde{z}_{i,t-1-K+k}\right|^p\right)^{1/p},$$

where the number of effective observations K can be determined using similar arguments for the RiskMetrics model (see Longestaey and Zangari 1996). While the time t stable covariation parameter between the i-th and the j-th returns is given by $\sigma_{ij,t/t-1}^2$ and

$$B_{ij,t/t-1}(p) = E_{t-1}\left(\tilde{z}_{i,t}\left(\tilde{z}_{j,t}\right)^{\langle p-1 \rangle}\right)A(p)$$

$$\simeq \left(A(p)(1-\lambda)\sum_{k=0}^{K}\lambda^{K-k}\tilde{z}_{i,t-1-K+k}\left(\tilde{z}_{j,t-1-K+k}\right)^{\langle p-1 \rangle}\right).$$

Under these assumptions any portfolio is defined as $x'z_t = \sqrt{x'Q_{t/t-1}x}X_t$ where $X_t \sim S_\alpha(1,0,0)$ and $Q_{t/t-1} = \left[\sigma_{ij,t/t-1}^2\right]$.

When we assume that $\varepsilon_{i,t}$ is independent by $\varepsilon_{i,s}$ for any $t \neq s$ and $i = 1, ..., n$, then for any integer T the centered continuously compounded return in the period $[t, t+T]$ is given by

$$\tilde{Z}_{i,t+T} = \sum_{s=1}^{T} \tilde{z}_{i,t+s} = \sum_{s=1}^{T} \sigma_{ii,t+s/t+s-1} \varepsilon_{i,t+s} \overset{d}{=} \left(\sum_{s=1}^{T} \sigma_{ii,t+s/t+s-1}^{\alpha} \right)^{\frac{1}{\alpha}} V$$

where $V \sim S_{\alpha}(1, 0, 0)$. Thus, the return $\tilde{Z}_{i,t+T} = \sigma_{ii,t+T/t} V$ is conditional α-stable distributed with dispersion parameter defined by:

$$\sigma_{ii,t+T/t}^{p} = E_t(|\tilde{Z}_{i,t+T}|^p) A(p) = E_t \left(\left(\sum_{s=1}^{T} \sigma_{ii,t+s/t+s-1}^{\alpha} \right)^{\frac{p}{\alpha}} \right) \tag{6.35}$$

for every $p \in (1, \alpha)$. Therefore, assuming the vector $\tilde{Z}_{t+T} = [\tilde{Z}_{1,t+T}, \ldots, \tilde{Z}_{n,t+T}]'$ conditionally α-stable sub-Gaussian distributed such that[5] for any integer $s \in [1, T]$,

$$\lim_{p \to \alpha} E_t \left(\tilde{z}_{i,t+s} \left(\sum_{q=1}^{T} \tilde{z}_{j,t+q} \right)^{\langle p-1 \rangle} \right) A(p)$$
$$= \lim_{p \to \alpha} A(p) E_t \left(\tilde{z}_{i,t+s} \left(\tilde{z}_{j,t+s} \right)^{\langle p-1 \rangle} \right), \tag{6.36}$$

then the other elements of dispersion matrix $Q_{t+T/t} = \left[\sigma_{ij,t+T/t}^2 \right]$ are given by:

$$B_{ij,t+T/t}(p) = E_t \left(\tilde{Z}_{i,t+T} \left(\tilde{Z}_{j,t+T} \right)^{\langle p-1 \rangle} \right) A(p)$$
$$= \sum_{q=1}^{T} E_t \left(\tilde{z}_{i,t+q} \left(\sum_{s=1}^{T} \tilde{z}_{j,t+s} \right)^{\langle p-1 \rangle} \right), A(p) \tag{6.37}$$

$$\sigma_{ij,t+T/t}^2 = B_{ij,t+T/t}(p) \sigma_{jj,t+T/t}^{2-p}.$$

Using the expectation operator at time t, we can write the forecast parameters over $s \geq 2$ periods as

$$\begin{bmatrix} E_t(\sigma_{ii,t+s/t+s-1}^p) \\ E_t \left(B_{ij,t+s/t+s-1}(p) \right) \end{bmatrix} = \begin{bmatrix} \lambda & 0 \\ 0 & \lambda \end{bmatrix} \begin{bmatrix} E_t(\sigma_{ii,t+s-1/t+s-2}^p) \\ E_t \left(B_{ij,t+s-1/t+s-2}(p) \right) \end{bmatrix}$$
$$+ \begin{bmatrix} 1-\lambda & 0 \\ 0 & 1-\lambda \end{bmatrix} \begin{bmatrix} E_t(E_{t+s-2} \left(|\tilde{z}_{i,t+s-1}|^p \right) A(p)) \\ E_t \left(E_{t+s-2} \left(\tilde{z}_{i,t+s-1} \left(\tilde{z}_{j,t+s-1} \right)^{\langle p-1 \rangle} \right) A(p) \right) \end{bmatrix}$$
$$= \begin{bmatrix} E_t(\sigma_{ii,t+s-1/t+s-2}^p) \\ E_t \left(B_{ij,t+s-1/t+s-2}(p) \right) \end{bmatrix}.$$

---

[5] Assumption (6.36) is a form of independence that we require, in order to describe a time rule relation for the calculation of VaR with stable sub-Gaussian distributions.

Moreover, relations (6.35 and 6.37) are also verified for the limit of $p \uparrow \alpha$. Therefore, as a consequence of (6.36)

$$\sigma_{ii,t+T/t}^{\alpha} = \sum_{s=1}^{T} E_t \left(\sigma_{ii,t+s/t+s-1}^{\alpha} \right) = TE_t \left(\sigma_{ii,t+1/t}^{\alpha} \right) = T\sigma_{ii,t+1/t}^{\alpha},$$

$$B_{ij,t+T/t}(\alpha) = \sum_{s=1}^{T} \lim_{p \to \alpha} E_t \left(E_{t+s-1} \left(\tilde{z}_{i,t+s} \left(\tilde{z}_{j,t+s} \right)^{\langle p-1 \rangle} \right) \right) A(p)$$

$$= \sum_{s=1}^{T} E_t \left(B_{ij,t+s/t+s-1}(\alpha) \right) = T B_{ij,t+1/t}(\alpha)$$

and

$$\sigma_{ij,t+T/t}^{2} = B_{ij,t+T/t}(\alpha)\sigma_{jj,t+T/t}^{2-\alpha}$$

$$= T B_{ij,t+1/t}(\alpha) T^{\frac{2-\alpha}{\alpha}} \sigma_{jj,t+1/t}^{2-\alpha} = T^{\frac{2}{\alpha}} \sigma_{ij,t+1/t}^{2}.$$

Therefore even for the stable version of EWMA we obtain a temporal rule for the volatility and the covariation matrix is $Q_{t+T/t} = \left[\sigma_{ij,t+T/t}^{2} \right] = T^{\frac{2}{\alpha}} Q_{t+1/t}$. Clearly, we can apply this temporal rule to value the Value at Risk (VaR) at time $t + 1$ and at time $t + T$ of a portfolio $\tilde{z}_{p,t} = \sum_{i=1}^{n} w_i \tilde{z}_{i,t}$ as in the RiskMetrics exponentially weighted moving average model. In fact the $(1-\theta)\%$ VaR in the period $[t, t+1]$ is gotten simply by multiplying the corresponding percentile, $k_{1-\theta,\alpha}$ of the standardized α-stable $S_\alpha(1, 0, 0)$ for the forecast volatility $\sigma_{p,t+1/t} = \sqrt{w' Q_{t+1/t} w}$, that is

$$VaR_{t+1} = k_{1-\theta,\alpha} \sigma_{p,t+1/t}. \tag{6.38}$$

The $(1 - \theta)\%$ VaR in the period $[t, t + T]$ is simply given by

$$VaR_{t+T} = k_{1-\theta,\alpha} T^{\frac{1}{\alpha}} \sigma_{p,t+1/t} = T^{\frac{1}{\alpha}} VaR_{t+1}.$$

In the above model we need to estimate three parameters: the index of stability α, the decay factor λ and the parameter p. In order to avoid a too complex algorithm, we assume that these parameters are a particular mean of the parameters α_i, λ_i, p_i valued for every distinct series of data. The decay factor λ_i and the parameter p_i are estimated minimizing the *mean absolute deviation error* on the historical series of data. Notice that $A(1)E(|\tilde{z}_{i,t}|) = \sigma_{ii,t/t-1}$. Substantially, the procedure is the same of Longestaey and Zangari (1996), i.e., we value for any series

$$(\lambda_i, p_i) = \arg \left(\min_{\overline{\lambda}, \overline{p}} \sum_{t=1}^{T} \left| |\tilde{z}_{i,t}| - \frac{\sigma_{ii,t/t-1}(\overline{\lambda}, \overline{p})}{A(1)} \right| \right). \tag{6.39}$$

However, our empirical results do not change excessively valuing $(\lambda_i, p_i) = \arg\left(\min_{\overline{\lambda}, \overline{p}} \sum_{t=1}^{T} \left|\left|\tilde{z}_{i,t}\right| - \sigma_{ii,t/t-1}(\overline{\lambda}, \overline{p})\right|\right)$. There has been an extensive discussion among academics and practitioners on what error measure to use when assessing post-sample prediction (see Ahlburg (1992), Armstrong and Collopy (1992), Fildes (1992)). We use the mean absolute deviation error because we need a risk measure independent of parameter p that can be valid for stable distributed returns with indexes of stability $\alpha_i > 1$. Observe that $p_i \in [1, \alpha_i)$ and $\lambda_i \in [0, 1]$. Thus, we solve the optimization problem (6.39) discretizing $\overline{p}_i$ and $\overline{\lambda}_i$ with the same steps $\Delta\overline{p} \simeq 0.05$ and $\Delta\overline{\lambda} \simeq 0.01$ for every i. The optimal parameters λ, α and p are defined as:

$$\widehat{\lambda} = \sum_{i=1}^{n} \phi_i \lambda_i; \quad \widehat{p} = \sum_{i=1}^{n} \phi_i p_i; \quad \widehat{\alpha} = \sum_{i=1}^{n} \phi_i \alpha_i,$$

where $\phi_i = \frac{\theta_i}{\sum_{k=1}^{n} \theta_k}$; $\theta_i = \dfrac{\sum_{k=1}^{n} \min_{\overline{\lambda}, \overline{p}} \sum_{t=1}^{T} \left|\left|\tilde{z}_{k,t}\right| - \frac{\sigma_{kk,t/t-1}(\overline{\lambda}, \overline{p})}{A(1)}\right|}{\min_{\overline{\lambda}, \overline{p}} \sum_{t=1}^{T} \left|\left|\tilde{z}_{i,t}\right| - \frac{\sigma_{ii,t/t-1}(\overline{\lambda}, \overline{p})}{A(1)}\right|}$ and α_i are estimated

with the maximum likelihood method. Under these assumptions the index of stability α and the corresponding percentile, $k_{1-\theta, \alpha}$, of the standardized α-stable $S_\alpha(1, 0, 0)$ change over time. This fact assures a further flexibility in the prediction of the future VaR. However, for large portfolios it could be convenient to fix the parameters α and p. For example we could consider $\alpha = \frac{1}{n} \sum_{k=1}^{n} \alpha_i$ and $p = \max(1, \alpha - 0.3)$.

Under the assumptions of the SEWMA model, when $E(x_t' z_t) = E(y_t' z_t)$ and the random dispersions of two admissible portfolios satisfy the relation $x_t' Q_{t/t-1} x_t \leq y_t' Q_{t/t-1} y_t$, every risk averse investor prefers the portfolio $x_t' z_t$ to $y_t' z_t$. Assume the vectors of returns $z_t = [z_{1,t}, ..., z_{n,t}]'$ $t = t_0, t_1, ..., t_{T-1}$ satisfy the hypothesis of the SEWMA model (i.e., they are α-stable sub-Gaussian conditional distributed). Then, for any given strategy $\{x_{t_i}\}_{0 \leq i \leq T-1}$, the final wealth

$$W_{t_T} = W_0 \prod_{k=0}^{T-1}(1 + z_{0,t_k}) + \sum_{i=0}^{T-2} x_{t_i}' p_{t_i} \prod_{k=i+1}^{T-1}(1 + z_{0,t_k}) + x_{t_{T-1}}' p_{t_{T-1}},$$

is α-stable conditional distributed with mean (6.16) and scale parameter $\sigma(W_{t_T})$ defined by

$$\sigma^\alpha(W_{t_T}) = \sum_{i=0}^{T-2} (x_{t_i}' Q_{t_{i+1}/t_i} x_{t_i})^{\alpha/2} \left(\prod_{k=i+1}^{T-1}(1 + z_{0,t_k})\right)^\alpha$$
$$+ \left(x_{t_{T-1}}' Q_{t_T/t_{T-1}} x_{t_{T-1}}\right)^{\alpha/2}.$$

Thus, we obtain a relation similar to (6.17) where, instead of the matrix Q_{t_j}, we have to use the matrix Q_{t_{j+1}/t_j} for any $t_j = t_0, t_1, ..., t_{T-1}$. Substituting opportunely the matrix Q_{t_j}, the admissible optimal strategies of any risk averse investor are given by (6.19). Since the matrix Q_{t_{j+1}/t_j} is generally unknown for $j > 0$ we could approximate the optimal strategies using the expected dispersion $\widehat{Q}_{t_{j+1}/t_j}$ on the period $[t_j, t_{j+1}]$ predicted at time t_0, that is

$$\widehat{Q}_{t_{j+1}/t_j} = E_{t_0}\left(Q_{t_{j+1}/t_j}\right) = \left(t_{j+1} - t_j\right)^{\frac{2}{\alpha}} Q_{t_0+1/t_0}.$$

Therefore, the analysis of the previous sections substantially does not change except for the interpretation of the parameters. If we accept this approximation we obtain deterministic solutions of multi-portfolio policies solving the optimization problem (6.18). However, in order to express a portfolio choice coherent with the previous model we can propose a scenario generation portfolio choice. Thus, at each time (stage) t_j we can generate S possible scenarios of vector $z_{t_j}^{(s)} = [z_{1,t_j}^{(s)}, ..., z_{n,t_j}^{(s)}]'$; $s = 1, ..., S$ with characteristic function $\Phi_{z_{t_j}}(u) = \exp\left(-\left(u'Q_{t_j/t_{j-1}}u\right)^{\alpha/2} + iu'E(z_{t_j})\right)$. Every stage is linked to the previous one by the decay factor which is computed by considering the previous scenario. Therefore, if the horizon of interest is the T period, then S^T alternative economic scenarios are generated. In fact, $x_{0,t_k} + \sum_{i=1}^{n} x_{i,t_k} = W_{t_k}$ and $W_{t_j} = \sum_{i=0}^{n} x_{i,t_{j-1}}(1 + z_{i,t_{j-1}})$, then at time t_k for $k > 0$ we have S^k possible strategies $\left\{x_{t_i}^{(s_1,...,s_i)}\right\}_{0 \le i \le k}$ and for every possible wealth $W_{t_k}^{(s_1,...,s_k)} = \sum_{i=0}^{n} x_{i,t_k}^{(s_1,...,s_k)}$ we will obtain S possible levels of wealth

$$W_{t_{k+1}}^{(s_1,...,s_k,s)} = \sum_{i=1}^{n} x_{i,t_k}^{(s_1,...,s_k)}(1 + z_{i,t_k}^{(s)}) + x_{0,t_k}^{(s_1,...,s_k)}(1 + z_{0,t_{k-1}}),$$

at the k-th step. Thus, if we suppose that no short sales are allowed, an investor with initial wealth W_{t_0} and utility function u that wants to maximize his/her expected utility of the final wealth W_{t_T} will choose the multi-portfolio policy solution of the following optimization problem for some given initial wealth W_0:

$$\begin{aligned}
&\max_{\left\{x_t^{(s_1,...,s_t)}\right\}} \quad E(u(W_{t_T})) \quad \text{subject to} \\
&\qquad\qquad t=0,1,...,T-1; \\
&\qquad\qquad s_1,...,s_{T-1}=1,...,S \\
&x_{i,t_{k-1}}^{(s_1,...,s_{k-1})} \ge 0; \quad W_{t_{k-1}}^{(s_1,...,s_{k-1})} = \sum_{i=0}^{n} x_{i,t_{k-1}}^{(s_1,...,s_{k-1})} \\
&W_{t_k}^{(s_1,...,s_{k-1},s)} = \sum_{i=1}^{n} x_{i,t_{k-1}}^{(s_1,...,s_{k-1})}(1 + z_{i,t_{k-1}}^{(s)}) + x_{0,t_{k-1}}^{(s_1,...,s_{k-1})}(1 + z_{0,t_{k-1}}) \\
&\text{for every } s_1, ..., s_{T-1}, s = 1, ..., S; \quad k = 1, 2, ..., T
\end{aligned} \tag{6.40}$$

where $x_{i,t_0}^{(s_1,...,s_0)} := x_{i,0}$; $W_{t_0}^{(s_1,...,s_0)} := W_0$ and we approximate the expected utility considering a uniform distribution of scenarios, i.e.,

$$E(u(W_{t_T})) \simeq \frac{1}{S^T} \sum_{s_1,...,s_T=1}^{S} u(W_{t_k}^{(s_1,...,s_{T-1},s_T)}). \tag{6.41}$$

Alernatively, we can choose the policy which maximizes the expected utility of final wealth (6.41) among the optimal portfolio policies of the efficient frontier given by the solutions of the following optimization problem for some given mean m and initial wealth W_0:

$$\min_{\substack{\left\{x_t^{(s_1,\ldots,s_t)}\right\} \\ t=0,1,\ldots,T-1 \\ ;s_1,\ldots,s_{T-1}=1,\ldots,S}} \sum_{i=0}^{T-2} \left(x_{t_i}'^{(s_1,\ldots,s_i)} Q_{t_{i+1}/t_i}^{(s_1,\ldots,s_i)} x_{t_i}^{(s_1,\ldots,s_i)}\right)^{\alpha/2}$$

$$\times \left(\prod_{k=i+1}^{T-1}(1+z_{0,t_k})\right)^{\alpha} + \left(x_{t_{T-1}}'^{(s_1,\ldots,s_{T-1})} Q_{t_T/t_{T-1}}^{(s_1,\ldots,s_{T-1})} x_{t_{T-1}}^{(s_1,\ldots,s_{T-1})}\right)^{\alpha/2}$$

$$W_0 \prod_{k=0}^{T-1}(1+z_{0,t_k}) + E\left(x_{t_{T-1}}'^{(s_1,\ldots,s_{T-1})} p_{t_{T-1}}^{(s_1,\ldots,s_{T-1})}\right) \qquad (6.42)$$

$$+ \sum_{i=0}^{T-2} E\left(x_{t_i}'^{(s_1,\ldots,s_i)} p_{t_i}^{(s_1,\ldots,s_i)}\right) \prod_{k=i+1}^{T-1}(1+z_{0,t_k}) = m$$

$$x_{i,t_k}^{(s_1,\ldots,s_k)} \geq 0;$$

for every $s_1, \ldots, s_{T-1} = 1, \ldots, S; \quad k = 0, 1, 2, \ldots, T-!$

Clearly, if we consider a market where short sales are allowed, we can find admissible solutions in a reasonable computational time, otherwise we could need too much time to find solutions that fit the efficient frontier. In fact if short sales are allowed, for every given m and W_0, the optimal strategies $\left\{x_{t_i}^{(s_1,\ldots,s_i)}\right\}_{0 \leq i \leq k}$ are those, among the S^T possible strategies, given by the formula (6.19) that minimize the dispersion $\sigma^{\alpha}(W_{t_T})$ of final wealth, i.e.,

$$\sum_{i=0}^{T-2} \left(x_{t_i}'^{(s_1,\ldots,s_i)} Q_{t_{i+1}/t_i}^{(s_1,\ldots,s_i)} x_{t_i}^{(s_1,\ldots,s_i)}\right)^{\alpha/2} \left(\prod_{k=i+1}^{T-1}(1+z_{0,t_k})\right)^{\alpha}$$

$$+ \left(x_{t_{T-1}}'^{(s_1,\ldots,s_{T-1})} Q_{t_T/t_{T-1}}^{(s_1,\ldots,s_{T-1})} x_{t_{T-1}}^{(s_1,\ldots,s_{T-1})}\right)^{\alpha/2}.$$

Generally, the estimator $\widetilde{Q}_{t_j/t_{j-1}}$ of matrix $Q_{t_j/t_{j-1}}$ is not symmetric, even if, most of the times, the alternative estimator $\dfrac{\left(\widetilde{Q}_{t_j/t_{j-1}}+\left(\widetilde{Q}_{t_j/t_{j-1}}\right)'\right)}{2}$ is a positive definite matrix. Thus, generally, in order to simulate the vector $z_{t_j}^{(s)} = [z_{1,t_j}^{(s)}, \ldots, z_{n,t_j}^{(s)}]'$ of the conditional α-stable sub-Gaussian return distributions, we use the dispersion matrix $\widehat{\Sigma} = \dfrac{\left(\widetilde{Q}_{t_j/t_{j-1}}+\left(\widetilde{Q}_{t_j/t_{j-1}}\right)'\right)}{2}$. As a matter of fact, we first generate the vector $G = [G_1, \ldots, G_n]'$ of the joint Gaussian distribution $G = N\left(0, \widehat{\Sigma}\right)$ using the Cholesky decomposition matrix (see Jobson (1992)). Then, the vector of centered returns (see Samorodnitsky, Taqqu (1994)) is given by:

$$\widetilde{z} = \sqrt{A}G$$

where $A \overset{d}{=} S_{\alpha/2}\left(2\left(\cos\left(\frac{\pi\alpha}{4}\right)\right)^{2/\alpha}, 1, 0\right)$ is an $\alpha/2$ stable random variable independent of the Gaussian vector G.

Similarly, we can extend the three-fund separation model previously considered. In particular we assume that the conditional distribution of the centered continuously compounded return vector $\widetilde{z}_{t+1} = [\widetilde{z}_{1,t+1}, \ldots, \widetilde{z}_{n,t+1}]'$ is jointly α-stable with characteristic function

$$\Phi_{\tilde{z}_{t+1}}(u) = \exp\left(-\left((u'Q_{t+1/t}u)^{\alpha/2} + |u'b\sigma_Y|^\alpha\right)\right.$$

$$\left. \times \left(1 - i\frac{|u'b\sigma_Y|^\alpha \, sgn(u'b)\beta_Y}{(u'Q_{t+1/t}u)^{\alpha/2} + |u'b\sigma_Y|^\alpha} \, \tan\left(\frac{\pi\alpha}{2}\right)\right)\right).$$

That is the centered continuously compounded returns $\tilde{z}_{i,t}$ are generated as follows:

$$\tilde{z}_{i,t+1} = b_i Y_{t+1} + \sigma_{ii,t+1/t}\varepsilon_{i,t+1} = \left(\sigma^\alpha_{ii,t+1/t} + |b_i\sigma_Y|^\alpha\right)^{\frac{1}{\alpha}} X_{i,t+1},$$

$$\sigma^p_{ii,t+1/t} = (1-\lambda)\left|\tilde{z}_{i,t} - b_i Y_t\right|^p A(p) + \lambda\sigma^p_{ii,t/t-1},$$

$$B_{ij,t+1/t}(p) = (1-\lambda)\left(\tilde{z}_{i,t} - b_i Y_t\right)\left(\tilde{z}_{j,t} - b_j Y_t\right)^{\langle p-1\rangle} A(p) + \lambda B_{ij,t/t-1}(p),$$

$$\sigma^2_{ij,t+1/t} = B_{ij,t+1/t}(p)\sigma^{2-p}_{jj,t+1/t},$$

where $A(p) = \dfrac{\Gamma(1-\frac{p}{2})\sqrt{\pi}}{2^p\Gamma(1-\frac{p}{\alpha})\Gamma(\frac{p+1}{2})}$, $Y_t \sim S_\alpha(\sigma_{Y_t}, \beta_{Y_t}, 0)$, $t = 1, 2, \dots$ are i.i.d. observations of the centered index return that we assume to be α-stable asymmetric ($\beta_Y \neq 0$) random variables independent of residual vectors

$$\tilde{z}_t - bY_t = [\sigma_{11,t/t-1}\varepsilon_{1,t}, \dots, \sigma_{nn,t/t-1}\varepsilon_{n,t}]'.$$

The conditional distribution of the residual vector is sub-Gaussian α-stable and for any i and t, $\varepsilon_{i,t} \sim S_\alpha(1, 0, 0)$ and $X_{i,t} \sim S_\alpha\left(1, \dfrac{|b_i\sigma_Y|^\alpha sgn(b_i)\beta_Y}{(\sigma^\alpha_{ii,t/t-1} + |b_i\sigma_Y|^\alpha)}, 0\right)$. λ is a parameter (decay factor) that regulates the weighting on past covariation parameters. The vector $b = [b_1, b_2, \dots, b_n]'$ is estimated considering the OLS estimator (6.23)[6]. The forecast time $t+1$ stable scale parameter of i-th residual is given by:

---

[6] However, we can also assume that the vector b evolves over time such that,

$$b_{t+1/t} = [b_{1,t+1/t}, b_{2,t+1/t}, \dots, b_{n,t+1/t}]',$$

where

$$\widehat{b}_{i,t+1/t} = \frac{\widehat{b}^{(1)}_{i,t+1/t}}{\widehat{b}^{(2)}_{t+1/t}}; \quad i = 1, \dots, n,$$

$$\widehat{b}^{(1)}_{i,t+1/t} = \sum_{k=0}^{K} Y_{t-K+k}\tilde{z}_{i,t-K+k}$$

$$= \widehat{b}^{(1)}_{i,t/t-1} + Y_t\tilde{z}_{i,t} - Y_{t-K-1}\tilde{z}_{i,t-K-1},$$

$$\widehat{b}^{(2)}_{t+1/t} = \sum_{k=0}^{K}(Y_{t-K+k})^2 = \widehat{b}^{(2)}_{i,t/t-1} + Y_t^2 - Y_{t-K-1}^2.$$

$$\sigma_{ii,t+1/t} = \left(E_t(|\tilde{z}_{i,t+1} - b_i Y_{t+1}|^p) A(p)\right)^{1/p}$$
$$\simeq \left(A(p)(1-\lambda) \sum_{k=0}^{K} \lambda^{K-k} |\tilde{z}_{i,t-K+k} - b_i Y_{t-K+k}|^p\right)^{1/p}.$$

While the time $t+1$ stable covariation parameter between the i-th and the j-th residual is defined by $\sigma_{ij,t+1/t}^2$ and

$$B_{ij,t+1/t}(p) = A(p) E_t\left((\tilde{z}_{i,t+1} - b_i Y_{t+1})(\tilde{z}_{j,t+1} - b_j Y_{t+1})^{\langle p-1 \rangle}\right) \simeq A(p)$$
$$\times (1-\lambda) \sum_{k=0}^{K} \left(\lambda^{K-k}(\tilde{z}_{i,t-K+k} - b_i Y_{t-K+k})(\tilde{z}_{j,t-K+k} - b_j Y_{t-K+k})^{\langle p-1 \rangle}\right).$$

Under these assumptions the forecast $(1-\theta)\%$ VaR of portfolio

$$\tilde{z}_{p,t} = w'\tilde{z}_t = \sum_{i=1}^{n} w_i \tilde{z}_{i,t}$$

in the period $[t, t+1]$ is given by the corresponding $(1-\theta)$ percentile, of the α-stable distribution $S_\alpha\left(\sigma_{p,t+1/t}, \beta_{p,t+1/t}, 0\right)$ where

$$\sigma_{p,t+1/t} = \left((w' Q_{t+1/t} w)^{\alpha/2} + |w' b \sigma_Y|^\alpha\right)^{1/\alpha}$$

is the forecast volatility and

$$\beta_{p,t+1/t} = \frac{|w' b \sigma_Y|^\alpha \, sgn(w'b) \beta_Y}{(w' Q_{t+1/t} w)^{\alpha/2} + |w' b \sigma_Y|^\alpha}$$

is the forecast skewness. As for the SEWMA model, we can obtain a time rule for this three-parameter model (see Lamantia, Ortobelli, Rachev (2003)). Besides, using the results of the previous sections, we can describe a three-parameter conditional portfolio choice model, where the parameters are interpreted via the above formulas.

All the above conditional models provide symmetric and asymmetric parametric stable models with a fixed index of stability α. Recently, Rachev, Schwartz and Khindanova (2000) have proposed a conditional model for the vector of centered continuously compounded returns $\tilde{z}_t = [\tilde{z}_{1,t}, ..., \tilde{z}_{n,t}]'$ which is assumed to be in the domain of attraction of an $(\alpha_1, ..., \alpha_n)$-stable law. In particular they assume that $\tilde{z}_{i,t} \overset{d}{=} S_{\alpha_i}(\gamma_i, \beta_i, 0)$ for any i and t and $\beta_i \neq \pm 1$. Next, we simply refer to this as the RSK model. Thus, every centered return $\tilde{z}_{i,t}$ can be seen as the sum of two independent stable random variables, i.e.,

$$\tilde{z}_{i,t} \overset{d}{=} \tilde{z}_{i,t}^{(1)} + \tilde{z}_{i,t}^{(2)}, \tag{6.43}$$

where $\tilde{z}_{i,t}^{(1)} \overset{d}{=} S_{\alpha_i}((k^{\alpha_i}+1)^{-\frac{1}{\alpha_i}} \gamma_i, 0, 0)$; $\tilde{z}_{i,t}^{(2)} \overset{d}{=} S_{\alpha_i}(k(k^{\alpha_i}+1)^{-\frac{1}{\alpha_i}} \gamma_i, (1+k^{-\alpha_i})\beta_i, 0)$ for some $k > 0$ belonging to the interval $\left[\max_{1 \leq i \leq n}\left(\frac{|\beta_i|}{1-|\beta_i|}\right)^{\frac{1}{\alpha_i}}, +\infty\right)$. Hence given the portfolio,

$$\widetilde{z}_{p,t} \overset{d}{=} \widetilde{z}_{p,t}^{(1)} + \widetilde{z}_{p,t}^{(2)} = \sum_{i=1}^{n} w_i \widetilde{z}_{i,t}^{(1)} + \sum_{i=1}^{n} w_i \widetilde{z}_{i,t}^{(2)}, \tag{6.44}$$

where $\widetilde{z}_{p,t}^{(1)} = \sum_{i=1}^{n} w_i \widetilde{z}_{i,t}^{(1)}$ and $\widetilde{z}_{p,t}^{(2)} = \sum_{i=1}^{n} w_i \widetilde{z}_{i,t}^{(2)}$. Then, for any $i = 1, ..., n$, we get $\widetilde{z}_{i,t}^{(1)} \overset{d}{=} \sqrt{X_{i,t}} G_{i,t}$, where

$$G_{i,t} \overset{d}{=} N(0, \sigma_{G_{i,t}}) \text{ and}$$

$$X_{i,t} \overset{d}{=} S_{\alpha_i/2} \left(\frac{2(k^{\alpha_i} + 1)^{-\frac{2}{\alpha_i}} \gamma_i^2}{\sigma_{G_{i,t}}^2} \left(\cos\left(\frac{\pi \alpha_i}{4} \right) \right)^{2/\alpha_i}, 1, 0 \right)$$

are two independent random variables. We also assume that vector

$$G = [G_{1,t}, ..., G_{n,t}]'$$

is jointly Gaussian distributed, i.e., $G_t \overset{d}{=} N\left(0, \Sigma_{G_t}\right)$ where $\Sigma_{G_t} = [\sigma_{ii,t}^2]$ is the forecast variance covariance matrix of vector G_t. The vector G_t is not observable. We assume that the dependence structure of Gaussian variables $G_{i,t}$ is "inherited" from the dependence structure of truncated values of stable variables where the truncation values for $\widetilde{z}_{i,t}$ are sufficiently large in order to value the effect of outliers. Thus, we can introduce and control the dependence among different returns with the first component of portfolio $\widetilde{z}_{p,t}^{(1)}$. In fact, the variance and covariance matrix is given by the following rules:

$$\sigma_{ii,t/t-1}^2 = \lambda \sigma_{ii,t-1/t-2}^2 + (1 - \lambda)\widetilde{z}_{i,t-1}^2, \tag{6.45}$$

$$\sigma_{ij,t/t-1}^2 = \lambda \sigma_{ij,t-1/t-2}^2 + (1 - \lambda)\widetilde{z}_{i,t-1}\widetilde{z}_{j,t-1}. \tag{6.46}$$

Under these assumptions we can simulate any portfolio of vector $\widetilde{z}_t = [\widetilde{z}_{1,t}, ..., \widetilde{z}_{n,t}]'$ in the domain of attraction of one $(\alpha_1, ..., \alpha_n)$-stable law.

6.5   Backtest of VaR models

This section recalls some of the empirical results obtained by Lamantia, Ortobelli and Rachev (2003). In particular we present an analysis through backtest in order to assess the reliability of the models proposed to compute VaR. This analysis consists of splitting the historical data into two parts:

- The first part is used to estimate the dispersion matrix.
- The second part is used to verify the hypothesis that the VaR, computed at the beginning of each period, correctly forecasts the realization of the actual profit/loss occurred at the end of the period.

We examined daily returns during the period 15/11/93–30/01/98, for a total of 1100 observations. Then, over a period of 900 days, we calculated the interval of confidence at $\theta = 95\%$ and $\theta = 99\%$ of some randomly selected portfolios of ten indexes (Goldman Sachs, S&P500, Dow Jones Industrials, Corn no.2, Reuters Commodities, Brent Crude, CAC 40, FTSE all Share, DAX 100, Nikkei 500). We basically determined how many times during the period taken into account the profits/losses fall outside the confidence interval. In particular, for $\theta = 95\%$ and $\theta = 99\%$, the expected number of observations outside the confidence interval must not exceed respectively 5% and 1%.

Tables 10 and 11 show the results of the backtest for the two levels of confidence respectively. A first analysis demonstrates that the RiskMetrics model — hypothesis of normality of the conditional returns — underestimates the number of observations that fall outside the forecast interval. This effect is more evident when the percentiles are low and it confirms that the empirical distribution tails are heavier. In particular, in Table 10 we can compare the backtest results between EWMA and SEWMA and the stable asymmetric models. In this comparison we assume that the SEWMA stability index α varies over time as described above. Recall that the formula for the portfolio's value-at-risk bands is given by $VaR_{t+1} = \pm k_{1-\theta,\alpha}\sigma_{p,t+1/t}$ where $k_{1-\theta,\alpha}$ is the percentile of the standardized α-stable $S_\alpha(1, 0, 0)$. Therefore, if α varies over time, the VaR forecasts are much more sensitive to the latest market information and for this reason the sub-Gaussian model presents good performances for low and high percentiles. Moreover, for the backtesting results of the stable asymmetric model we use the DAX 30 series as index return Y and we fix the parameters α and p of the model respectively with $\alpha = \alpha_Y = 1.7557$ and $p = 1.45$. In this way the stable asymmetric model appears less flexible than the SEWMA model, even if it presents better performances than RiskMetrics.

Finally, in order to compare EWMA and RSK model performance we recall in Table 11 some backtest results obtained by Consiglio, Massabò and Ortobelli (2001). As for the previous models we observe that the stable Paretian hypothesis presents better performances than the Gaussian one in the VaR forecasts.

Among the alternative models for the VaR calculation, stable models are more reliable than that of RiskMetrics, in particular for confidence interval $\theta = 99\%$. The advantage of using stable models as an alternative to the normal one is reduced when the percentiles are higher than 5%. In this case, the percentage which has been realized is almost equal to that expected, except for the stable sub-Gaussian model that overestimates the losses.

6.6   Option pricing with stable distributions

The previous section dealt with models for the unconditional and conditional return distribution; this section discusses further stable models of option pricing.

A large part of modern finance has been concerned with modeling the evolution over time of return processes. Typically using subordinated processes, it is possible to capture empirically observed anomalies that contradict the classic log-normality

TABLE 10. The following Table obtained by Lamantia, Ortobelli, Rachev (2003) summarizes the backtesting results of RiskMetrics (Ewma), Sewma and stable asymmetric (SA) models for a level of confidence of $\theta = 95\%$ and $\theta = 99\%$

	Ewma θ=95%	Sewma θ=95%	SA θ=95%	Ewma θ=99%	Sewma θ=99%	SA θ=99%
Port 1	4.52%	1.13%	4.46%	1.63%	0.00%	1.11%
Port 2	5.40%	1.25%	5.91%	1.76%	0.00%	1.11%
Port 3	5.40%	0.63%	4.79%	1.76%	0.00%	0.89%
Port 4	6.65%	0.75%	6.69%	2.38%	0.13%	0.89%
Port 5	6.52%	1.13%	6.47%	2.13%	0.00%	1.11%
Port 6	4.64%	0.63%	4.57%	1.88%	0.00%	0.67%
Port 7	4.89%	0.50%	4.68%	1.76%	0.00%	0.89%
Port 8	6.27%	1.13%	4.91%	2.26%	0.13%	1.11%
Port 9	5.65%	0.88%	4.79%	1.88%	0.00%	1.11%
Port 10	4.77%	1.38%	5.02%	2.01%	0.00%	0.78%
Port 11	6.15%	1.00%	4.91%	2.51%	0.13%	1.23%
Port 12	5.90%	0.75%	6.13%	1.38%	0.00%	1.00%
Port 13	4.52%	1.00%	5.24%	1.63%	0.00%	1.00%
Port 14	5.02%	1.38%	4.57%	2.13%	0.13%	1.00%
Port 15	5.52%	0.13%	7.02%	1.51%	0.00%	0.56%
Port 16	5.14%	1.13%	4.57%	2.01%	0.13%	1.00%
Port 17	5.65%	1.25%	5.24%	2.38%	0.00%	0.78%
Port 18	5.90%	0.88%	7.02%	2.38%	0.00%	1.34%
Port 19	6.02%	1.63%	5.91%	2.38%	0.13%	1.23%
Port 20	5.40%	0.38%	6.35%	2.26%	0.00%	1.23%
Mean	5.50%	0.95%	5.46%	2.00%	0.04%	1.00%

assumption of asset prices. That is, we substitute the physical or calendar time for an intrinsic or operational time which provides tail effects as observed in the market. This means: if $W = \{W(t), t \geq 0\}$ is a stochastic process and $T = \{T(t), t \geq 0\}$ is a non-negative stochastic process, a new process $Z = \{Z(t) = W(T(t)), t \geq 0\}$ may be formed and it is defined as *subordinated* to W by the intrinsic time process T. The intrinsic time process $T = \{T(t), t \geq 0\}$ represents the cumulative trading volume process which measures the cumulative volume of all the transactions up to the physical time t. Whereas the process W represents the noise process introduced in the intrinsic time scale. When W is a standard Wiener process and the intrinsic time process T is the deterministic physical time (that is $T(t) = t$) we obtain the classic log-normal model (see Samuelson (1955) and Osborne (1959)). While subordinated models with random intrinsic time are leptokurtic with heavier tails than the normal distributions tails. When the intrinsic time process has non-negative stationary independent increments, then the subordinated process Z also has stationary independent increments (see Feller (1966)) and therefore it is a martingale.

TABLE 11. Summary obtained by Consiglio, Massabò, Ortobelli (2002) of the backtesting results of RiskMetrics and RSK models on some Italian stocks and portfolios. The stock series include 1044 observations from 01/01/96 to 28/09/00, for a total of 1044 observations. They have computed the interval of confidence of the last 977 days, for $\theta = 95\%$ and $\theta = 99\%$

	RSK $\theta=99\%$	Ewma $\theta=99\%$	RSK $\theta=95\%$	Ewma $\theta=95\%$
ALITALIA	0.10%	3.69%	7.57%	7.06%
BULGARI	0.00%	3.38%	5.94%	6.86%
EDISON	0.51%	2.15%	5.94%	7.27%
MEDIASET	0.00%	2.05%	4.20%	6.96%
MEDIOBANCA	0.92%	2.76%	7.37%	6.04%
OLIVETTI	0.10%	2.87%	5.63%	5.83%
ROLO BANCA	0.31%	3.17%	5.43%	6.65%
SAIPEM	0.10%	2.87%	4.81%	5.83%
TELECOM	0.41%	1.84%	4.40%	5.73%
TIM	1.95%	1.95%	6.86%	6.55%
PORT A	0.21%	2.15%	5.02%	5.94%
PORT B	0.31%	2.15%	6.14%	6.76%
PORT C	0.51%	2.46%	7.06%	6.55%
AVERAGE	0.42%	2.58%	5.87%	6.46%

Generally, we assume frictionless[7] markets where the log price process Z is subordinated to a standard Wiener process W by the independent intrinsic time process T. Thus, the asset price process $S(t)$ assumes the stochastic form

$$S(t) = S(t_0) \exp \left\{ \int_{t_0}^{t} \mu(s)ds + \int_{t_0}^{t} \rho(s)dT(s) + \int_{t_0}^{t} \sigma(s)dW(T(s)) \right\},$$

where the drift in the physical time scale $\mu(s)$, the drift in the intrinsic time scale $\rho(s)$ and the volatility $\sigma(s)$ are generally assumed to be constant. The appeal of processes subordinated to a standard Wiener process W by an intrinsic time process T with non-negative stationary independent increments is also due to the option pricing formula which follows from the classic Black and Scholes (B&S) formula in a frictionless complete market and from a risk minimizing strategy in incomplete markets[8]. In the Rachev–Hurst–Platen stable subordinated model the unique continuous martingale measure, making sense in a discrete time setting, is used, but a priori it

---

[7] Hedging strategies in the presence of taxes and transaction costs for subordinated processes can be studied adopting arguments similar to those used for log normal processes (see, among others, Gilster and Lee (1984), Leland (1985), Dybvig and Ross (1986), Ross (1987)).

[8] In incomplete markets non-redundant claims exist carrying an intrinsic risk. Then in order to value a contingent claim, we need to apply a risk-minimizing strategy that exists and

does not derive from a risk minimizing strategy even if the markets are incomplete (see Rachev and Mittnik (2000)). Typically, the value at time t_0 of a European call option with exercise price K and time to maturity t is given by:

$$C_t = S(t_0) F_+ \left(\ln \left(\frac{S(t_0)}{K_{r,t_0,t}} \right) \right) - K_{r,t_0,t} F_- \left(\ln \left(\frac{S(t_0)}{K_{r,t_0,t}} \right) \right), \qquad (6.47)$$

where $F_\pm(x) = \int_0^{+\infty} \Phi \left(\frac{x \pm \frac{1}{2} y}{\sqrt{y}} \right) d F_Y(y)$, Φ is the cumulative distribution of a standardized normal $N(0, 1)$, F_Y is the cumulative distribution of $Y = \int_{t_0}^t \sigma(s) dT(s)$; $K_{r,t_0,t} = K \exp \left(- \int_{t_0}^t r(s) ds \right)$ is the discounted exercise price, and the right-continuous with left-hand limits time-dependent function $r(t)$ defines the short term interest rate.

We recall that the Mandelbrot and Taylor subordinated processes Z, can be characterized by a simple analytical expression. Mandelbrot and Taylor (1967) required the intrinsic time process T to have stationary independent increments $T(t+s) - T(t) \overset{d}{=} S_{\frac{\alpha}{2}}(cs^{2/\alpha}, 1, 0)$ for all $s, t \geq 0$, $\alpha \in (0, 2)$ and $c > 0$. Where $\alpha/2$ is the index of stability, $cs^{\alpha/2}$ is the scale parameter, the skewness parameter is equal to 1 and the location parameter is equal to zero, that is T is a maximal positively skewed $\alpha/2$ stable Lévy process. It can be shown that for $\alpha \to 2$ the intrinsic time process T asymptotically approaches the physical time which leads us to the log-normal process. Under these assumptions the subordinated process Z is a symmetric α-stable motion with stationary independent increments $Z(t + s) - Z(t) \overset{d}{=} S_\alpha(\nu s^{\frac{1}{\alpha}}, 0, 0)$ for all $s, t > 0$, where $\nu = \frac{\sigma \sqrt{c}}{\sqrt{2}(\cos(\frac{\pi \alpha}{4}))^{\frac{1}{\alpha}}}$. If we consider the volatility σ constant, then the random variable Y in equation 6.47 is $Y = \sigma^2(T(t) - T(t_0)) = \lambda V$ where $\lambda = c\sigma^2(t - t_0)^{2/\alpha}$ and $V = S_{\frac{\alpha}{2}}(1, 1, 0)$. Hence, assuming that $c = 2 \left(\cos \left(\frac{\pi \alpha}{4} \right) \right)^{\frac{2}{\alpha}}$ it follows that $Z(t) \overset{d}{=} S_\alpha(\sigma t^{\frac{1}{\alpha}}, 0, 0)$. Thus, we can value the index of stability α and the volatility σ using the maximum likelihood estimates for daily returns and the daily volatility $\hat{\sigma}$ can be converted to an annual volatility $\hat{\sigma}_{ann} = \hat{\sigma} \times 365^{1/\hat{\alpha}}$. Moreover, considering the density function f_V of the $\alpha/2$ stable random variable V, we obtain with the transformation $v = \frac{u}{(1-u)^3}$ (that implies $u \in (0, 1)$ and

$$u = 1 + \sqrt[3]{-\frac{1}{2v} + \sqrt{\frac{1}{4v^2} + \frac{1}{27v^3}}} + \sqrt[3]{-\frac{1}{2v} - \sqrt{\frac{1}{4v^2} + \frac{1}{27v^3}}})$$

$$F_\pm(x) = \int_0^{+\infty} \Phi \left(\frac{x \pm \frac{1}{2} \lambda v}{\sqrt{\lambda v}} \right) f_V(v) dv$$

$$= \int_0^1 \Phi \left(\frac{x \pm \frac{1}{2} \lambda \frac{u}{(1-u)^3}}{\sqrt{\lambda \frac{u}{(1-u)^3}}} \right) f_V \left(\frac{u}{(1-u)^3} \right) \frac{1+2u}{(1-u)^4} du.$$

is unique in the martingale case (see Hofmann, Platen and Schweizer (1982), Föllmer and Sondermann (1986), Föllmer and Schweizer (1989)).

Therefore, $F_{\pm}(x)$ can now be numerically integrated over the finite interval $[0, 1]$. In this sense the efficiency of the stable subordinated model is strictly linked to the integral approximation. In particular, by comparing the integral calculation using different methodologies in the numerical integration, we observe that with the extended midpoint rule (see Press, Teukolsky, Vetterling and Flannery (1992)) and using the program STABLE to generate values of the density $f_V(v)$, we err by about 2×10^{-4} in the valuation of $F_{\pm}\left(\ln\left(\frac{S(t_0)}{K_{r,t_0,t}}\right)\right)$.

We observe that the stable option prices are generally greater than those obtained from the classic B&S model. We can identify at least two reasons: first in incomplete markets we cannot always replicate the option value thus the price has to consider this additional risk; while in the stable subordinated model the option price takes into account the risk due to the heavy tails of continuously compounded returns. Even if for subordinated models we cannot obtain the classic differential equation $r(t_0)f = r(t_0)S\frac{\partial f_t}{\partial S} + \frac{\partial f_t}{\partial t} + \frac{1}{2}\sigma^2 S^2 \frac{\partial^2 f_t}{\partial S^2}$, there exist other relations among Greek letters of a European option f with exercise price K and time to maturity t. Thus, as for the classic log normal distributional approach, we can monitor and manage the variation in the derivative asset price with respect to the parameters which enter in the option formula (i.e., the Greek letters). For example, assuming that the log price process evolves in the same way as a Mandelbrot and Taylor subordinated process, we obtain the relation

$$r(t_0)f = r(t_0)S\Delta_f + \Theta_f + \frac{\sigma}{\alpha(t - t_0)}\Lambda_f,$$

where the volatility σ is constant, Δ_f, Θ_f and Λ_f are respectively the *delta* $\frac{\partial f_t}{\partial S}$ the *theta* $\frac{\partial f_t}{\partial t}$ and the *vega* $\frac{\partial f_t}{\partial \sigma}$ of the option f. In particular, when the option f is an European call, we obtain the following *delta* $\Delta_c = \frac{\partial C_t}{\partial S} = F_+\left(\ln\left(\frac{S(t_0)}{K_{r,t_0,t}}\right)\right)$, *gamma*

$$\Gamma_c = \frac{\partial^2 C_t}{\partial S^2} = \frac{1}{S(t_0)} \int_0^{+\infty} \frac{1}{\sqrt{2\pi\lambda v}} \exp\left(-\left(\frac{\ln\left(\frac{S(t_0)}{K_{r,t_0,t}}\right) + \frac{1}{2}\lambda v}{\sqrt{2\lambda v}}\right)^2\right) f_V(v)dv$$

theta

$$\Theta_c = \frac{\partial C_t}{\partial t} = -r(t_0)K_{r,t_0,t}F_-\left(\ln\left(\frac{S(t_0)}{K_{r,t_0,t}}\right)\right)$$

$$-S(t_0)\int_0^{+\infty} \frac{\sigma\sqrt{cv}}{\alpha(t-t_0)^{\frac{\alpha-1}{\alpha}}\sqrt{2\pi}} \exp\left(-\left(\frac{\ln\left(\frac{S(t_0)}{K_{r,t_0,t}}\right) + \frac{1}{2}\lambda v}{\sqrt{2\lambda v}}\right)^2\right) f_V(v)dv$$

and *Vega*

$$\Lambda_c = \frac{\partial C_t}{\partial \sigma} = S(t_0) \int_0^{+\infty} \frac{(t - t_0)^{\frac{1}{\alpha}}\sqrt{cv}}{\sqrt{2\pi}}$$

$$\times \exp\left(-\left(\frac{\ln\left(\frac{S}{K_{r,t_0,t}}\right) + \frac{1}{2}\lambda v}{\sqrt{2\lambda v}}\right)^2\right) f_V(v)dv.$$

In addition, we can introduce a new Greek letter that is $\chi_f = \frac{\partial f}{\partial \alpha}$ and it describes the evolution of contingent claim prices as a function of the index of stability[9].

In order to compare the effects of the stable subordinated model and the classic Black–Scholes model, we describe under the two distributional hypothesis the dynamic hedging scheme applied to the S&P 500. In particular, we assume that a financial institution in data 31/10/97 has sold a European call option on 10000 shares of the S&P 500. The financial institution is faced with the problem of hedging its exposure. The S&P 500 price in data 31/10/97 was 914.62 \$, and we estimate the parameters of the models using daily data from 1/3/95 to 30/10/97. We obtain for the B&S model a weekly volatility $\sigma = 0.018559$ and for the stable subordinated model $\sigma = 0.012078$ and $\alpha = 1.7052$. We assume that the time to maturity is 13 weeks and we consider a 6% p.a. risk-free rate. Therefore the cost of hedging the option is given by

$$\sum_{i=1}^{n} \Delta(t_i) dS(t_i) \exp\left(\int_{t_i}^{t} r(s)ds\right) = \sum_{i=1}^{n} dC(t_i) \exp\left(\int_{t_i}^{t} r(s)ds\right),$$

when discounted to the beginning of the period must be close to, but not exactly the same as the option price. We call *effective cost of the delta hedging strategy* the difference between the discounted cost of the delta hedging strategy and the option cost. We want to compare the effective costs and the costs of the delta hedging strategies with stable or Gaussian models using the ex-post daily prices of S&P 500 and assuming several exercise prices. For example, if we apply a weekly dynamic delta hedging strategy considering an exercise price 970\$, that is the option closes in the money, we obtain Table 12 for the B&S model and for the stable subordinated model. As we can see in Table 12 at the end the cost of the stable delta hedging strategy (that is

---

[9] In the case of the European call we obtain

$$\chi_c = \frac{\partial C_t}{\partial \alpha} = -S(t_0) \int_0^{+\infty} \exp\left(-\left(\frac{\ln\left(\frac{S(t_0)}{K_{r,t_0,t}}\right) + \frac{1}{2}\lambda v}{\sqrt{2\lambda v}}\right)^2\right)$$

$$\times \frac{\sqrt{v}\left(\frac{\pi}{2\alpha}\tan\left(\frac{\pi\alpha}{4}\right) + \frac{2\ln((t-t_0)\cos\left(\frac{\pi\alpha}{4}\right))}{\alpha^2}\right)}{2\sqrt{\pi}} f_V(v)dv$$

$$+ \int_0^{+\infty} \left(S(t_0)\Phi\left(\frac{\ln\left(\frac{S(t_0)}{K_{r,t_0,t}}\right) + \frac{1}{2}\lambda v}{\sqrt{\lambda v}}\right)\right.$$

$$\left. - K_{r,t_0,t}\Phi\left(\frac{\ln\left(\frac{S(t_0)}{K_{r,t_0,t}}\right) - \frac{1}{2}\lambda v}{\sqrt{\lambda v}}\right)\right) \frac{\partial f_V(v)}{\partial \alpha} dv.$$

Moreover we could obtain a more pricise expression for $\frac{\partial f_V(v)}{\partial \alpha}$ using the density formula in Nolan (1997).

TABLE 12. Scheme of weekly delta hedging strategies (with both the Black and Scholes model and the stable subordinated model) considering an exercise price of 970$, that is option closes in the money

		Black and Scholes Model				
					Cost of Cumul. Int.	
Data	S&P 500 Price	Delta	Shares purch.	shares purch.	costs (incl.Int.)	cost
31/10/1997	914.62	0.267333	2673.33	2445084	2445084	2821.25
7/11/1997	927.51	0.326644	593.11	550115.6	2998021	3459.26
14/11/1997	928.35	0.317034	-96.10	-89217.7	2912262	3360.3
21/11/1997	963.09	0.541468	2244.33	2161500	5077123	5858.22
28/11/1997	955.4	0.47686	-646.08	-617264	4465717	5152.75
5/12/1997	983.79	0.681182	2043.22	2010102	6480971	7478.04
12/12/1997	953.39	0.435374	-2458.08	-2343508	4144941	4782.62
19/12/1997	946.78	0.360189	-751.85	-711836	3437888	3966.79
26/12/1997	936.46	0.245674	-1145.15	-1072391	2369464	2734
2/1/1998	975.04	0.611227	3655.53	3564286	5936484	6849.79
9/1/1998	927.69	0.103182	-5080.45	-4713079	1230255	1419.53
16/01/1998	961.51	0.407534	3043.52	2926375	4158049	4797.75
23/01/1998	957.59	0.266858	-1406.76	-1347099	2815748	3248.94
30/01/1998	980.28	1	7331.42	7186843	10005840	0
				Cost of hedging	305840	

		Stable Model				
					Cost of Cumul. Int.	
Data	S&P 500 Price	Delta	Shares purch.	shares purch.	costs (incl.Int.)	cost
31/10/1997	914.62	0.23985	2398.50	2193718	2193718	2531.21
7/11/1997	927.51	0.287018	471.68	437489.7	2633739	3038.93
14/11/1997	928.35	0.280085	-69.33	-64365.1	2572413	2968.17
21/11/1997	963.09	0.47221	1921.25	1850337	4425719	5106.6
28/11/1997	955.4	0.417348	-548.62	-524150	3906674	4507.7
5/12/1997	983.79	0.61034	1929.92	1898633	5809815	6703.63
12/12/1997	953.39	0.385247	-2250.93	-2146012	3670506	4235.2
19/12/1997	946.78	0.320851	-643.96	-609688	3065054	3536.6
26/12/1997	936.46	0.223122	-977.29	-915191	2153399	2484.69
2/1/1998	975.04	0.563066	3399.44	3314588	5470472	6312.08
9/1/1998	927.69	0.102598	-4604.69	-4271722	1205062	1390.46
16/01/1998	961.51	0.371473	2688.75	2585261	3791714	4375.05
23/01/1998	957.59	0.231663	-1398.10	-1338803	2457285	2835.33
30/01/1998	980.28	1	7683.37	7531854	9991974	0
				Cost of hedging	291974	

TABLE 13. Effective costs and costs of dynamic weekly and daily delta hedging strategies considering the Black and Scholes model and the stable subordinated model

	Weekly	Weekly	Daily	Daily
Exercise price (in the money)	Stable effective cost	Gaussian effective cost	Stable Effective Cost	Gaussian effective cost
960	149058.9	231079.6	39452.1	152387.9
965	135043.3	222424.3	31215.3	134605.4
970	111354.1	204851	18368.4	113852.2
975	77915.7	177886.9	-205.27	90101.23
980	35921.17	141875.5	-30353	63381.18
Exercise price (out the money)	Stable effective cost	Gaussian effective cost	Stable Effective Cost	Gaussian effective cost
985	33797.82	144366.2	-30524	80270.71
990	29869.32	143127.2	-38226	65127.78
995	23169.26	137067.1	-45747	59105.99
1000	15066.89	127767.5	-50652	52295.8
1005	6523.959	116571.9	-54171	45720.78
	Weekly	Weekly	Daily	Daily
Exercise price (in the money)	Stable effective cost	Gaussian effective cost	Stable Effective Cost	Gaussian effective cost
960	149058.9	231079.6	39452.1	152387.9
965	135043.3	222424.3	31215.3	134605.4
970	111354.1	204851	18368.4	113852.2
975	77915.7	177886.9	-205.27	90101.23
980	35921.17	141875.5	-30353	63381.18
Exercise price (out the money)	Stable effective cost	Gaussian effective cost	Stable Effective Cost	Gaussian effective cost
985	33797.82	144366.2	-30524	80270.71
990	29869.32	143127.2	-38226	65127.78
995	23169.26	137067.1	-45747	59105.99
1000	15066.89	127767.5	-50652	52295.8
1005	6523.959	116571.9	-54171	45720.78

equal to the cumulative cost 9991974 \$ minus the exercise price 9700000 \$) will be lower than the cost of the B&S one even if the stable call price is greater than B&S call price. Therefore, the effective cost of the B&S model is much greater than the stable one. In addition, we could say that this is not a unique case. In fact, in Table 13 we see that generally the stable delta hedging strategy is less expensive than the Gaussian one considering daily or weekly hedging strategies. Thus, independently of the option price we apply, it is more cost-effective using the stable delta hedging strategy than the Gaussian one. Clearly, the effective cost of a stable delta hedging strategy is always lower than the B&S one. In particular, the effective cost of stable subordinated hedging strategy could be lower than zero (see Table 13) because we do not take into consideration the transaction costs and the stable option price is always greater than the B&S one. We obtain similar results by changing the maturity and the riskless return. This analysis underlines the limits of B&S model. The best approximation of a stable subordinated model reduces the costs of hedging strategies and the effective costs of the classic hedging strategies. Thus, for any financial institution it is more cost-effective to apply a delta hedging strategy using a stable subordinated model than the classic B&S one. Even if, further ex-post studies and comparisons would be necessary to confirm these results, the more protective behavior of the stable option pricing formula led us to reject the classic log-normal hypothesis and to propose the log-stable model of asset prices.

Many other subordinated models have been proposed in writing. In particular we recall the Clark model (see Clark (1973)), the log-student t model (see Praetz (1972), Blattberg and Gonedes (1974)), the log Laplace model (see Mittnik and Rachev, (1993)), the hyperbolic model (see Barndorff-Nielsen (1994), Eberlein and Keller (1995)). These subordinated models were recently analyzed and compared by Hurst, Platen and Rachev (1997)[10]. For a general approach to contingent claim pricing based on a diffusion model for asset returns with stochastic and path dependent volatilities we could refer to Hofmann, Platen and Schweizer (1982), Duan (1995), Kallsen and Taqqu (1998).

6.7   Conclusions

In this paper we have shown that many of the models used in modern finance can be generalized assuming stable distributions for the underlined random variables and that the generalized models not only are theoretically justifiable and empirically testable, but they generally present better performances than the respective Gaussian models.

In the first analysis, we study, analyze and discuss portfolio choice models considering returns with heavy-tailed distributions. The first distributional model considered: the case of sub-Gaussian α-stable distributed returns permits a mean risk

---

[10] For alternative approaches to option pricing with stable distributions that do not use subordinated processes see McCulloch (1996) Rachev and Mittnik (2000) and the references therein.

analysis pretty similar to the Markowitz–Tobin mean variance one. As a matter of fact, this model admits the same analytical form for the efficient frontier, but the parameters in the two models have a different meaning. Therefore, the most important difference is given by the way of estimating the parameters. Moreover we extend the model to a discrete time analysis and we obtain multi-portfolio policies for α-stable sub-Gaussian distributed return. In order to present heavy-tailed models that consider the asymmetry of the returns, we study a discrete time three fund separation model where the portfolios are in the domain of attraction of a stable law. We generalize the analysis to the case of $k + 1$-fund separation model in the domain of attraction of a stable law. For all models we explicate the efficient frontier for the risk averse investors and we show how to estimate all parameters. Moreover we propose a conditional extension of the previous models in order to value the risk of a given portfolio.

In the second analysis, the comparison made between the dynamic stable non-Gaussian and the normal approach to the allocation problems has indicated that the stable non-Gaussian allocation presents better performances than the normal one. To be precise, the stable approach not only approximates better the asset returns, but, differently from the normal one, considers the component of risk due to the heavy tails. Therefore, we found that the main differences between the Gaussian and the stable non-Gaussian approaches imply important differences in the problems of allocation. These results are empirically confirmed if we compare portfolio choices obtained considering portfolio returns that follows an ARMA(1,1) model with stable or Gaussian distributed residuals. In addition we observed that the distribution of residuals is asymmetric and leptokurtic and the hypothesis of normality is usually rejected under statistical testing.

In order to compare alternative models for the VaR calculation we have submitted each model to a backtest analysis. We use some of the most representative index returns of the international market and, using the relative exchange rates, we converted their values into USD. The obtained results confirm that when the percentiles are below 5%, the hypothesis of normality of the conditional return distribution determines intervals of confidence whose forecast ability is low. Among the alternative models we proposed, the α-stable densities are reliable in the VaR calculation.

We conclude this analysis by presenting and comparing option pricing models. In particular, we review option pricing models for subordinated return distributions and we compare dynamic hedging strategies obtained from the classic Black and Scholes model and the stable subordinated one. We can see that by considering different exercise prices the costs of hedging and the effective costs are lower for the stable subordinated model than for the B&S one. Thus, financial institutions could find it more cost-effective to apply a delta hedging strategy using a stable subordinated model rather than the classic B&S one.

References

[1] Ahlburg, D.A.(1992): A comment on error measures, *International Journal of Forecasting* 8, 99–111.

[2] Akgiray, V. (1989): Conditional heteroskedasticity in time series of stock returns: evidence and forecast, *Journal of Business* 62, 55–80.

[3] Armstrong, J.S., and F. Collopy (1992): Error measures for generalizing about forecasting methods: Empirical comparisons, *International Journal of Forecasting*, n.8, 69–80.

[4] Barndorff-Nielsen O.E. (1994): Gaussian inverse Gaussian processes and the modeling of stock returns, *Technical Report*, Aarhus University.

[5] Bawa, V. S. (1976): Admissible portfolio for all individuals, *Journal of Finance* 31, 1169–1183.

[6] Bawa, V. S. (1978): Safety-first stochastic dominance and optimal portfolio choice, *Journal of Financial and Quantitative Analysis*, 255–271.

[7] Blattberg, R.C. and N.J. Gonedes (1974): A comparison of the stable and student distributions as statistical models for stock prices, *Journal of Business* 47, 244–280.

[8] Boender, G.C.E.(1997): A hybrid simulation/optimization scenario model for asset/liability management, *European Journal of Operation Research* 99, 126–135.

[9] Bollerslev, T. (1986): Generalized autoregressive conditional heteroskedasticity, *Journal of Econometrics* 31, 307–327.

[10] Box G.E.P. and G.M. Jenkins (1976): Time series analysis: forecasting and control, 2nd ed. San Francisco: Holden-Day

[11] Brokwell, P.J. and R.A. Davis (1991): Time series: theory and methods, 2nd ed. New York: Springer.

[12] Cadenillas A. and S.R. Pliska (1999): Optimal trading of a security when there are taxes and transaction costs, *Finance and Stochastics* 3, 137–165.

[13] Campbell, J. (1987): Stock returns and the term structure, *Journal of Financial Economics* 18, 373–399.

[14] Chamberlein, G. (1983): A characterization of the distributions that imply mean variance utility functions, *Journal of Economic Theory* 29, 975–988.

[15] Chambers, J., S.J. Mallows and B. Stuck (1976): A method for simulating stable random variables, *Journal of the American Statistical Association* 71, 340–344.

[16] Chen, N. F. and J., Ingersoll (1983): Exact pricing in linear factor models with finitely many assets: a note, *Journal of Finance* 38, 985–988.

[17] Cheng, B. and S., Rachev (1995): Multivariate stable futures prices, *Mathematical Finance* 5, 133–153.

[18] Clark, P.K. (1973): A subordinated stochastic process model with finite variance for speculative prices, *Econometrica* 41, 135–155.

[19] Connor, G. (1984): A unified beta pricing theory, *Journal of Economic Theory* 34, 13–31.

[20] Consiglio, A., I. Massabo and S. Ortobelli (2001): Non-Gaussian distribution for VaR calculation: an assessment for the Italian Market, *Proceeding IFAC SME 2001 Symposium*.

[21] Cvitanic, J., and I., Karatzas (1992): Convex duality in constrained portfolio optimization, *Annals of Applied Probability* 2, 767–818.

[22] Davis, M.H.A. and A.R. Norman (1990): Portfolio selection with transaction cost, *Mathematical Operation Research* 15, 676–713.

[23] Davis, R. (1996): Gauss-Newton and M-estimation for ARMA processeswith infinite variance, *Stochastic Processes and Applications* 63, 75–95.

[24] Davis, R.A. and S.I Resnick (1986): Limit theory for the sample covariance and correlation functions of moving averages, *Annals of Statistics* 14, 533–558.

[25] Davis, R.A. and S.I Resnick (1996): Limit theory for bilinear processes with heavy tailed noise, *Annals of Applied Probability* 6, 1191–1210.

[26] Davis, R.A., T. Mikosch and B. Basrak (1999): Sample ACF of multivariate stochastic recurrence equations with applications to GARCH, *Technical Report*, University of Groningen.

[27] De Haan, L., S.I. Resnick, H. Rootzen and C.G. Vries (1989): Extremal behavior of solutions to a stochastic difference equation with applications to ARCH processes, *Stochastic Processes and Applications* 32, 213–224.

[28] Duan, J.C. (1995): The GARCH option pricing model, *Mathematical Finance* 5, 13–32.

[29] Dutta P. K. (1994): Bankruptcy and expected utility maximization, *Journal of Economic Dynamics and Control* 18, 539–560.

[30] Dybvig, P. (1985): Acknowledgment: Kinks on the mean variance frontier, *Journal of Finance* 40, 245.

[31] Dybvig, P. (1983): An explicit bound on individual assets' deviations from APT pricing in a finite economy, *Journal of Financial Economics* 12, 483–496.

[32] Dybvig, P. and S. Ross (1986): Tax clienteles and asset pricing, *Journal of Finance* 41, 751–762.

[33] Eberlein, E. and K. Keller (1995):Hyperbolic distributions in Finance, *Bernoulli* 1, 281–299.

[34] Engle R.F. (1982): Autoregressive conditional heteroskedasticity with estimates of the variance of U.K. Inflation, *Econometrica* 50, 987–1008.

[35] Fama, E. (1963): Mandelbrot and the stable paretian hypothesis *Journal of Business* 36, 420–429.

[36] Fama, E. (1965a): The behavior of stock market prices *Journal of Business* 38, 34–105.

[37] Fama, E. (1965b): Portfolio analysis in a stable paretian market *Management Science* 11, 404–419.

[38] Fama, E. and K., French (1988): Permanent and temporary components of stock prices, *Journal of political Economy*, 96, 246–273.

[39] Feller, W. (1966): An introduction to probability theory and its applications II. New York, Wiley.

[40] Fildes, R. (1992): The evaluation of extrapolative forecasting methods, *International Journal of Forecasting* 8, 81–98.

[41] Föllmer, H. and D., Sondermann (1986): Hedging of non-redundant contingent claims, in *Contributions to Mathematical Economics*, W. Hildenbrand and A. Mas Aollell eds. 205–223.

[42] Föllmer, H and M., Schweizer (1989): Microeconomic approach to diffusion models for stock prices, *Mathematical Finance* 3, 1–23.

[43] French, K., G., Schwert and R., Stambaugh (1987): Expected stock returns and volatility, *Journal of Financial Economics* 19, 3–29.

[44] Gamrowski, B. and S., Rachev (1999): A testable version of the Pareto-stable CAPM, *Mathematical and computer modeling* 29, 61–81.

[45] Gamrowski, B. and S., Rachev (1994): Stable models in testable asset pricing, in *Approximation, probability and related fields*, New York: Plenum Press.

[46] Giacometti, R. and S., Ortobelli (2003): "Risk measures for asset allocation models", in Chapter 6 "New Risk Measures in Investment and Regulation" Elsevier Science Ltd., 69–86.

[47] Gilster, J. and W. Lee (1984): The effect of transaction costs and different borrowing and lending rates on the option pricing model: a note, *Journal of Finance* 39, 1215–1222.

[48] Götzenberger, G., S., Rachev and E., Schwartz (2001): Performance measurements: the stable paretian approach, to appear in *Applied Mathematics Reviews*, Vol. 1, World Scientific Publ. 2000, 329–406.

[49] Grinblatt, M. and S., Titman (1983): Factor pricing in a finite economy, *Journal of Financial Economics* 12, 497–508.

[50] Hardin, Jr. (1984): Skewed stable variable and processes, *Technical Report* 79, Center for Stochastic Processes at the University of North Carolina, Chapel Hill.

[51] Hartvig, N.V. J.L. Jensen, and J. Pedersen (2001): A class of risk neutral densities with heavy tails, *Finance and Stochastics 5, 115–128.*

[52] He, H., and N. Pearson (1993): Consumption and portfolio policies with incomplete markets and short sale constraints: The infinite dimensional case, *Journal of Economic Theory* 54, 259–304.

[53] Hofmann, N., E., Platen, M. Schweizer (1982): Option pricing under incompleteness and stochastic volatility, *Technical Report,* Department of Mathematics, University of Bonn

[54] Huberman, G. (1982): A simple approach to arbitrage pricing theory, *Journal of Economic Theory* 28, 183–191.

[55] Hurst, S.H., E. Platen and S. Rachev (1997): Subordinated market index model: a comparison, *Financial Engineering and the Japanese Markets* 4, 97–124.

[56] Ingersoll, J. Jr. (1987): Theory of financial decision making, Totowa: Rowman & Littlefield.

[57] Janicki, A. and A., Weron (1994): Simulation and chaotic behavior of stable stochastic processes, New York: Marcel Dekker.

[58] Jobson, J.D. (1992): Applied Multivariate Data Analysis, Heidelberg:.Springer-Verlag.

[59] Kallsen, J. and M. Taqqu (1998): Option pricing in ARCH-Type models, *Mathematical Finance* 8, 13–26.

[60] Kesten H. (1973): Random difference equations and renewal theory for products of random matrices, *Acta Mathematica* 131, 207–248.

[61] Khindanova, I., S., Rachev and E., Schwartz (2001): Stable modeling of Value at Risk, *Mathematical and Computer Modelling* 34, 1223–1259.

[62] Klebanov, L.B., Rachev S., Szekely G. (1999): Pre-limit theorems and their applications, *Acta Applicandae Mathematicae* 58, 159–174.

[63] Klebanov, L.B., Rachev S., Safarian G. (2000): Local pre-limit theorems and their applications to finance, *Applied Mathematics Letters* 13, 70–73.

[64] Korn R. (1998): Portfolio optimization with strictly positive transaction costs and impulse control, *Finance and Stochastics* 2, 85–114.

[65] Kraus, A. and R., Litzenberger (1976): Skewness preference and the valuation of risk assets, *Journal of Finance* 31, 1085–1100.

[66] Lamantia, F., S., Ortobelli, and S., Rachev (2003):Value at risk with stable distributed returns, *Technical Report,* University of Bergamo, to appear in *Annals of Operation Research.*

[67] Leland H. (1985): Option pricing and replication with transaction costs, *Journal of Finance* 40, 1283–1301.

[68] Levy, H. (1992): Stochastic dominance and expected utility: survey and analysis, *Management Science* 38, 555–593.

[69] Li, D. and W.L., Ng (2000): Optimal dynamic portfolio selection: multiperiod mean-variance formulation, *Mathematical Finance* 10, 387–406.

[70] Lo, A. (1991): Long term memory in stock market prices, *Econometrica* 59, 1279–1313.

[71] Longestaey, J. and P. Zangari (1996). *RiskMetrics -Technical Document.* J.P. Morgan, Fourth edition, New York.

[72] Mandelbrot, B. (1963a): New methods in statistical economics, *Journal of Political Economy* 71, 421–440.

[73] Mandelbrot, B. (1963b): The variation of certain speculative prices, *Journal of Business* 26, 394–419.

[74] Mandelbrot, B. (1967): The variation of some other speculative prices, *Journal of Business* 40, 393–413.

[75] Mandelbrot, B. and M., Taylor (1967): On the distribution of stock price differences, *Operations Research* 15, 1057–1062.

[76] Markowitz, H. (1959): Portfolio selection; efficient diversification of investment, New York: Wiley.

[77] McCulloch J.H. (1996): Financial applications of stable distributions, in *Handbook of Statistics* 14, G.S. Maddala and C.R. Rao eds., Elsevier Science.

[78] Mittnik, S. and S., Rachev (1993): Modeling asset returns with alternative stable distributions, *Econometric Reviews* 12, 261–330.

[79] Mittnik, S., S., Rachev and M., Paolella (1996): Integrated stable GARCH processes, *Technical Report* Institute of Statistics and Econometrics, Christian Albrechts University at Kiel

[80] Milne F. (1988): Arbitrage and diversification a general equilibrium asset economy, *Econometrica* 56, 815–840.

[81] Mikosh, T., T. Gadrich, C. Kluppelberg, and R.J. Adler (1995): Parameter estimation for ARMA models with infinite variance innovations, *The Annals of Statistics,* 23, 305–326.

[82] Mikosh, T., and C. Starica (1998): Limit Theory for the sample autocorrelations and extremes of a GARCH(1,1) process, *Tecnical Report*, University of Groningen.

[83] Morton A., and S.R., Pliska (1995): Optimal portfolio management with fixed transaction costs, *Mathematical Finance* 5, 337–356.

[84] Mossin, J. (1966): Equilibrium in a capital asset market, *Econometrica* 34, 768–783.

[85] Nolan, J. (1997): Numerical computation of stable densities and distribution functions, *Communications in Statistics Stochastic Models* 13, 759–774.

[86] Ortobelli, S. (2001): The classification of parametric choices under uncertainty: analysis of the portfolio choice problem, *Theory and Decision* 51, 297–327.

[87] Ortobelli, S. I., Huber, M., Höchstötter, and S., Rachev (2001): A comparison among Gaussian and non-Gaussian portfolio choice models, *Proceeding IFAC SME 2001 Symposium.*

[88] Ortobelli, S. and S. Rachev (2001): Safety first analysis and stable paretian approach to portfolio choice theory. *Mathematical and Computer Modelling* 34: 1037–1072.

[89] Ortobelli S, I., Huber, S. Rachev and E. Schwartz (2002): Portfolio choice theory with non-Gaussian distributed returns Chapter 14 in the *Handbook of Heavy Tailed Distributions in Finance*. North Holland Handbooks of Finance (Series Editor W. T. Ziemba).

[90] Ortobelli, S., I., Huber, E. Schwartz (2002): Portfolio selection with stable distributed returns, *Mathematical Methods of Operations Research* 55, 265–300.

[91] Ortobelli, S., A., Biglova, I., Huber, B., Racheva and S., Stoyanov (2003) Portfolio choice with heavy tailed distributions, *Technical Report 22*, University of Bergamo, to appear in *Journal of Concrete and Applicable Mathematics.*

[92] Osborne, M. F. (1959): Brownian motion in the stock market, *Operation Research* 7, 145–173.

[93] Owen, J. and R., Rabinovitch (1983): On the class of elliptical distributions and their applications to the theory of portfolio choice, *Journal of Finance* 38, 745–752.

[94] Panorska, A., S. Mittnik and S., Rachev (1995): Stable GARCH models for financial time series, *Applied Mathematics Letters* 815, 33–37.

[95] Praetz, P. (1972): The distribution of share price changes, *Journal of Business* 45, 49–55.

[96] Press W.H., S.A., Teukolsky, W.T. Vetterling, B.P. Flannery (1992): Numerical recipes in C: the art of scientific computing, 2nd ed. New York: Cambridge University Press.

[97] Pyle, D. and S., Turnovsky (1970): Safety first and expected utility maximization in mean standard deviation portfolio selection, *Review of Economic Statistics* 52, 75–81.

[98] Pyle, D. and S., Turnovsky (1971): Risk aversion in change-constrained portfolio selection, *Management Science* 18, 218–225

[99] Rachev, S. (1991): Probability metrics and the stability of stochastic models, New York: Wiley.

[100] Rachev, S.T., S. Ortobelli, S. Schwartz, and E. Schwartz (2002): The problem of optimal asset allocation with stable disctributed returns, *Technical Report*, University of California at Santa Barbara, to appear in *Stochastic Processes and Functional Analysis: Recent Advanced*, A volume in honor of M.M. Rao, Marcel Dekker Inc.

[101] Rachev, S. H., Xin (1993): Test on association of random variables in the domain of attraction of multivariate stable law, *Probability and Mathematical Statistics*, 14, 125–141.

[102] Rachev, S.T., S., Ortobelli, S., and E., Schwartz, (2002): "The problem of optimal asset allocation with stable distributed returns", *Technical Report*, University of California at Santa Barbara, "Stochastic Processes and Functional Analysis" Marcel Dekker Editor.

[103] Rachev, S., E., Schwartz and I., Khindanova (2000): Stable modeling of credit risk, *Technical Report*, Anderson School of Management, Department of Finance.

[104] Ross, S. (1975): Return, risk and arbitrage, in *Studies in Risk and Return*; Irwin Friend and J. Bicksler, eds. Cambridge: Mass.: Ballinger Publishing Co.

[105] Ross, S. (1976): The arbitrage theory of capital asset pricing, *Journal of Economic Theory* 13, 341–360.

[106] Ross, S. (1978): Mutual fund separation in financial theory-the separating distributions, *Journal of Economic Theory* 17, 254–286.

[107] Ross, S. (1987): Arbitrage and martingale with taxation, *Journal of Political Economy 95, 371–393*.

[108] Rothschild, M. and J., Stiglitz (1970): Increasing risk: I. definition, *Journal of Economic Theory* 2, 225–243.

[109] Roy. A.D. (1952): Safety-first and the holding of assets, *Econometrica* 20, 431–449.

[110] Roy, S. (1995): Theory of dynamic portfolio choice for survival under uncertainty, *Mathematical Social Sciences* 30, 171–194

[111] Samorodnitsky, G. and M.S., Taqqu (1994): Stable non Gaussian random processes: stochastic models with infinite variance, New York: Chapman and Hall.

[112] Samuelson, P.A. (1955): Brownian motion in stock market, *Unpublished Manuscript*.

[113] Shaked, M. and G., Shanthikumar (1994): Stochastic orders and their applications, New York: Academic Press Inc. Harcourt Brace & Company.

[114] Simaan, Y. (1993): Portfolio selection and asset pricing -Three parameter framework, *Management Science* 5, 568–577.

[115] Tesler, L.G. (1955/6): Safety first and hedging, *Review of Economic Studies* 23, 1–16.

[116] Tokat, Y., S., Rachev and E., Schwartz (2003): The stable non-Gaussian asset allocation: a comparison with the classical Gaussian approach, *Journal of Economic Dynamics and Control*, 27, 937–969.

[117] Young, M.R. (1998): A minimax portfolio selection rule with linear programming solution, *Management Science*, 44, 673–683.

[118] Zolatorev, V. M. (1986): One-dimensional stable distributions, Amer. Math. Soc. Transl. of Math. Monographs 65, Providence: RI. Transl. of the original 1983 Russian.

7

Optimal Quantization Methods and Applications to Numerical Problems in Finance

Gilles Pagès

Huyên Pham

Jacques Printems

ABSTRACT We review optimal quantization methods for numerically solving nonlinear problems in higher dimensions associated with Markov processes. Quantization of a Markov process consists in a spatial discretization on finite grids optimally fitted to the dynamics of the process. Two quantization methods are proposed: the first one, called marginal quantization, relies on an optimal approximation of the marginal distributions of the process, while the second one, called Markovian quantization, looks for an optimal approximation of transition probabilities of the Markov process at some points. Optimal grids and their associated weights can be computed by a stochastic gradient descent method based on Monte Carlo simulations. We illustrate this optimal quantization approach with four numerical applications arising in finance: European option pricing, optimal stopping problems and American option pricing, stochastic control problems and mean-variance hedging of options and filtering in stochastic volatility models.

7.1 Introduction

Optimal quantization of random vectors consists in finding the best possible approximation (in L^p) of an $\mathbb{R}^d$-valued random vector X by a measurable function $\varphi(X)$ where φ takes at most N values in $\mathbb{R}^d$. This is a very old story which starts in the early 1950s. The idea was to use a finite number N of codes (or "quantizers") to transmit efficiently a continuous stationary signal. Then it became essential to optimize the geometric location of these quantizers for a given distribution of the signal and to evaluate the resulting error. In a more mathematical form, the problem is to find a measurable function φ^* (if one exists) such that

$$\|X - \varphi^*(X)\|_p = \inf \left\{ \|X - \varphi(X)\|_p, \ \varphi : \mathbb{R}^d \to \mathbb{R}^d, \ |\varphi(\mathbb{R}^d)| \leq N \right\}$$

and then to evaluate $\|X - \varphi^*(X)\|_p$, especially when N goes to infinity. These problems have been extensively investigated in information theory and signal processing (see [12]). However, from a computational point of view, optimal quantization remained essentially limited to one-dimensional signals, the optimization process, essentially deterministic, becoming intractable for multi-dimensional signals. The drastic cut down of massive Monte Carlo simulation cost on computers made possible the implementation of alternative procedures based on probabilistic ideas (see the $CLVQ$ algorithm below). This gave birth to many applications and extensions in various fields like automatic classification, data analysis and artificial neural networks. Let us mention e.g., the self-organizing maps introduced by Kohonen in the early 1980s (see [15]). More recently, this leads to consider optimal quantization as a possible spatial discretization method to solve multi-dimensional (discrete time) problems arising in numerical probability. An important motivation to tackle these questions comes from finance since most problems arising in that field are naturally multi-dimensional.

First, an application to numerical integration in medium dimension ($1 \leq d \leq 4$) was developed and analyzed in [18]. A second step consisted in applying optimal quantization to solve nonlinear problems related to a (discrete time) Markovian dynamics. A first example was provided by discrete time optimal stopping problems (by the way of American option pricing), still in a multi-dimensional setting (see [2], [4] and [5]). From a probabilistic point of view, the nonlinearity usually appears through functionals of conditional expectations that need to be computed. From a computational point of view, the quantization approach leads to some tree algorithms in which, at every time step is associated a grid of quantizers, assumed to be optimal in some sense for the Markov chain. Then, investigating various fields of applications like stochastic control or nonlinear filtering, it turned out that it could be useful to specialize the way one quantizes Markov chains according to the nature of the encountered problem. This gave rise to two variants of the quantization: the *marginal quantization* introduced in [2] that focused on the optimization of the marginal distributions of the Markov chain and the *Markovian quantization* introduced in [19] that enhances the approximation of the conditional distributions at some points. Both approaches are presented here with some applications to finance, along with some further developments (1st-order schemes).

The paper is organized as follows: Section 7.2 is devoted to general background on optimal vector quantization of random vectors. First, the main properties concerning the existence of an optimal quantization and its rate of convergence toward 0 as its size goes to infinity are recalled. Then, numerical methods to get optimal quantizers and their associated weights are described. A first application to numerical integration is presented which points out in a simple setting the main features of this spatial discretization method. In Section 7.3, we present the two methods used so far to quantize Markov chains, called *marginal* and *Markovian* quantization methods. Both methods are applied to compute expectation of functionals $\phi_0(X_0) \ldots \phi(X_n)$ of the Markov chain. Then, the main theoretical and computational features of both methods are discussed. In Section 7.4, three main applications to finance are described including some numerical illustrations: American option pricing, stochastic control and filtering of stochastic volatility. Finally, in Section 7.5 we explain by an example how one can design some first-order schemes based on optimal quadratic quantization that significantly improve the rate of convergence of the above methods.

Throughout the paper, $|\xi|$ will denote the usual canonical Euclidean norm of $\xi \in \mathbb{R}^d$. Let $(a_n)_{n\geq 1}$ and $(b_n)_{n\geq 1}$ be two sequences of real numbers: we denote by $a_n \sim b_n$ (strong equivalence as n goes to infinity) the fact that $a_n = u_n b_n$ with $\lim\limits_n u_n = 1$.

7.2 Optimal quantization of a random vector

7.2.1 Existence and asymptotics of optimal quantization

The basic idea of quantization is to replace an $\mathbb{R}^d$-valued random vector $X \in L^p(\Omega, \mathbb{P})$ by a random vector taking at most N values in order to minimize the induced L^p-error i.e., one wishes to solve the minimizing problem error

$$\min\left\{\|X - Y\|_p,\ Y : \Omega \to \mathbb{R}^d,\ \text{measurable}\ ,\ |Y(\Omega)| \leq N\right\}.$$

Let $Y : \Omega \to \mathbb{R}^d$ be such a random vector and let $\Gamma = Y(\Omega)$. Then, consider a closest neighbor rule projection $\text{Proj}_\Gamma : \mathbb{R}^d \to \Gamma$ and set,

$$\widehat{X}^\Gamma := \text{Proj}_\Gamma(X).$$

One easily checks that $\|X - \widehat{X}^\Gamma\|_p \leq \|X - Y\|_p$. Assume $|\Gamma| = N$ and $\Gamma = \{x^1, \dots, x^N\}$. Closest neighbor rule projections Proj_Γ are in one-to-one correspondence with *Voronoi tessellations* of Γ, that is with Borel partitions $C_1(\Gamma), \dots, C_N(\Gamma)$ of $\mathbb{R}^d$ satisfying:

$$C_i(\Gamma) \subset \left\{\xi \in \mathbb{R}^d : |\xi - x^i| = \min_{x^j \in \Gamma} |\xi - x^j|\right\}, \quad i = 1, \dots, N.$$

Then, one may set $\text{Proj}_\Gamma(\xi) := \sum_{i=1}^N x^i \mathbf{1}_{C_i(\Gamma)}(\xi)$ so that

$$\widehat{X}^\Gamma = \sum_{i=1}^N x^i \mathbf{1}_{C_i(\Gamma)}(X). \tag{7.1}$$

In the sequel, the exponent Γ in $\widehat{X}^\Gamma$ will be often dropped.

The L^p-error induced by this projection — called L^p-quantization error — is given by $\|X - \widehat{X}^\Gamma\|_p$. It clearly depends on the grid Γ; in fact, one easily derives from (7.1):

$$\|X - \widehat{X}^\Gamma\|_p^p = \mathbb{E}\left[\min_{1\leq i\leq N} |X - x^i|^p\right], \tag{7.2}$$

for $\Gamma = \{x^1, \ldots, x^N\}$. So, if one identifies a grid Γ of size N with the N-tuple $x := (x^1, \ldots, x^N) \in (\mathbb{R}^d)^N$ or any of its permutations, the pth power of the L^p-quantization error is a symmetric function Q_N^p defined on N-tuples with pairwise distinct components by

$$Q_N^p(x^1, \ldots, x^N) := \int q_N^p(x, \xi) \, \mathbb{P}_X(d\xi) \, (\mathbb{P}_X \text{ is for the distribution of } X)$$

where $q_N^p(x, \xi) := \min_{1 \le i \le N} |x^i - \xi|^p$, $x = (x^1, \ldots, x^N) \in (\mathbb{R}^d)^N$, $\xi \in \mathbb{R}^d$.

(The function $\sqrt[p]{q_N^p}$ is sometimes called *local quantization error*.) The extension of the function Q_N^p on the whole $(\mathbb{R}^d)^N$ is obvious.

Then two questions naturally arise: does this function reach a minimum? How does this minimum behave as N goes to infinity? They have been investigated for a long time as part of quantization theory for probability distributions, first in information theory and signal processing in the 1950s and, more recently in probability for both numerical or theoretical purposes (see [13, 18]). They make up the core of optimal quantization. We will now shortly recall these main results. For a comprehensive approach, one may consult [13] and the references therein.

First, the size N being settled, the function $\sqrt[p]{Q_N^p}$ is Lipschitz continuous and does reach a minimum (although Q_N^p does not go to infinity as $\max_{1 \le i \le N} |x^i| \to \infty$). If $|X(\Omega)| \ge N$, then any N-tuple that achieves the minimum has pairwise distinct components i.e., defines a grid Γ^* of size N satisfying

$$\|X - \widehat{X}^{\Gamma^*}\|_p = \min \left\{ \|X - Y\|_p, \; Y \text{ random vector in } \mathbb{R}^d, \; |Y(\Omega)| \le N \right\}. \quad (7.3)$$

If $|X(\Omega)|$ is infinite, this minimum (strictly) decreases to 0 as N goes to infinity. Its rate of convergence is ruled by the so-called Zador theorem, completed by several authors: Zador, Bucklew & Wise (see [8]) and finally Graf & Luschgy in [13].

Theorem 1. *(see [13]) Assume that* $\mathbb{E}|X|^{p+\varepsilon} < +\infty$ *for some* $\varepsilon > 0$. *Then*

$$\lim_N \left(N^{\frac{p}{d}} \min_{|\Gamma| \le N} \|X - \widehat{X}^{\Gamma}\|_p^p \right) = J_{p,d} \left(\int_{\mathbb{R}^d} g^{\frac{d}{d+p}}(\xi) \, d\xi \right)^{1 + \frac{p}{d}} \quad (7.4)$$

where $\mathbb{P}_X(d\xi) = g(\xi) \lambda_d(d\xi) + \nu(d\xi)$, $\nu \perp \lambda_d$ (λ_d *Lebesgue measure on* $\mathbb{R}^d$). *The constant* $J_{p,d}$ *corresponds to the case of the uniform distribution on* $[0, 1]^d$.

Remark 1. In higher dimension, the true value of $J_{p,d}$ is unknown except in one dimension where $J_{p,1} = \frac{1}{2^p(p+1)}$. However, one shows that $J_{2,2} = \frac{5}{18\sqrt{3}}$ and that $J_{p,d} \sim \left(\frac{d}{2\pi e}\right)^{\frac{p}{2}}$ as d goes to infinity (see [13] for some proofs and other results using non-Euclidean norms).

This theorem says that $\min_{|\Gamma| \le N} \|X - \widehat{X}^{\Gamma}\|_p \sim C_{\mathbb{P}_X, p, d} N^{-\frac{1}{d}}$ as $N \to \infty$. This is in accordance with the rates $O(N^{-1/d})$ easily obtained with product quantizer of size $N = m^d$ for the uniform distribution $U([0, 1]^d)$ over the unit hypercube $[0, 1]^d$. In fact, even in that very specific setting, these product quantizers are not optimal quantizers for $U([0, 1]^d)$ (except if $d = 1$). In fact *optimal quantization provides for every $N \geq 1$ the "best fitting" grid of size N for a given distribution $\mathbb{P}_X$. This grid corresponds to the real constant $C_{\mathbb{P}_X, p, d}$ when N goes to infinity.*

7.2.2 How to get optimal quantization?

At this stage, the next question clearly is: how to get numerically an optimal N-grid with a minimal L^p-quantization error? Historically, the first attempt to solve this optimization problem — when $p = 2$ and $d = 1$ — is the so-called "Lloyd's methods I". This iterative procedure acts on the grids as follows: let Γ^0 be a grid of size N. Then set by induction

$$\Gamma^{s+1} = \mathbb{E}\left[X \mid \mathrm{Proj}_{\Gamma^s}(X)\right](\Omega) = \left(\mathbb{E}\left[X \mid X \in C_i(\Gamma^s)\right]\right)_{1 \leq i \leq N}, \quad s \in \mathbb{N}.$$

One shows that $\{\|X - \mathrm{Proj}_{\Gamma^s}(X)\|_2, s \in \mathbb{N}\}$ is a nonincreasing sequence and that, under some appropriate assumptions (see [14]), $\mathrm{Proj}_{\Gamma^s}(X)$ converges toward some random vector $\widehat{X}$ taking N values as s goes to infinity. Moreover, $\widehat{X}$ satisfies the *stationary quantizer* property

$$\widehat{X} = \mathbb{E}[X \mid \widehat{X}] \tag{7.5}$$

and is the only solution to the original optimization problem $\mathrm{argmin}\,\{\|X - Y\|_2, |Y(\Omega)| \leq N\}$.

When the dimension d is greater than 1, the convergence may fail. When some convergence holds, the limit $\widehat{X}$ is still stationary but has no reason to minimize the quadratic quantization error. In a general setting, this algorithm has two main drawbacks: it is a purely "local" procedure which does not explore the whole state space, and, furthermore, it becomes numerically intractable in its original form since it requires the computation of d-dimensional integrals $\int_C \ldots d\mathbb{P}_X$. When the random vector X *is simulatable*, one can randomize the Lloyd's methods I by using a Monte Carlo simulation to compute the above integrals. This version is sometimes used as a final step of the optimization procedure to "refine" locally the results obtained by other methods like that described below.

We will describe a procedure which partially overcomes these drawbacks, based on another property of the L^p-quantization error function Q_N^p: its smoothness. Let us temporarily identify a grid $\Gamma := \{x^1, \ldots, x^N\}$ of size N with the N-tuple $x = (x^1, \ldots, x^N)$ and let us denote the Voronoi cell of x^i by $C_i(x)$ instead of $C_i(\Gamma)$.

Proposition 1. *([18]) Let $p > 1$. The function Q_N^p is continuously differentiable at any N-tuple $x \in (\mathbb{R}^d)^N$ having pairwise distinct components and a $\mathbb{P}_X$-negligible Voronoi tessellation boundary $\cup_{i=1}^N \partial C_i(x)$. Its gradient $\nabla Q_N^p(x)$ is obtained by formal differentiation:*

$$\nabla Q_N^p(x) = \mathbb{E}\left[\nabla_x q_N^p(x, X)\right], \tag{7.6}$$

$$\text{where} \quad \nabla_x q_N^p(x, \xi) = \left(\frac{\partial q_N^p}{\partial x^i}(x, \xi)\right)_{1 \le i \le N}$$

$$:= p\left(\frac{x^i - \xi}{|x^i - \xi|}|x^i - \xi|^{p-1}\mathbf{1}_{C_i(x)}(\xi)\right)_{1 \le i \le N} \tag{7.7}$$

with the convention $\frac{0}{|0|} = 0$. If $\mathbb{P}_X$ is continuous the above formula (7.6) still holds for $p = 1$. (Note that $\nabla_x q_N^p(x, \xi)$ has exactly one non-zero component $i(x, \xi)$ defined by $\xi \in C_{i(x,\xi)}(x)$.)

One shows (see [13], p. 38) that any N-tuple $x^* \in \text{argmin} Q_N^p$ satisfies the "boundary" assumption of Proposition 1 so that $\nabla Q_N^p(x^*) = 0$.

The integral representation (7.6) of ∇Q_N^p strongly suggests, as soon as independent copies of X can be easily simulated on a computer, to implement a *stochastic gradient algorithm* (or *descent*). It is a stochastic procedure recursively defined by

$$\Gamma^{s+1} = \Gamma^s - (\delta_{s+1}/p)\nabla_x q_N^p(\Gamma^s, \xi^{s+1}) \tag{7.8}$$

where the initial grid Γ^0 has N pairwise distinct components, $(\xi^s)_{s \ge 1}$ is an i.i.d. sequence of $\mathbb{P}_X$-distributed random vectors, and $(\delta_s)_{s \ge 1}$ a non-increasing sequence of $(0, 1)$-valued step parameters satisfying the usual conditions:

$$\sum_s \delta_s = +\infty \quad \text{and} \quad \sum_s \delta_s^2 < +\infty. \tag{7.9}$$

Note that (7.8) *a.s.* grants by induction that Γ^s has pairwise distinct components. In an abstract framework (see e.g., [9] or [17]), under some appropriate assumptions, a stochastic gradient descent associated to the integral representation of a so-called potential function *a.s.* converges toward *a local minimum of this potential function* (Q_N^p in our problem). Although these assumptions are not fulfilled by Q_N^p, the encountered theoretical problems can be partially overcome (see [18] for some a.s. convergence results in one dimension or when $\mathbb{P}_X$ is compactly supported). Practical implementation does provide satisfactory results (a commonly encountered situation with gradient descent procedures). Some estimates of the *companion parameters* ($\mathbb{P}_X$-weights of the cells and L^r-quantization errors, $r \le p$) can be obtained as by-product of the procedure. This is discussed below.

STATIONARY QUANTIZERS (BACK TO): When $p = 2$, standard computations show that Equation $\nabla Q_N^2(x) = 0$ is simply the stationary quantizer property: if Γ is the corresponding grid, then $\widehat{X}^\Gamma$ satisfies Equation (7.5). This identity has interesting applications (see the next two paragraphs below). It also implies that, for every $p \in [1, +\infty]$, $\|\widehat{X}^\Gamma\|_p \le \|X\|_p$.

Note that non-optimal quantizers may be stationary: when $\mathbb{P}_X = \mathbb{P}_{X^1} \otimes \cdots \otimes \mathbb{P}_{X^d}$ is a product measure, any "product quantizer" made up with optimal — or even stationary — quantizers of its marginal distributions $\mathbb{P}_{X^i}$ is stationary. It can also be the case of any *local minima* of Q_N^2 which are the natural targets of the above stochastic gradient descent algorithm.

PRACTICAL ASPECTS OF THE OPTIMIZATION, COMPANION PARAMETERS: Formula (7.8) can be developed as follows if one sets $\Gamma^s := \{x^{1,s}, \ldots, x^{N,s}\}$,

COMPETITIVE PHASE :

$$\text{Select } i(s+1) := i(\Gamma^s, \xi^{s+1}) \in \text{argmin}_i |x^{i,s} - \xi^{s+1}|. \tag{7.10}$$

LEARNING PHASE :

$$\begin{cases} x^{i(s+1),s+1} := x^{i(s+1),s} - \delta_{s+1} \frac{x^{i(s+1),s} - \xi^{s+1}}{|x^{i(s+1),s} - \xi^{s+1}|} |x^{i(s+1),s} - \xi^{s+1}|^{p-1}, \\ x^{i,s+1} := x^{i,s}, \qquad i \neq i(s+1). \end{cases} \tag{7.11}$$

The competitive phase (7.10) corresponds to selecting the closest point in Γ^s i.e., $i(s+1)$ such that $\xi^{s+1} \in C_{i(s+1)}(\Gamma^s)$. The learning phase (7.11) consists in updating the closest neighbor and leaving still other components of the grid Γ^s.

Furthermore, it is established in [18] (see also [2]) that, if $X \in L^{p+\varepsilon}$ ($\varepsilon > 0$), then the sequences $(Q_N^{r,s})_{s \geq 1}$, $1 \leq r \leq p$, and $(\pi_i^s)_{t \geq 1}$, $1 \leq i \leq N$, of random variables recursively defined by

$$Q_N^{r,s+1} := Q_N^{r,s} - \widetilde{\delta}_{s+1}(Q_N^{r,s} - |x^{i(s+1),s} - \xi^{s+1}|^r), \quad Q_N^{r,0} := 0, \tag{7.12}$$

$$\pi_i^{s+1} := \pi_i^s - \widetilde{\delta}_{s+1}(\pi_i^s - 1_{\{i=i(s+1)\}}), \quad \pi_i^0 := 1/N, \quad 1 \leq i \leq N \tag{7.13}$$

where $\widetilde{\delta}_s := 1/s$, $s \geq 1$, satisfy on the event $\{\Gamma^s \to \Gamma^*\}$

$$Q_N^{r,s} \xrightarrow{a.s.} Q_N^r(\Gamma^*), \ 0 < r \leq p, \text{ and } \pi_i^s \xrightarrow{a.s.} \mathbb{P}_X(C_i(\Gamma^*)), \ 1 \leq i \leq N, \text{ as } s \to \infty.$$

In fact, an easy adaptation of the proof shows that other choices are possible for the "companion" gain parameter sequence $\widetilde{\delta}$. In fact, (7.12) and (7.13) hold as soon as $\widetilde{\delta}_s \in (0, 1)$, $\sum_{s \geq 1} \widetilde{\delta}_s = +\infty$ and $\sum_{s \geq 1} \widetilde{\delta}_s^{\frac{p+\varepsilon}{r} \wedge 2} < +\infty$. These two companion procedures are essentially costless since they are based on computational steps of the grid optimization procedure itself and they yield the parameters of numerical interest (weights of the Voronoi cells, L^r-quantization errors of Γ^*, $0 < r \leq p$) for the grid Γ^*. Note that this holds whatever the limiting grid Γ^* is: this means that the procedure is consistent.

The *quadratic case* $p = 2$ is the most commonly implemented for applications and is known as the *Competitive Learning Vector Quantization* (CLVQ) algorithm. Then one considers $(0, 1)$-valued step parameters δ_s so that Γ^{s+1} lives in the convex hull of Γ^s and ξ^{s+1} and the cooperative procedure (7.11) becomes a simple homothety centered at ξ^{s+1} with ratio $1 - \delta_{s+1}$. These features have a stabilizing effect on the procedure: one checks on simulations that the CLVQ algorithm does behave better than its non-quadratic counterpart. The numerical aspects of the $CLVQ$ algorithm are deeply investigated in [21], especially when X is a d-dimensional normal vector (or is uniformly distributed over $[0, 1]^d$). In particular the optimal choice for the gain parameter sequence and the induced (weak) rate of convergence of the $CLVQ$ are discussed: one usually sets $\delta_s := \frac{A}{B+s}$ (and $\widetilde{\delta}_s := \delta_s$) with $A \leq B$; then, for some special distributions like $U([0, 1])$, a Central Limit Theorem at rate $\sqrt{s}$ can be established for the $CLVQ$. It seems empirically that this rate is standard (when $X \in L^{2+\varepsilon}$). Note that the heuristics given in [21] to specify the real constants A and B are in fact valid for more general distributions. Concerning rates, see also [1].

Figure 1 shows an optimal grid for the bivariate standard Normal distribution with 500 points. It is obtained by the $CLVQ$ procedure described above.

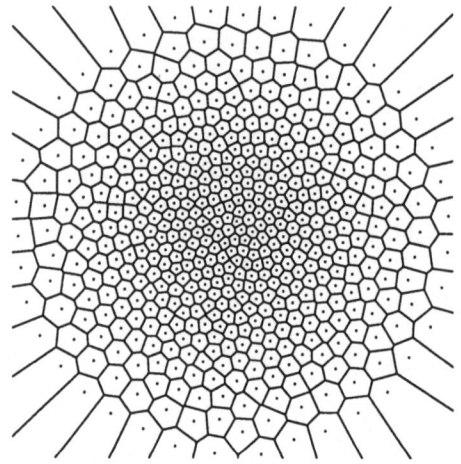

FIGURE 1. A L^2-optimal 500-quantization grid for the Normal distribution $\mathcal{N}(0; I_2)$ with its Voronoi tessellation.

7.2.3   Application to numerical integration

Consider a simulatable $\mathbb{R}^d$-valued *integrable* random vector X with probability distribution $\mathbb{P}_X$. The quantization method for numerical integration consists in approximating the probability distribution $\mathbb{P}_X$ by $\mathbb{P}_{\widehat{X}}$, the distribution of (any of) its closest neighbor rule projection(s) $\widehat{X} = \mathrm{Proj}_\Gamma(X)$ on a grid $\Gamma = \{x^1, \dots, x^N\}$:

$$\mathbb{P}_{\widehat{X}} = \sum_{i=1}^{N} \hat{p}^i \, \delta_{x^i}.$$

So, $\mathbb{P}_{\widehat{X}}$ is a discrete probability distribution whose weights $\hat{p}^i$ are defined by

$$\hat{p}^i = \mathbb{P}[\widehat{X} = x^i] = \mathbb{P}_X[C_i(\Gamma)], \qquad 1 \le i \le N,$$

where δ_{x^i} is the Dirac mass at x^i and $C_i(\Gamma) = \mathrm{Proj}_\Gamma^{-1}(x^i)$ denotes the Voronoi cells of $x_i \in \Gamma$. Then, one approximates the expectation of a Lipschitz continuous function ϕ on $\mathbb{R}^d$ w.r.t. $\mathbb{P}_X$, i.e.,

$$E[\phi(X)] = \int_{\mathbb{R}^d} \phi(\xi)\,\mathbb{P}_X(d\xi),$$

$$\text{by} \qquad E[\phi(\widehat{X})] = \int_{\mathbb{R}^d} \phi(\xi)\,\mathbb{P}_{\widehat{X}}(d\xi) = \sum_{i=1}^{N} \hat{p}^i \phi(x^i).$$

THE LIPSCHITZ CASE: When ϕ is simply Lipschitz continuous, the induced error is then simply measured by:

$$|E[\phi(X)] - E[\phi(\widehat{X})]| \le [\phi]_{Lip} \|X - \widehat{X}\|_1, \tag{7.14}$$

$$\le [\phi]_{Lip} \|X - \widehat{X}\|_p \quad (\text{if } X \in L^p, \ p \ge 1). \tag{7.15}$$

Optimal grids (of size N) which minimize the L^1-quantization error then provide a $O\left(N^{-1/d}\right)$ rate. Such a grid, its associated weights $\hat{p}_i$ and the induced L^1-quantization error can be computed by the algorithm described above (implemented with $p = 1$) if X has a continuous distribution. However, it often happens that, essentially for stability, one prefers to implement the algorithm in the quadratic case $p = 2$ ($CLVQ$). Then, it produces in practice an — at least locally — optimal *quadratic* grid Γ^*. Then, it also produces as a by-product, the companion parameters of the grid Γ^*: its $\mathbb{P}_X$-weights $(\hat{p}_i^*)_{1 \le i \le n}$ and any L^r-quantization error $\|X - \widehat{X}^{\Gamma^*}\|_r$, $1 \le r < p$ if $X \in L^p$ (some slightly more stringent assumption is required on step sequence (δ_s) to get the L^r-quantization error estimates). Some extensions of (7.14) to locally Lipschitz continuous functions can be found in [11].

THE LIPSCHITZ DERIVATIVE CASE: Assume now that X is at least *square integrable* and that function ϕ is *continuously differentiable with a Lipschitz continuous differential* $D\phi$. Furthermore, assume that $\widehat{X} = \widehat{X}^{\Gamma^*}$ where Γ^* is an *optimal quadratic* grid. By Taylor's formula, we have

$$|\phi(X) - (\phi(\widehat{X}) + D\phi(\widehat{X}).(X - \widehat{X}))| \le [D\phi]_{Lip} |X - \widehat{X}|^2$$

so that

$$\left|E[\phi(X)] - E[\phi(\widehat{X})] - E[D\phi(\widehat{X}).(X - \widehat{X})]\right| \le [D\phi]_{Lip} \|X - \widehat{X}\|_2^2,$$

$$\le [D\phi]_{Lip} \|X - \widehat{X}\|_p^2 \ (\text{if } X \in L^p, \ p \ge 2).$$

Now, $\widehat{X}$ is in particular a stationary quantizer, hence it satisfies (7.5) so that

$$E[D\phi(\widehat{X}).(X - \widehat{X})] = E\left[D\phi(\widehat{X}).E[X - \widehat{X} \mid \widehat{X}]\right] = 0,$$

$$\text{and} \quad |E[\phi(X)] - E[\phi(\widehat{X})]| \le [D\phi]_{Lip} \|X - \widehat{X}\|_2^2 = O\left(N^{-2/d}\right). \tag{7.16}$$

THE CONVEX CASE: When ϕ is a convex function and $\widehat{X}$ is a stationary quantizer satisfying $\widehat{X} = E[X \mid \widehat{X}]$, we have by Jensen's inequality:

$$E\left[\phi(\widehat{X})\right] = E\left[\phi\left(E[X \mid \widehat{X}]\right)\right] \le E[\phi(X)], \tag{7.17}$$

so that $E[\phi(\widehat{X})]$ is always a lower bound for $E[\phi(X)]$.

7.2.4   A first numerical Test (European option approximation)

The aim of this section is to test the optimal quantizers that we obtained by the numerical methods described in subsection 7.2.2 in dimension $d = 4$. Simultaneously, we aim to illustrate the performances of vector quantization for numerical integration. That is why we carry out a short comparison between quantization methods and Monte Carlo methods on a simple numerical integration problem.

Recall that the Strong Law of Large Numbers implies that, given a sequence $(Z_k)_{k \geq 1}$ of independent copies of a random vector Z with normal distribution $\mathcal{N}(0; I_d)$,

$$\mathbb{P}(d\omega)\text{-a.s.} \quad \frac{f(Z_1(\omega)) + \cdots + f(Z_N(\omega))}{N} \xrightarrow{N \to +\infty} \mathbb{E}(f(Z))$$

$$= \int_{\mathbb{R}^d} f(z) \exp(-|z|^2/2) \frac{dz}{(2\pi)^{d/2}}$$

for every $f \in L^1(\mathbb{R}^d, \mathbb{P}_Z)$. The Monte Carlo method consists in generating on a computer a path $(Z_k(\omega))_{k \geq 1}$ to compute the above Gaussian integral. The Law of the Iterated Logarithm says that, if $f(Z) \in L^2$, this convergence *a.s.* holds at a

$$\sigma(f(Z)) \sqrt{\frac{\log \log N}{N}}$$

rate where $\sigma(f(Z))$ is the standard deviation of $f(Z)$.

When f is twice differentiable, this is to be compared to the error bound provided by (7.16) when using a quadratic optimal N-quantizer $x^* := (x_1^*, \ldots, x_N^*)$, namely

$$[Df]_{Lip} Q_N^2(x^*) \approx \left(J_{2,d}(1 + 2/d)^{1+d/2} [Df]_{Lip} \right) N^{-2/d}.$$

Consequently, the dimension $d = 4$ appears as the (theoretical) critical dimension for the numerical integration of such functions by quantization for a given computational complexity (quantization formulae involving higher order differentials yield better rates): one assumes that the optimal quantizers have been formerly computed and that the computation times of a (Gaussian) random number and of a weight are both negligible w.r.t. the computation time of a value $f(z)$ of f.

The test is processed in each selected dimension d with five random variables $g_i(Z)$, $Z = (Z^1, \ldots, Z^d) \sim \mathcal{N}(0; I_d)$, $i = 0, 1, 2, 3, 4$ where the g_i's are five functions with compact support such that

- g_0 is an indicator function of a (bounded) interval (hence discontinuous),

- g_1 is convex and Lipschitz continuous,

- g_2 is convex and twice differentiable,

- g_3 is the difference of two convex functions and Lipschitz continuous,

- g_4 is the difference of two convex functions and twice differentiable.

The test functions are borrowed from classical option pricing in mathematical finance: one considers d traded assets $S^1, \ldots, S^d$, following a d-dimensional Black & Scholes dynamics. We assume that these assets are independent (this is not very realistic but corresponds to the most unfavorable case for quantization). We assume as well that $S_0^1 = s_0 > 0$, $i = 1, \ldots, d$ and that the d assets share the same volatility $\sigma^i = \sigma > 0$. At maturity $T > 0$, we then have:

$$S_T^i = s_0 \exp\left((r - \frac{\sigma^2}{2})T + \sigma\sqrt{T}Z^i\right), \qquad i = 1, \ldots, d.$$

One considers, still at time T, the geometric index $I_T = \left(S_T^1 \ldots S_T^d\right)^{\frac{1}{d}}$ satisfying

$$I_T = I_0 \exp\left((r - \frac{\sigma^2}{2d})T + \frac{\sigma\sqrt{T}}{\sqrt{d}}\frac{Z^1 + \cdots + Z^d}{\sqrt{d}}\right)$$

with

$$I_0 = s_0 \exp\left(-\frac{\sigma^2(d-1)}{2d}T\right).$$

Then, one specifies the random variables $g_i(Z)$ for $i = 1$ and $i = 3$ as follows

$$g_1(Z) = e^{-rT}(K_1 - I_T)_+ \qquad \text{(Put}(K_1, T) \text{ payoff)},$$
$$g_3(Z) = e^{-rT}(K_2 - I_T)_+ - e^{-rT}(K_1 - I_T)_+, \quad K_1 < K_2,$$
$$\text{(Put-Spread}(K_1, K_2, T) \text{ payoff)}.$$

The random variables are the payoffs of a Put option with strike price K_1 and a put-spread option with strike prices $K_1 < K_2$ respectively, both on the geometric index I_T. Some closed forms for the premia $\mathbb{E}[g_1(Z)]$ and $\mathbb{E}[g_2(Z)]$ are given by the Black & Scholes formula:

$$\mathbb{E}[g_1(Z)] = \pi(I_0, K_1, r, \sigma, T) \text{ and } \mathbb{E}[g_3(Z)] = \psi(I_0, K_1, K_2, r, \sigma, T) \ (7.18)$$
$$\text{with } \pi(x, K, r, \sigma, T) = Ke^{-rT}\text{erf}(-d_2) - I_0 \text{erf}(-d_1),$$
$$d_1 = \frac{\log(x/K) + (r + \frac{\sigma^2}{2d})T}{\sigma\sqrt{T/d}}, \qquad d_2 = d_1 - \sigma\sqrt{T/d}$$
$$\text{and} \qquad \psi(x, K_1, K_2, r, \sigma, T) = \pi(x, K_2, r, \sigma, T) - \pi(x, K_1, r, \sigma, T).$$

Then, one sets

$$g_2(Z) = e^{-rT/2}\pi(I_{\frac{T}{2}}, K_1, r, \sigma, T/2) \text{ and } g_4(Z)$$
$$= e^{-rT/2}\psi(I_{\frac{T}{2}}, K_1, K_2, r, \sigma, T/2).$$

The random variables $g_2(Z)$ and $g_4(Z)$ have the distributions of the (discounted) premia at time $T/2$ of the put(K_1, T) and of the put-spread(K_1, K_2, T) respectively. Functions g_2 and g_4 are C^∞ and using the martingale property of the discounted premia yields

$$\mathbb{E}\, g_2(Z) = \pi(I_0, K_1, r, \sigma, T) \quad \text{and} \quad \mathbb{E}\, g_4(Z) = \psi(I_0, K_1, K_2, r, \sigma, T). \quad (7.19)$$

Finally we specify g_0 as the "hedge function at maturity" of the put-spread option:

$$g_0(Z) = -e^{-rT} 1_{\{I_T \in [K_1, K_2]\}}. \quad (7.20)$$

The numerical specifications of the functions g_i's are as follows:

$$s_0 = 100, \quad K_1 = 98, \quad K_2 = 102, \quad r = 5\%, \quad \sigma = 20\%, \quad T = 2.$$

• NUMERICAL RESULTS IN FOUR DIMENSIONS: The comparison with the Monte Carlo estimator

$$\widehat{g_i(Z)}_N = \frac{1}{N} \sum_{k=1}^{N} g_i(Z_k), \qquad Z_k \text{ i.i.d., } Z_k \sim \mathcal{N}(0; I_d) \quad (7.21)$$

of $\mathbb{E}[g_i(Z)]$ is presented in the last column on the right: we first computed (a proxy of) the standard deviation $\sigma(\widehat{g_i(Z)}_N)$ of the above estimator (7.21) by an $N = 10\,000$ trial Monte Carlo simulation. Then, in order to measure the error induced by the quantization in the scale of the MC estimator Standard Deviation, we wrote down the *ratio* $\frac{absolute\ error}{\sigma(\widehat{g_i(Z)}_N)}$.

The results in Table 1 illustrate a widely observed phenomenon when integrating functions by quantization: differences of convex functions behave better than convex functions (this is obviously due to (7.17)), and Lipschitz derivative functions behave better than Lipschitz continuous functions (as predicted by (7.16)). The whole tests set suggests that the convexity feature is prominent.

• GRAPHICAL COMPARISON IN DIMENSIONS $d = 3, 4, 5$: We focus here on the convex C^2 function g_2. We wish to emphasize the dimension effect (keeping unchanged the other specifications). So, we depict in Figure 2, in dimension $d = 3, 4, 5$ (in a log-log scale), both the absolute error and the standard deviation $\sigma(\widehat{g_2(Z)}_N)$ of its Monte Carlo estimator as a function of N (the dotted lines are the induced least square regressions).

TABLE 1. Quantization versus Monte Carlo in four dimensions.

$d = 4$ & $N = 6540$	B&S Reference value	Quantized value	Relative error	$\sigma(\widehat{g_i(Z)}_N)$	$\dfrac{absolute\ error}{\sigma(\widehat{g_i(Z)}_N)}$
$\mathbb{E}\, g_0(Z)$	-0.093	$-0,091$	2.40%	0.034	0.064
$\mathbb{E}\, g_1(Z)$	2.077	2.047	1.44%	0.054	0.548
$\mathbb{E}\, g_2(Z)$	2.077	2.061	0.77%	0.033	0.482
$\mathbb{E}\, g_3(Z)$	1.216	1.213	0.26%	0.021	0.015
$\mathbb{E}\, g_4(Z)$	1.216	1.215	0.08%	0.012	0.001

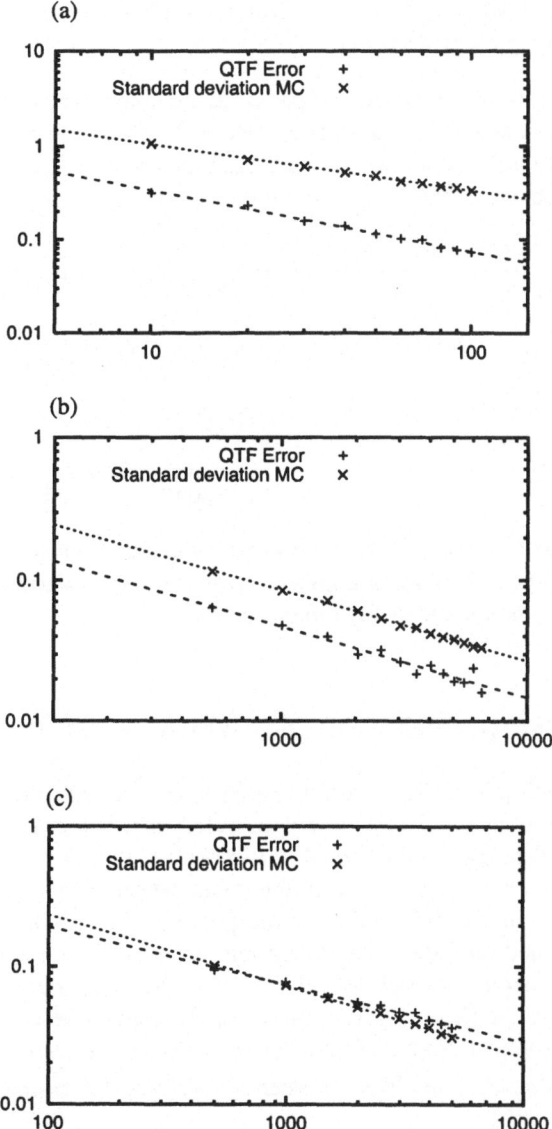

FIGURE 2. Linear regression in log-log scale of $N \mapsto |\mathbb{E}g_2(\widehat{Z}) - \widehat{\mathbb{E}g_2(Z)}_N|$. In a) $d = 3$; b) $d = 4$; c) $d = 5$.

Concerning the dimensionality effect, the theoretical rates for the error bounds ($N^{-1/d}$ in the Lipschitz case and $N^{-2/d}$ for Lipschitz differential case) are confirmed: when $d \leq 4$, quantization overperforms more and more the Monte Carlo method as N increases. When $d > 4$, quantization overperforms up to a critical number N_d of points (for a given trust level in the MC method). More detailed numerical results are presented in [21].

Remark 2. In the above tests, we compared quantization *versus* Monte Carlo for the computation of a *single* integral. If one looks for a uniform error bound over an infinite class of Lipschitz continuous functions, the conclusion can be quite different: thus, with the notation of the former subsection 7.2.3,

$$
\sup_{f, [f]_{Lip} \leq 1} \left| \frac{f(Z_1) + \cdots + f(Z_N)}{N} - \int_{\mathbb{R}^d} f(\zeta) \mathbb{P}_Z(d\zeta) \right|
$$

$$
\geq \int_{\mathbb{R}^d} \min_{1 \leq i \leq N} |Z_i - \zeta| \mathbb{P}_Z(d\zeta)
$$

$$
\geq Q_N^1(Z_1, \ldots, Z_N)
$$

$$
> \min_{(\mathbb{R}^d)^N} Q_N^1 \sim \frac{c_Z}{N^{\frac{1}{d}}}, \quad (c_Z > 0).
$$

This means that for every fixed N the worst behaviour of the Monte Carlo method on 1-Lipschitz functions induces a greater error than that obtained by optimal L^1-quantization. This holds true in any dimension d.

7.3   Optimal quantization of a Markov chain

We consider an $\mathbb{R}^d$-valued $(\mathcal{F}_k)$-Markov chain $(X_k)_{0 \leq k \leq n}$, with probability transition $P_k(x, dx')$ (from time $k - 1$ to k) and with initial distribution μ for X_0. The joint distribution of $(X_k)_{0 \leq k \leq n}$, is then equal to $\mu(dx_0) P_1(x_0, dx_1) \ldots P_n(x_{n-1}, dx_n)$.

In this section, we are interested in the quantization of this Markov chain, i.e., an approximation of the distribution of the process (X_k) by the distribution of a process $(\widehat{X}_k)$ valued on finite grids taking into account the probabilistic feature of the process. The naive approach would consist in the quantization of the $\mathbb{R}^{(n+1)d}$-valued random vector $(X_0, \ldots, X_n)$ following the method described in Section 2. However, by Theorem 1, for a total number N of points in such a "time-space" grid, the L^p-quantization error would be of order $N^{-\frac{1}{nd}}$. This is of course very slow when n is large.

Instead, we propose an approach based on the fact that a Markov chain is completely characterized by its initial distribution and its transitions probabilities. The idea is then to "quantize" the initial distribution of X_0 and the conditional probabilities of X_k given X_{k-1}. We propose two different quantization methods which shall provide a better rate of convergence of order $n^{1+1/d}/N^{1/d}$. The first approach, based on a quantization at each time k of the random variable X_k, was introduced in [2] and is called *marginal quantization*. The second one that enhances the preservation of the dynamics, namely the Markov property, was introduced in [19] and is called *Markovian quantization*.

7.3.1   Marginal quantization

At each time k and given a grid $\Gamma_k = \{x^1, \dots, x^{N_k}\}$ of N_k points in $\mathbb{R}^d$, associated to a Voronoi tessellation $C_1(\Gamma_k), \dots, C_{N_k}(\Gamma_k)$, we define:

$$\widehat{X}_k = \mathrm{Proj}_{\Gamma_k}(X_k), \quad k = 0, \dots, n. \tag{7.22}$$

Hence, in the marginal approach, the emphasis is put on the accuracy of the distribution approximations: if at every time k, the grid Γ_k is L^p-optimal, then $\widehat{X}_k$ is the best possible L^p-approximation of X_k by a random variable taking $N_k := |\Gamma_k|$ points. Notice that since the projection on the closest neighbor is not injective, the process $(\widehat{X}_k)_k$ constructed in (7.22) is not a Markov chain. However, if we define the probability transition matrices $[\hat{p}_k^{ij}]$ at times $k = 1, \dots, n$ by:

$$\hat{p}_k^{ij} = \mathbb{P}\left[\widehat{X}_k = x_k^j \mid \widehat{X}_{k-1} = x_k^i\right] = \frac{\hat{\beta}_k^{ij}}{\hat{p}_{k-1}^i}, \quad i = 1, \dots, N_{k-1}, \; j = 1, \dots, N_k,$$

where

$$\hat{p}_{k-1}^i = \mathbb{P}[\widehat{X}_{k-1} = x_{k-1}^i] = \mathbb{P}[X_{k-1} \in C_i(\Gamma_{k-1})],$$
$$\hat{\beta}_k^{ij} = \mathbb{P}[\widehat{X}_{k-1} = x_{k-1}^i, \; \widehat{X}_k = x_k^j] = \mathbb{P}[X_{k-1} \in C_i(\Gamma_{k-1}), \; X_k \in C_j(\Gamma_k)],$$

then it is well known that there exists a Markov chain $(\widehat{X}_k^c)$ with initial distribution $\hat{p}_0$ and probability transition matrices $[\hat{p}_k^{ij}]$ at times $k = 1, \dots, n$. The marginal quantization method consists in approximating the distribution of the Markov chain $(X_k)_{0 \le k \le n}$ by that of the Markov chain $(\widehat{X}_k^c)_{0 \le k \le n}$: by construction, the conditional distribution of $\widehat{X}_{k+1}^c$ given $\widehat{X}_k^c$ is equal to the conditional distribution of $\widehat{X}_{k+1}$ given $\widehat{X}_k$, and the distribution of $\widehat{X}_0^c$ is equal to the distribution of $\widehat{X}_0$. We will evaluate the rate of approximation (in distribution) of $\widehat{X}^c$ toward X on functions of the form $(x_0, x_1, \dots, x_n) \mapsto \phi_0(x_0)\phi_1(x_1)\dots\phi_n(x_n)$, where $\phi_0, \dots, \phi_n$ are bounded Lipschitz continuous functions on $\mathbb{R}^d$. First, notice that both quantities $\mathbb{E}[\phi_0(X_0)\phi_1(X_1)\dots\phi_n(X_n)]$ and $\mathbb{E}[\phi_0(\widehat{X}_0^c)\phi_1(\widehat{X}_1^c)\dots\phi_n(\widehat{X}_n^c)]$ follow a dynamic programming formula induced by the Markov property. Namely

$$\mathbb{E}\left[\phi_0(X_0)\dots\phi_n(X_n)\right] = \mathbb{E}\left[v_0(X_0)\right] \text{ and } \mathbb{E}\left[\phi_0(\widehat{X}_0^c)\dots\phi_n(\widehat{X}_n^c)\right]$$
$$= \mathbb{E}\left[\widehat{v}_0(\widehat{X}_0^c)\right]$$

where $\widehat{v}_0(\widehat{X}_0^c)$ and $v_0(X_0)$ satisfy

$$v_n(X_n) = \phi_n(X_n),$$
$$v_{k-1}(X_{k-1}) = \phi_{k-1}(X_{k-1})\mathbb{E}[v_k(X_k) \mid X_{k-1}], \qquad k = 1, \dots, n \tag{7.23}$$
$$\widehat{v}_n(\widehat{X}_n^c) = \phi_n(\widehat{X}_n^c),$$
$$\widehat{v}_{k-1}(\widehat{X}_{k-1}^c) = \phi_{k-1}(\widehat{X}_{k-1}^c)\mathbb{E}[\widehat{v}_k(\widehat{X}_k^c) \mid \widehat{X}_{k-1}^c], \qquad k = 1, \dots, n. \tag{7.24}$$

This will be the key to evaluate the error induced by approximating the first expectation term by the second one. Furthermore, the dynamic programming formula for $\widehat{X}^c$, once written "in distribution", provides a simple numerical algorithm to compute $\mathbb{E}\left[\widehat{v}_0(\widehat{X}_0^c)\right]$:

$$\widehat{v}_n(x_n^i) = \phi_n(x_n^i), \qquad \forall x_n^i \in \Gamma_n,$$

$$\widehat{v}_{k-1}(x_{k-1}^i) = \phi_{k-1}(x_{k-1}^i)\mathbb{E}[\widehat{v}_k(\widehat{X}_k) \mid \widehat{X}_{k-1} = x_{k-1}^i]$$

$$= \phi_{k-1}(x_{k-1}^i) \sum_{j=1}^{N_k} \hat{p}_k^{ij}\, \widehat{v}_k(x_k^j), \quad \forall x_{k-1}^i \in \Gamma_{k-1}, \quad k = 1, \ldots, n.$$

$$\mathbb{E}\left[\widehat{v}_0(\widehat{X}_0^c)\right] = \mathbb{E}\left[\widehat{v}_0(\widehat{X}_0)\right] = \sum_{i=1}^{N_0} \hat{p}_0^i\, \widehat{v}_0(x_0^i). \tag{7.25}$$

We rely on the following Lipschitz assumption on the transitions P_k of the Markov chain (X_k).

(A1)   For any $k = 1, \ldots, n$, the probability transition P_k is Lipschitz with ratio $[P_k]_{Lip}$, i.e., for any Lipschitz function ϕ on $\mathbb{R}^d$, with ratio $[\phi]_{Lip}$, we have:

$$\left| \int_{\mathbb{R}^d} \phi(x') P_k(x, dx') - \int_{\mathbb{R}^d} \phi(x') P_k(\widehat{x}, dx') \right|$$

$$\leq [P_k]_{Lip}[\phi]_{Lip}|x - \widehat{x}|, \quad \forall x, \widehat{x} \in \mathbb{R}^d.$$

Then we set $[P]_{Lip} = \max_{k=1,\ldots,n}[P_k]_{Lip}$. Let

$$BL_1(\mathbb{R}^d) = \Big\{ \phi : \mathbb{R}^d \to \mathbb{R},$$

$$\phi \text{ is bounded by 1 and } \phi \text{ is Lipschitz with ratio } [\phi]_{Lip} \leq 1 \Big\}.$$

Theorem 2. *Let $p \geq 1$. Under (A1), we have the error estimation in the marginal quantization method: for any functions $\phi_k \in BL_1(\mathbb{R}^d)$, $k = 0, \ldots, n$,*

$$\left| \mathbb{E}[\phi_0(X_0)\ldots\phi_n(X_n)] - \mathbb{E}[\phi_0(\widehat{X}_0^c)\ldots\phi_n(\widehat{X}_n^c)] \right|$$

$$\leq \sum_{k=0}^{n} \left(1 + (2 - \delta_{2,p})\frac{[P]_{Lip}^{n-k+1} - 1}{[P]_{Lip} - 1}\right) \|\Delta_k\|_p, \quad (7.26)$$

where $\|\Delta_k\|_p = \|X_k - \widehat{X}_k\|_p$ is the L^p-quantization error at time k of X_k. In (7.26), we make the usual convention that $\frac{1}{u-1}(u^m - 1) = m$ if $u = 1$ and $m \in \mathbb{N}$.

Proof. We set $\|\phi\|_{sup} = \max_{k=0,\ldots,n} \|\phi_k\|_{sup} \leq 1$ and $[\phi]_{Lip} = \max_{k=0,\ldots,n}[\phi_k]_{Lip} \leq 1$. From (7.23), a standard backward induction shows that

$$\|v_k\|_{sup} \leq \|\phi\|_{sup}^{n+1-k} \quad \text{and} \quad [v_k]_{Lip} \leq [P]_{Lip}\|\phi\|_{sup}[v_{k+1}]_{Lip} + \|\phi\|_{sup}^{n-k}[\phi]_{Lip}$$

so that
$$[v_k]_{Lip} \leq \|\phi\|_{sup}^{n-k}[\phi]_{Lip}\frac{[P]_{Lip}^{n-k+1} - 1}{[P]_{Lip} - 1}.$$

For any bounded Borel function f on $\mathbb{R}^d$, we set
$$P_k f(x) = \mathbb{E}[f(X_k)|X_{k-1} = x], \quad x \in \mathbb{R}^d,$$
$$\widehat{P}_k f(x) = \mathbb{E}[f(\widehat{X}_k)|\widehat{X}_{k-1} = x], \quad x \in \Gamma_{k-1},$$

for $k = 1, \ldots, n$. Hence, by (7.23) and (7.25), we have
$$\left\|v_k(X_k) - \widehat{v}_k(\widehat{X}_k)\right\|_p \leq \left\|v_k(X_k) - \mathbb{E}[v_k(X_k)|\widehat{X}_k]\right\|_p$$
$$+ \left\|\mathbb{E}\left[\left(\phi_k(X_k) - \phi_k(\widehat{X}_k)\right)P_{k+1}v_{k+1}(X_k)\big|\widehat{X}_k\right]\right\|_p \quad (7.27)$$
$$+ \left\|\phi_k(\widehat{X}_k)\mathbb{E}\left[P_{k+1}v_{k+1}(X_k) - \widehat{P}_{k+1}\widehat{v}_{k+1}(\widehat{X}_k)\big|\widehat{X}_k\right]\right\|_p.$$

On one hand, notice that, for every $p \geq 1$,
$$\|v_k(X_k) - \mathbb{E}(v_k(X_k)|\widehat{X}_k)\|_p \leq \|v_k(X_k) - v_k(\widehat{X}_k)\|_p + \|v_k(\widehat{X}_k)$$
$$- \mathbb{E}(v_k(X_k)|\widehat{X}_k)\|_p \quad (7.28)$$
$$\leq 2\|v_k(X_k) - v_k(\widehat{X}_k)\|_p$$
$$\leq 2[v_k]_{Lip}\|\Delta_k\|_p. \quad (7.29)$$

When $p = 2$, the very definition of the conditional expectation as an orthogonal projection shows that the above inequality holds without the 2 factor. On the other hand, using that conditional expectation (given $\widehat{X}_k$) is a L^p-contraction and that $\widehat{X}_k$ is $\sigma(X_k)$-measurable yields
$$\left\|\mathbb{E}\left[\left(\phi_k(X_k) - \phi_k(\widehat{X}_k)\right)P_{k+1}v_{k+1}(X_k)\big|\widehat{X}_k\right]\right\|_p \leq [\phi]_{Lip}\|v_{k+1}\|_{sup}\|\Delta_k\|_p$$
$$\leq [\phi]_{Lip}\|\phi\|_{sup}^{n-k}\|\Delta_k\|_p \quad (7.30)$$

and
$$\left\|\phi_k(\widehat{X}_k)\left(\mathbb{E}\left[P_{k+1}v_{k+1}(X_k) - \widehat{P}_{k+1}\widehat{v}_{k+1}(\widehat{X}_k)\big|\widehat{X}_k\right]\right)\right\|_p$$
$$\leq \|\phi\|_{sup}\|v_{k+1}(X_{k+1}) - \widehat{v}_{k+1}(\widehat{X}_{k+1})\|_p. \quad (7.31)$$

Plugging inequalities (7.29), (7.30) and (7.31) in (7.27) leads to the backward induction formula
$$\|v_k(X_k) - \widehat{v}_k(\widehat{X}_k)\|_p \leq ((2 - \delta_{2,p})[v_k]_{Lip} + [\phi]_{Lip}\|\phi\|_{sup}^{n-k})\|\Delta_k\|_p$$
$$+ \|\phi\|_{sup}\|v_{k+1}(X_{k+1}) - \widehat{v}_{k+1}(\widehat{X}_{k+1})\|_p$$
$$\leq [\phi]_{Lip}\|\phi\|_{sup}^{n-k}\left(1 + (2 - \delta_{2,p})\frac{[P]_{Lip}^{n-k+1} - 1}{[P]_{Lip} - 1}\right)\|\Delta_k\|_p$$
$$+ \|\phi\|_{sup}\|v_{k+1}(X_{k+1}) - \widehat{v}_{k+1}(\widehat{X}_{k+1})\|_p$$

with $v_n = \widehat{v}_n \equiv \phi_n$. This yields the expected result after some standard computations. $\qquad \square$

7.3.2 Markovian quantization

Here, we suppose that the dynamics of the $(\mathcal{F}_k)$ Markov chain $(X_k)_k$ is given in the form:

$$X_k = F_k(X_{k-1}, \varepsilon_k), \quad k = 1, \dots, n \qquad (7.32)$$

(starting from some initial state X_0), where $(\varepsilon_k)_k$ is a sequence of identically distributed $\mathcal{F}_k$-measurable random variables in $\mathbb{R}^m$, such that ε_k is independent of $\mathcal{F}_{k-1}$, and F_k is some measurable function on $\mathbb{R}^d \times \mathbb{R}^m$. Given a sequence of grids $\Gamma_k = \{x^1, \dots, x^{N_k}\}$ of N_k points in $\mathbb{R}^d$, associated to a Voronoi tessellation $C_1(\Gamma_k), \dots, C_{N_k}(\Gamma_k), k = 0, \dots, n$, we define the process $(\widehat{X}_k)_k$ by:

$$\widehat{X}_k = \mathrm{Proj}_{\Gamma_k}(F_k(\widehat{X}_{k-1}, \varepsilon_k)), \quad k = 1, \dots, n, \qquad (7.33)$$

and $\widehat{X}_0 = \mathrm{Proj}_{\Gamma_0}(X_0)$. By construction, the process $(\widehat{X}_k)_k$ is still a Markov chain w.r.t. the same filtration $(\mathcal{F}_k)$. Its probability transition matrix $[\hat{p}_k^{ij}]$ at times $k = 1, \dots, n$ reads:

$$\hat{p}_k^{ij} = \mathbb{P}\left[\widehat{X}_k = x_k^j \mid \widehat{X}_{k-1} = x_k^i\right] = \frac{\hat{\beta}_k^{ij}}{\hat{p}_{k-1}^i},$$
$$i = 1, \dots, N_{k-1}, \; j = 1, \dots, N_k, \qquad (7.34)$$

where

$$\hat{p}_{k-1}^i = \mathbb{P}[\widehat{X}_{k-1} = x_{k-1}^i]$$
$$= \begin{cases} \mathbb{P}[F_k(\widehat{X}_{k-2}, \varepsilon_{k-1}) \in C_i(\Gamma_{k-1})], & \text{if } k \geq 2 \\ \mathbb{P}[\widehat{X}_0 \in C_i(\Gamma_0)] & \text{if } k = 1, \end{cases} \qquad (7.35)$$

and
$$\hat{\beta}_k^{ij} = \mathbb{P}[\widehat{X}_{k-1} = x_{k-1}^i, \; \widehat{X}_k = x_k^j],$$
$$\begin{cases} \mathbb{P}[F(\widehat{X}_{k-2}, \varepsilon_{k-1}) \in C_i(\Gamma_{k-1}), \; F_k(\widehat{X}_{k-1}, \varepsilon_k) \in C_j(\Gamma_k)], & \text{if } k \geq 2, \\ \mathbb{P}[\widehat{X}_0 \in C_i(\Gamma_0), \; F(\widehat{X}_0, \varepsilon_1) \in C_j(\Gamma_k)], & \text{if } k = 1. \end{cases}$$

We still intend to estimate the approximation of (X_k) by the Markov quantized process $(\widehat{X}_k)$ along functions $(x_0, x_1, \dots, x_n) \mapsto \phi_0(x_0)\phi_1(x_1)\dots\phi_n(x_n)$, $\phi_0, \dots,$ where ϕ_n are bounded Lipschitz continuous functions on $\mathbb{R}^d$. This time, the quantized process $(\widehat{X}_k)$ itself being a Markov chain, one may compute directly $\mathbb{E}[\phi_0(\widehat{X}_0)\phi_1(\widehat{X}_1)\dots\phi_n(\widehat{X}_n)]$. This quantity can be obtained as the final result of a backward dynamic programming formula formally identical to (7.25) but where the coefficients $[\hat{p}_k^{ij}]$ and $\hat{p}_0^i$ are given by (7.34) and (7.35) i.e., are based on the Markov chain $(\widehat{X}_k)_{0 \leq k \leq n}$ described in (7.33).

We will rely now on a pathwise Lipschitz assumption on the Markov chain $(X_k)_{0 \leq k \leq n}$:

(A1′)   For any $k = 1, \ldots , n$, there exists some positive constant $[F_k]_{Lip}$ such that:

$$\| F_k(x, \varepsilon_k) - F_k(\widehat{x}, \varepsilon_k)\|_1 \leq [F_k]_{Lip}\,|x - \widehat{x}|, \quad \forall x, \widehat{x} \in \mathbb{R}^d.$$

We then set $[F]_{Lip} = \max_{k=1,\ldots,n}[F_k]_{Lip}$.

Theorem 3. *Under* **(A1′)**, *we have the error estimation in the Markov quantization method: for any functions* $\phi_k \in BL_1(\mathbb{R}^d)$, $k = 0, \ldots , n$,

$$\left|\mathbb{E}\left[\phi_0(X_0)\ldots\phi_n(X_n)\right] - \mathbb{E}\left[\phi_0(\widehat{X}_0)\ldots\phi_n(\widehat{X}_n)\right]\right| \leq \sum_{k=0}^{n} \frac{[F]_{Lip}^{n-k+1} - 1}{[F]_{Lip} - 1}\|\Delta_k\|_1,$$

$$\text{(7.36)}$$

$$\text{where } \|\Delta_k\|_1 = \| F_k(\widehat{X}_{k-1}, \varepsilon_k) - \widehat{X}_k\|_1 \qquad \text{(7.37)}$$

is the L^1-*quantization error at time* k *of* $F_k(\widehat{X}_{k-1}, \varepsilon_k)$. *In* (7.36), *we make the usual convention that* $\frac{1}{u-1}(u^m - 1) = m$ *if* $u = 1$ *and* $m \in \mathbb{N}$.

Proof. Set $\|\phi\|_{sup} = \max\limits_{k=0,\ldots,n} \|\phi_k\|_{sup} \leq 1$ and $[\phi]_{Lip} = \max\limits_{k=0,\ldots,n} [\phi_k]_{Lip} \leq 1$. We also denote

$$L_k = \prod_{j=0}^{k} \phi_j(X_j) \text{ and } \widehat{L}_k = \prod_{j=0}^{k} \phi_j(\widehat{X}_j).$$

We then have

$$L_k - \widehat{L}_k = \left(\phi_k(X_k) - \phi_k(\widehat{X}_k)\right) L_{k-1} + (L_{k-1} - \widehat{L}_{k-1})\phi_k(\widehat{X}_k).$$

From the boundedness and Lipschitz conditions on ϕ_k, we deduce that

$$\left|L_k - \widehat{L}_k\right| \leq \|\phi\|_{sup}^{k}[\phi]_{Lip}\left|X_k - \widehat{X}_k\right| + \|\phi\|_{sup}\left|L_{k-1} - \widehat{L}_{k-1}\right|,$$

for all $k = 1, \ldots , n$. By a straightforward backward induction, we get

$$|L_n - \widehat{L}_n| \leq \|\phi\|_{sup}^{n}[\phi]_{Lip} \sum_{k=0}^{n} |X_k - \widehat{X}_k|. \qquad \text{(7.38)}$$

On the other hand, from the definitions (7.32) and (7.33) of X_k and $\widehat{X}_k$, and (7.37) of Δ_k, we obviously get for any $k \geq 1$:

$$\|X_k - \widehat{X}_k\|_1 \leq \| F_k(X_{k-1}, \varepsilon_k) - F_k(\widehat{X}_{k-1}, \varepsilon_k)\|_1 + \|\Delta_k\|_1.$$

By Assumption **(A1′)** and since ε_k is independent of $\mathcal{F}_{k-1}$, we then obtain:

$$\|X_k - \widehat{X}_k\|_1 \leq [F_k]_{Lip}\|X_{k-1} - \widehat{X}_{k-1}\|_1 + \|\Delta_k\|_1.$$

Recalling that $\|X_0 - \widehat{X}_0\|_1 = \|\Delta_0\|_1$, we deduce by backward induction that:

$$\forall k \in \{0, \ldots, n\}, \quad \|X_k - \widehat{X}_k\|_1 \leq \sum_{j=0}^{k} [F]_{Lip}^{k-j} \|\Delta_j\|_1. \tag{7.39}$$

Finally, using (7.38) and (7.39), one completes the proof noting that

$$\left| \mathbb{E}\left[\phi_0(X_0) \ldots \phi_n(X_n)\right] - \mathbb{E}\left[\phi_0(\widehat{X}_0) \ldots \phi_n(\widehat{X}_n)\right] \right| \leq \|L_n - \widehat{L}_n\|_1$$

$$\leq \|\phi\|_{sup}^n [\phi]_{Lip} \sum_{k=0}^{n} \sum_{j=0}^{k} [F]_{Lip}^{k-j} \|\Delta_j\|_1$$

$$\leq \|\phi\|_{sup}^n [\phi]_{Lip} \sum_{j=0}^{n} \left(\sum_{k=0}^{n-j} [F]_{Lip}^{k} \right) \|\Delta_j\|_1. \qquad \square$$

7.3.3   Comparison of both methods

THEORETICAL ASPECTS: The marginal and the Markovian quantization processes were assigned two different objectives. The marginal quantization process is originally designed to optimize the marginal distribution approximation at every time step $k = 0, \ldots, n$, namely

$$\|\widehat{X}_k - X_k\|_p = \min\left\{\|Y - X_k\|_p, \ |Y(\Omega)| \leq N_k\right\}$$

(with in mind some algorithmic stability properties of the grid optimization). Then, at every time k, the conditional distribution $\mathcal{L}(X_{k+1} \mid X_k = x) = P(x, dy)$ for a point $x \in C_i(\Gamma_k)$ (i.e., x in the cell of x_k^i) is approximated by

$$\mathcal{L}(X_{k+1} \mid X_k = x) \approx \mathcal{L}(\mathrm{Proj}_{\Gamma_{k+1}}(X_{k+1}) \mid \mathrm{Proj}_{\Gamma_k}(X_k) = x)$$

$$= \frac{1}{\mathbb{P}_{X_k}(C_i(\Gamma_k))} \int_{C_i(\Gamma_k)} P_{k+1}(x, dy) \circ \mathrm{Proj}_{\Gamma_{k+1}}^{-1} \mathbb{P}_{X_k}(dx). \tag{7.40}$$

This induces a loss of the Markov property.

In contrast, the Markovian quantization is designed at every time $k = 0, \ldots, n-1$ to optimize the approximation of the transition $P_{k+1}(x, dy) = F_{k+1}(x, \mathbb{P}_{\varepsilon_{k+1}})(dy)$ of the chain at the points of the quantizing grid $x_k^i \in \Gamma_k$ (in $L^p(\mathbb{P}_{\widehat{X}_k})$), namely

$$\|\widehat{X}_{k+1} - F_{k+1}(\widehat{X}_k, \varepsilon_{k+1})\|_p = \min\{\|Y - F_{k+1}(\widehat{X}_k, \varepsilon_{k+1})\|_p,$$

$$|Y(\Omega)| \leq N_{k+1}\}.$$

In this approach, for every $x \in C_i(\Gamma_k)$, one approximates

$$\mathcal{L}(X_{k+1} \mid X_k = x) \approx \mathcal{L}(\widehat{X}_{k+1} \mid \widehat{X}_k = x_k^i) = P_{k+1}(x_k^i, dy) \circ \mathrm{Proj}_{\Gamma_{k+1}}^{-1}. \qquad (7.41)$$

Then the Markov property w.r.t. the filtration of $(X_k)_{0 \le k \le n}$ is preserved.

In the marginal quantization, the conditional distributions are not approximated by a specific optimization process, but by averaging the transition w.r.t. the marginal distribution over the Voronoi tessellation of the best possible grid. In the Markovian approach, the conditional distributions are obtained by an optimization procedure which minimizes the error induced at the points of the grid.

One may notice by looking at the a priori estimates (7.26) and (7.36) that, provided that Assumption $(\mathbf{A'1})$ is satisfied, both approaches lead to quite similar a priori error bound structures: they differ by the Lipschitz constants $[P]_{Lip}$ in the marginal quantization and $[F]_{Lip}$ in the Markovian quantization on one hand and by some multiplicative factor (in favor of the Markovian quantization) on the other hand. It is easy to prove that the inequality $[F]_{Lip} \le [P]_{Lip}$ always holds and in many "regular" models (like Lipschitz mixing models, Gaussian Euler schemes of diffusions, etc), the inequality stands as an equality. On the other hand, the multiplicative factor appearing in the marginal quantization is likely to be an artifact of the method of proof. Overall, the assets and drawbacks of both methods essentially annihilate each other.

Then, how to discriminate between the two quantization methods? One first difference lies in the proofs: the general a priori error bounds like (7.36) are significantly easier to get by Markovian approach and so far, provided slightly lower theoretical constants.

When F is the Euler scheme of a diffusion process over $[0, T]$ with Lipschitz coefficients, then $[P]_{Lip} \le [F]_{Lip} = 1 + c_T/n$, where the time step is T/n. Then, in both methods, if one assigns $N/(n + 1)$ elementary quantizers to each grid Γ_k and assumes this grid is optimal, inequalities (7.26) and (7.36) lead to the structure of an a priori global error bound, namely

$$\forall n,\, N \ge 1, \qquad \mathrm{Error}(n, N) = O\left(\frac{n}{(N/n)^{1/d}}\right) = O\left(\frac{n^{1+1/d}}{N^{1/d}}\right).$$

In fact, without any further assumption on the probability density functions of the $\mathcal{L}(X_k)$'s, the above bound is only heuristic since it is impossible to control the rates of convergence in the asymptotics of the n optimal quantization errors. So far, this control turned out to be possible with marginal quantization under some *domination-scaling property* (see e.g., [2] for American option pricing but has no rigorous counterpart with Markovian quantization (see [19] for such a situation). The preservation of the Markov property maybe induces a greater degeneracy of the "innovation process": thus, there is more randomness in $\mathrm{Proj}_{\Gamma_k}(X_k)$ where X_k follows (7.32) than in $\mathrm{Proj}_{\Gamma_k}(F(\widehat{X}_{k-1}, \varepsilon_k))$ in (7.33).

So, when the choice is possible, it seems to be essentially motivated by the constraints of the problem: thus, the Markovian quantization, being a Markov chain w.r.t. the filtration of the original chain X_k, seems more appropriate for control problems (for which it was originally designed ...) whereas marginal quantization yield more satisfactory results in optimal stopping problems (for which it was originally designed ...). But once again, it may be only an artifact.

Let us mention however that the marginal quantization requires only some weak convergence Lipschitz assumption on the chain (namely $[P]_{Lip} < +\infty$) while the Markovian quantization requires some L^p-pathwise Lipschitz assumption (namely $[F]_{Lip} < +\infty$). It may happen that the first approach turns out to be the only available one because

$$[P]_{Lip} < +\infty = [F]_{Lip}.$$

This is, for example, the case for Markovian dynamics like

$$X_{k+1} = F(X_k, \varepsilon_{k+1}) \quad \text{with} \quad F(x, \varepsilon) := \text{sign}(x - \varepsilon) G(x, \varepsilon),$$

where $(\varepsilon_k)_k$ is an i.i.d. sequence, $\mathbb{P}_{\varepsilon_1}(du) = g(u)\,\lambda_q(du)$ (λ_q lebesgue measure on $\mathbb{R}^q$) and $(x, \varepsilon) \mapsto G(x, \varepsilon)$ is Lipschitz continuous in x uniformly in ε with ratio $[G]_{Lip}$. Then, one shows that

$$[P]_{Lip} = [G]_{Lip} < +\infty \text{ whereas } x \mapsto F(x, \varepsilon_1) = \text{sign}(x - \varepsilon_1) G(x, \varepsilon_1)$$

is not continuous.

COMPUTATIONAL ASPECTS: Although, both dynamic programming formulae are formally identical and the fact that, in both cases, the grid optimization phase consists in processing a chain of stochastic gradient descents, one for each time step, the optimization phases are radically different for the marginal and the Markovian quantization processes. Since these procedures have been extensively described in [2] and [19], we refer to these papers for details of implementation.

We wish to discuss here what make them different. First, they lead to different optimal grids with different transition matrices (using the same set of grids to process the marginal and the Markovian methods would provide two different sets of transition matrices).

In the *marginal quantization*, the optimization consists in two steps

— Computation for every $k = 0, \ldots, n$ of grids Γ_k^* which minimize over all grids of size N_k, the L^p-quantization error $\|\Delta_k\|_p$ i.e., solving

$$\text{argmin}_{\Gamma_k}\{\|X_k - \text{Proj}_{\Gamma_k}(X_k)\|_p, |\Gamma_k| \le N_k\} \tag{7.42}$$

— Estimation of the companion parameters i.e., the resulting transition matrices $[\hat{p}_k^{ij}]$ and the quantization errors $\|X_k - \text{Proj}_{\Gamma_k^*}(X_k)\|_p$.

At every step k, the optimization problem (7.42) *only depends on the distribution of X_k*. The main consequence is that if one looks carefully at the recursive stochastic algorithm described in Section 2.2., the optimization of the grid Γ_k at the kth time step only depends on the simulation of a large number M of independent copies of X_k. So if one simulates on a computer M independent paths of the whole chain $(X_k)_{0 \le k \le n}$, all the grids can be optimized independently by simply implementing procedures (7.10), (7.11).

The estimation of the companion parameters can be carried out "on line" as described in the algorithm of Section 2.2 using (7.12) and (7.13). It may be more efficient to carry on the companion parameter estimation after the grid optimization is achieved: once the optimal grids are settled, the companion parameter estimation procedure becomes a standard Monte Carlo simulation.

At a first glance, in the *Markovian quantization*, the two steps look similar. However, since $\widehat{X}_k = \text{Proj}_{\Gamma_k}(F(\widehat{X}_{k-1}, \varepsilon_k))$, the L^p-optimization problem for the kth grid Γ_k^* reads

$$\text{argmin}_{\Gamma_k}\{\|F(\widehat{X}_{k-1}, \varepsilon_k) - \text{Proj}_{\Gamma_k}(F(\widehat{X}_{k-1}, \varepsilon_k))\|_p, \ |\Gamma_k| \leq N_k\}. \qquad (7.43)$$

Consequently, *the optimization of the grids Γ_k at time k does depend on the distribution of $\widehat{X}_{k-1}$*, i.e., essentially upon Γ_{k-1}^*. This means that the grid optimization phase of a quantized Markov chain is deeply recursive: any optimization default at time k is propagated at times $\ell \geq k$, inducing a great instability of the global optimization process.

This provides an interpretation for a usually observed phenomenon: numerical grid optimization works much better with marginal quantization than Markovian quantization. It is in accordance with the idea that it is more difficult to estimate accurately conditional distributions than marginal ones.

7.4   Some applications in finance

7.4.1   *Optimal stopping problems and pricing of multi-dimensional American options*

We consider a multidimensional diffusion $X = (X^1, \ldots, X^d)^*$ governed by:

$$dX_t = b(X_t)dt + \sigma(X_t)dW_t, \quad X_0 = x_0, \qquad (7.44)$$

where b, σ are functions on $\mathbb{R}^d$ valued in $\mathbb{R}^d$ and $\mathbb{R}^{d \times m}$, satisfying usual growth and Lipschitz conditions, and W is an m-dimensional standard Brownian motion on a filtered probability space $(\Omega, \mathcal{F}, \mathbb{F} = (\mathcal{F}_t)_t, \mathbb{P})$.

Given a reward process $(g(t, X_t))_{t \in [0,T]}$, where g is some continuous function on $[0, T] \times \mathbb{R}^d$, Lipschitz continuous in x, we consider the optimal stopping problem:

$$V_t = \text{ess sup}_{\tau \in \mathcal{T}_{t,T}} \mathbb{E}[g(\tau, X_\tau)|\mathcal{F}_t]. \qquad (7.45)$$

Here $\mathcal{T}_{t,T}$ denotes the set of stopping times valued in $[t, T]$ and V is called the Snell envelope of $(g(t, X_t))_{t \in [0,T]}$.

We first approximate this continuous-time optimal stopping problem by a discrete-time optimal stopping problem where the set of possible stopping times is valued in $\{kT/n : k = 0, \ldots, n\}$ for n large. When the diffusion X is not simulatable, we approximate it by a discretization scheme, and we denote by $\overline{X}_k$ this approximation at time $t_k = kT/n$ of X. For example, in the case of an Euler scheme with step T/n, we have:

$$\overline{X}_0 = x_0, \qquad \overline{X}_{k+1} = \overline{X}_k + b(\overline{X}_k)\frac{T}{n} + \sigma(\overline{X}_k)\sqrt{\frac{T}{n}}\varepsilon_{k+1}$$

$$=: F(\overline{X}_k, \varepsilon_{k+1}), \quad k = 0, \dots, n-1,$$

where $\varepsilon_{k+1} = (W_{t_{k+1}} - W_{t_k})/\sqrt{h}$ is a centered Gaussian random variable in $\mathbb{R}^m$ with variance I_m, independent of $\overline{\mathcal{F}}_k := \mathcal{F}_{t_k}$. The process $(\overline{X}_k)$ is a Markov chain w.r.t. the filtration $(\overline{\mathcal{F}}_k)$. The associated discrete-time optimal stopping problem is:

$$\overline{V}_k = \text{ess sup}_{\tau \in \overline{\mathcal{T}}_{k,n}} \mathbb{E}\left[g(\tau T/n, \overline{X}_\tau)|\overline{\mathcal{F}}_k\right], \qquad (7.46)$$

where $\overline{\mathcal{T}}_{k,n}$ denotes the set of stopping times (with respect to the filtration $(\overline{\mathcal{F}}_k)$) valued in $\{j : j = k, \dots, n\}$.

We have the classical time discretization error estimation:

$$\max_{k=0,\dots,n} \|V_{kT/n} - \overline{V}_k\|_p \leq \frac{C_{b,\sigma}}{\sqrt{n}}$$

In fact, if g is slightly more regular, namely *semi-convex* and if one replaces the Euler scheme by the diffusion itself sampled at times kT/n, the above bound holds with $\frac{C_{b,\sigma}}{n}$.

It is well known that the Snell envelope $(\overline{V}_k)_k$ of $(g(t_k, \overline{X}_k))_k$ satisfies $\overline{V}_k = \overline{v}_k(\overline{X}_k), k = 0, \dots, n$, where the Borel functions $\overline{v}_k$ on $\mathbb{R}^d$ are given by the backward dynamic programming formula:

$$\overline{v}_n(x) = g(T, x), \forall x \in \mathbb{R}^d,$$

$$\overline{v}_k(x) = \max\left(g(t_k, x), \mathbb{E}[\overline{v}_{k+1}(\overline{X}_{k+1})|\overline{X}_k = x]\right),$$

$$\forall x \in \mathbb{R}^d, k = 0, \dots, n-1.$$

This backward formula remains intractable for numerical computations since it requires one to compute at each time step $k = 0, \dots, n$, conditional expectations of $\overline{X}_{k+1}$ given $\overline{X}_k = x$ at any point $x \in \mathbb{R}^d$ of the state space of $(\overline{X}_k)$.

The quantization approach for solving this problem is to first approximate the Markov chain $(\overline{X}_k)$ by a quantized Markov chain as described in Section 3. This means that at each time $t_k, k = 0, \dots, n$, we are given an (optimal) grid $\Gamma_k = \{x_k^1, \dots, x_k^{N_k}\}$ of N_k points in $\mathbb{R}^d$, and we approximate the distribution $\overline{X}_0$ by the distribution of $\widehat{X}_0 = \text{Proj}_{\Gamma_0}(\overline{X}_0)$, and the conditional distribution of $\overline{X}_{k+1}$ given $\overline{X}_k$ by the conditional distribution of $\widehat{X}_{k+1}$ given $\widehat{X}_k$: for $k \geq 1$, $\widehat{X}_k$ is defined by $\widehat{X}_k = \text{Proj}_{\Gamma_k}(\overline{X}_k)$ in the marginal quantization method, while $\widehat{X}_k = \text{Proj}_{\Gamma_k}(F(\widehat{X}_{k-1}, \varepsilon_k))$ in the Markovian quantization method. We then approximate the functions $\overline{v}_k$ by the functions $\widehat{v}_k$ defined on $\Gamma_k, k = 0, \dots, n$, by the backward dynamic programming formula or *optimal quantization tree descent*:

$$\widehat{v}_n(x_n^i) = g(T, x_n^i), \quad \forall x_n^i \in \Gamma_n, \tag{7.47}$$

$$\widehat{v}_k(x_k^i) = \max\left(g(t_k, x_k^i), \mathbb{E}[\widehat{v}_{k+1}(\widehat{X}_{k+1})|\widehat{X}_k = x_k^i]\right) \tag{7.48}$$

$$= \max\left(g(t_k, x_k^i), \sum_{j=1}^{N_{k+1}} \widehat{p}_{k+1}^{ij} \widehat{v}_{k+1}(x_{k+1}^j)\right),$$

$$x_k^i \in \Gamma_k, \ k = 0, \ldots, n-1, \tag{7.49}$$

where $\widehat{p}_{k+1}^{ij} = \mathbb{P}\left[\widehat{X}_{k+1} = x_{k+1}^j | \widehat{X}_k = x_k^i\right]$.

Then one gets an approximation of the process $(\overline{V}_k)$ by the process $(\widehat{V}_k)_k$, with $\widehat{V}_k = \widehat{v}_k(\widehat{X}_k)$. Namely, if the diffusion is uniformly elliptic, with coefficients b and σ either bounded Lipschitz continuous or $C_b^\infty(\mathbb{R}^d)$ and if the obstacle function g is Lipschitz over $[0, T] \times \mathbb{R}^d$, then the following error estimation holds for an L^p-optimal marginal quantization (see [2]):

$$\max_{0 \le k \le n} \|\overline{V}_k - \widehat{V}_k\|_p \le C_{b,\sigma,T,p} \frac{n^{1+1/d}}{N^{1/d}}, \tag{7.50}$$

where $N = \sum_{k=0}^n N_k$ is the total number of points to be dispatched among all grids Γ_k.

This estimate strongly relies on the sub-Gaussian upper-bound for the probability density of the diffusion density. The same bound holds if one substitutes the diffusion sampled at times $t_k, k = 0, \ldots, n$ to its Euler scheme.

7.4.1.1   Numerical illustration

As a numerical illustration, we consider a 2-dimensional uncorrelated Black–Scholes model with geometric dividends, i.e., for $x = (x^1, \ldots, x^{2d})^* \in \mathbb{R}^{2d}$, $b(x) = -(\mu^1 x^1, \ldots, \mu^{2d} x^{2d})$, $\sigma(x)$ is a $2d \times 2d$ diagonal matrix with ith diagonal term $\sigma_i x^i$, where $\sigma_i, i = 1, \ldots, d$ are constant volatilities. We assume that the short-term interest rate is zero. The American option price at time t of a payoff function $(g(X_t))$ is given by:

$$V_t = \text{ess sup}_{\tau \in \mathcal{T}_{t,T}} \mathbb{E}[g(X_\tau)|\mathcal{F}_t], \tag{7.51}$$

which is computed by the above algorithm. We consider an American 2-dimensional exchange option characterized by the payoff

$$g(t, x) = \max\left(x^1 \ldots x^d - x^{d+1} \ldots x^{2d}, 0\right)$$

with the market parameters

$$x_0^1 \ldots x_0^d = 36, \ x_0^{d+1} \ldots x_0^{2d} = 40, \ \sigma_i = 20d^{-\frac{1}{2}}\%, \ \mu^1 = 5\%, \ \mu^2 = \ldots = \mu^{2d} = 0.$$

Our reference price is obtained by a specific difference method devised in [22] for two dimensions. We reproduce in Figures 3 and 4 for $2d = 4$ and 6 the graphs

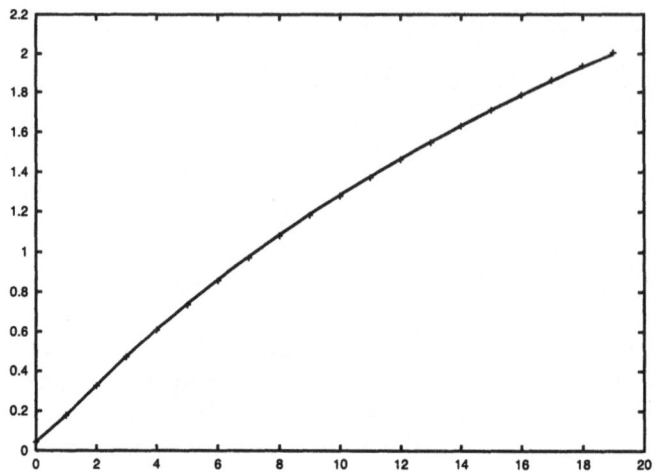

FIGURE 3. American exchange option in dimension 4 (out-of-the-money case). The reference price is depicted by a line and the quantized price by the cross line.

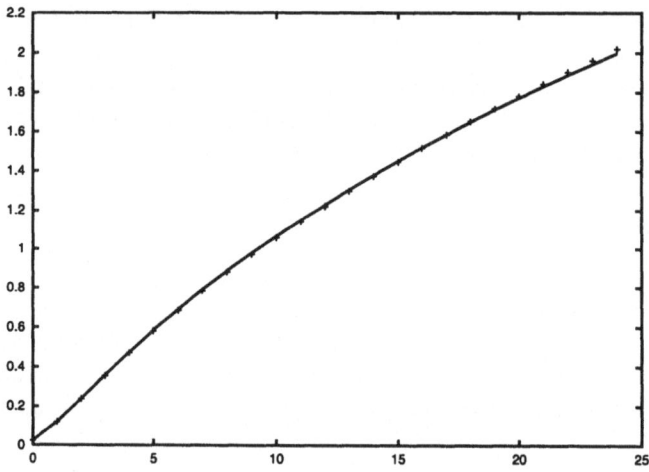

FIGURE 4. American exchange option in dimension 6 (out-of-the-money case). The reference price is depicted by a line and the quantized price by the cross line.

$\theta \mapsto \widehat{V}_0(\theta)$ where $\widehat{V}_0(\theta)$ denotes the premium at time 0 of the above American option when time to maturity θ runs over $\{k/n, \ k = 1, \dots, n\}$. The numerical parameters are settled as follows: $n = 25$, k time discretization steps when $\theta = k/n$, and

$$N_{25} = 750 \quad \text{if} \quad 2d = 4 \quad \text{and} \quad N_{25} = 1\,000 \quad \text{if} \quad 2d = 6.$$

The sizes N_k of the grid Γ_k are specified following the dispatching rule given in [2].

7.4.2   A stochastic control problem: mean-variance hedging of options

7.4.2.1   Error bounds using the Markovian quantization

We consider the following portfolio optimization problem. The dynamics of the controlled process is governed by:

$$dX_t = b(X_t)\,dt + \sigma(X_t)\,dW_t,$$
$$dY_t = \alpha_t^* dX_t, \qquad\qquad Y_0 = y_0,$$

where b, σ are functions on $\mathbb{R}^d$ valued in $\mathbb{R}^d$ and $\mathbb{R}^d \times m$, satisfying usual growth and Lipschitz conditions, and W is an m-dimensional standard Brownian motion on a filtered probability space $(\Omega, \mathcal{F}, \mathbb{F} = (\mathcal{F}_t)_t, \mathbb{P})$. The control process $\alpha = (\alpha_t)_t$ is an $\mathbb{F}$-adapted process valued in some subset A of $\mathbb{R}^d$. We denote by $\mathcal{A}$ the set of such control processes. Here, $X = (X^1, \dots, X^d)^*$ represents the dynamics of risky assets and/or volatility, Y is the (self-financed) wealth process of an investor who can trade α_t shares of risky assets at time t, and starting from some initial capital y_0. The set A models the constraints on the portfolio held by the investor. For example, if $A = \mathbb{R} \times \{0\}^{d-1}$, this means that the investor can trade only in the first asset.

We are now given an option written on the risky assets, i.e., a payoff function in the form $g(X_T)$, for some Lipschitz continuous function g on $\mathbb{R}^d$, that one wants to hedge with the available risky assets, and according to a quadratic criterion. In other words, one has to solve the stochastic control problem:

$$v(t, x, y) = \inf_{\alpha \in \mathcal{A}} \mathbb{E}\left[(g(X_T) - Y_T)^2 \Big|\right.$$
$$\left.(X_t, Y_t) = (x, y)\right], \quad (t, x, y) \in [0, T] \times \mathbb{R}^d \times \mathbb{R}. \quad (7.52)$$

We first approximate the continuous-time control problem (7.52) by a discrete-time control problem at times $t_k = kT/n$, $k = 0, \dots, n$ for n large. We consider for (X_t) an approximation by the Euler scheme with step $h = T/n$. The approximation $\overline{X}_k$ of X_{t_k} is then defined by:

$$\overline{X}_0 = X_0 \qquad \text{and} \qquad \overline{X}_{k+1} = \overline{X}_k + b(\overline{X}_k)\frac{T}{n} + \sigma(\overline{X}_k)\sqrt{\frac{T}{n}}\,\varepsilon_{k+1}$$
$$=: F(\overline{X}_k, \varepsilon_{k+1}), \quad k = 0, \dots, n-1,$$

where $\varepsilon_{k+1} = (W_{t_{k+1}} - W_{t_k})/\sqrt{h}$ is a centered Gaussian random variable in $\mathbb{R}^m$ with variance I_m, independent of $\overline{\mathcal{F}}_k := \mathcal{F}_{t_k}$. The process $(\overline{X}_k)$ is a Markov chain w.r.t. the filtration $(\overline{\mathcal{F}}_k)$. We denote by $\overline{\mathcal{A}}$ the set of all $\{\mathcal{F}_k, k = 0, \dots, n-1\}$-adapted processes $\overline{\alpha} = \{\overline{\alpha}_k, k = 0, \dots, n-1\}$ valued in A. Given $\overline{\alpha} \in \overline{\mathcal{A}}$, we consider the approximation $(\overline{Y}_k)$ of the controlled process (Y_t) at times (t_k), and defined by:

$$\overline{Y}_0 = y_0 \qquad \text{and} \qquad \overline{Y}_{k+1} = \overline{Y}_k + \overline{\alpha}_k^*(\overline{X}_{k+1} - \overline{X}_k)$$
$$=: G(\overline{Y}_k, \overline{X}_k, \overline{\alpha}_k, \overline{X}_{k+1}), \quad k = 0, \dots, n-1.$$

We then consider the stochastic control problem in discrete-time:

$$\overline{v}_k(x, y) = \inf_{\overline{\alpha} \in \overline{\mathcal{A}}} \mathbb{E}\left[\left(g(\overline{X}_n) - \overline{Y}_n\right)^2 \Big| (\overline{X}_k, \overline{Y}_k) = (x, y)\right], \qquad (7.53)$$

for all $k = 0, \ldots, n$ and $(x, y) \in \mathbb{R}^d \times \mathbb{R}$. The convergence from the discrete-time control problem to the continuous one may be proved either by probabilistic arguments (see [16]) or by viscosity solutions approach (see [7]):

$$\overline{v}_k(x, y) \to v(t, x, y),$$

for all $(x, y) \in \mathbb{R}^d \times \mathbb{R}$, as n goes to infinity and $t_k \to t$.

The functions $\overline{v}_k$ satisfy the dynamic programming formula:

$$\overline{v}_n(x, y) = (g(x) - y)^2, \quad (x, y) \in \mathbb{R}^d \times \mathbb{R},$$
$$\overline{v}_k(x, y) = \inf_{a \in A} \mathbb{E}\left[\overline{v}_{k+1}(\overline{X}_{k+1}, \overline{Y}_{k+1}) \big| (\overline{X}_k, \overline{Y}_k) = (x, y)\right],$$

$$k = 0, \ldots, n - 1, \quad (x, y) \in \mathbb{R}^d \times \mathbb{R}.$$

From a numerical viewpoint, this backward formula remains intractable since we have to compute at each time step, conditional expectations of $(\overline{X}_{k+1}, \overline{Y}_{k+1})$ given $(\overline{X}_k, \overline{Y}_k) = (x, y)$ at every point (x, y) of the state space $\mathbb{R}^d \times \mathbb{R}$. With respect to optimal stopping problems, we have in addition to calculate an infimum of these conditional expectations over the possible values of the control set A.

The starting point in the quantization approach for solving (7.53) is to discretize the controlled $(\overline{\mathcal{F}}_k)$-Markov chain $(\overline{X}_k, \overline{Y}_k)_k$ by a controlled Markov chain $(\widehat{X}_k, \widehat{Y}_k)_k$ valued in a finite state space. Here, recall that $(\overline{X}_k)_k$ is an uncontrolled process while $(\overline{Y}_k)_k$ is a one-dimensional controlled process. We shall then consider two different spatial discretizations for $(\overline{X}_k)_k$ and $(\overline{Y}_k)_k$. Moreover, we also want to keep the Markov property of the controlled quantized Markov chain w.r.t. the same filtration $(\overline{\mathcal{F}}_k)$. This means that we want to approximate the control problem (7.53) by another control problem where the controls are still adapted w.r.t. the filtration $(\overline{\mathcal{F}}_k)$. More precisely, we shall discretize the d-dimensional process $(\overline{X}_k)$ on an optimal grid $\Gamma_k = \{x_k^1, \ldots, x_k^{N_k}\}$ at each time k and define a Markovian quantization of $(\overline{X}_k)$ by:

$$\widehat{X}_0 = \mathrm{Proj}_{\Gamma_0}(X_0)$$

and $\qquad \widehat{X}_{k+1} = \mathrm{Proj}_{\Gamma_{k+1}}\left(F(\widehat{X}_k, \varepsilon_{k+1})\right), \quad k = 0, \ldots, n - 1.$

The controlled one-dimensional process $(\overline{Y}_k)$ is discretized using a regular orthogonal grid of $\mathbb{R}$, namely $\Gamma^Y = (2\delta)\mathbb{Z} \cap [-R, R]$, and we then define:

$$\widehat{Y}_0 = y_0$$

and $\qquad \widehat{Y}_{k+1} = \mathrm{Proj}_{\Gamma^Y}\left(G(\widehat{Y}_k, \widehat{X}_k, \widehat{\alpha}_k, \widehat{X}_{k+1})\right), \quad k = 0, \ldots, n - 1.$

Therefore, $(\widehat{X}_k, \widehat{Y}_k)$ is a controlled Markov chain w.r.t. $(\overline{\mathcal{F}}_k)$. We then consider the stochastic control problem in discrete-time:

$$\widehat{v}_k(x, y) = \inf_{\overline{\alpha} \in \overline{\mathcal{A}}} \mathbb{E}\left[\left(g(\widehat{X}_n) - \widehat{Y}_n\right)^2 \Big| (\widehat{X}_k, \widehat{Y}_k) = (x, y) \right], \tag{7.54}$$

for all $k = 0, \ldots, n$ and $(x, y) \in \Gamma_k \times \Gamma^Y$. By the dynamic programming principle, functions $\widehat{v}_k$ are computed recursively by:

$$\widehat{v}_n(x, y) = (g(x) - y)^2, \quad (x, y) \in \Gamma_n \times \Gamma^Y,$$
$$\widehat{v}_k(x, y) = \inf_{a \in A} \mathbb{E}\left[\widehat{v}_{k+1}(\widehat{X}_{k+1}, \widehat{Y}_{k+1}) \Big| (\widehat{X}_k, \widehat{Y}_k) = (x, y) \right],$$
$$k = 0, \ldots, n-1, \quad (x, y) \in \Gamma_k \times \Gamma^Y.$$

From an algorithmic point of view, this reads:

$$\widehat{v}_n(x_n^i, y) = (g(x_n^i) - y)^2, \quad \forall x_n^i \in \Gamma_n, \ \forall y \in \Gamma^Y,$$

$$\widehat{v}_k(x_k^i) = \inf_{a \in A} \sum_{j=1}^{N_{k+1}} \widehat{p}_{k+1}^{ij} \widehat{v}_{k+1}\left(x_{k+1}^j, \mathrm{Proj}_{\Gamma^Y}\left(G(y, x_k^i, a, x_{k+1}^j) \right) \right)$$

$$\forall x_k^i \in \Gamma_k, \ \forall y \in \Gamma^Y, \quad k = 0, \ldots, n-1,$$

where $\quad \widehat{p}_{k+1}^{ij} = \mathbb{P}\left[\widehat{X}_{k+1} = x_{k+1}^j \big| \widehat{X}_k = x_k^i \right].$

It is proved in [19] that the estimation error for the value functions by this quantization method is measured by:

$$\mathbb{E}\left| \overline{v}_0(\overline{X}_0, y_0) - \widehat{v}_0(\overline{X}_0, y_0) \right|$$

$$\leq C_1(1 + |y_0|)\left(\frac{1}{\sqrt{n}} \sum_{k=0}^{n}(n-k)\|\Delta_k\|_2 + \sum_{k=0}^{n} \|\Delta_k\|_2 \right)$$

$$+ C_2 n\delta + C_3(1 + |y_0|^{\overline{p}}) \frac{n}{R^{\overline{p}-1}},$$

for all $\overline{p} > 1$ and $y_0 \in \mathbb{R}$. Here, C_1, C_2 and C_3 are positive constants depending on the coefficients of the diffusion process X and

$$\|\Delta_k\|_2 = \|F(\widehat{X}_{k-1}, \varepsilon_k) - \widehat{X}_k\|_2$$

is the L^2-quantization error at date k in the Markovian approach.

7.4.2.2   Numerical illustrations

As a numerical illustration, we consider the two following models:

A stochastic volatility model (2-dim X-process) Let $X = (X^1, X^2)$ be governed by:

$$dX_t^1 = X_t^1 X_t^2 dW_t^1,$$

$$dX_t^2 = -\eta(X_t^2 - \overline{\sigma})dt + \beta dW_t^2, \quad X_0^2 \sim \mathcal{N}(\overline{\sigma}; \frac{\beta^2}{2\eta}),$$

where (W^1, W^2) is a standard two-dimensional Brownian motion. Here X^1 represents the price process of one risky asset and X^2 is the (stationary) stochastic volatility process of the risky asset. The investor trades only in the risky asset X^1, i.e., $A = \mathbb{R} \times \{0\}$, and he wants to hedge a put option on this asset, i.e., $g(x) = (K - x^1)_+$ for $x = (x^1, x^2)$.

By projecting $(K - X_T^1)_+$ on the set of stochastic integrals w.r.t. S, we have by Itô's formula:

$$(K - X_T^1)_+ = \mathbb{E}[(K - X_T^1)_+ \mid X_0^2] + \int_0^T \alpha_s^{opt} dX_s^1 + Z_T$$

where $\alpha_t^{opt} = \frac{\partial P}{\partial s}(t, X_t^1, X_t^2)$, $Z_T = \beta \int_0^T \frac{\partial P}{\partial s}(t, X_t^1, X_t^2)dW_t^2$ with $P(t, x^1, x^2) = \mathbb{E}[(K - X_T^1)_+ \mid (X_t^1, X_t^2)) = (x^1, x^2)]$.

Then, the function $y_0 \mapsto \mathbb{E}[v_0(y_0, x_0^1, X_0^2)]$ reaches its minimum at $y_{\min} = \mathbb{E}[(K - X_T^1)_+)]$. So, the optimal control is always α^{opt} regardless of y. Since the volatility process X_t^2 is independent of W^1, we notice by Jensen's inequality that

$$\mathbb{E}[(K - X_T^1)_+] \geq \text{Put}_{B\&S}(x_0^1, K, T, \overline{\sigma}).$$

PARAMETERS FOR NUMERICAL IMPLEMENTATION :
- Model parameters: $T = 1, \overline{\sigma} = 20\%, \eta = 0.5, \beta = 0.05, x_0^1 = K = 100$.
- Time discretization: $n = 25$.
- Spatial discretization (quantization grid parameters):
  — Grid Γ^X: $2\delta = \frac{1}{20}, n_X = 50$ (i.e., $|\Gamma^X| = 2 \times 100 + 1$), centered at $I_0 = 7.96$.
  — Grids Γ_k: Total numbers of points used to produce the $n = 25$ grids that quantize the Euler scheme of (S, σ), $N = 5\,750$ $(N_{25} = 299)$.
  – Optimization of the grids using $M = 10^6$ independent trials of the Euler scheme.
- Approximation of the optimal control: dichotomy method on $A = [-1, 0]$.

NUMERICAL RESULTS: Figure 5 below depicts a quantization of $X_T = (X_T^1, X_T^2)$ using $N_{25} := 299$ points obtained as a result of an optimization process described above

Figure 6 and Figure 7 display the computed graph of $y \mapsto \mathbb{E}[\widehat{v}_0(y, x_0^1, \widehat{X}_0^2)]$ and the value of the optimal control α_0^{opt} at $t = 0$ respectively. The global shape of the graph is parabolic and reaches its minimum at $y_{min} = 8.06$. This is to be compared with the premium provided by a direct Monte Carlo simulation, namely 8.00. The optimal control is nearly constant and its value at $y_{min} = 8.06$, $\alpha_0^{mv}(y_{min}) = -0.38$, is satisfactory w.r.t. the theoretical value estimated by Monte Carlo (-0.34).

A put on a Black & Scholes geometric asset (4-dim X-process) We consider a Black & Scholes model $X = (X^1, \ldots, X^4)$ in four dimensions: $b = 0$ and $\sigma(x)$ is

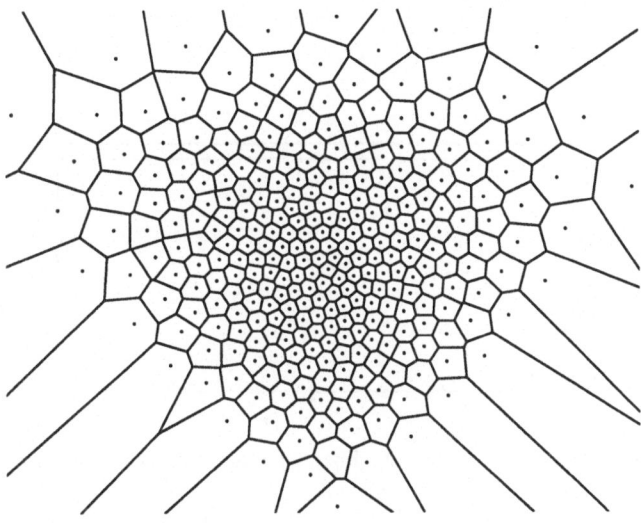

FIGURE 5. L^2-optimal 300-quantizer of $(\log(X_T^1), X_T^2)$ with its Voronoi tessellation.

the 4×4 diagonal matrix with i^{th} diagonal term $\sigma_i(x) := \sigma_i x^i$. The payoff function to be hedged is a geometric put option on $J_t = X_t^1 \ldots X_t^d$:

$$g(X_T) = (K - J_T)_+.$$

The investor is allowed to trade only in the first asset X^1 hence $A = \mathbb{R} \times \{0\}^3$.

So, the mean variance hedging problem of the investor at time $t = 0$ is

$$v(0, x_0, y_0) = \min_{\alpha \in \mathcal{A}} \mathbb{E}\left[\left(Y_T^{y_0, \alpha} - (K - J_T)_+ \right)^2 \right] \tag{7.55}$$

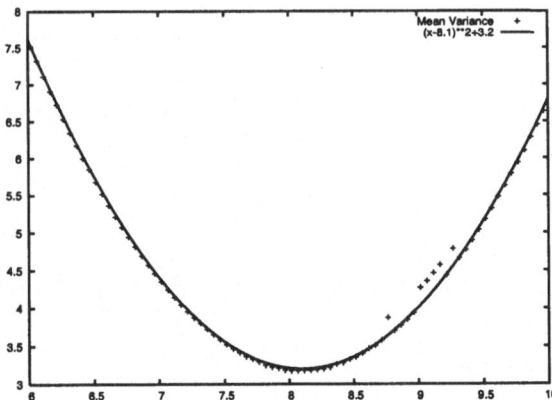

FIGURE 6. Graph of $y \mapsto \mathbb{E}[\widehat{v}_0(y, x_0^1, \widehat{X}_0^2)]$ (dot line), closest parabola (thick line).

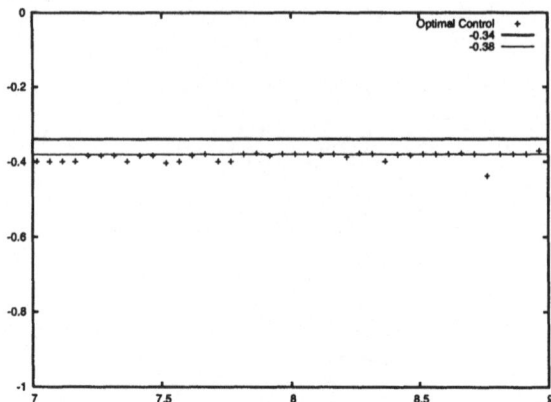

FIGURE 7. Quantized optimal control (dot line). Optimal control computed by a Monte Carlo simulation.

where x_0 is the initial vector value risky asset. Itô's formula then classically yields

$$(K - J_T)_+ = \mathbb{E}[(K - J_T)_+] + \int_0^T H_t \, dJ_t$$

$$\text{with } H_t = \frac{\partial P}{\partial x}(T - t, J_t, K, \sigma), \ t \in [0, T),$$

where $P(\theta, x, K, \sigma)$ denotes the price of a one-dimensional European put option with residual maturity θ, asset price x, strike price K, constant volatility σ. It follows that

$$\mathbb{E}\left[(Y_T^{y_0,\alpha} - (K - J_T)_+)^2\right] = \left(y_0 - P(T, J_0, K, \sigma)\right)^2$$

$$+ \sigma_1^2 \int_0^T \mathbb{E}\,(J_s H_s - \alpha_s X_s^1)^2 \, ds$$

$$+ \sum_{i=2}^d \sigma_i^2 \int_0^T \mathbb{E}\,(J_s H_s)^2 \, ds.$$

Hence, the solution of (7.55) is given by

$$v_0(x_0, y_0) = \left(y_0 - P(T, J_0, K, \sigma)\right)^2 + \sum_{i=2}^d \sigma_i^2 \int_0^T \mathbb{E}\,(J_s H_s)^2 \, ds$$

using the optimal control

$$\alpha_t^{opt} = \frac{J_t H_t}{S_t^1} = X_t^2 \ldots X_t^d \frac{\partial P}{\partial x}(T - t, J_t, K, \sigma).$$

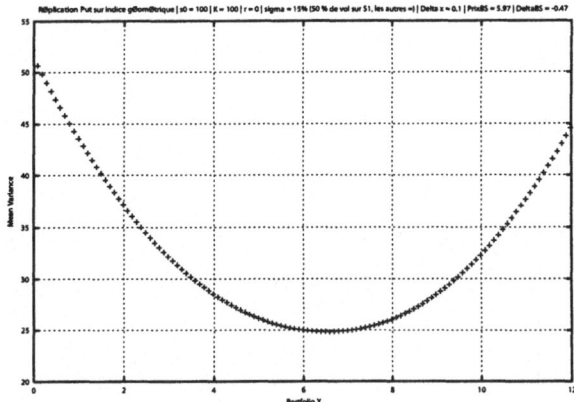

FIGURE 8. Graph of $x_0 \mapsto v_0(x_0, s_0)$ (dot line), closest parabola (thick line).

In the above model, the non-correlation assumption of the assets may look not very realistic but corresponds to the most difficult case to solve for quantization since it corresponds in some way to a "full d-dimensional problem".

PARAMETERS FOR NUMERICAL IMPLEMENTATION:

• Model parameters: $d = 4$, $T = 1$, $\sigma = 15\%$, $\sigma_1 = \sigma/\sqrt{2}$ and $\sigma_i = \sigma/\sqrt{2(d-1)}$, $i = 2, \ldots, d$, $X_0^i = (100)^{1/d}$, $i = 1, \ldots, d$, $K = 100$.

• Spatial discretization (quantization grid parameters):

— Grid Γ^Y: $2\delta := \frac{1}{10}$, $n_Y := 2\,00$ (i.e., $|\Gamma^X| = 2 \times 2\,00 + 1$), centered at $I_0 = 5.97$ ($B\&S$ premium of the put option with volatility σ).

— Grids Γ_k: Total numbers of points used to produce the $n = 20$ grids that quantize the geometric Brownian motion S: $N = 22\,656$ ($N_{20} := 1\,540$).

— Optimization of the grids using $M := 10^6$ independent trials of the Euler scheme.

• Approximation of the optimal control: dichotomy method on $A \subset [-2, 0]$.

NUMERICAL RESULTS: Figure 8 and Figure 9 below display the computed graphs $x_0 \mapsto v_0(x_0, s_0)$ and $y_0 \mapsto \alpha_0^{opt}(y_0, I_0)$. The global shape of the graph is parabolic, reaches its minimum (equal to 25.82) at $y_{min} = 6.27$ (the true value is 5.97, $B\&S$ premium for the put). The optimal control is satisfactorily constant as expected; its value at y_{min} is $\alpha_0^{opt}(y_{min}) = -0.48$ (true value is -0.47).

7.4.3 Filtering of stochastic volatility models

We consider the following filtering model. The signal (X_k) is an $\mathbb{R}^d$-valued Markov chain given by:

$$X_k = F(X_{k-1}, \varepsilon_k), \quad k \in \mathbb{N}^*, \tag{7.56}$$

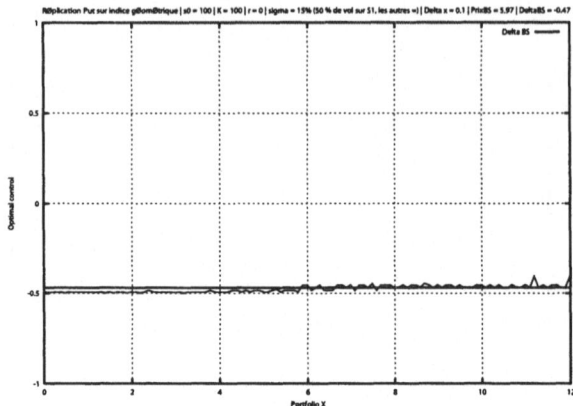

FIGURE 9. Quantized optimal control with theoretical optimal control (straight line).

where $(\varepsilon_k)_k$ is a sequence of i.i.d. random variables valued in $\mathbb{R}^m$, and F is some measurable function on $\mathbb{R}^d \times \mathbb{R}^m$. The initial distribution of X_0 is known equal to μ. The observation process valued (Y_k) valued in $\mathbb{R}^q$ takes the form:

$$Y_k = G(X_k, \eta_k), \quad k \in \mathbb{N}, \tag{7.57}$$

where $(\eta_k)_k$ is a sequence of i.i.d. random variables in $\mathbb{R}^l$, independent of $(\varepsilon_k)_k$, and G is a measurable function on $\mathbb{R}^d \times \mathbb{R}^l$. We assume that for every $x \in \mathbb{R}^d$, the random variable $G(x, \eta_1)$ admits a bounded density $y \mapsto g(x, y)$ w.r.t. the Lebesgue measure on $\mathbb{R}^q$.

We are interested in the computation at some time $n \geq 1$, of the conditional distribution $\Pi_{y,n}$ of the signal X_n given the observations $(Y_0, \ldots, Y_n)$ fixed to $y = (y_0, \ldots, y_n)$. In other words, we wish to calculate the conditional expectations

$$\Pi_{y,n} f = \mathbb{E}[\, f(X_n)| (Y_0, \ldots, Y_n) = y] \tag{7.58}$$

for all reasonable functions f on $\mathbb{R}^d$. From the Markov property of the pair (X_k, Y_k) and Bayes formula, we have the following expression for $\Pi_{y,n}$:

$$\Pi_{y,n} f = \frac{\pi_{y,n} f}{\pi_{y,n} \mathbf{1}} \qquad \text{where} \qquad \pi_{y,n} f = \mathbb{E}\left[f(X_n) \prod_{k=0}^{n} g(X_k, y_k) \right], \tag{7.59}$$

for any $f \in \mathcal{B}(\mathbb{R}^d)$, the set of bounded measurable functions on $\mathbb{R}^d$. This can be derived by noting that the function $y = (y_0, \ldots, y_n) \in (\mathbb{R}^q)^{n+1} \mapsto \pi_{y,n} \mathbf{1} = \mathbb{E}[\prod_{k=0}^{n} g(X_k, y_k)]$ is actually equal to the density ϕ_{n+1} of $(Y_0, \ldots, Y_n)$ w.r.t. to the Lebesgue measure on $(\mathbb{R}^q)^{n+1}$.

In the sequel, the observations are fixed to $y = (y_0, \ldots, y_n)$ and we write π_n for $\pi_{y,n}$ and $\Pi_n = \Pi_{y,n}$.

The computation of the unnormalized filter $\pi_{y,n}$ is based on the following inductive formula:

$$\pi_k f = \pi_{k-1} H_k f, \quad k = 1, \ldots, n,$$

where H_k is the transition kernel given by:

$$H_k f(x) = \mathbb{E}\left[f(X_k) g(X_k, y_k) | X_{k-1} = x \right].$$

Hence, the inductive formula of the unnormalized filter relies on successive computations of conditional expectations of X_{k+1} given X_k. Notice that with regard to the problems of optimal stopping or stochastic control problems, we have here an infinite-dimensional problem, since we have to calculate these conditional expectations for any Borel bounded functions on $\mathbb{R}^d$. For solving numerically this problem, we are then suggested to approximate the conditional distributions of X_k given X_{k-1} for any $k = 1, \ldots, n$ by a quantization approach as described in Section 3. We are then given, at each time $k = 0, \ldots, n$, an (optimal) grid $\Gamma_k = \{x_k^1, \ldots, x_k^{N_k}\}$ of N_k points in $\mathbb{R}^d$, and we approximate the distribution μ of X_0 by the distribution of $\widehat{X}_0$ $= \text{Proj}_{\Gamma_0}(X_0)$, and the conditional distribution of X_k given X_{k-1} by the conditional distribution of $\widehat{X}_k$ given $\widehat{X}_{k-1}$: for $k \geq 1$, $\widehat{X}_k$ is defined by $\widehat{X}_k = \text{Proj}_{\Gamma_k}(X_k)$ in the marginal quantization method, while $\widehat{X}_k = \text{Proj}_{\Gamma_k}(F(\widehat{X}_{k-1}, \varepsilon_k))$ in the Markovian quantization method.

We then approximate the transition kernel H_k by the transition matrix $\widehat{H}_k$ defined by:

$$\widehat{H}_k^{ij} = \mathbb{E}\left[1_{\widehat{X}_k = x_k^j} g(\widehat{X}_k, y_k) \Big| \widehat{X}_{k-1} = x_{k-1}^i \right]$$
$$= \hat{p}_k^{ij} g(x_k^j, y_k), \quad i = 1, \ldots, N_{k-1}, \ j = 1, \ldots, N_k.$$

Here, $(\hat{p}_k)_k$ is the probability transition matrix of $(\widehat{X}_k)_k$, i.e.,

$$\hat{p}_{k+1}^{ij} = \mathbb{P}\left[\widehat{X}_{k+1} = x_{k+1}^j | \widehat{X}_k = x_k^i \right],$$

and $\hat{p}_0 = (\hat{p}_0^i)_{i=1,\ldots,N_0}$ is the probability distribution of $\widehat{X}_0$, i.e., $\hat{p}_0^i = \mathbb{P}[\widehat{X}_0 = x_0^i]$. The unnormalized filter π_n is then approximated by the discrete probability measure $\hat{\pi}_n$ on Γ_n:

$$\hat{\pi}_n = \sum_{\ell=1}^{N_n} \hat{\pi}_n^\ell \delta_{x_n^\ell},$$

where $(\hat{\pi}_k^\ell)$, $k = 0, \ldots, n$, $\ell = 1, \ldots, N_k$, are computed inductively by:

$$\hat{\pi}_0 = \hat{p}_0 \quad \text{and} \quad \hat{\pi}_k^j = \sum_{1 \leq i \leq N_{k-1}} \widehat{H}_k^{ij} \hat{\pi}_{k-1}^i, \quad k = 1, \ldots, n, \quad j = 1, \ldots, N_k.$$

The normalized filter Π_n is finally approximated by the discrete probability measure $\widehat{\Pi}_n$ on Γ_n:

$$\widehat{\Pi}_n = \frac{\hat{\pi}_n}{\sum_{\ell=1}^{N_n} \hat{\pi}_n^\ell}.$$

Under the Lipschitz assumption (**A1′**) on the scheme F, and assuming also that the function $g(x, y)$ is Lipschitz in x, uniformly in y, with ratio $[g]_{Lip}$, we have the following estimation error for the approximate filter (see [20]): for any $f \in BL_1(\mathbb{R}^d)$,

$$
|\Pi_{y,n} f - \widehat{\Pi}_{y,n} f| \leq \frac{\|g\|_\infty^{n+1}}{\phi_{n+1}(y)} \sum_{k=0}^{n} C_{n,k} \|\Delta_k\|_2,
$$

where $C_{n,k}(f) = [F]_{Lip}^{n-k+1} + 2 \dfrac{[g]_{Lip}}{\|g\|_\infty} \left(\dfrac{[F]_{Lip} + 1}{[F]_{Lip} - 1} ([F]_{Lip}^{n-k+1} - 1) + 1 \right)$,

$\|\Delta_0\|_2 = \|X_0 - \widehat{X}_0\|_2$, $\|\Delta_k\|_2 = \|X_k - \widehat{X}_k\|_2$ in the marginal quantization method, and $\|\Delta_k\|_2 = \|F(\widehat{X}_{k-1}, \varepsilon_k) - \widehat{X}_k\|_2$ in the Markovian quantization, $k = 1, \ldots, n$, are the L^2-quantization errors.

Remark 3. ● In the marginal quantization method, the constant $[F]_{Lip}$ may be replaced by the constant $[P]_{Lip}$. Note that in regular examples studied here, we have $[P]_{Lip} = [F]_{Lip}$.

● The uniform Lipschitz condition in y of $x \mapsto g(x, y)$ may be relaxed into a nonuniform Lipschitz condition in the form: $|g(x, y) - g(x', y)| \leq [g]_{Lip}(y)(1 + |x| + |x'|)|x - x'|$, with in this case a more complex estimation error term.

Numerical illustrations : 1. *The (familiar) Kalman–Bucy model*:

$$
\begin{aligned}
X_{k+1} &= A X_k + \varepsilon_k \quad \in \mathbb{R}^d, \\
Y_k &= B X_k + \eta_k \quad \in \mathbb{R}^q,
\end{aligned}
$$

for $k \in \mathbb{N}$, and X_0 is normally distributed with mean $m_0 = 0$ and variance Σ_0. Here A and B are matrices of appropriate dimensions, and $(\varepsilon_k)_k$, $(\eta_k)_k$ are independent centered Gaussian processes, $\varepsilon_k \rightsquigarrow \mathcal{N}(0, I_d)$ and $\eta_k \rightsquigarrow \mathcal{N}(0, \Lambda)$. In this case, we have

$$
g(x, y) = \frac{1}{(2\pi)^{d/2} \det(\Lambda^{\frac{1}{2}})} \exp\left(-\frac{1}{2} \left| \Lambda^{-\frac{1}{2}}(y - Bx) \right|^2 \right).
$$

Of course, the filter $\Pi_{y,n}$ is explicitly known, see e.g., [10]: it is a Gaussian distribution of mean m_n and variance Σ_n given by the inductive equations:

$$
\Sigma_{k+1}^{-1} = I_d - A \left(A'A + \Sigma_k^{-1} \right)^{-1} A' + B' \Lambda^{-1} B,
$$

$$
\left(\Sigma_{k+1}^{-1} m_{k+1} \right), = A \left(A'A + \Sigma_k^{-1} \right)^{-1} \left(\Sigma_k^{-1} m_k \right) + B' \Lambda^{-\frac{1}{2}} y_{k+1}.
$$

We will illustrate the numerical scheme in dimension $d = 3$. Here A and B are

$$
A = \begin{bmatrix} 0.8 & 0 & 0 \\ 0 & 0.5 & 0 \\ 0 & 0 & 0.2 \end{bmatrix} \quad \text{and} \quad B = I_3.
$$

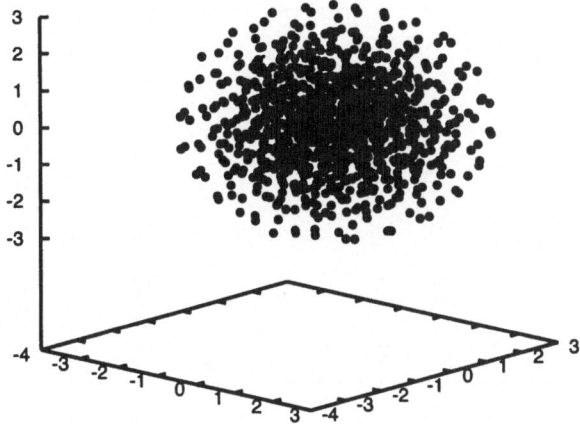

FIGURE 10. A L^2 optimal 1 000-quantizer of $\mathcal{N}(0, I_3)$. L^2-quantization error is equal to 0.233.

We take $\Lambda = (0.5)^2 I_3$. The variance Σ_0 is such that (X_k) is stationary. In this case, we can work with a single grid (1000 points). In Figure 10 is represented the 1000-optimal quantizer used for $\mathcal{N}(0, I_3)$. Computations are carried out with its Σ_0-rescaled version (which is a non-optimal but straightforwardly accessible and quite efficient quantizer for $\mathcal{N}(0, \Sigma_0)$). The number n of observations is equal to 20. We compute the conditional expectations $\mathbb{E}[f(X_n) \mid Y_0, \ldots, Y_n]$ with $f(x) = x$ (the conditional mean) and $f(x) = x \cdot {}^t x$ (the conditional variance). The quantized version of the conditional mean is denoted by $\widehat{m}_n$ and that given by the Kalman filter by m_n. We take the same convention for the conditional variance Σ_n. We represent in Figure 11 the errors $\|m_k - \widehat{m}_k\|$ and $\|\Sigma_k - \widehat{\Sigma}_k\|$ plotted w.r.t. $k \in \{0, \ldots, 20\}$. Finally, Figure 12 depicts the three components of the conditional mean in its Kalman filter version and its quantized version. These figures show that in this setting the $3d$ Kalman filter is well captured by the quantization method.

2. *A stochastic volatility model arising in financial time series*: Let S_k, $k \in \mathbb{N}$, be a positive process describing the stock prices in time, and define $Y_k = \ln S_{k+1} - \ln S_k$, the *log*-returns of the stock prices. A standard stochastic volatility model (SVM) is given by

$$Y_k = \sigma(X_k)\eta_k \quad \in \mathbb{R} \quad \text{with} \quad X_k = \rho \, X_{k-1} + \varepsilon_{k-1} \quad \in \mathbb{R} \qquad (7.60)$$

where ρ is a real constant, $\sigma(.)$ is a positive Borel function on $\mathbb{R}$ and $(\varepsilon_k)_k$, $(\eta_k)_k$ are independent Gaussian processes. We consider dynamics (7.60) as a time discretization Euler scheme with step size $\Delta t = 1/n$, of a continuous-time Ornstein–Uhlenbeck stochastic volatility model :

$$dX_t = -\lambda X_t dt + \tau dW_t, \quad 0 \le t \le 1.$$

We then assume that

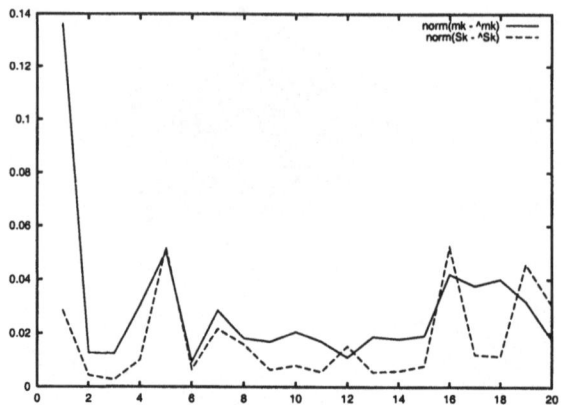

FIGURE 11. Errors $\|m_n - \widehat{m_n}\|$ (line) and $\|\Sigma_n - \widehat{\Sigma_n}\|$ (dot line) plotted w.r.t. n.

$$\rho = 1 - \lambda \Delta t, \quad \varepsilon_k \rightsquigarrow \mathcal{N}(0, \tau^2 \Delta t) \quad \text{and} \quad \eta_k \rightsquigarrow \mathcal{N}(0, \Delta t),$$

for some positive parameters λ and τ. Typical examples of SVM are specified with $\sigma(x) = |x| + \gamma$, $\sigma(x) = x^2 + \gamma$, or $\sigma(x) = \exp(x)$ for some positive constant γ. The filtering problem consists in estimating the volatility $\sigma(X_n)$ at step n given the observations of the prices $(Y_0, \ldots, Y_n)$. Here,

$$g(x, y) = \frac{1}{\sqrt{2\pi \Delta t}\, \sigma(x)} \exp\left(-\frac{y^2}{2\sigma^2(x)\Delta t}\right).$$

The values of the parameters in our simulation are for $(\lambda, \tau, \Delta t) = (1, 0.5, 1/250)$. The Gaussian distribution of X_0 is specified so that the sequence $(X_k)_k$ is stationary i.e., $X_0 \sim \mathcal{N}(0, \Sigma_0^2)$ with $\Sigma_0 = \tau\sqrt{\Delta t/(1 - \rho^2)} \approx \tau/\sqrt{2\lambda} = 0.35\ldots$.

There are two types of models involved here:

$$(ABS) \equiv \sigma(X_k) = \gamma + |X_k| \quad \text{and} \quad (EXP) \equiv \sigma(X_k) = \overline{\sigma}\exp(X_k),$$

with the values $(\gamma, \overline{\sigma}) = (0.05, 0.2)$.

We represent in Figure 13 the stock price simulation according (EXP) together with the simulation of the volatility σ_n and its mean conditionally to Y_k, $0 \le k \le n = 250$. Idem in Figure 14 with (ABS).

We represent in Figure 15, the conditional variance of the volatility w.r.t. the observations in the two models. Since we are here in a nonlinear model, we cannot compare our results with an explicit filter, but we can see that the filter captures well the dynamic of the stochastic volatility.

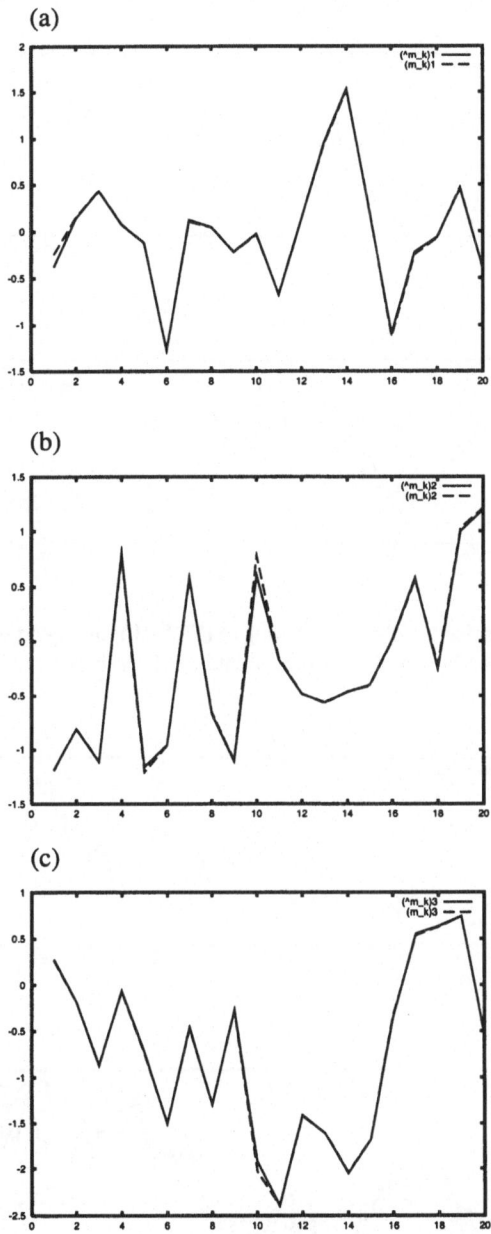

FIGURE 12. Components of $\widehat{m_k}$ (Quantized filter) (line) and of m_k (Kalman filter) (dot line). (a) x, (b) y and (c) z.

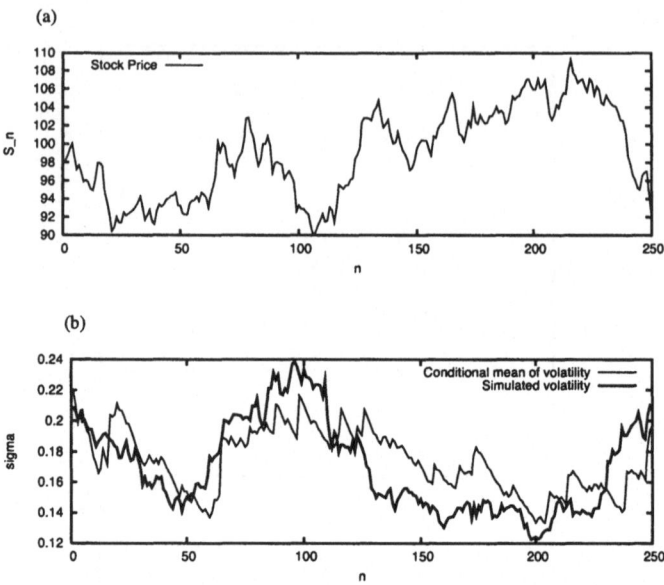

FIGURE 13. a) Stock price simulated according to (EXP). b) Simulated volatility according to (EXP) (Thick line), Conditional mean of the volatility (Thin line).

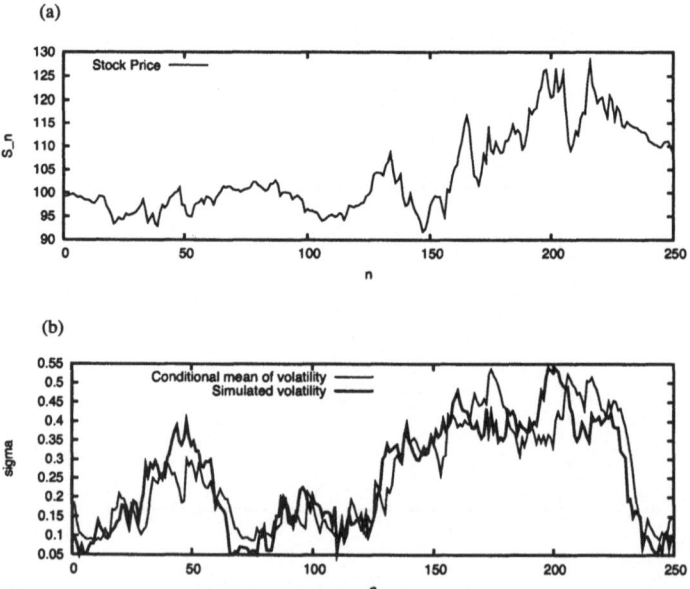

FIGURE 14. a) Stock price simulated according to (ABS). b) Simulated volatility according to (ABS) (Thick line), Conditional mean of the volatility (Thin line).

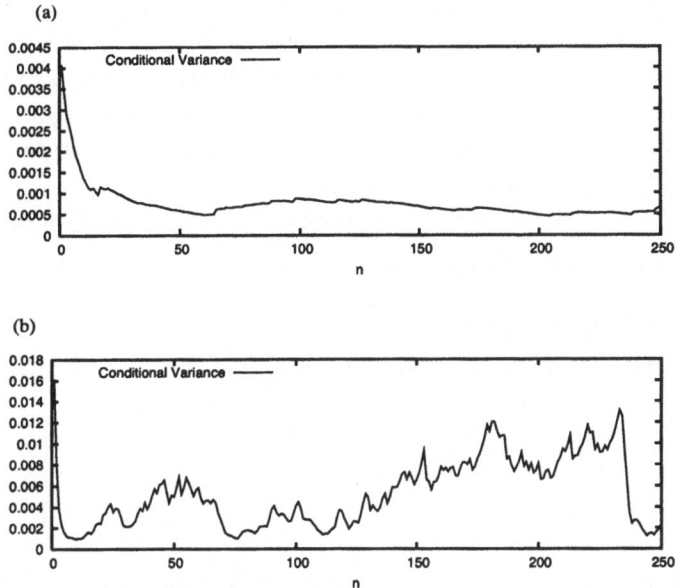

FIGURE 15. a) Conditional Variance of the Volatility according to (EXP). b) Conditional Variance of the Volatility (ABS).

7.5 Toward higher order schemes in quantization methods

The aim of this section is to present in a slightly different setting the first order scheme introduced in [6] and successfully tested on the pricing of Exchange options in a d-dimensional Black & Scholes model ($d = 2, 4$ and 6). One comes back to the expectation computation

$$\mathbb{E}\left[\phi_0(X_0)\phi_1(X_1)\ldots\phi_n(X_n)\right]$$

along the path of a Markov chain investigated in Section 7.3. The idea is to try taking advantage of the specificity of the stationary quantizers, as for the numerical integration of smooth functions (see Section 7.2.3).

We deal here with marginal quantization and the approach is partially heuristic. However, to enhance the quantization aspects we will essentially focus on a smooth setting where the ϕ_k functions are smooth, say C_b^2 (twice differentiable with bounded existing derivatives). We will shortly comment below on how it can be somewhat relaxed.

Assume that $(X_k)_{0 \leq k \leq n}$ is a homogeneous Markov chain with a transition $P(x, dy)$ satisfying on C_b^2 functions:

$$\|D(Pf)\|_\infty \leq K\|Df\|_\infty \quad \text{and} \quad \|D^2(Pf)\|_\infty \leq K(\|Df\|_\infty + \|D^2f\|_\infty) \quad (7.61)$$

for some real constant $K > 0$ (note that then P is Lipschitz with constant K as well). Such an assumption is satisfied e.g., by the transition P_t of a (simulatable) diffusion having C_b^2 coefficients b and σ at any time t or by the transition of its Euler scheme. Then, the transition P clearly maps C_b^2 into itself and one shows by induction that the functions v_k defined by 7.23 all lie in C_b^2 and that their first two derivatives can be controlled using K, $\|D\phi_k\|_\infty$ and $\|D^2\phi_k\|_\infty$.

The key result to design a first-order scheme is the following proposition.

Proposition 2. *Assume Assumption (7.61) and set by induction*

$$\widetilde{v}_n(\widehat{X}_n) = \phi_n(\widehat{X}_n), \tag{7.62}$$

$$\widetilde{v}_{k-1}(\widehat{X}_{k-1}) = \phi_{k-1}(\widehat{X}_{k-1})\mathbb{E}\left(\widetilde{v}_k(\widehat{X}_k) + Dv_k(\widehat{X}_k).(X_k - \widehat{X}_k) \mid \widehat{X}_{k-1}\right),$$
$$1 \le k \le n. \tag{7.63}$$

Set $L := \max\limits_{0 \le k \le n} (\|\phi_k\|_\infty, \|D\phi_k\|_\infty, \|D^2\phi_k\|_\infty)$.
Then for every $k \in \{0, \dots, n\}$,

$$\left\|\mathbb{E}(v_k(X_k) \mid \widehat{X}_k) - \widetilde{v}_k(\widehat{X}_k)\right\|_1 \le L^{n+1-k} \sum_{\ell=k}^{n} c_\ell^n(K)\|X_\ell - \widehat{X}_\ell\|_2^2$$

$$\tag{7.64}$$

where

$$c_\ell^n(K) =$$

$$\begin{cases} \dfrac{1}{K-1}\left(K^{n+1-\ell}\,(6(n-\ell)+1) - 6\dfrac{K^{n+1-\ell}-1}{K-1} + \dfrac{K+9}{2}\right) & 1 \le \ell \le n-1 \\[2ex] 1/2, & \ell = n \end{cases}$$

when $K \ne 1$ *and*

$$c_\ell^n(1) = 3(n-\ell)^2 + 4(n-\ell) + \frac{3}{2}, \quad 0 \le \ell \le n-1, \qquad c_n^n(1) = 1/2.$$

In particular,

$$|\mathbb{E}\, v_0(X_0) - \mathbb{E}\, \widetilde{v}_0(\widehat{X}_0)| \le L^{n+1} \sum_{\ell=0}^{n} c_\ell^n(K)\|X_\ell - \widehat{X}_\ell\|_2^2.$$

How to use this result to design a first order scheme? First, one reads (7.63) in distribution i.e.,

$$\widetilde{v}_{k-1}(x_{k-1}^i) = \phi_{k-1}(x_{k-1}^i)\left(\sum_{j=1}^{N_k} \widehat{p}_k^{ij}\widetilde{v}_k(x_k^j) + Dv_k(x_k^j).\widehat{\chi}_k^{ij}\right)$$

where the $\mathbb{R}^d$-valued *correcting vectors* $\widehat{\chi}_k^{ij}$ are defined by

$$\hat{\chi}_k^{ij} := \mathbb{E}\left[(X_k - \widehat{X}_k)\mathbf{1}_{\{\widehat{X}_k = x_k^j\}} \mid \widehat{X}_{k-1} = x_{k-1}^i\right]. \tag{7.65}$$

The key point for numerical application is to note that these correcting vectors can easily be estimated like the former companion parameters $\hat{p}_k^{ij}$'s, either on line during the grid optimization phase or using a Monte Carlo simulation once the grids are settled.

The second step is mostly heuristic so far: the weak link in (7.63) is of course that the differential Dv_k is involved in the computation of $\tilde{v}_{k-1}$ and this function is not numerically accessible since we precisely intend approximating the functions v_k. Note that if Dv_{k-1} had been involved in (7.63), the scheme would have been definitely intractable. In its present form, several approaches can be considered.

It often happens, e.g., for diffusions or Euler schemes with smooth coefficients, that $D(Pf) = Q(Df)$ where Q is an integral kernel. This kernel Q can be approximated by quantization as well so that one can rely on the backward induction formula

$$Dv_k = D\phi_k \, Pv_{k+1} + \phi_k \, Q(Dv_{k+1})$$

to design an algorithm that approximates the differentials Dv_k. In many cases one can also write $D(Pf) = \widetilde{Q}(f)$ using an appropriate integration by parts ("à la Malliavin" when the Euler scheme is concerned, see [6]). Then the computation of $D(Pv_{k+1})$ is directly connected to the function v_{k+1}. Another approach can be to use some approximation by convolution: one approximates Dv_k by $(D\varphi_\varepsilon) * \tilde{v}_k$ where $(\varphi_\varepsilon)_{\varepsilon>0}$ is e.g., a Gaussian unit approximation. The practical task is then to tune the band width ε. When the functions ϕ_k are not smooth enough, one uses the regularizing properties of the Markov semi-group if any, although this induces a loss in the rate of convergence.

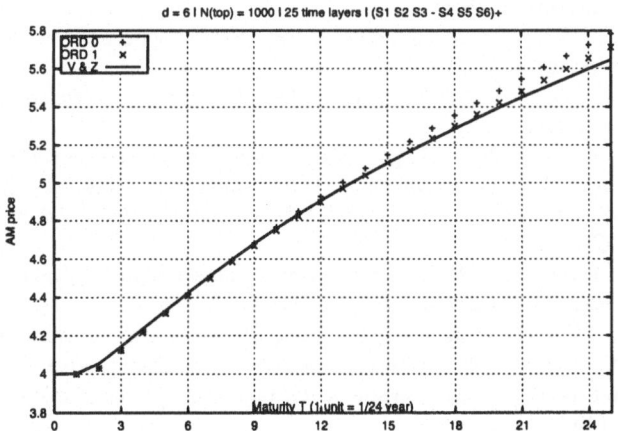

FIGURE 16. American exchange option in dimension 6 (*in-the-money case*). The reference price is depicted by a line and the quantized prices (order 0 and order 1) by cross lines.

We present in Figure 16 a graph that emphasizes the improvement provided by this first-order *quantization tree descent* versus the original one for pricing 6-dimensional American exchange options, as described in section 7.4.1. We consider an "in-the-money" case consisting in setting $x_0^1 \dots x_0^3 = 40$ and $x_0^4 \dots x_0^6 = 36$, all other parameters being unchanged. For more details we refer to [6].

ACKNOWLEDGMENT : We thank Harald Luschgy for fruitful comments.

References

[1] Bally V. (2002): The Central Limit Theorem for a non-linear algorithm based on quantization, forthcoming in *Proceedings of the Royal Society*.

[2] Bally V., Pagès G. (2000): A quantization algorithm for solving discrete time multi-dimensional optimal stopping problems, pre-print LPMA-628, Laboratoire de Probabilités & Modèles Aléatoires, Universités Paris 6 & 7 (France), to appear in *Bernoulli*.

[3] Bally V., Pagès G. (2003): Error analysis of the quantization algorithm for obstacle problems, *Stochastic Processes and their Applications*, **106**, $n^0 1$, 1–47.

[4] Bally V., Pagès G., Printems J. (2001): A stochastic quantization method for nonlinear problems, *Monte Carlo Methods and Applications*, **7**, $n^0 1$-2, 21–34.

[5] Bally V., Pagès G., Printems J. (2002): A quantization method for pricing and hedging multi-dimensional American style options, pre-print LPMA-753, Laboratoire de Probabilités & Modèles Aléatoires, Université Paris 6 & 7 (France), to appear in *Mathematical Finance*.

[6] Bally V., Pagès G., Printems J. (2003): First order schemes in the numerical quantization method, *Mathematical Finance*, **13**, n^0 1, 1–16.

[7] Barles G., Souganidis P. (1991): Convergence of approximation schemes for fully nonlinear second-order equations, *Asymptotics Analysis*, **4**, 271–283.

[8] Bucklew J., Wise G. (1982): Multidimensional Asymptotic Quantization Theory with r^{th} Power distortion Measures, *IEEE Transactions on Information Theory, Special issue on Quantization*, **28**, n^0 2, 239–247.

[9] Duflo, M. (1997): *Random Iterative Models*, Coll. Applications of Mathematics, **34**, Springer-Verlag, Berlin, 1997.

[10] Elliott R., Aggoun L. and J. Moore (1995): *Hidden Markov Models, Estimation and Control*, Springer Verlag.

[11] Fort J.C., Pagès G. (2002): Asymptotics of optimal quantizers for some scalar distributions, *Journal of Computational and Applied Mathematics*, **146**, 253–275.

[12] Gersho A., Gray R. (eds.) (1982): *IEEE Transactions on Information Theory, Special issue on Quantization*, **28**.

[13] Graf S., Luschgy H. (2000): *Foundations of Quantization for Probability Distributions*, Lecture Notes in Mathematics $n^0 1730$, Springer, Berlin.

[14] Kieffer J. (1982): Exponential rate of Convergence for the Lloyd's Method I, *IEEE Transactions on Information Theory, Special issue on Quantization*, **28**, $n^0 2$, 205–210.

[15] Kohonen T. (1982): Analysis of simple self-organizing process, *Biological Cybernetics*, **44**, 135–140.

[16] Kushner H.J., Dupuis P. (2001): *Numerical methods for stochastic control problems in continuous time*, 2nd edition, Applications of Mathematics, **24**, Stochastic Modelling and Applied Probability, Springer-Verlag, New York.

[17] Kushner H.J., Yin G.G. (1997): *Stochastic Approximation Algorithms and Applications*, Springer, New York.

[18] Pagès G. (1997): A space vector quantization method for numerical integration, *Journal of Computational and Applied Mathematics*, **89**, 1–38.

[19] Pagès G., Pham H. (2001): A quantization algorithm for multidimensional stochastic control problems, pre-print LPMA-697, Laboratoire de Probabilités et Modèles Aléatoires, Universités Paris 6 & 7 (France).

[20] Pagès G., Pham H. (2002): Optimal quantization methods for nonlinear filtering with discrete-time observations, pre-print LPMA-778, Laboratoire de Probabilités et modèles aléatoires, Universités Paris 6 & 7 (France).

[21] Pagès G., Printems J. (2003): Optimal quadratic quantization for numerics: the Gaussian case, *Monte Carlo Methods and Applications*, **9**, $n^0 2$.

[22] Villeneuve S., Zanette A. (2002) Parabolic A.D.I. methods for pricing american option on two stocks, *Mathematics of Operation Research*, **27**, $n^0 1$, 121–149.

8

Numerical Methods for Stable Modeling in Financial Risk Management

Stoyan Stoyanov

Borjana Racheva-Jotova

ABSTRACT The seminal work of Mandelbrot and Fama, carried out in the 1960s, suggested the class of α-stable laws as a probabilistic model of financial assets returns. Stable distributions possess several properties which make plausible their application in the field of finance — heavy tails, excess kurtosis, domains of attraction. Unfortunately working with stable laws is very much obstructed by the lack of closed-form expressions for probability density functions and cumulative distribution functions. In the current paper we review statistical and numerical techniques which make feasible the application of stable laws in practice.

8.1 Introduction

The distributional assumption for financial asset returns is central in economics and finance both from theoretical and practical viewpoints. For example the problem of portfolio risk management, important results in optimal portfolio theory and option pricing heavily depend on the hypothesis of the distributional form of asset returns. Many important concepts are developed on the presumption that the distributional form is Gaussian. On the other hand, the Gaussian distribution does not possess the properties which empirical research shows to be typical of financial time series — heavy tails, asymmetry and excess kurtosis.

In the pioneering work of Mandelbrot and Fama in the 1960s, the normal assumption is rejected and a more general family of distributions is proposed — the class of stable laws. The Gaussian distribution is a member of the class of stable laws. Moreover the stable non-Gaussian distributions possess the empirical properties of financial time series — they are heavy-tailed and allow for modeling of skewness and excess kurtosis. Stable non-Gaussian laws are also called stable Paretian (because of the Pareto-like power decay of the tail of the distribution), or Lévy stable (after the name of Paul Lévy who carried out the fundamental research of characterizing the family of non-Gaussian stable laws). Subsequent studies, after the initial ones conducted in the 1960s, supported the stable Paretian hypothesis.

As pointed out in [17], stable distributions are attractive because they have an important desirable property — domains of attraction. Loosely speaking, according to this property, if a distribution is in the domain of attraction of a stable law, it has properties which are close to those of the specified stable law. The domain of attraction is completely determined by the tail behavior of the distribution. As a result, it is reasonable to adopt the stable law as the "idealized" model if the true distribution has the appropriate tail behavior.

Another attractive feature is the stability property. Stable laws have an important shape parameter which governs the properties of the distribution. It is called the index of stability and is denoted by α. Because of the significance of the index of stability, stable distributions are also called α-stable. According to the stability property, appropriately centralized and normalized sums of independent identically distributed (iid) α-stable random variables is again α-stable. This feature is desirable in portfolio theory because it follows directly that a portfolio of assets with α-stable returns has again α-stable returns.

A well-known property of stable non-Gaussian distributions is that, due to the power decay of the tails, they do not possess a finite second moment. Certainly the application of infinite-variance distributions as theoretical models of bounded variables, such as financial assets returns, seems inappropriate. Moreover any empirical distribution has a finite variance, hence it may seem that infinite variance distributions are inapplicable in any context. Nevertheless there is ample empirical evidence that the probability of large deviations of the changes in stock market prices is so great that any statistical theory based on finite-variance distributions is impossible to predict accurately. As it is remarked in [13] and [8], the sum of a large number of these variables is often dominated by one of the summands which is a theoretical property of infinite-variance distributions. Hence an infinite-variance distribution may be an appropriate probabilistic model.

The problem of parameter estimation of stable distributions was first tackled by Mandelbrot then by Fama and Roll, [9] and [10]. This is a non-trivial task because, with a few exceptions, there are no closed-form expressions for the probability density functions (pdf) and cumulative distribution functions (cdf). For example the classical maximum likelihood method, in this case, depends on numerical approximations of the density and could be extremely time-consuming. Moreover standard estimation techniques based on asymptotic results which rely on a finite second moment are irrelevant.

Mathematical models with application of stable laws in finance, economics and other areas can be found in [1], [7], [17] and [20]. We continue with a rigorous definition of the family of α-stable distributions. Then we focus on the numerical issues in the approximation of probability density and distribution functions. The last section considers some well-known approaches to the estimation of stable parameters, namely quantile methods, chf based methods and maximum likelihood.

8.2 Definition and basic properties

8.2.1 Definition and parametrizations

There are several equivalent ways to define the class of α-stable distributions. The first definition identifies the stability property.

Definition 1. *A random variable X is said to have stable distribution if for any $n \geq 2$, there is a positive number C_n and a real number D_n such that*

$$X_1 + X_2 + \cdots + X_n \overset{d}{=} C_n X + D_n$$

where $X_1, X_2, \ldots, X_n$ are independent copies of X and $\overset{d}{=}$ means equality in distribution.

The second definition states that stable distributions are the only distributions that can be obtained as limits of properly normalized sums of iid random variables.

Definition 2. *A random variable X is said to have a stable distribution if it has a domain of attraction, i.e., if there is a sequence of iid random variables $Y_1, Y_2, \ldots$ and sequences of positive numbers $\{d_n\}$ and real numbers $\{a_n\}$ such that*

$$\frac{Y_1 + Y_2 + \cdots + Y_n}{d_n} + a_n \overset{d}{\to} X$$

where $\overset{d}{\to}$ denotes convergence in distribution.

The third definition specifies the characteristic function of stable laws.

Definition 3. *A random variable X is said to have a stable distribution if there are parameters $0 < \alpha \leq 2, \sigma > 0, -1 \leq \beta \leq 1, \mu \in \mathbb{R}$ such that its characteristic function (chf) has the form*

$$\varphi(t) = E e^{itX} = \begin{cases} \exp\{-\sigma^\alpha |t|^\alpha (1 - i\beta \frac{t}{|t|} \tan(\frac{\pi\alpha}{2})) + i\mu t\}, & \alpha \neq 1, \\ \exp\{-\sigma |t|(1 + i\beta \frac{2}{\pi} \frac{t}{|t|} \ln(|t|)) + i\mu t\}, & \alpha = 1, \end{cases} \tag{8.1}$$

where $\frac{t}{|t|} = 0$ if $t = 0$.

Proofs of the equivalence between the three definitions can be found in [20].

The parameter α is the index of stability, β is a skewness parameter, σ is a scale parameter and μ is a location parameter. Since stable distributions are uniquely determined by the four parameters, the common notation is $S_\alpha(\sigma, \beta, \mu)$. If X is said to have stable distribution, we write $X \sim S_\alpha(\sigma, \beta, \mu)$. One can easily notice that if $\beta = 0$ and $\mu = 0$, the chf becomes real-valued, hence the random variable is symmetric. If X belongs to the class of symmetric α-stable distributions, we write $X \sim S\alpha S$. If $X \sim S\alpha S$, then the chf of X has the simple form

$$\varphi(t) = \exp\{-\sigma^\alpha |t|^\alpha\}.$$

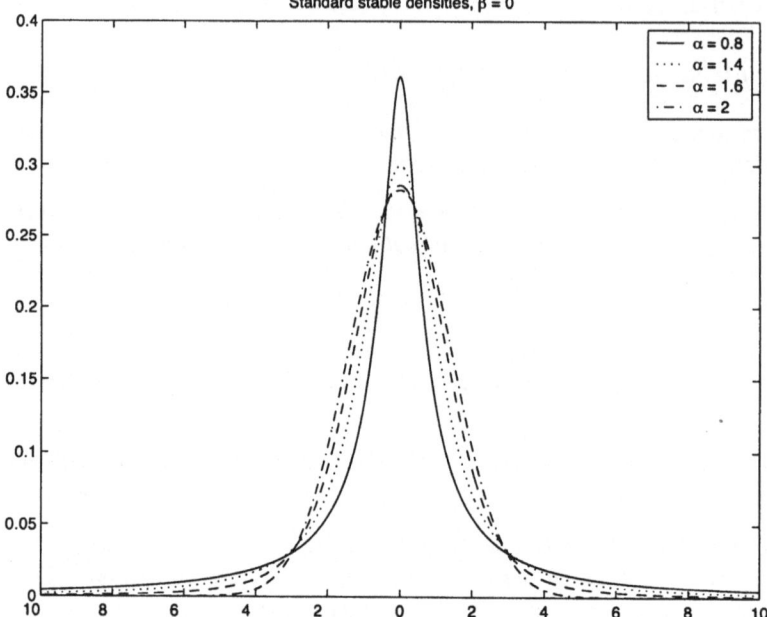

FIGURE 1. Standard symmetric α-stable densities ($\sigma = 1$, $\mu = 0$) for varying α.

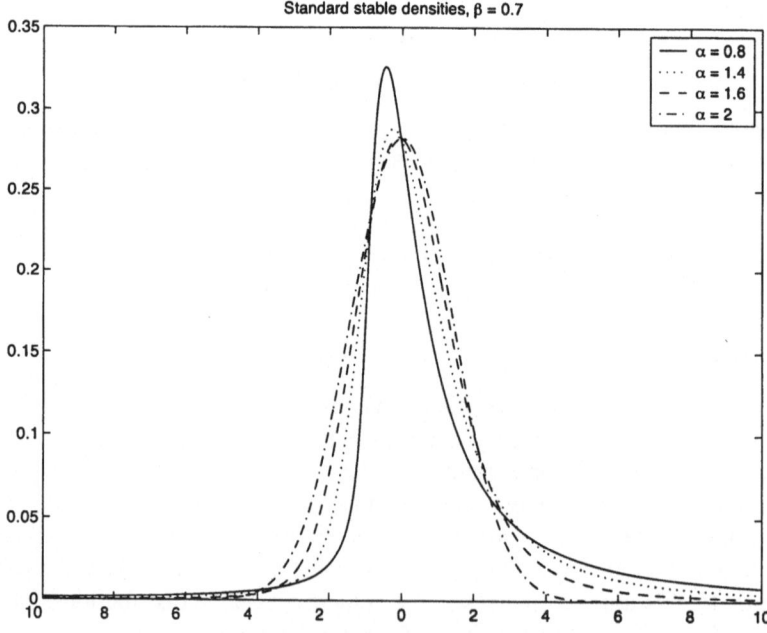

FIGURE 2. Standard α-stable densities ($\sigma = 1$, $\mu = 0$) for varying α, with $\beta = 0.7$.

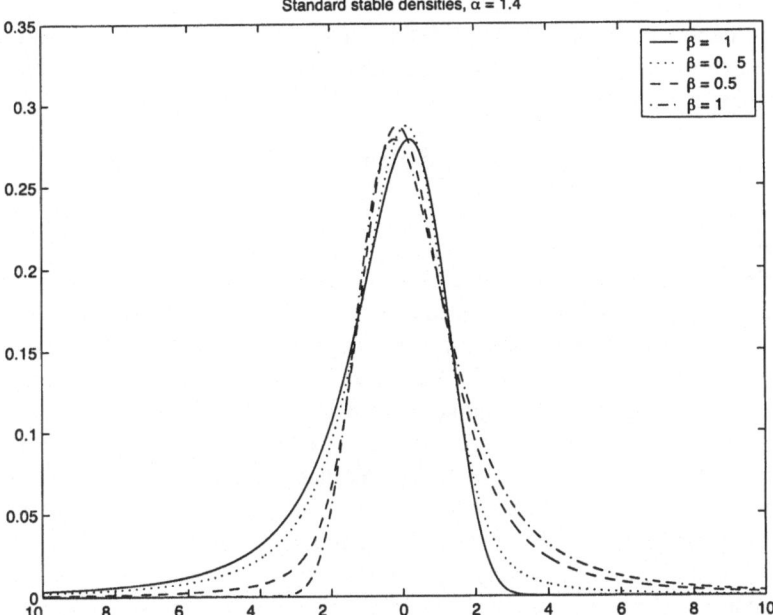

FIGURE 3. Standard 1.4-stable densities ($\sigma = 1$, $\mu = 0$) for varying β.

If $\alpha = 2$, we arrive at the chf of the Gaussian distribution, that is if $X \sim S_2(\sigma, \beta, \mu)$, then

$$\varphi(t) = E e^{itX} = \exp\{-\sigma^2 |t|^2 + i\mu t\}.$$

Hence X has the Gaussian distribution with mean equal to μ and variance equal to $2\sigma^2$: $X \sim N(\mu, 2\sigma^2)$. Note that in this case β is irrelevant. Nevertheless the Gaussian distribution is usually associated with $\beta = 0$.

The parametrization in Definition 3 is one possible way to define the characteristic function. It has the advantage that the parameters are easy to interpret in terms of shape and location. However there is a serious disadvantage when it comes to numerical or statistical work — it is discontinuous at $\alpha = 1$ and $\beta \neq 0$, i.e., if $\alpha \to 1, \beta \to \beta^* \neq 0, \sigma \to \sigma^*$ and $\mu \to \mu^*$, then the limit function is not the chf of the stable random variable $S_1(\sigma^*, \beta^*, \mu^*)$. This drawback is not an inherent property of the class of α-stable laws and appears because of the special form of the classical chf given in Definition 3. As noted in [18] and [20], it is possible to change the parametrization in order to have convergence in distribution when $\alpha \to \alpha^*, \beta \to \beta^*, \sigma \to \sigma^*$ and $\mu \to \mu^*$. An alternative parametric representation of the chf equipped with this property is the following:

$$\varphi(t) = \begin{cases} \exp\{-|\sigma t|^\alpha + i\sigma t\beta(|\sigma t|^{\alpha-1} - 1)\tan(\frac{\pi\alpha}{2}) + i\mu_1 t\}, & \alpha \neq 1, \\ \exp\{-|\sigma t| + i\sigma t\beta\frac{2}{\pi}\ln|\sigma t| + i\mu_1 t\}, & \alpha = 1, \end{cases} \quad (8.2)$$

where $0 < \alpha \le 2, -1 \le \beta \le 1, \sigma > 0$ and $\mu_1 \in R$.

Let us denote the parametrization defined in equation (8.1) as P_0 and the continuous one defined in (8.2) as P_1. The relation between P_0 and P_1 is given in terms of the parameters μ and μ_1

$$\mu_1 = \begin{cases} \mu + \beta\sigma \tan\frac{\pi\alpha}{2}, & \alpha \ne 1, \\ \mu, & \alpha = 1. \end{cases} \tag{8.3}$$

Obviously P_0 is different from P_1 only when $\beta \ne 0$. An attractive feature of the continuous parametrization is that it is not necessary to consider the case $\alpha = 1$ separately, i.e., it can be defined by means of the limit $\alpha \to 1$. Convergence in distribution follows because of the one-to-one relationship between the cumulative distribution functions (cdfs) and the chfs. Moreover it is preferable to have a continuous parametrization in statistical and numerical work.

Yet another parametrization appears to be more suitable in the derivation of some analytic properties of stable laws. It comes out that it is more convenient for the coefficient $(1 - i\beta\frac{t}{|t|} \tan(\frac{\pi\alpha}{2}))$ to be in the form e^{iy}. Such transformation is not possible if $\alpha = 1$ but for the case $\alpha \ne 1$ it is easy to establish. The following parametrization P_2 is proposed by Zolotarev and is more appropriate for the examination of the analytic properties. Moreover certain approximations of the probability density function and cumulative distribution function discussed in Section 8.3 are derived by taking advantage of parametrization P_2. According to P_2, the natural logarithm of the chf of a standardized stable distribution has the representation:

$$\ln(\varphi(t)) = -|t|^\alpha \exp\left\{-i\frac{\pi}{2}\beta_2 K(\alpha)sign(t)\right\} \tag{8.4}$$

where $K(\alpha) = \alpha - 1 + sign(1 - \alpha)$.

Note that, to simplify the expression, equation (8.4) is the chf of a standardized stable distribution, i.e., with scale parameter equal to 1 and location parameter equal to zero. It is easily checked that if the chf of a standardized random variable X is in P_2 form and if we set

$$\beta_2 = \frac{2}{\pi K(\alpha)} \arctan\left(\beta \tan\frac{\pi\alpha}{2}\right) \tag{8.5}$$

and

$$c = \left(\cos\left(\frac{\pi}{2}\beta_2 K(\alpha)\right)\right)^{1/\alpha} = \left(1 + \beta^2 \tan^2\frac{\pi\alpha}{2}\right)^{-1/2\alpha}, \tag{8.6}$$

then the natural logarithm of the chf of the random variable X/c admits the representation

$$\ln(\varphi(t)) = -|t|^\alpha\left(1 - i\beta sign(t) \tan\frac{\pi\alpha}{2}\right).$$

It follows that the chf of X/c is in parametrization P_0, hence equations (8.5) and (8.6) establish the relation between P_2 and P_0.

The three parametrizations defined in this section do not comprise all possible parametric representations of the chf of stable laws. In [20] there are more examples which appear to be appropriate in different situations. Therefore, to avoid confusion, it is necessary to specify which particular parametrization is used when exploring certain properties. We shall write $X \sim S_\alpha(\sigma, \beta, \mu; P_1 \text{ or } P_2)$ to indicate that X has stable distribution and that the parametric representation of the chf is in P_1 or P_2 form. Otherwise the notation assumes the classical parametrization P_0.

8.2.2   Basic properties

The basic properties we shall consider are easier to establish when working with P_0. Most of them follow directly from the particular form of equation (8.1). The proofs can be found in [18].

Proposition 1. *Let X_1 and X_2 be independent random variables such that $X_1 \sim S_\alpha(\sigma_1, \beta_1, \mu_1)$ and $X_2 \sim S_\alpha(\sigma_2, \beta_2, \mu_2)$. Then $X_1 + X_1 \sim S_\alpha(\sigma, \beta, \mu)$, with*

$$\sigma = (\sigma_1^\alpha + \sigma_2^\alpha)^{\frac{1}{\alpha}}, \quad \beta = \frac{\beta_1 \sigma_1^\alpha + \beta_2 \sigma_2^\alpha}{\sigma_1^\alpha + \sigma_2^\alpha}, \quad \mu = \mu_1 + \mu_2.$$

Proposition 2. *Let $X \sim S_\alpha(\sigma, \beta, \mu)$ and $a \in \mathbb{R}$. Then*

$$X + a \sim S_\alpha(\sigma, \beta, \mu + a).$$

Proposition 3. *Let $X \sim S_\alpha(\sigma, \beta, \mu)$ and $a \in \mathbb{R}, a \neq 0$. Then*

$$aX \sim S_\alpha(|a|\sigma, sign(a)\beta, a\mu), \quad \alpha \neq 1$$

$$aX \sim S_1(|a|\sigma, sign(a)\beta, a\mu - \frac{2}{\pi}(\ln(|a|)\sigma\beta)), \quad \alpha = 1.$$

The first three properties identify σ and μ as a scale and a shift parameter respectively.

Proposition 4. *For any $0 < \alpha < 2$, if $X \sim S_\alpha(\sigma, \beta, 0)$, then $-X \sim S_\alpha(\sigma, -\beta, 0)$.*

We shall use the standard notation for the cumulative distribution function and probability density function: $P(X < x) = F(x; \alpha, \beta)$ specifies the cdf and $f(x; \alpha, \beta) = F'(x; \alpha, \beta)$ denotes the pdf of a random variable $X \sim S_\alpha(1, \beta, 0)$. The fact that we consider only standardized random variables is not limiting. By Propositions 2 and 3 it follows that if $X \sim S_\alpha(\sigma, \beta, \mu)$, then $(X - \mu)/\sigma \sim S_\alpha(1, \beta, 0)$. Furthermore because of the symmetry introduced by Proposition 4, it is sufficient to examine only the cdf and the pdf of $X \sim S_\alpha(1, \beta, 0)$, with $\beta \geq 0$ and then take advantage of the expressions:

- if $\beta < 0$:

$$f(x; \alpha, \beta) = f(-x; \alpha, -\beta), \quad F(x; \alpha, \beta) = 1 - F(-x; \alpha, -\beta); \qquad (8.7)$$

- if $\sigma \neq 1$ or $\mu \neq 0$:

$$f(x) = \frac{1}{\sigma} f(\frac{x - \mu}{\sigma}; \alpha, \beta), \quad F(x) = F(\frac{x - \mu}{\sigma}; \alpha, \beta), \tag{8.8}$$

where $f(x)$ and $F(x)$ are the pdf and the cdf of $X \sim S_\alpha(\sigma, \beta, \mu)$ respectively.

Propositions 1 and 3 imply that if we have a portfolio of assets the returns of which are independent α-stable random variables, then the portfolio return has again α-stable distribution since it equals the weighted average of assets returns. More precisely, let r_p denote the portfolio return, r_i denote the return of the i-th asset and ω_i denote the weight of the i-th portfolio item, then

$$r_p = \sum_{i=1}^{n} \omega_i r_i$$

where $r_i \sim S_\alpha(\sigma_i, \beta_i, \mu_i)$, with $\alpha \neq 1$, $\omega_i > 0$ for all $i = 1, 2 \ldots n$ and $\sum_{i=1}^{n} \omega_i = 1$.

As a consequence of Propositions 1 and 3, it follows that $r_p \sim S_\alpha(\sigma, \beta, \mu)$, with $\alpha \neq 1$ and

$$\sigma = \left(\sum_{i=1}^{n} (\omega_i \sigma_i)^\alpha \right)^{1/\alpha}, \quad \beta = \frac{\sum_{i=1}^{n} \beta_i (\omega_i \sigma_i)^\alpha}{\sum_{i=1}^{n} (\omega_i \sigma_i)^\alpha}, \quad \mu = \sum_{i=1}^{n} \omega_i \mu_i.$$

The case $\alpha = 1$ is examined in the same way. The parameter β is a skewness parameter because of the next proposition.

Proposition 5. $X \sim S_\alpha(\sigma, \beta, \mu)$ *is symmetric if and only if* $\beta = 0$ *and* $\mu = 0$. *It is symmetric about* μ *if and only if* $\beta = 0$.

The distribution is said to be skewed to the right if $\beta > 0$ and to the left if $\beta < 0$. It is said to be totally skewed to the right if $\beta = 1$ and totally skewed to the left if $\beta = -1$.

Stable distributions can be used as a theoretical model when empirical data is heavy-tailed. As we have mentioned in the introduction, empirical studies confirm that financial time series possess this property. The application of stable laws in this aspect is motivated by the fact that the tail of the stable law approaches zero as a power function. This is what is called "Pareto-like" behavior of the tail because of the same power decay of the tail of the Pareto distribution. The next proposition provides a rigorous description of the tail behavior.

Proposition 6. *Let* $X \sim S_\alpha(\sigma, \beta, \mu)$ $0 < \alpha < 2$. *Then*

$$\lim_{\lambda \to \infty} \lambda^\alpha P(X > \lambda) = C_\alpha \frac{1 + \beta}{2} \sigma^\alpha,$$

$$\lim_{\lambda \to \infty} \lambda^\alpha P(X < -\lambda) = C_\alpha \frac{1 - \beta}{2} \sigma^\alpha,$$

where

$$C_\alpha = \left(\int_0^\infty x^{-\alpha} \sin(x) dx \right)^{-1} = \begin{cases} \frac{1-\alpha}{\Gamma(2-\alpha)\cos(\pi\alpha/2)}, & \alpha \neq 1, \\ 2/\pi, & \alpha = 1. \end{cases}$$

The tail behavior of the Gaussian distribution as $\lambda \to \infty$ is specified by

$$P(X < -\lambda) = P(X > \lambda) \sim \frac{1}{2\sqrt{\pi}\sigma\lambda} e^{-\frac{\lambda^2}{4\sigma^2}}.$$

The difference in the asymptotic behavior of the tails of α-stable distributions, with $\alpha < 2$ and the Gaussian distribution motivates the distinction between Gaussian and non-Gaussian stable distributions. The latter are also called Pareto stable or Lévy stable as mentioned in the introduction.

The power decay of the tail of Pareto stable distributions implies that they do not possess a finite second moment. More exactly

Proposition 7. *Let* $X \sim S_\alpha(\sigma, \beta, \mu)$ *and* $0 < \alpha < 2$. *Then*

$$E|X|^p < \infty, \quad 0 < p < \alpha,$$

$$E|X|^p = \infty, \quad \alpha \leq p.$$

As a consequence, if $\alpha \leq 1$, then the corresponding α-stable distribution does not have a finite first absolute moment. Therefore statistical techniques valid for the Gaussian distribution are not applicable for the stable Paretian distributions.

The shift parameter μ has the following nice property if $\alpha > 1$:

Proposition 8. *If* $1 < \alpha \leq 2$, *the location parameter* μ *equals the mathematical expectation of* $X \sim S_\alpha(\sigma, \beta, \mu)$.

8.3   Probability density and distribution functions

According to Definition 2, α-stable distributions are the only ones that have domains of attraction. Naturally one would expect that these distributions would play a central role in statistical theory and the applications. One representative, the Gaussian distribution, is really of primary importance which is due to the fact that all distributions with finite second moment belong to its domain of attraction — certainly they comprise most of the interesting distributions for the applications. Undoubtedly the basic difficulty with the application of α-stable laws in practice is the lack of closed-form expressions for the pdf and the cdf, with the following exceptions:

* The Gaussian distribution $S_2(\sigma, 0, \mu) = N(\mu, 2\sigma^2)$, the cdf $F(x) = P(X \leq x)$ and the pdf $f(x)$ are well known:

$$F(x) = \Phi\left(\frac{x-\mu}{\sigma\sqrt{2}}\right), \quad f(x) = \frac{1}{2\sigma\sqrt{\pi}}e^{-\frac{(x-\mu)^2}{4\sigma^2}}, \quad x \in \mathbb{R}$$

where $\Phi(x)$ denotes the cdf of $N(0,1)$ distribution:

$$\Phi(x) = \frac{1}{\sqrt{2\pi}} \int_{-\infty}^{x} e^{-\frac{u^2}{2}} du. \tag{8.9}$$

- The Cauchy distribution $S_1(\sigma, 0, \mu)$, the cdf and the pdf are given by the following equations:

$$F(x) = \frac{1}{2} + \frac{1}{\pi}\arctan\left(\frac{x-\mu}{\sigma}\right), \quad f(x) = \frac{\sigma}{\pi((x-\mu)^2 + \sigma^2)}, \quad x \in \mathbb{R}.$$

- The Lévy distribution $S_{1/2}(\sigma, 1, \mu)$, the cdf and the pdf are given by:

$$F(x) = 2\left(1 - \Phi\left(\sqrt{\frac{\sigma}{x-\mu}}\right)\right), \quad f(x) = \sqrt{\frac{\sigma}{2\pi}}\frac{1}{(x-\mu)^{3/2}}e^{-\frac{\sigma}{2(x-\mu)}}, \quad x \geq \mu.$$

- Because of Proposition 4, there exist closed-form expressions for the cdf and the pdf of the symmetric to the Lévy distribution: $S_{1/2}(\sigma, -1, \mu)$.

The task of developing efficient numerical algorithms to overcome the lack of closed formulae is not trivial and many efforts have been concentrated on it. We shall continue with a brief summary of some of the main approaches.

8.3.1 Infinite series expansions

Fama and Roll [9] were the first to publish tabulated cdfs and percentiles of symmetric stable distributions. They took advantage of the two infinite series expansions of Bergström [3]. The series expansions are derived from the chf in parametrization P_2. We summarize the basic result for the case $\alpha \neq 1$ in the following:

Theorem 1. *Let $X \sim S_\alpha(1, \beta, 0; P_2)$. The density $f(x; \alpha, \beta; P_2)$ has the following representation in terms of the infinite series expansions:*

$$f(x; \alpha, \beta; P_2) = \frac{1}{\pi}\sum_{k=1}^{\infty}(-1)^{k-1}\frac{\Gamma(1+k/\alpha)}{k!}x^{k-1}\sin\left[\frac{k\pi}{2\alpha}B\right] \tag{8.10}$$

and

$$f(x; \alpha, \beta; P_2) = \frac{1}{\pi x^{1+\alpha}}\sum_{k=1}^{\infty}\frac{\Gamma(1+k\alpha)}{k!}(-x^{-\alpha})^{k-1}\sin\left[\frac{k\pi}{2}B\right] \tag{8.11}$$

where $B = (\alpha + \beta_2 K(\alpha))$.

The series (8.10) is absolutely convergent for $x > 0$ and $1 < \alpha \leq 2$, and is an asymptotic expansion for $x \to 0$, when $0 < \alpha < 1$. In contrast to (8.10), (8.11) is absolutely convergent for $x > 0$ and $0 < \alpha < 1$, and is an asymptotic expansion for $x \to \infty$, when $1 < \alpha \leq 2$.

Infinite series expansions of the cdf are derived by term-by-term integration of the expansions of the densities and are provided in the next

Theorem 2. *Let $X \sim S_\alpha(1, \beta, 0; P_2)$. The cumulative distribution function $F(x; \alpha, \beta; P_2)$ has the following representation in terms of the infinite series expansions:*

$$F(x; \alpha, \beta; P_2) = A + \frac{1}{\pi\alpha} \sum_{k=1}^{\infty}(-1)(-c)^k \frac{\Gamma(k/\alpha)}{k!} x^k \sin\left[\frac{k\pi}{2\alpha}B\right] \qquad (8.12)$$

and

$$F(x; \alpha, \beta; P_2) = 1 - \frac{1}{\pi} \sum_{k=1}^{\infty}(-1)^{k-1} c^{-k\alpha} \frac{\Gamma(k\alpha)}{k!} x^{-\alpha k} \sin\left[\frac{k\pi}{2}B\right] \qquad (8.13)$$

where $A = F(0; \alpha, \beta; P_2) = \frac{1}{2}\left(1 - \frac{\beta_2 K(\alpha)}{\alpha}\right)$ and $B = (\alpha + \beta_2 K(\alpha))$.

The series (8.12) is absolutely convergent for $x > 0$ and $1 < \alpha \leq 2$, and is an asymptotic expansion for $x \to 0$, when $0 < \alpha < 1$. In contrast to (8.12), (8.13) is absolutely convergent for $x > 0$ and $0 < \alpha < 1$, and is an asymptotic expansion for $x \to \infty$, when $1 < \alpha \leq 2$.

It is possible to translate the series expansions into any of the specified parametric forms by taking advantage of expressions (8.3) or (8.5) and (8.6). Note that the formulae in the two theorems are valid only for $x > 0$. However this is not a restriction because we can use equations (8.7) and (8.8).

DuMouchel [5] remarks that the infinite series expansions are efficient only for either small or large values of $|x|$. Thus they can be used to approximate the tail of the distribution. For the middle range of $|x|$, another type of approximation can be developed. For example in [5] it is combined with the method of the Fast Fourier Transform (FFT) algorithm and in [4] — with a polynomial type approximation.

8.3.2   The FFT approach

In theory, distribution and density functions are derived from the chf via the Fourier inversion integral. So perhaps the most straightforward method to compute stable cdfs and pdfs is direct numerical integration of the inversion formula:

$$f(x) = \frac{1}{2\pi} \int_{-\infty}^{\infty} e^{-itx} \varphi(t) dt \qquad (8.14)$$

or

$$F\left(x + \frac{h}{2}\right) - F\left(x - \frac{h}{2}\right) = \frac{1}{2\pi} \int_{-\infty}^{\infty} e^{-itx} \varphi(t) \frac{2}{t} \sin\frac{ht}{2} dt; \quad h > 0 \qquad (8.15)$$

where $\varphi(t)$ is the chf of the distribution. First we shall consider the problem of evaluating the density function following the approach in [17].

If it is necessary to compute the density function for a large number of x values, much efficiency is gained if the FFT algorithm is employed. This is particularly appropriate if we aim at evaluating the likelihood function. The advantage is that the FFT approach is computationally efficient and the main disadvantage is that we have the density function evaluated on an equally spaced grid. As a consequence, we have to interpolate for intermediate values and if the argument is outside the grid, we need to employ another method only for the tail approximation. For this purpose we can use an infinite series expansion or the integral representations of Zolotarev.

The main idea is to calculate the integral in (8.14) for the grid of equally spaced x values

$$x_k = (k - 1 - \frac{N}{2})h, \quad k = 1, ..., N.$$

That is, we arrive at the equation

$$f(x_k) = \int_{-\infty}^{\infty} e^{-i2\pi\omega(k-1-\frac{N}{2})h}\varphi(2\pi\omega)d\omega. \tag{8.16}$$

Since the integral in (8.16) is convergent, we can choose a large enough upper and a small enough lower bound to approximately compute (8.16) through the Riemann sum:

$$f(x_k) \approx s \sum_{n=1}^{N} \varphi\left(2\pi s\left(n - 1 - \frac{N}{2}\right)\right)e^{-i2\pi(k-1-\frac{N}{2})(n-1-\frac{N}{2})sh}, \quad k = 1, ..., N. \tag{8.17}$$

In this particular example, the upper and the lower bounds equal $\frac{sN}{2}$ and $-\frac{sN}{2}$ respectively. The integrand is evaluated for the equally spaced grid $n - 1 - \frac{N}{2}$, $n = 1, ..., N$ with distance s between them. The choice of the integral bounds is not arbitrary. If $s = (hN)^{-1}$, we have the following expression for the density:

$$f(x_k) \approx \frac{1}{hN} \sum_{n=1}^{N} \varphi\left(2\pi \frac{1}{hN}\left(n - 1 - \frac{N}{2}\right)\right)e^{-i2\pi(k-1-\frac{N}{2})(n-1-\frac{N}{2})\frac{1}{N}},$$

$$k = 1, ..., N.$$

Having rearranged the terms in the exponent, finally we arrive at

$$f(x_k) \approx \frac{(-1)^{k-1+\frac{N}{2}}}{hN} \sum_{n=1}^{N}(-1)^{n-1}\varphi\left(\frac{2\pi}{hN}\left(n - 1 - \frac{N}{2}\right)\right)e^{\frac{-i2\pi(n-1)(k-1)}{N}},$$

$$k = 1, ..., N. \tag{8.18}$$

The discrete FFT is a numerical method developed for calculation of sequences such as $f(x_k)$ in (8.18) given the sequence

$$(-1)^{n-1} \varphi\left(\frac{2\pi}{hN}\left(n - 1 - \frac{N}{2}\right)\right), \quad n = 1, ..., N. \tag{8.19}$$

So by applying the discrete Fourier transform for the sequence (8.19) we approximately compute the density values $f(x_k)$. The benefit of this approach is that the FFT algorithm needs $N \log_2 N$ arithmetical operations. In comparison, the direct computation of the integral needs N^2. Obviously the FFT approach reduces the computational burden enormously when N is a large number.

It should be noted that the approximation error can be divided into three types:

- error arising from the interchange of the infinite integral bounds in (8.16) with finite ones;
- error arising from the approximation of (8.16)with the Riemann sum (8.17);
- error arising in the interpolation for intermediate values if the function argument is not a grid node;

The parameters of the FFT method are N (the number of summands in the Riemann sum) and h (the grid spacing). They should be carefully chosen as there is a trade-off between accuracy and computational burden. From the construction of (8.17), it follows that, to reduce the approximation error, N should be as large as possible. A peculiarity of the numerical method is that efficiency is gained if N is expressed as 2^q. In addition the length of the integration region in terms of the original variable t is $2\pi/h$. Hence h should be as small as possible to increase accuracy. Numerical experiments show [17] that if $q = 13$ and $h = 0.01$ the magnitude of approximation error is 10^{-6} which is satisfactory for most of the applications. In [17] it is mentioned that the first fifteen summands in the infinite series expansion (8.11) produce the same error. Hence they can be combined with the FFT method to approximate the tail of the distribution.

If the purpose of the approximation is estimation of the likelihood function, it is preferable to use a continuous parametrization. Furthermore, taking advantage of (8.7) and (8.8), we can always approximate the normalized case and then estimate the pdf or cdf for any choice of σ and μ. Because of the symmetry of the grid nodes, further efficiency can be achieved in the computations of (8.19), if we use the following relationship valid for any chf:

$$\varphi(-t) = \overline{\varphi(t)} \tag{8.20}$$

where $\overline{z}$ means the complex conjugate of z.

Due to (8.20), it is possible to compute the chf values only for the positive nodes, then change the sign of their imaginary parts and achieve the chf values for the symmetric with respect to zero, negative grid nodes.

Concerning the cdf, there are two potential approaches:

- on the one hand, it is possible to use (8.15) and derive an expression similar to (8.18);
- on the other hand, we can work directly with the evaluated pdf according to (8.18) and;

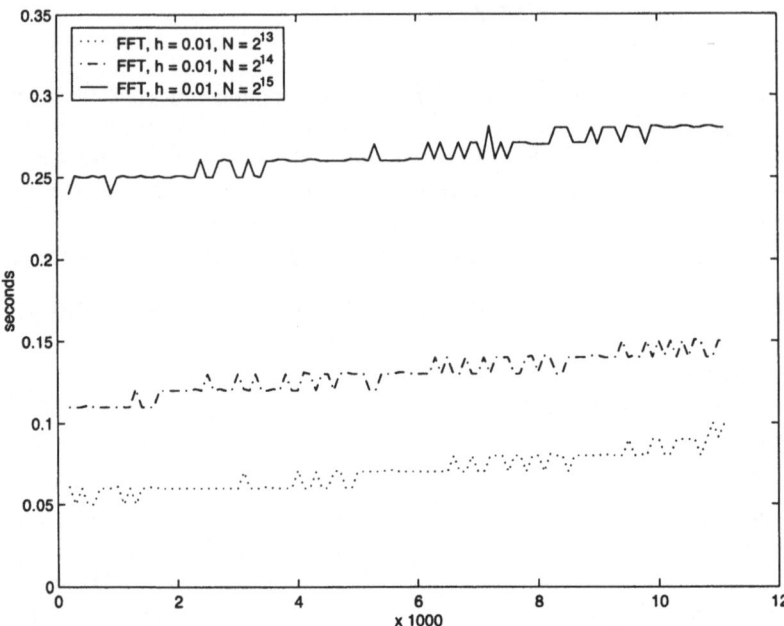

FIGURE 4. Time (sec) necessary to evaluate the pdf of 1.6-symmetric stable distribution as a function of the number of pdf arguments (horizontal axis) and for different choices of N and h.

$$F(x) = \int_{-\infty}^{x} f(u)du = h \sum_{\{k:u_k \leq x\}} f(u_k).$$

Figure 4 shows the computational burden of evaluating the pdf according to the FFT method with different numbers of pdf arguments and different choices of the parameters N and h. Clearly the computational time necessary for interpolation for intermediate values is negligible with respect to the computational time for the discrete FFT — the three curves rise very slowly as the number of pdf arguments increases. The platform is PC, Windows XP, Pentium III 733MHz.

8.3.3   Polynomial-type approximations

By Proposition 6, the tails of α-stable laws exhibit power decay. It is tempting to verify if it is possible to employ polynomials with negative powers to approximate the tail of the distribution and polynomials with positive powers for the central part around the mode.

A straightforward approach is to choose

$$f(y; \alpha, \beta, 1, 0) \approx \sum_{k=k_1}^{k_K} \sum_{l=l_1}^{l_L} \sum_{m=m_1}^{m_M} c_{klm} \alpha^k \beta^l y^m. \tag{8.21}$$

The coefficients c_{klm} can be computed by constructing a linear system of equations in which the relevant pdf values $f(y; \alpha, \beta, 1, 0)$ are calculated with the FFT-based method or the integral representations of Zolotarev.

A more involved approach is proposed in [4] and [17]. The domain of the pdf is partitioned into several regions and different approximation schemes are specified for the different regions; the tails of the distribution are approximated with a Bergström series expansion. The coefficients c_{klm} are calibrated with a two-step procedure which relies on least-squares techniques (for more details see [4] and [17]). The authors report that 10 000 pdf evaluations take approximately one second on a 300MHz Pentium II PC with 64MB RAM. Approximate estimation shows that the computational burden is comparable to that of the approach of McCulloch discussed in Section 8.3.5. Of course, it should be noted that the computational efficiency is conditional on the particular implementation of the approximation.

8.3.4   The integral representations of Zolotarev

The pdf and the cdf of stable laws can be very accurately evaluated with the help of integral representations derived by Zolotarev in [19] in parametrization P_2. Du-Mouchel remarks in [5] that they can be potentially used to compute the pdf and the cdf and John Nolan is the first to apply them in the software STABLE. His approach is described in [15]. Zolotarev's result is proved in [15] in terms of the continuous parametrization P_1 and in [20] it is given in terms of P_2. Following [15], the pdf and cdf of stable laws can be expressed as

$$f(x; \alpha, \beta, P_1) = c_2(x; \alpha, \beta) \int_{-\theta_0}^{\frac{\pi}{2}} g(\theta; x, \alpha, \beta) \exp(-g(\theta; x, \alpha, \beta)) d\theta, \quad (8.22)$$

$$F(x; \alpha, \beta, P_1) = c_1(\alpha, \beta) + c_3(\alpha) \int_{-\theta_0}^{\frac{\pi}{2}} \exp(-g(\theta; x, \alpha, \beta)) d\theta, \quad (8.23)$$

where

$$c_1(\alpha, \beta) = \begin{cases} \frac{1}{\pi}\left(\frac{\pi}{2} - \theta_0\right), & \alpha < 1, \\ 0, & \alpha = 1, \\ 1, & \alpha > 1, \end{cases}$$

$$c_2(x; \alpha, \beta) = \begin{cases} \frac{\alpha}{\pi|\alpha-1|(x-\zeta)}, & \alpha \neq 1, \\ \frac{1}{2|\beta|}, & \alpha = 1, \end{cases}$$

$$c_3(\alpha) = \begin{cases} \frac{sign(1-\alpha)}{\pi}, & \alpha \neq 1, \\ \frac{1}{\pi}, & \alpha = 1, \end{cases}$$

$$g(\theta; x, \alpha, \beta) = \begin{cases} (x - \zeta)^{\frac{\alpha}{\alpha-1}} V(\theta; \alpha, \beta), & \alpha \neq 1, \\ e^{-\frac{\pi x}{2\beta}} V(\theta; \alpha, \beta), & \alpha = 1, \end{cases}$$

and

$$\zeta = \zeta(\alpha, \beta) = \begin{cases} -\beta \tan \frac{\pi\alpha}{2}, & \alpha \neq 1, \\ 0, & \alpha = 1, \end{cases}$$

$$\theta_0 = \theta_0(\alpha, \beta) = \begin{cases} \frac{1}{\alpha} \arctan(\beta \tan \frac{\pi\alpha}{2}), & \alpha \neq 1, \\ \frac{\pi}{2}, & \alpha = 1, \end{cases}$$

$$V(\theta; \alpha, \beta) = \begin{cases} (\cos \alpha\theta_0)^{\frac{1}{1-\alpha}} \left(\frac{\cos \theta}{\sin \alpha(\theta_0+\theta)} \right)^{\frac{\alpha}{\alpha-1}} \frac{\cos(\alpha\theta_0+(\alpha-1)\theta)}{\cos \theta}, & \alpha \neq 1, \\ \frac{2}{\pi} \left(\frac{\frac{\pi}{2}+\beta\theta}{\cos \theta} \right) \exp\left(\frac{1}{\beta} \left(\frac{\pi}{2} + \beta\theta \right) \tan \theta \right), & \alpha = 1. \end{cases} \quad (8.24)$$

Equations (8.22) and (8.23) are correct for $x > \zeta$. The case $x < \zeta$ can be treated by taking advantage of the relations

$$f(x; \alpha, \beta, P_1) = f(-x; \alpha, -\beta, P_1)$$

and

$$F(x; \alpha, \beta, P_1) = 1 - F(-x; \alpha, -\beta, P_1)$$

which are the same as (8.7). The parameter ζ appears because of the change from P_2 to P_1.

It is possible to show that the integrand in (8.22) is composed of two monotone segments, it is equal to zero at the lower and the upper bound of the integral and has a unique maximum for that value of θ for which $g(\theta; x, \alpha, \beta) = 1$. As Nolan remarks in [15], for large values of $|x|$ the integrand in (8.22) has a sharp spike which can be an issue in the numerical integration. Figure 5 is a nice illustration of this point. As a result, for computational purposes it is more suitable to split the integration region into two parts:

$$f_L(x; \alpha, \beta, P_1) = c_2(x; \alpha, \beta) \left[\int_{-\theta_0}^{t} g(\theta; x, \alpha, \beta) \exp(-g(\theta; x, \alpha, \beta)) d\theta \right]. \quad (8.25)$$

and

$$f_R(x; \alpha, \beta, P_1) = c_2(x; \alpha, \beta) \left[\int_{t}^{\frac{\pi}{2}} g(\theta; x, \alpha, \beta) \exp(-g(\theta; x, \alpha, \beta)) d\theta \right], \quad (8.26)$$

where

$$t = \left\{ \theta : \ln(g(\theta; x, \alpha, \beta)) = 0 \right\} \quad (8.27)$$

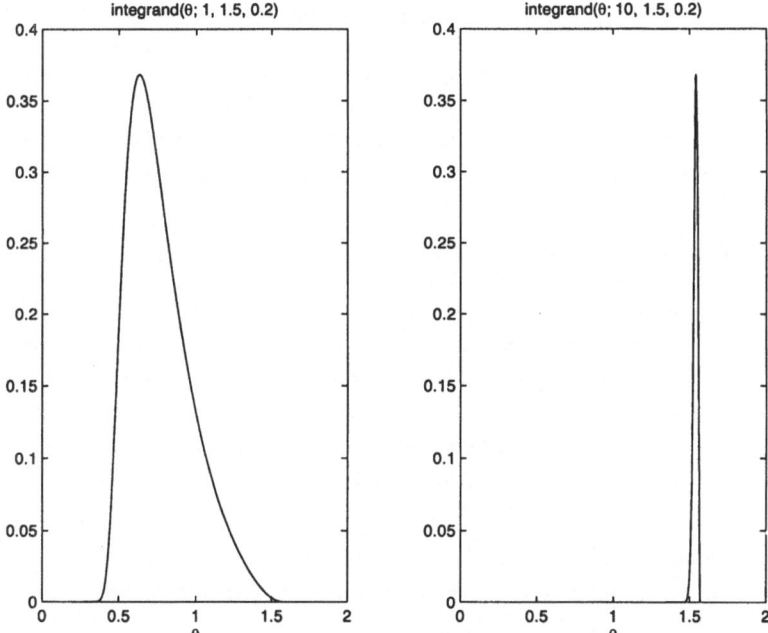

FIGURE 5. The integrand in (8.22) for $\alpha = 1.5$, $\beta = 0.2$ and two different values of the pdf argument: $x = 1$ and $x = 10$

and the pdf is the sum of (8.25) and (8.26):

$$f(x; \alpha, \beta, P_1) = f_L(x; \alpha, \beta, P_1) + f_R(x; \alpha, \beta, P_1).$$

The use of the logarithm in (8.27) is permitted because $g(\theta; x, \alpha, \beta)$ is positive (see [15] and the references therein). Moreover the logarithm transforms the product in (8.24) into a sum which is numerically more stable.

In contrast to (8.22), the expression for the cdf (8.23) is by far less problematic. The integrand is strictly monotonic (increasing or decreasing depending on the sign of β), at one of the integral bounds it equals 1 and at the other it equals zero.

The benefits of pdf and cdf estimation using Zolotarev's integral representations can be summarized as follows:

- the integral bounds are finite in contrast to the infinite ones in the Fourier inversion integral;
- the integrands are not periodic, they are bounded, continuous and composed of monotonic segments;
- it is possible to approximate the pdf and the cdf for any x with no interpolation as in the FFT method.

It is important to remark that the accuracy and the computational burden significantly depend on the choice of quadrature to perform the numerical integration. In case we choose an adaptive quadrature, the speed of the pdf evaluation will depend on the particular value of x, because for large x the integrand is not well behaved for numerical work and more iterations will be necessary before convergence. Apart from this point, Nolan notes in [15] that the program STABLE encounters numerical difficulties when $|\alpha - 1| < 0.02$ or when α is close to zero, because the integrands change very rapidly.

8.3.5 Approximations of McCulloch for SαS distributions

McCulloch suggested in [1] an approach for efficient evaluation of the pdf and the cdf of $S\alpha S$ laws. The approximations work well for $\alpha \in [0.92, 2]$. They are neither extremely accurate nor compact but represent an interesting approach that could be further improved or extended for the asymmetric case. Since we are working with symmetric distributions the parametrization is not an issue because both P_1 and P_2 are continuous when $\beta = 0$. We consider the cdf and the pdf on the positive half-line.

The main idea is summarized as follows:

- The domain $[0, \infty]$ is suitably transformed into the interval $[0, 1]$ which is bounded, hence better for numerical work. The transform $z_\alpha(x) : [0, \infty) \to [0, 1]$ is chosen to have the special form:

$$z_\alpha(x) = 1 - (1 + a_\alpha x)^{-\alpha}$$

because of the power decay of the tail of stable laws.

- In the new interval $[0, 1]$ the tail of the $S\alpha S$ distribution is represented as a linear combination of the tails of Cauchy and Gaussian distribution plus a residual:

$$\overline{F}(z; \alpha) = (2 - \alpha)\overline{F}(x_1(z); 1) + (\alpha - 1)\overline{F}(x_2(z); 2) \\ + R_\alpha(z), \quad z \in [0, 1]$$

where $\overline{F}(z; \alpha) = 1 - F(z; \alpha)$ and $x_\alpha(z)$ is the inverse of $z_\alpha(x)$:

$$x_\alpha(z) = \frac{(1 - z)^{-1/\alpha} - 1}{a_\alpha}.$$

- Quintic splines are used to approximate the residual $R_\alpha(z)$. The spline coefficients are calibrated via Zolotarev's integral representations and are given in [1]. Finally the cdf approximation has the following form:

$$\hat{F}(z; \alpha) = 1 - (2 - \alpha)\left(\frac{1}{2} - \frac{1}{\pi} \arctan x_1(z)\right) \\ - (\alpha - 1)\left[1 - \Phi\left(\frac{x_2(z)}{\sqrt{2}}\right)\right] - \hat{R}_\alpha(z), \quad z \in [0, 1]$$

where $\Phi(x)$ is defined in equation (8.9) and $\hat{R}_\alpha(z)$ is the quintic spline approximation of the residual $R_\alpha(z)$.

Because of the special form of the approximation, we have the following limit behavior of the residual for any $\alpha \in (0, 2]$:

$$\lim_{z \to 0} R_\alpha(z) = \lim_{z \to 1} R_\alpha(z) = 0$$

which corresponds to higher accuracy level in the tails.

The approximation $\hat{f}(x; \alpha)$ of the pdf is received after differentiation of the expression derived for the cdf. It should be noted that the spline coefficients given in [1] are computed to minimize the error in the estimation of the likelihood function. As a result, the approximations are suitable for maximum likelihood estimation of stable parameters. The precision for $\alpha \in [0.92, 2]$ is:

$$\max_{\alpha, x} |\hat{F}(x; \alpha) - F(x; \alpha)| = 2.2 \times 10^{-5},$$

$$\max_{\alpha, x} |\hat{f}(x; \alpha) - f(x; \alpha)| = 6.6 \times 10^{-5}.$$

In Figure 6 we compare the computational burden of McCulloch's approach and the FFT method. If $N = 2^{13}$ and $h = 0.01$, the accuracy level of the FFT method is roughly the same as the precision of McCulloch's approach. It follows that the latter is faster if we would like to estimate the pdf at less than 2000 points. Such a

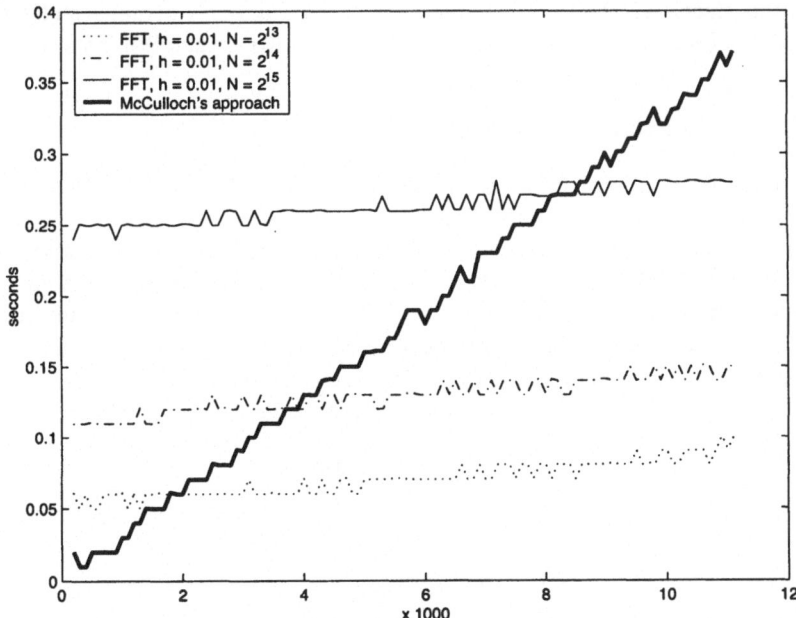

FIGURE 6. McCulloch's approach compared to the FFT method in terms of the computational time necessary to evaluate the pdf of 1.6-symmetric stable distribution as a function of the number of pdf arguments.

comparison is useful for the purposes of maximum likelihood estimation of stable parameters — if the sample consists of less than 2000 observations, then it would be more efficient to use McCulloch's approximations. All computations are carried out on a PC, Windows XP, Pentium III 733MHz.

8.4   Parameter estimation

Generally speaking, parameter estimation techniques for the class of stable laws fall into three categories — quantile methods, characteristic function based methods and maximum likelihood. The approaches from the first type use predetermined empirical quantiles to estimate stable parameters. For example the method of Fama and Roll [9], [10] for symmetric α-stable distributions and its modified version of McCulloch [14] for the skewed case belong to this group.

Chf based methods include the method of moments approach suggested by Press [16] and regression-type procedures proposed by Koutrouvelis [12] and Kogon and Williams [1]. Simulation studies available in the literature ([1], [2]), show the superiority of the regression-type estimation over the quantile methods.

The validity of maximum likelihood estimation (MLE) theory was demonstrated by DuMouchel [6]. The comparison studies between MLE and the quantile method of McCulloch in [17] recommend the maximum likelihood estimator.

In this paper we shall review and compare McCulloch's quantile method, the method of moments, the regression-type estimator of Kogon and Williams and MLE.

8.4.1   Quantile method of McCulloch

The estimation procedure proposed by McCulloch in [14] is a generalization of the quantile method of Fama and Roll in [9], [10] for the symmetric case. The estimates of stable parameters in parametrization P_0 are consistent and asymptotically normal if $0.6 < \alpha \leq 2$. We shall adopt the standard notation for theoretical and empirical quantiles, namely x_p is the p-th quantile if $F(x_p) = p$, where $F(x)$ is the cdf of a random variable and given a sample of observations $x_1, x_2, \ldots, x_n$, then $\hat{x}_p$ is the sample quantile if $F_n(\hat{x}_p) = p$, where $F_n(x)$ is the sample cdf.

According to [14], let us define two functions of theoretical quantiles :

$$v_\alpha = \frac{x_{0.95} - x_{0.05}}{x_{0.75} - x_{0.25}},$$

$$v_\beta = \frac{x_{0.95} + x_{0.05} - 2x_{0.50}}{x_{0.95} - x_{0.05}}.$$

The functions v_α and v_β have this special form because by expression (8.8) it appears that they do not depend on the scale and the location parameter, i.e.,

$$\begin{vmatrix} v_\alpha = \phi_1(\alpha, \beta), \\ v_\beta = \phi_2(\alpha, \beta). \end{vmatrix} \tag{8.28}$$

Employing equation (8.7), we have that $F(-x_p; \alpha, -\beta) = F(x_{1-p}; \alpha, \beta)$ and therefore we have the relations:

$$\begin{aligned} \phi_1(\alpha, \beta) = \phi_1(\alpha, -\beta), \\ \phi_2(\alpha, \beta) = -\phi_2(\alpha, -\beta). \end{aligned} \tag{8.29}$$

The system of equations (8.28) can be inverted and the parameters α and β can be expressed as functions of the quantities v_α and v_β:

$$\begin{vmatrix} \alpha = \psi_1(v_\alpha, v_\beta), \\ \beta = \psi_2(v_\alpha, v_\beta). \end{vmatrix} \tag{8.30}$$

Replacing v_α and v_β in equations (8.30) with their sample counterparts $\hat{v}_\alpha$ and $\hat{v}_\beta$:

$$\hat{v}_\alpha = \frac{\hat{x}_{0.95} - \hat{x}_{0.05}}{\hat{x}_{0.75} - \hat{x}_{0.25}},$$

$$\hat{v}_\beta = \frac{\hat{x}_{0.95} + \hat{x}_{0.05} - 2\hat{x}_{0.50}}{\hat{x}_{0.95} - \hat{x}_{0.05}}$$

yields estimators $\hat{\alpha}$ and $\hat{\beta}$:

TABLE 1. $\alpha = \psi_1(v_\alpha, v_\beta) = \psi_1(v_\alpha, -v_\beta)$

		v_β						
		0	0.1	0.2	0.3	0.5	0.7	1
	2.439	2	2	2	2	2	2	2
	2.5	1.916	1.924	1.924	1.924	1.924	1.924	1.924
	2.6	1.808	1.813	1.829	1.829	1.829	1.829	1.829
	2.7	1.729	1.73	1.737	1.745	1.745	1.745	1.745
	2.8	1.664	1.663	1.663	1.668	1.676	1.676	1.676
	3	1.563	1.56	1.553	1.548	1.547	1.547	1.547
	3.2	1.484	1.48	1.471	1.46	1.448	1.438	1.438
v_α	3.5	1.391	1.386	1.378	1.364	1.337	1.318	1.318
	4	1.279	1.273	1.266	1.25	1.21	1.184	1.15
	5	1.128	1.121	1.114	1.101	1.067	1.027	0.973
	6	1.029	1.021	1.014	1.004	0.974	0.935	0.874
	8	0.896	0.892	0.887	0.883	0.855	0.823	0.769
	10	0.818	0.812	0.806	0.801	0.78	0.756	0.691
	15	0.698	0.695	0.692	0.689	0.676	0.656	0.595
	25	0.593	0.59	0.588	0.586	0.579	0.563	0.513

TABLE 2. $\beta = \psi_2(v_\alpha, v_\beta) = -\psi_2(v_\alpha, -v_\beta)$

					v_β			
		0	0.1	0.2	0.3	0.5	0.7	1
	2.439	0	2.16	1	1	1	1	1
	2.5	0	1.592	3.39	1	1	1	1
	2.6	0	0.759	1.8	1	1	1	1
	2.7	0	0.482	1.048	1.694	1	1	1
	2.8	0	0.36	0.76	1.232	2.229	1	1
	3	0	0.253	0.518	0.823	1.575	1	1
	3.2	0	0.203	0.41	0.632	1.244	1.906	1
v_α	3.5	0	0.165	0.332	0.499	0.943	1.56	1
	4	0	0.136	0.271	0.404	0.689	1.23	2.195
	5	0	0.109	0.216	0.323	0.539	0.827	1.917
	6	0	0.096	0.19	0.284	0.472	0.693	1.759
	8	0	0.082	0.163	0.243	0.412	0.601	1.596
	10	0	0.074	0.174	0.22	0.377	0.546	1.482
	15	0	0.064	0.128	0.191	0.33	0.478	1.362
	25	0	0.056	0.112	0.167	0.285	0.428	1.274

$$\left| \begin{array}{l} \hat{\alpha} = \psi_1(\hat{v}_\alpha, \hat{v}_\beta), \\ \hat{\beta} = \psi_2(\hat{v}_\alpha, \hat{v}_\beta). \end{array} \right. \tag{8.31}$$

The functions $\psi_1(.)$ and $\psi_2(.)$ are tabulated in Tables 1 and 2. It should be noted that because of property (8.29), we have that:

$$\psi_1(v_\alpha, v_\beta) = \psi_1(v_\alpha, -v_\beta),$$
$$\psi_2(v_\alpha, v_\beta) = -\psi_2(v_\alpha, -v_\beta).$$

In other words, the sign of $\hat{v}_\beta$ determines the sign of β.

Since there are no closed-form expressions for the functions $\psi_1(.)$ and $\psi_2(.)$, we compute estimates of α and β from the statistics $\hat{v}_\alpha$ and $\hat{v}_\beta$ using the values in Tables 1 and 2 and linear interpolation for intermediate values. If it happens that $\hat{v}_\alpha$ is below 2.439, $\hat{\alpha}$ should be set equal to 2 and $\hat{\beta}$ equal to zero. Table 2 contains values larger than 1 for more precise interpolation. If $\hat{\beta} > 1$, it should be reduced to 1.

McCulloch provides an estimator for the scale parameter σ which is very similar to the estimator given by Fama and Roll. Let us first define v_σ as:

$$v_\sigma = \frac{x_{0.75} - x_{0.25}}{\sigma} = \phi_3(\alpha, \beta).$$

The function $\psi_3(\alpha, \beta)$ is given in Table 3. Employing the same arguments that led us to equations (8.29) yields the relation $\phi_3(\alpha, \beta) = \phi_3(\alpha, -\beta)$. The estimator $\hat{\sigma}$ is received after replacing α and β with the estimates found according to equations (8.31):

TABLE 3. $v_\sigma = \phi_3(\alpha, \beta) = \phi_3(\alpha, -\beta)$

		β			
		0	0.25	0.5	0.75	1
	0.5	2.588	3.073	4.534	6.636	9.144
	0.6	2.337	2.635	3.542	4.808	6.247
	0.7	2.189	2.392	3.004	3.844	4.775
	0.8	2.098	2.244	2.676	3.265	3.912
	0.9	2.04	2.149	2.461	2.886	3.356
	1	2	2.085	2.311	2.624	2.973
	1.1	1.98	2.04	2.205	2.435	2.696
	1.2	1.965	2.007	2.125	2.294	2.491
α	1.3	1.955	1.984	2.067	2.188	2.333
	1.4	1.946	1.967	2.022	2.106	2.211
	1.5	1.939	1.952	1.988	2.045	2.116
	1.6	1.933	1.94	1.962	1.997	2.043
	1.7	1.927	1.93	1.943	1.961	1.987
	1.8	1.921	1.922	1.927	1.936	1.947
	1.9	1.914	1.915	1.916	1.918	1.921
	2	1.908	1.908	1.908	1.908	1.908

$$\hat{\sigma} = \frac{\hat{x}_{0.75} - \hat{x}_{0.25}}{\phi_3(\hat{\alpha}, \hat{\beta})}.$$

Estimation of the location parameter μ is a more involved affair because of the discontinuity of the parametric representation of the chf P_0 when $\alpha \to 1$ and $\beta \neq 0$. First we estimate the shifted location parameter ζ defined by:

$$\zeta = x_{0.50} + \sigma \, sign(\beta)\phi_4(\alpha, \beta) \tag{8.32}$$

where $\phi_4(\alpha, \beta)$ is tabulated in Table 4 and has the property $\phi_4(\alpha, \beta) = \phi_4(\alpha, -\beta)$. The location parameter μ is related to ζ according to:

$$\mu = \begin{cases} \zeta - \beta\sigma \tan \frac{\pi\alpha}{2}, & \alpha \neq 1, \\ \zeta, & \alpha = 1. \end{cases} \tag{8.33}$$

Replacing the parameters in equations (8.32) and (8.33) with their sample counterparts yields the estimator $\hat{\mu}$:

$$\hat{\zeta} = \hat{x}_{0.50} + \hat{\sigma} \, sign(\hat{\beta})\phi_4(\hat{\alpha}, \hat{\beta})$$

and

$$\hat{\mu} = \begin{cases} \hat{\zeta} - \hat{\beta}\hat{\sigma} \tan \frac{\pi\hat{\alpha}}{2}, & \hat{\alpha} \neq 1, \\ \hat{\zeta}, & \hat{\alpha} = 1. \end{cases}$$

TABLE 4. $\phi_4(\alpha, \beta) = \phi_4(\alpha, -\beta)$

		β			
		0	0.25	0.5	0.75	1
	0.5	0	-0.061	-0.279	-0.659	-1.198
	0.6	0	-0.078	-0.272	-0.581	-0.997
	0.7	0	-0.089	-0.262	-0.52	-0.853
	0.8	0	-0.096	-0.25	-0.469	-0.742
	0.9	0	-0.099	-0.237	-0.424	-0.652
	1	0	-0.098	-0.223	-0.383	-0.576
	1.1	0	-0.095	-0.208	-0.346	-0.508
	1.2	0	-0.09	-0.192	-0.31	-0.447
α	1.3	0	-0.084	-0.173	-0.276	-0.39
	1.4	0	-0.075	-0.154	-0.241	-0.335
	1.5	0	-0.066	-0.134	-0.206	-0.283
	1.6	0	-0.056	-0.111	-0.17	-0.232
	1.7	0	-0.043	-0.088	-0.132	-0.179
	1.8	0	-0.03	-0.061	-0.092	-0.123
	1.9	0	-0.017	-0.032	-0.049	-0.064
	2	0	0	0	0	0

It should be observed that a significant advantage of the method considered is the lack of heavy computations. On the personal homepage of McCulloch (http://www.econ.ohio-state.edu/jhm/jhm.html) a FORTRAN implementation of the algorithm is publicly available.

8.4.2   Chf based methods

The characteristic function based methods rely on the sample chf for parameter estimation. The sample chf is defined as:

$$\hat{\varphi}(t) = \frac{1}{n} \sum_{j=1}^{n} e^{itx_j}, \quad t \in \mathbb{R} \tag{8.34}$$

where $x_1, x_2, \ldots, x_n$ is a sample of independent, identically distributed (iid) observations on a random variable X. Since $|\hat{\varphi}(t)| \leq 1$, all moments of the random variable $\hat{\varphi}(t)$ are finite and, according to equation (8.34), for any t it is the sample mean of the iid random variables e^{itx_j}. As a consequence, from the law of large numbers, it can be inferred that the sample chf is a consistent estimator of the chf $\varphi_X(t) = \mathbb{E}e^{itX}$, $t \in \mathbb{R}$ of a random variable X.

8.4.2.1   The method of moments

Press [16] suggested a simple and straightforward approach to estimation of parameters of stable laws, which was called the method of moments. His approach is based on certain transformations of the chf in parametrization P_0. From the parametric representation (8.1) it follows that

$$|\varphi(t)| = \exp(-\sigma^\alpha |t|^\alpha), \quad t \in \mathbb{R} \tag{8.35}$$

and therefore $-\ln|\varphi(t)| = \sigma^\alpha |t|^\alpha$ for any real t.

8.4.2.1.1   Case $\alpha \neq 1$.

If we choose t_1 and t_2 such that $t_1 \neq t_2 \neq 0$, we have the following system of two equations:

$$\left|\begin{array}{l} -\ln|\varphi(t_1)| = \sigma^\alpha |t_1|^\alpha, \\ -\ln|\varphi(t_2)| = \sigma^\alpha |t_2|^\alpha, \end{array}\right.$$

which can be solved for α and σ. Replacing the chf for its sample equivalent $\hat{\varphi}(t)$ yields the estimators $\hat{\alpha}$ and $\hat{\sigma}$:

$$\hat{\alpha} = \frac{\ln \frac{\ln|\hat{\varphi}(t_1)|}{\ln|\hat{\varphi}(t_2)|}}{\ln|\frac{t_1}{t_2}|} \tag{8.36}$$

and

$$\ln\hat{\sigma} = \frac{\ln|t_1|\ln(-\ln|\hat{\varphi}(t_2)|) - \ln|t_2|\ln(-\ln|\hat{\varphi}(t_1)|)}{\ln\left|\frac{\hat{\varphi}(t_1)}{\hat{\varphi}(t_2)}\right|}. \tag{8.37}$$

Estimation of the skewness and the location parameter requires more efforts. Let us first denote the imaginary part of the logarithm of the chf in P_0 as $u(t)$:

$$u(t) = \Im(\ln\varphi(t)) = \mu t + \sigma^\alpha |t|^\alpha \beta \, sign(t) \tan\frac{\pi\alpha}{2}.$$

Then if we choose two non-zero values t_3 and t_4 such that $t_3 \neq t_4$ we can write a system of two equations:

$$\left|\begin{array}{l} \frac{u(t_3)}{t_3} = \mu + \sigma^\alpha |t_3|^{\alpha-1}\beta\tan\frac{\pi\alpha}{2}, \\ \frac{u(t_4)}{t_4} = \mu + \sigma^\alpha |t_4|^{\alpha-1}\beta\tan\frac{\pi\alpha}{2}. \end{array}\right.$$

It is possible to solve the system for β and μ and again replacing α, σ and $u(t)$ with their sample counterparts yields the required estimators. Since

$$\hat{\varphi}_\xi(t) = \left(\frac{1}{n}\sum_{j=1}^{n}\cos tx_j\right) + i\left(\frac{1}{n}\sum_{j=1}^{n}\sin tx_j\right)$$

and taking advantage of the properties of complex numbers we achieve the estimator $\hat{u}(t)$:

$$\tan \hat{u}(t) = \frac{\sum_{j=1}^{n} \sin t x_j}{\sum_{j=1}^{n} \cos t x_j}.$$

Finally for $\hat{\beta}$ and $\hat{\mu}$ we have:

$$\hat{\beta} = \frac{\frac{\hat{u}(t_4)}{t_4} - \frac{\hat{u}(t_3)}{t_3}}{\left[|t_4|^{\hat{\alpha}-1} - |t_3|^{\hat{\alpha}-1}\right]\hat{\sigma}^{\hat{\alpha}} \tan \frac{\pi\hat{\alpha}}{2}} \qquad (8.38)$$

and

$$\hat{\mu} = \frac{|t_4|^{\hat{\alpha}-1}\frac{\hat{u}(t_3)}{t_3} - |t_3|^{\hat{\alpha}-1}\frac{\hat{u}(t_4)}{t_4}}{|t_4|^{\hat{\alpha}-1} - |t_3|^{\hat{\alpha}-1}}. \qquad (8.39)$$

8.4.2.1.2   *Case* $\alpha = 1$.

If $\alpha = 1$, equation (8.35) allows us to construct the estimator $\hat{\sigma}$ directly:

$$\hat{\sigma} = -\frac{\ln |\varphi(t_1)|}{t_1}$$

where $t_1 \neq 0$. Similar arguments as in the case $\alpha \neq 1$ lead us to:

$$\hat{\beta} = \frac{\frac{\hat{u}(t_3)}{t_3} - \frac{\hat{u}(t_4)}{t_4}}{\frac{2}{\pi}\hat{\sigma} \ln |\frac{t_4}{t_3}|},$$

$$\hat{\mu} = \frac{\ln |t_4|\frac{\hat{u}(t_3)}{t_3} - \ln |t_3|\frac{\hat{u}(t_4)}{t_4}}{\ln |t_4| - \ln |t_3|},$$

where $t_3 \neq t_4$ and both are non-zero.

The estimators of stable parameters are consistent since they are based on $\hat{\varphi}(t)$, $\Re\hat{\varphi}(t)$ and $\Im\hat{\varphi}(t)$ which are consistent estimators of $\varphi(t)$, $\Re\varphi(t)$ and $\Im\varphi(t)$ by the law of large numbers. The question which still remains is the best way to choose $t_1, \ldots, t_4$, since obviously the derived estimators are not invariant of their choice. Koutrouvelis in his simulation studies in [11] uses the values $t_1 = 0.2$, $t_2 = 0.8$, $t_3 = 0.1$ and $t_4 = 0.4$, which are selected for the normalized case ($\sigma = 1$, $\mu = 0$). Because of the following property of the chf of an arbitrary random variable X:

$$\varphi_{\sigma X + \mu}(t) = e^{it\mu}\varphi_X(\sigma t)$$

it is clear that for different σ and μ we shall have to choose different values for $t_1, \ldots, t_4$ to achieve equal performance, i.e., the values determined for the normalized case will not be equally "good" for a non-normalized case. For this reason, if we aim at estimation of stable parameters by the method of moments, we need first to find initial estimates of the scale and the location parameter and to normalize the sample. Without incurring significant additional computational burden, initial estimates could be computed with the help of a quantile method. For such purposes Koutrouvelis [11] uses the method of Fama and Roll [9], [10] despite the bias in the estimate of σ even in the symmetric case. We shall adopt his approach in our computations.

To summarize, the algorithm for estimating stable parameters by the method of moments, given a sample of iid observations $x_1, x_2, \ldots, x_n$, is as follows:

1. Compute initial estimates $\hat{\sigma}_0$ and $\hat{\mu}_0$ of σ and μ respectively, according to:

$$\hat{\sigma}_0 = \frac{\hat{x}_{0.72} - \hat{x}_{0.28}}{1.654}$$

   and $\hat{\mu}_0$ equals the 50% truncated sample average — the mean of the middle 50% of the ordered observations.

2. Normalize the sample with the initial estimates:

$$x_k' = (x_k - \hat{\mu}_0)/\hat{\sigma}_0, \quad k = 1, 2, \ldots, n.$$

3. Using the normalized sample $x_1', x_2', \ldots, x_n'$, calculate $\hat{\alpha}, \hat{\beta}, \hat{\sigma}_1$ and $\hat{\mu}_1$ according to equations (8.36), (8.38), (8.37) and (8.39) respectively.

4. Compute the final estimates $\hat{\sigma}$ and $\hat{\mu}$:

$$\hat{\sigma} = \hat{\sigma}_0 \hat{\sigma}_1, \qquad \hat{\mu} = \hat{\sigma}_0 \hat{\mu}_1 + \hat{\mu}_0.$$

8.4.2.2   The regression-type estimator of Kogon–Williams

Regression-type estimators are also based on the sample chf. It is possible to derive simple expressions, linear with respect to stable parameters, and construct estimators using the least squares technique. Kogon and Williams suggest such a procedure in [1] with the chf being parametrized according to the continuous parametrization P_1 defined in equation (8.2). Their approach is similar to that of Koutrouvelis in [11] and [12].

The linear equations follow directly from the convenient form of the logarithm of the chf:

$$\ln[-\Re(\ln \varphi(t))] = \alpha \ln \sigma + \alpha \ln |t|, \tag{8.40}$$

$$\Im(\ln \varphi(t)) = \mu_1 t + \beta \sigma t (|\sigma t|^{\alpha-1} - 1) \tan \frac{\pi \alpha}{2}. \tag{8.41}$$

Estimators of the stable parameters can be constructed using the method of least squares after replacing the chf for the sample chf. Certainly here we face the same problem as in the method of moments — the sample chf should be evaluated for certain values of the argument. Koutrouvelis gives tables in [11] and [12] which relate the values of the sample chf argument to the value of the index of stability α and the sample size. The major advantage of the procedure in [1] is that the provided values of the sample chf argument are invariant of any other parameters. Having conducted numerous experiments, Kogon and Williams report in [1] that the most suitable choice is $t_k = \{0.1 + 0.1k, \quad k = 0, 1, \ldots, 9\}$ — 10 equally spaced points in the interval [0.1, 1]. Undoubtedly the sample should be normalized before applying the method of least squares, otherwise the optimal selection of the sample chf arguments would depend on the scale and the modified location parameter. For preliminary estimation of σ and μ, it is suggested to use the quantile method of McCulloch.

The algorithm is as follows:

1. Given a sample of iid observations $x_1, x_2, \ldots, x_n$ first we find preliminary estimates σ_0 and μ_{01} utilizing the quantile method of McCulloch and we normalize the observations:

$$x'_j = \frac{x_j - \hat{\mu}_{01}}{\hat{\sigma}_0}, \quad j = 1, 2, \ldots, n.$$

2. Next we consider the regression equation constructed from equation (8.40):

$$y_k = b + \alpha w_k + \epsilon_k, \quad k = 0, 1, \ldots, 9$$

where $y_k = \ln[-\Re(\ln \hat{\varphi}(t_k))]$, $w_k = \ln |t_k|$, $t_k = \{0.1 + 0.1k, \quad k = 0, 1, \ldots, 9\}$ and ϵ_k denotes the error term. We find $\hat{\alpha}$ and $\hat{b}$ according to the method of least squares using the normalized sample $x'_1, x'_2, \ldots, x'_n$. The estimator $\hat{\sigma}_1$ of the scale parameter of the normalized sample is:

$$\hat{\sigma}_1 = \exp\left(\frac{\hat{b}}{\hat{\alpha}}\right).$$

3. Estimators $\hat{\beta}$ and $\hat{\mu}_{11}$ of the skewness parameter and the modified location parameter respectively are derived from the second regression equation based on (8.41):

$$z_k = \mu_{11} t_k + \beta v_k + \eta_k, \quad k = 0, 1, \ldots, 9$$

where $z_k = \Im(\ln \hat{\varphi}(t_k))$, $v_k = \hat{\sigma}_1 t_k (|\hat{\sigma}_1 t_k|^{\hat{\alpha}-1} - 1) \tan \frac{\pi \hat{\alpha}}{2}$, $t_k = \{0.1 + 0.1k, k = 0, 1, \ldots, 9\}$ and η_k is the error term.

4. The final estimators $\hat{\sigma}$ and $\hat{\mu}_1$ proceed from:

$$\hat{\sigma} = \hat{\sigma}_0 \hat{\sigma}_1, \quad \hat{\mu}_1 = \hat{\mu}_{01} + \hat{\sigma}_0 \hat{\mu}_{11}.$$

If we aim at estimating the location parameter μ, we need to take advantage of the connection between the two parametric forms P_0 and P_1:

$$\hat{\mu} = \hat{\mu}_1 - \hat{\beta} \hat{\sigma} \tan \frac{\pi \hat{\alpha}}{2}.$$

In [1] there is a huge Monte Carlo study in which the method of Kogon–Williams is compared to the approach of Koutrouvelis [11], [12]. The result is that from a computational viewpoint the former is more efficient. It is definitely superior to the latter when α is close to zero and $\beta \neq 0$. The approach of Koutrouvelis outperforms that of Kogon–Williams only in the estimation of β.

8.4.3 *Maximum likelihood*

The method of maximum likelihood is very attractive because of the good asymptotic properties of the estimates, provided that the likelihood function obeys certain general conditions. The likelihood function is defined as:

$$L(x_1, x_2, \ldots x_n | \theta) = \prod_{k=1}^{n} f(x_k | \theta)$$

where $x_1, x_2, \ldots x_n$ is a sample of iid observations of a random variable X, $f(x|\theta)$ is the pdf of X, and θ is a vector of parameters. In the case of stable distributions, $\theta = (\alpha, \beta, \sigma, \mu)$. Maximum likelihood estimates are found by searching for those parameter values which maximize the likelihood function, or equivalently, the log-likelihood function:

$$\hat{\theta}_n = \arg \max_{\theta} \log(L(x_1, x_2, \ldots x_n | \theta)). \tag{8.42}$$

DuMouchel studied the applicability of maximum likelihood theory in the case of α-stable distributions in [6] by verifying whether the likelihood function complies with a set of conditions that guarantee the validity of the theory. The theorem proved in the paper and adapted to parametrization P_0 is the following

Theorem 3. *When sampling from a stable distribution, $\hat{\theta}_n$, the maximum likelihood estimate for $\theta = (\alpha, \beta, \sigma, \mu)$ based on the first n observations, restricted so that $\hat{\alpha}_n$, the estimate of α, satisfies $\hat{\alpha}_n > \epsilon$, ϵ arbitrarily small and positive, is consistent and asymptotically normal as long as θ_0, the true value of θ is in the interior of the parameter space (that is the cases $\alpha_0 \leq \epsilon$, $\alpha_0 = 2$ and $|\beta| = 1$ are excluded) and the additional case $(\alpha_0 = 1, \beta_0 \neq 0)$ is excluded.*

Clearly if we intend to derive expressions for MLE analytically, we need to have closed-form expressions for the pdfs of stable laws. Such expressions are not known to exist in the general case and the problem of MLE of stable parameters should be attacked numerically, i.e., we have to numerically search for the solution of problem (8.42) in which the pdf is approximated. In our comparison studies we use the FFT-approach combined with Bergström series expansion for tail approximation.

8.5   Conclusions

An important issue in financial risk management is the distributional presumption for assets returns. In the current paper we have considered α-stable distributions as candidates for a probabilistic model and we have reviewed the most common numerical and statistical methods that make feasible their application. The lack of closed-form expressions for the density and the distribution function raises some practical issues for instance in quantile computation or parameter estimation. We have discussed several approaches with a focus on computational burden and accuracy. We arrived at the following conclusions:

- Bergström series expansions are efficient for small or large values of $|x|$;
- The most efficient global approaches to pdf approximation for a large number of x values are the FFT and polynomial-type methods. The former should be combined with a method for tail approximation. In the symmetric case, McCulloch's method is faster than the former under certain conditions;
- Zolotarev's integral representations are advantageous when high precision is necessary. The computational time is conditional on the choice of quadrature.

We have also discussed the problem of parameter estimation. Simulation studies indicate that the maximum likelihood estimator has the best performance followed by the regression-type estimators and then by the quantile estimators (for more details, see [1], [2], [17] and the references therein). The method of moments has the worst performance out of the approaches discussed. From a computational viewpoint, the most complicated approach is MLE in which the speed of calculations strongly depends on the density approximation method. By far less computational burden is entailed by the estimator of Kogon–Williams followed by the quantile method of McCulloch and the method of moments.

References

[1] R. J. Adler, R. Feldman and M. Taqqu, (Eds.) *A Practical Guide to Heavy Tails: Statistical Techniques and Applications*, Birkhäuser, Boston, Basel, Berlin, 1998.

[2] V. Akgiray, C. G. Lamoureux, Estimation of Stable-law Parameter: A Comparative Study, *Journal of Business and Economic Statistics*, 7, 85–93, (1989).

[3] H. Bergström, On Some Expansions of Stable Distributions, *Arkiv for Matematik* II, 375–378 (1952).

[4] T. Doganoglu, S. Mittnik, An Approximation Procedure for Asymmetric Stable Paretian Densities, *Computational Statistics*, 13, 463–475 (1998).

[5] W. H. DuMouchel, Stable Distributions in Statistical Inference, Ph.D. dissertation, Department of Statistics, Yale University (1971).

[6] W. H. DuMouchel, On the Asymptotic Normality of the Maximum-Likelihood Estimate when Sampling from a Stable Distribution, *Annals of Statistics*, 1, 948–957 (1973).

[7] P. Embrechts, C. Klüppelberg, T. Mikosch, *Modeling Extremal Events for Insurance and Finance*, Springer, Berlin, Heidelberg, New York, 1997.

[8] E. Fama, The behavior of stock market prices, *J. Bus. Univ. Chicago*, 38, 34–105, (1965).

[9] E. Fama, R. Roll, Some Properties of Symmetric Stable Distributions, *Journal of the American Statistical Association*, 63, 817–836, (1968).

[10] E. Fama, R. Roll, Parameter Estimates for Symmetric Stable Distributions, *Journal of the American Statistical Association*, 66, 331–338, (1971) .

[11] I. A. Koutrouvelis, Regression-type Estimation of the Parameters of Stable Laws, *Journal of the American Statistical Association*, 75, 918–928, (1980).

[12] I. A. Koutrouvelis, An Iterative Procedure for the Estimation of the Parameters of Stable Laws, *Communications in Statistics. Simulation and Computation*, 10, 17–28, (1981).

[13] B. Mandelbrot, The variation of certain speculative prices, *J. Bus. Univ. Chicago*, 26, 394–419, (1963).

[14] J. H. McCulloch, Simple Consistent Estimators of Stable Distribution Parameters, *Communications in Statistics. Simulation and Computation*, 15, 1109–1136, (1986).

[15] J. P. Nolan, Numerical Computation of Stable Densities and Distribution Functions, *Stochastic Models*, 13, 759–774, (1997).

[16] S. J. Press, Estimation in Univariate and Multivariate Stable Distribution, *Journal of the American Statistical Association*, 67, 842–846, (1972b).

[17] S.T. Rachev and S. Mittnik *Stable Paretian Models in Finance*, John Wiley & Sons Ltd, 2000.

[18] G. Samorodnitsky, M. S. Taqqu, *Stable Non-Gaussian Random Processes, Stochastic Models with Infinite Variance*, Chapman and Hall, New York, London, 1994.

[19] V. M. Zolotarev, On the Representations of Stable Laws by Integrals, *Selected Translations in Mathematical Statistics and Probability*, 6, 84–88, American Mathematical Society, Providence, Rhode Island, 1964.

[20] V. M. Zolotarev, *One-Dimensional Stable Distributions*, Nauka (in Russian), Moscow, 1983.

9

Modern Heuristics for Finance Problems: A Survey of Selected Methods and Applications

Frank Schlottmann

Detlef Seese

ABSTRACT   The high computational complexity of many problems in financial decision-making has prevented the development of time-efficient deterministic solution algorithms so far. At least for some of these problems, e.g., constrained portfolio selection or non-linear time series prediction problems, the results from complexity theory indicate that there is no way to avoid this problem. Due to the practical importance of these problems, we require algorithms for finding optimal or near-optimal solutions within reasonable computing time. Hence, heuristic approaches are an interesting alternative to classical approximation algorithms for such problems. Over the last years many interesting ideas for heuristic approaches were developed and tested for financial decision-making. We present an overview of the relevant methodology, and some applications that show interesting results for selected problems in finance.

9.1   Introduction

It is one of the goals of computational finance to develop methods and algorithms to support decision making. Unfortunately, many problems of practical or theoretical interest are too complex to be solvable exactly by a deterministic algorithm in reasonable computing time, e.g., using a method that applies a simple closed-form analytical expression. Such problems require approximation procedures which provide sufficiently good solutions while requiring less computational effort compared to an exact algorithm. Heuristic approaches are a class of algorithms which have been developed to fulfil these requirements in many problem contexts. Today, there are many different paradigms for heuristic approaches which have been tested in several fields of application. Particularly in the last ten years, a growing number of applications in the area of finance were investigated. Comparing the tremendous variability of the different heuristic approaches, it is often difficult to decide which heuristic method should be applied to a given financial problem setting. To support the right choice among different heuristic approaches we gather their basic characteristics with a special focus on financial applications. In the following text we mainly introduce the core methodology of different heuristics and present many references which cover theoretical aspects, e.g., convergence properties or parameter choice, and a selection of successful applications in finance. Before discussing the modern heuristic approaches, we will first point out a formal view of complexity and a classical local search algorithm in the following section.

9.2   Complexity of finance problems and local search algorithms

A widely accepted formal definition of complex problems is the computational intractability of problems which are hard to solve from an algorithmic perspective, i.e., we require huge computational resources to compute the exact solution of a problem having input size n (e.g., an exponential number 2^n of necessary calculations for n given input variables of the considered problem). Besides other complexity classifications, the theory of **NP**-*completeness* yields a well-defined formalisation of such complex problems, see e.g., Garey & Johnson [1] for a detailed coverage of this topic and a large collection of **NP**-*complete* problems. Until now, there is no known algorithm which requires only a polynomial number of computational steps depending on the input size n for an arbitrarily chosen problem that belongs to the class of **NP**-*complete* problems. See e.g., Papadimitriou [2] for the formal definitions of computational complexity and further implications.

Many problems in finance belong to the class of **NP**-*complete* problems, since they have a combinatorial structure which is equivalent (with respect to polynomial-time reductions) to well-known **NP**-*complete* problems, e.g., constrained portfolio selection and related questions of asset allocation are equivalent to **NP**-*complete* knapsack problems. Cf. Seese & Schlottmann [3] for such complexity results.

Therefore, we require approximation algorithms that yield sufficiently good solutions for complex finance problems and consume only polynomial computational resources measured by the size of the respective problem instance (e.g., number of independent variables). For some complex problem settings and under certain assumptions, particularly linearity or convexity of target functions in optimization problems, there are analytical approximation algorithms which provide a fast method of finding solutions having a guaranteed quality of lying within an ϵ-region around the globally best solution(s). If the considered problem instance allows the necessary restrictions for the application of such algorithms, these are the preferred choice, see Ausiello et al. [4] for such considerations. However, some applications in finance require non-linear, non-convex functions (e.g., valuation of exotic option contracts), and sometimes we know only the data (parameters) but not the functional dependency between them (bankruptcy prediction problems for instance), so there is nevertheless a need for methods that search for good solutions in difficult problem settings while spending only relatively small computational cost. This is the justification for heuristic approaches.

Almost all heuristics that are discussed in this survey belong to *local search algorithms*. This means that for a given current solution x_i to the problem which is to be solved, a promising solution candidate that can be derived by small modifications of x_i is to be chosen to continue the search for the globally best solutions. The set of solutions which are close to x_i will be called the local *neighbourhood* $N(x_i)$ of x_i in the following text. The exact definition of $N(x_i)$ is problem-specific, and an adequate choice for a given problem is often crucial for the success of a certain algorithm. Most heuristics require an initial solution from which they start their search for improvement. In many cases, a random initialisation of this initial solution yields the best outcome of the respective problem solving method on average, i.e., over all possible instances for a fixed problem class.

A standard local search algorithm is *Hill Climbing (HC)* which basically works as shown in Algorithm 1. The algorithm moves from the current solution x_i to a solution x_j from the neighbourhood of x_i if and only if x_j is better than x_i concerning the problem to be solved, otherwise the algorithm stops and returns x_i as the candidate for the globally best solution (in the following text, φ denotes a sample target function that is to be maximized by the respective algorithm).

Algorithm 1. *Hill Climbing*

Input: Initial solution x_j
$i := 1$ *(iteration counter)*
Repeat
 Set current solution $x_i := x_j$
 Generate neighbourhood $N(x_i)$
 Choose the best solution $x_j \in N(x_i)$
 If $\varphi(x_j) > \varphi(x_i)$ Then
  TerminateSearch := False
  $i := i + 1$
 Else
  TerminateSearch := True
Until TerminateSearch = True
Terminate
Output: Best solution found x_i

There are many examples of such HC procedures for solving finance problems in the standard literature, so we do not consider them here in detail. Usually, analytically tractable problems can be solved using Newton's method or other gradient-based approaches. On the other hand, if the problem to be solved contains non-linear, non-convex functions and/or integer constraints, such a local search procedure that decides about the next move solely based on local information in the neighbourhood of x_i, can get stuck in suboptimal solutions without discovering the globally best (optimal) solution throughout the search process. The methods described in the following sections try to avoid this problem by using certain strategies, e.g., for choosing the next solution x_j to be investigated or for deciding about the termination of the search process. Beside the more specific references appearing in the text, general overviews of the concepts presented below can e.g., be found in Reeves [5], Osman & Kelly [6], Aarts & Lenstra [7], Fogel & Michalewicz [8], Pham & Karaboga [9] and Nelles [10] which cover a variety of methods and mostly non-financial applications. A more finance-related, recent survey on a subset of the methods discussed below is Chen's book [11].

9.3   Simulated Annealing

The basic working principle of *Simulated Annealing (SA)* is an analogy to conducting thermodynamical annealing processes in a heat bath to obtain low-energy states for a solid (cf. Kirkpatrick et al. [12] and Cerny [13]). A simulation algorithm for such a thermodynamical process was introduced in 1953 by Metropolis et al. [14], and the so-called *Metropolis criterion* proposed in their work is also the central part of the SA heuristic that is shown in Algorithm 2 ($Z_i \in [0, 1]$ are assumed to be independently and identically distributed uniform random variates).

Algorithm 2. *Simulated Annealing*

Input: *Initial solution x_j, parameter value $T_1 \in \mathbb{R}$*
$i := 1$ *(iteration counter)*
$x_{best} := x_j$ *(best solution found so far)*
Repeat
    Set current solution $x_i := x_j$
    Randomly choose a solution $x_j \in N(x_i)$
    If $\varphi(x_j) > \varphi(x_{best})$ **Then**
        $x_{best} := x_j$
    If $\varphi(x_j) > \varphi(x_i)$ **Then**
        TerminateSearch := False
    Else
        If $Z_i < e^{-\frac{\varphi(x_i)-\varphi(x_j)}{T_i}}$ **Then**
            TerminateSearch := False
            $i := i + 1$
            Adapt T_i according to predefined rule (cooling schedule)
        Else
            TerminateSearch := True
Until *TerminateSearch = True*
Terminate
Output: *Best solution found x_{best}*

The parameter T_i which represents the temperature in the heat bath conducts the annealing process. In analogy to the Metropolis criterion, the SA algorithm not only accepts a better solution x_j with probability 1, but also a deterioration of the current solution with probability $e^{-\frac{\varphi(x_i)-\varphi(x_j)}{T_i}}$. For an optimal low-energy state of the solid to be annealed, an appropriate cooling schedule is required for T_i, and this is also crucial if the SA heuristic is applied to finance problems. Aarts et al. [16] provide a thorough analysis of the convergence properties of the SA heuristic using Markov chains, some guidelines for choosing appropriate cooling schedules and many references to applications of the SA heuristic. Here, the main convergence results are shortly summarized by the following theorem (for more details see [16]):

Theorem 1. *Assuming an appropriate cooling schedule for T_i in the SA algorithm, x_i converges to a globally optimal solution with probability 1 for $i \to \infty$ (i is the number of iterations of the loop in Algorithm 2).*

Regrettably, this does neither hold for a finite number of iterations, nor for an arbitrary cooling schedule. Nevertheless, there are many reports of successful SA applications in the literature, particularly in other areas than finance. A sample finance application was e.g., built by Chang et al. [17] who applied different heuristics including SA to constrained and unconstrained portfolio selection problems. Their goal was to identify the mean-variance efficient frontier (cf. e.g., Markowitz [18]) for given sets of alternative investments. Besides the good results for smaller problems from the Hang Seng, DAX, FTSE or S & P stock market indices, the SA heuristic found a good approximation of a constrained efficient frontier for a draw of 10 from 225 alternative investments from the Nikkei stock market index in less than 10 minutes on a single workstation computer.

9.4   Threshold Accepting

The *Threshold Accepting (TA)* heuristic was introduced in 1990 by Dueck & Scheurer [19] as a simplification of Simulated Annealing. Instead of applying the Metropolis criterion to decide about the acceptance of a deterioration of the current solution (cf. Algorithm 2), TA uses a straightforward threshold parameter T_i for the maximum deterioration of the current solution that is accepted without terminating the algorithm. Like the temperature cooling schedule in SA, an adaptation rule for T_i is necessary to ensure proper convergence, and the choice of this adaptation rule is a crucial point when applying TA to a problem. Moreover, it has to be emphasized that the success of TA depends heavily on the proper modelling of a neighbourhood $N(x_i)$ for any possible solution x_i. There are both deterministic and non-deterministic variants of TA depending on the adaptation of T_i.

Algorithm 3. *Threshold Accepting*

Input: *Initial solution x_j, initial parameter $T_1 \in \mathbb{R}, T_1 > 0$,*
        iteration limit $i_{max} \in \mathbb{N}$
$i := 1$ *(iteration counter)*
$x_{best} := x_j$ *(best solution found so far)*
Repeat
        Set current solution $x_i := x_j$
        Randomly choose a solution $x_j \in N(x_i)$
        If** $\varphi(x_j) > \varphi(x_{best})$ **Then
            $x_{best} := x_j$
        If** $\varphi(x_i) - \varphi(x_j) < T_i$ **Then
            TerminateSearch := False
            $i := i + 1$
            Adapt T_i according to predefined rule
        Else
            TerminateSearch := True
***Until** $i > i_{max}$ **Or** TerminateSearch = True*
Terminate
Output: *Best solution found x_{best}*

The convergence results for TA are similar to the results for SA (cf. Theorem 1), i.e., they use Markov chain results to derive convergence properties of the algorithm. A good source for many aspects of TA including theoretical aspects as well as econometric applications is Winker [23].

Both Dueck & Winker [20] and Gilli & Kellezi [21] applied TA to portfolio selection problems. Their respective objective was to find the optimal asset allocation that is efficient concerning the aggregate portfolio risk as well as the aggregate portfolio return for a mean-variance approach or a mean/downside risk approach. The studies incorporated the successful application of TA to analytically intractable optimization problems.

Gilli & Kellezi [22] reported the application of TA to an index tracking problem, where the goal of a so-called *passive investor* was to minimize the difference between his tracking portfolio and a given market index (a stock index like Dow Jones Industrial Average for instance) under constraints, e.g., rebalancing and transaction cost, round lots etc. Even for a large number of 528 assets, the TA heuristic found a good approximation for the optimal tracking portfolio within less than one minute on a Personal Computer.

9.5 Tabu Search

The main ideas of *Tabu Search (TS)* were formulated by Glover [24] and Hansen [25]. This heuristic is different from the other approaches discussed in this survey because it explicitly keeps track of the activities that were performed in the problem solving process so far and tries to conduct this process towards relevant solutions by forbidding certain activities that have already been performed. To achieve this, the definition of a *move* is the central concept in TS. A move from a current solution x_i is the operator that yields a neighbourhood $N(x_i)$, i.e., the (usually small) modification applied to x_i to obtain $N(x_i)$. TS keeps a *Tabu List* of maximum length k that contains the most recently performed moves throughout the problem solving process. These moves are not considered for the next move to be performed, i.e., they are tabu (cf. taboo). An obvious advantage of memorizing recently performed moves instead of recently investigated solutions is that the former method usually requires less memory. The length k of the Tabu List is a crucial parameter for a given problem. If the list is too short the search for optimal solutions might run into a loop (recently visited solutions are revisited again and again), and if the list is too long the algorithm may not find optimal solutions that require the repeated application of certain moves. Algorithm 4 shows an overview of the TS local search method.

Algorithm 4. *Tabu Search*

Input: Initial solution x_j, iteration limit $i_{max} \in \mathbb{N}$
$i := 1$ *(iteration counter)*
$x_{best} := x_j$ *(best solution found so far)*
$T := \emptyset$ *(tabu list)*
Repeat
    Set current solution $x_i := x_j$
    Generate neighbourhood $N(x_i)$ for x_i using a move $m_i \notin T$
    Choose the best solution $x_j \in N(x_i)$
    If $\varphi(x_j) > \varphi(x_{best})$ **Then**
        $x_{best} := x_j$
    $i := i + 1$
    Update T according to predefined rule
Until $i > i_{max}$
Terminate
Output: Best solution found x_{best}

The update procedure for T in Algorithm 4 usually consists of the operation $T := T \cup \{m_i\}$, and the least recently used move m_j is removed from the list by $T := T \setminus \{m_j\}$ (First-In-First-Out principle). In addition, different strategies (*aspiration* of tabu criteria, and *intensification* as well as *diversification* concerning search regions) are discussed in the literature to enhance the performance of the algorithm and to avoid the above mentioned problems of a too restricted search for better solutions. See e.g., Glover & Laguna [26] for a detailed coverage of these and further details of TS. A short introduction and a selection of non-finance TS applications are e.g., given by Hertz et al. [27].

Glover et al. [28] reported a successful application of TS to a multi-period asset allocation problem using a time-dependent mean-variance approach with non-convex constraints which allow for modelling taxes, transaction cost etc. For the absence of taxes and transaction cost in a problem containing 8 assets and 20 time periods, the results of the TS algorithm were nearly identical to an analytical ϵ-approximative approach for the efficient frontier. When these constraints were included to obtain an analytically intractable problem structure, the TS algorithm also found an approximation of the mean-variance efficient frontier within 17 minutes on a workstation.

The study of Chang et al. [17], which has already been cited in the Simulated Annealing section, also contains results of a TS application for mean-variance portfolio selection problems. TS showed a similar performance like Simulated Annealing in the study, being slightly better in some cases and slightly worse in other concerning the quality of the solutions found, and having a similar runtime of e.g., about 10 minutes for an approximation of the mean-variance efficient frontier in a constrained portfolio selection problem for 10 assets from the Nikkei-225 stock market index.

9.6   Evolutionary Computation

Some problem solving mechanisms like selection, reproduction and mutation which can be observed in natural environments and populations are the basic working principles for *Evolutionary Computation (EC)*. The history of EC started in the 1950s (see DeJong et al. [29] for an overview of such early work). In the 1960s, Fogel et al. [30] developed the concepts of *Evolutionary Programming*, which was followed later by Koza's introduction of *Genetic Programming (GP)* [31, 32, 33], Holland [34] proposed the *Genetic Algorithm (GA)* and Rechenberg [35] and Schwefel [36] introduced the *Evolution Strategies (ES)* (cf. also [29]). Since the 1990s these evolutionary approaches have attracted many research activities due to their interesting theoretical properties as well as their successful applications. We will concentrate on the GA and GP methodology because of their dominance concerning finance applications. A good source for many aspects of EC are Baeck et al. [37, 38] and the annual conference proceedings of GECCO (e.g., [39]).

9.6.1   Genetic Algorithms

GAs are randomised heuristic search algorithms reflecting the Darwinian *survival of the fittest* principle that can be observed in many natural evolution processes. A GA works on a set of n potential solutions to a problem rather than on a single solution. The current set of solutions being processed by a GA at each time step t of the algorithm is called *population* or *generation* $P(t) = \{x, y, ...\}$, $|P(t)| = n$, and each $x \in P(t)$ is called an *individual*. To apply a GA to a problem, the decision variables have to be transformed into *gene strings*, i.e., each element from the decision variable space D has to be transformed into a string consisting of digits or characters in the search space of the algorithm S by applying a 1-1 function $g : D \rightarrow S$. The original representation $x \in D$ is called *phenotype*, the genetic counterpart $g(x) \in S$ is called *genotype*. For the sake of simplicity, we will not distinguish between x and $g(x)$ in the following text. We will denote the length of the gene string x by $length(x)$. A simple GA scheme is shown in Algorithm 5.

Algorithm 5. *Genetic Algorithm*

Input: Iteration limit $t_{max} \in \mathbb{N}$
$t := 0$ (population counter)
Generate initial population $P(t)$
Evaluate $P(t)$
Repeat
    Select individuals from $P(t)$
    Recombine selected individuals
    Mutate recombined individuals
    Create offspring population $P'(t)$
    Evaluate $P'(t)$
    Generate $P(t+1)$
    $t := t + 1$
Until *$t > t_{max}$*
Output: *$P(t)$*

The initial population $P(0)$ can be generated e.g., by random initialisation of every individual. For the evaluation of each individual $x \in P(t)$ the GA requires a so-called *fitness function* f which is defined either on the set of possible genotypes S or on the set of phenotypes D. Usually, the fitness function is defined on D and takes real values, i.e., $f : D \to \mathbb{R}$. This function expresses the quality of the solution represented by the individual. It is problem specific and therefore, it has to be designed according to the problem parameters and constraints. During the evolution process the GA selects individuals for reproduction from the current population $P(t)$ according to their fitness value, i.e., the probability of surviving or creating offspring for the next population $P(t+1)$ is higher for individuals having higher fitness values. The intention of *selection* is to guide the evolution process towards the most promising solutions. A common method is *tournament selection*, where the fitness values $f(x)$, $f(y)$ of two randomly drawn individuals $x, y \in P(t)$ are compared, and the winner is determined by the rule $f(x) > f(y) \Rightarrow x$ survives, $f(x) < f(y) \Rightarrow y$ survives, $f(x) = f(y) \Rightarrow x, y$ survive. For an overview and a comparison of selection methods, see part 3 of Baeck et al. [37].

Since the GA's task is to explore the search space S to find globally optimal and feasible solutions, e.g., $x^* = \max_{x \in S'} f(x)$ in a constrained maximisation problem where $S' \subseteq S$ specifies the space of feasible solutions, the selected individuals from each $P(t)$ are modified using *genetic operators*, sometimes called *variation operators* (cf. Fogel & Michalewicz [8], p. 173). A typical variation operator for recombination is the one-point crossover, i.e., the gene strings of two selected individuals are cut at a randomly chosen position and the resulting tail parts are exchanged with each other to produce two new offspring. This variation operator is applied to the selected individuals using a problem specific crossover probability $Prob_{cross}$. Common values are $Prob_{cross} \in (0.6, 1)$. The main goal of this operator is to conduct the simulated evolution process through the search space S. Most GAs implement a second variation operator called *mutation*. In analogy to natural mutation, this operator randomly changes the genes of selected individuals with the parameter $Prob_{mut}$ per gene to allow the invention of new, previously undiscovered solutions in the population. Its second task is the prevention of the GA stalling in local optima because of high selection pressure since there is always a positive probability to leave a local optimum if $Prob_{mut} > 0$. Usually, $Prob_{mut}$ is set small, e.g., $Prob_{mut} \leq \frac{1}{length(x)}$. For a survey of different variation operators see e.g., part 6 of Baeck et al. [37]. The creation of $P(t + 1)$ is usually performed by selecting the n best individuals either from the offspring population $P'(t)$ or from the joint population containing both the parent and the offspring individuals $P(t) \cup P'(t)$.

The evolution of individuals during the GA's search process can be modelled by Markov chains (cf. e.g., Rudolph [40]). There are different results concerning the convergence of the GA population towards global optimal solutions depending on the fitness function, properties of the search space etc., see e.g., Muehlenbein [41], Droste et al. [42], Vose [43], Wegener [44]. Despite the theoretical value of these results, they can only provide some rules of thumb for proper genetic modelling of most application-oriented problems, for choosing adequate fitness functions, selection mechanisms, variation operators and parameters like n, $Prob_{cross}$, $Prob_{mut}$. These choices are not necessarily independent from each other, e.g., the choice of the fitness function is important for the selection mechanism and vice versa. As a consequence, there is currently no GA fitting each application-oriented context. Therefore, most application studies focus on empirical evaluations of problem specific GAs.

A very natural application of GAs is the modelling of a group of individual entities, e.g., traders in financial markets, to observe the emerging macro-level output (e.g., asset prices) from individual decisions. Such approaches benefit from the fact that the GA provides a built-in adaptation mechanism through its evolution process which can be used to model individuals that try to improve their financial decisions by processing historical information. An introduction to such approaches in economics is given by Riechmann [45]. The study by Rieck [46] and the Santa Fe artificial stock market (see e.g., Tayler [47]) are two examples for early studies that analyzed asset prices resulting from individual decisions made by artificial traders which are improved by a GA. Meanwhile, there is a large number of similar approaches which analyze e.g., asset price time series properties (cf. LeBaron et al. [48]), the Efficient Market Hypothesis (cf. Coche [49], Farmer & Lo [50]) and further questions related to real-world asset markets.

Another common field of GA applications in finance is the discovery and the classification of patterns in financial data, e.g., for evaluation of counterparties in credit scoring models (cf. e.g., Walker et al. [51]), for detecting insider trading (cf. e.g., Mott [52]) or for the identification of successful trading strategies in stock markets (cf. e.g., Frick et al. [53]). The book by Bauer [54] covers many aspects of such approaches. An interesting application in this context is Tsang & Lajbcygier's [56] modified GA framework for the discovery of successful foreign exchange trading strategies in historical data. It uses a split population that is divided into a group of individuals which is modified using a higher mutation probability, and another group of individuals which is changed with a low mutation probability. The modified GA found strategies yielding higher mean returns while having similar return standard deviations than a standard GA, but both approaches showed advantages over the other in certain test criteria.

9.6.2   Genetic Programming

The concept of GP is very similar to the GA paradigm, since it uses the same elements like selection, reproduction and mutation in its core and implements the same scheme like Algorithm 5. However, there is a significant difference concerning the genetic representation of the individuals in the considered population as e.g., pointed out by Koza et al. [33], p. 31:

Genetic programming is an extension of the genetic algorithm in which the genetic population contains computer programs.

Rather than being genetic representations of single solutions, the genotypes in a GP algorithm represent programs that are candidates for solving the considered problem. The most widely used representation is the *Automatically Defined Function* by Koza [32] where a program is represented by a syntax tree containing variables and operators that can be used to formulate mathematical expressions which are similar to those that can be formulated in the programming language LISP. These expressions are evaluated by setting the variables' contents to different input parameter sets which are instances of the problem to be solved. The output of this evaluation yields a fitness value for the program concerning the problem to be solved. For a further discussion of different representations see e.g., Langdon & Poli [57], p. 9 ff.

Therefore, at each step t the population $P(t)$ contains different programs each of which is evaluated by running it on a number of instances (i.e., input parameter values) for the given problem type and observing its performance (cf. the fitness evaluation in a GA). The methods to prove convergence of $P(t)$ to globally optimal solutions (i.e., programs which solve the given problem instances well concerning pre-defined performance criteria), are similar to the GA convergence analysis methods and partially, yield even the same convergence properties like GAs. For a detailed discussion of these topics see also [57].

An obvious application of GP is the approximation of a priori unknown functional dependencies: Given different sets of input parameter values and their associated output values, the goal is to find a function that describes the dependency of the output on the input approximately. This kind of GP application was e.g., used by Keber [58] to find closed-form approximations for valuing American put options on non-dividend paying stocks instead of using finite differences (cf. Brennan & Schwarz [59]) or the tree approach by Cox et al. [60]. Keber compared the results of a GP-based search for an approximation formula to frequently quoted approximation procedures in the literature which differ from the two numerical approaches cited above. The best formula found by his GP approach was quite similar in its structure to some of the existing approximations, but it outperformed the other approximations concerning the numerical accuracy of the resulting option values both on a standard test data set used by several other studies and on another large sample of theoretical American put option prices.

In a series of papers (see [61] for further references), Li & Tsang developed a GP approach to support investment decisions. They used a decision tree containing rules like 'IF Condition Is Met THEN Buy Asset' which was improved by GP using daily asset price data of 10 US stocks. They compared the results on the test data to the results of both a linear classification approach and problem-specific Artificial Neural Networks obtained by another study and found that the GP approach yielded better results (i.e., higher excess returns) on their test data set. For a related problem setting, where a GP algorithm was used to solve a multi-period constrained asset allocation problem for an Italian pension fund see Baglioni et al. [62].

Similar to the studies cited in the GA section which model individual decision-making and adaptation of rules in a dynamic environment, there are also applications of the GP methodology to study properties of artificial financial markets. See e.g., Chen & Kuo [63] and Chen [11] for a more detailed coverage of this subject.

9.7   Artificial Neural Networks

An *Artificial Neural Network (ANN)* is a computing model which is built in analogy to the structure of the human brain. A remarkable work which introduced such a computing model based on artificial neurons was published by McCulloch and Pitts in 1943 [64]. But it took about 40 years until the ANN approach (sometimes called *connectionist* approach) reached a wide-spread interest during the 1980s. Since then, many successful real-world applications have been reported, from which we will cite some finance-related examples below after a brief introduction to some methodic details.

An ANN consists of $m \in \mathbb{N}$ artificial *neurons*, where each neuron $j \in \{1, ..., m\}$ represents a small computation unit performing basic functionality, e.g., taking the values of a vector x containing $s_j \in \mathbb{N}$ variables $x_i \in \mathbb{R}$, $i \in \{1, ..., s_j\}$ as input, which are modified by an *activation function* a_j, and calculating its output, denoted by $output_j$, from the sum of the activation function values $a_j(x_i)$ according to a specified *output function* o_j as follows:

$$output_j := o_j(\sum_{i=1}^{s_j} a_j(x_i)). \tag{9.1}$$

In many applications, the functions a_j are linear functions of the type $f(x) := w_{ij}x_i$ and the o_j are non-linear functions like $f(x) = \tanh(x)$ (*tangens hyperbolicus*) or $f(x) = \frac{1}{1-e^{-x}}$ (*logistic function*). Each w_{ij} is called *weight* of the input value i for neuron j.

To obtain more computational power, the neurons can be connected pairwise by directed links, e.g., connecting the output of neuron i with the input of neuron j. From a graph theoretic point of view, each neuron in an ANN can be represented by a vertex and each link between two neurons can be modelled by a directed edge between their vertices. There are different kinds of ANN paradigms depending on the theoretical and/or application context. We focus on the *Multi-Layer-Perceptron*

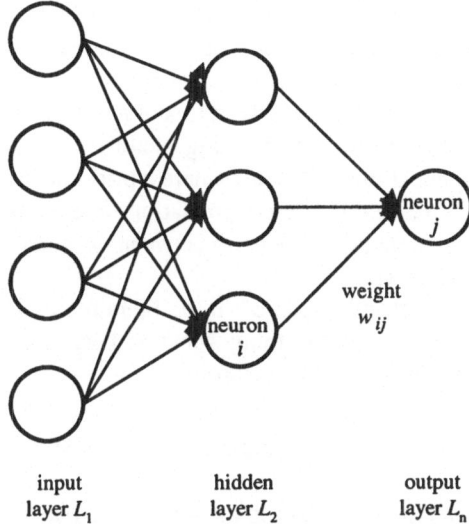

input
layer L_1

hidden
layer L_2

output
layer L_n

FIGURE 1. Multi-Layer-Perceptron Network

(MLP) paradigm here since it is a widely used neural computing model in finance applications. A discussion of major ANN computing models is e.g., given by Schalkoff [65]. Arbib's handbook [66] contains a large collection of papers concerning all kinds of ANNs. Kohonen's book [67] is the standard reference for the *Self Organizing Map* ANN paradigm which we do not cover here, while Deboeck & Kohonen [68] contains some interesting financial applications of this paradigm. A survey of recent work concerning this ANN model is given in Seiffert & Jain [69].

An MLP network M usually consists of $n \geq 3$ layers $L_1, L_2, ..., L_n$, each containing at least one neuron. The first layer L_1 is called *input layer*, the last layer L_n is called *output layer*. All possible layers between the input and the output layers are called *hidden layers*. The connections between the neurons build a simple structure: The output of each neuron in layer L_k is connected to the input of each neuron in its succeeding layer L_{k+1} except for all neurons in L_n. There are no other links in this ANN topology. An example is shown in Figure 1.

A single computation step in an MLP network M works as follows: Each component value of an input vector $x = (x_i)_{i=1,...,|L_1|}$ is put directly into the corresponding neuron i of the input layer L_1. The output value from every input neuron i resulting by evaluation of $o_i(x_i)$ is propagated to each neuron j of the first hidden layer L_2. Each neuron $j \in L_2$ calculates its output value from these input values according to equation (9.1). These results are propagated to the next layer and so on. Finally, the output values of the neurons in layer L_n are the *total output* $M(x)$ of the MLP network for the given input vector x. Due to this computing method, the MLP is a *feed forward* network.

To use this computation scheme for an approximation of a given target function g, a proper *learning* method *(training)* is needed. Training an MLP network means starting from a random initialisation of the weights w_{ij} which have to be adapted in a way so that for any *training vector* x the total output of the network $M(x)$ gets very close to the known result of $g(x)$, i.e., $|M(x) - g(x)| < \epsilon$ for a given $\epsilon > 0$. It is interesting to note here that g is not needed in functional form, since it is sufficient to know the correct output $M(x)$ for each training vector x. This is particularly important for *data mining* applications in finance where a priori unknown information has to be learned from data sets and for other applications like nonparametric prediction problems, where the functional dependencies between exogeneous and endogeneous variables are a priori unknown.

There are different methods discussed in the literature to adapt the weights according to an error function during training. Most of these methods use the gradient of the quadratic error function (9.2) for neuron j:

$$err_j(x_i) = \frac{1}{2}(output_j(x_i) - t_j(x_i))^2 \tag{9.2}$$

where x_i is the input vector and $t_j(x_i)$ is the desired output from neuron j for this input. Probably due to its simplicity, the most popular method for training an MLP net using this scheme is *backpropagation*. It works basically as follows: A training vector x is presented, the net computes its output $M(x)$, and the error is computed by applying (9.2) to all output neurons. Afterwards, the gradient values of this error function are computed for the weights of the incoming links of the output neurons, and these weights are adapted by

$$\Delta w_{ij} = -\eta \frac{\partial}{\partial w_{ij}} err_j \tag{9.3}$$

where $\eta \in (0, 1)$ is a global parameter of the MLP net called *learning rate*. This process is repeated from the output neuron layer L_n to the preceding hidden neuron layer L_{n-1}, and so on until the input neuron layer is reached (see [70] for further details). As the name of this training method states, the network error is propagated backwards through the MLP network. It is easy to see that backpropagation and related gradient descent methods can particularly run into potential local minima of the error function or oscillate in steep valleys of the error function, therefore a number of modifications are discussed in the literature to restrict these problems, e.g., by using second-order derivatives or conjugate gradient approaches, see e.g., Hecht-Nielsen [71] for an overview.

After training of an appropriately constructed MLP network, the network should be able to 'know' the functional dependence between an input vector x and the resulting output $g(x)$. Therefore, the network can be used for *generalisation*, i.e., a number of new input vectors x not used during the training process can be presented to the network and the network outputs $M(x)$ will be considered as an approximation of the usually unknown values of $g(x)$. This is comparable to the application of non-linear regression models, e.g., for time series prediction.

To obtain a good generalisation capability of a MLP, the problem of *overfitting* has to be avoided: Many instances show that the generalisation capabilities of trained

MLPs will be sub-optimal if the MLP memorises the training vectors perfectly. There are different strategies to overcome this situation, e.g., by an early termination of the learning process before final convergence of the weights. For a discussion of this problem and its avoidance see e.g., [65], p. 195ff.

An interesting result concerning the general approximation capabilities of MLP networks is the following theorem by Hornik [72]:

Theorem 2. *Given a measure μ, a distance $\epsilon > 0$ and assuming the presence of one hidden layer using bounded, non-constant activation functions a_j, there is an MLP which can ϵ-approximate any function $f \in L^P(\mu)$, where $L^P(\mu) := \{f : \int_{\mathbb{R}^n} |f(x)|^P d\mu(x) < \infty\}$.*

There are many other results proving the approximation capabilities of ANNs for other restrictions on the target functions, the activation or output functions (see Anthony and Bartlett [73] for a detailed theoretical analysis). A little disadvantage of all these proofs is that they do not provide exact construction guidelines for ANNs in applications. So the topology of the network, particularly the number of necessary hidden layer neurons which has a strong influence on the approximation properties and further parameters have to be chosen e.g., by empirical tests or by experiences from similar cases in each application context. However, a general guideline for adequate ANN modelling and validation in finance applications is provided by Zapranis and Refenes in [74].

Due to the difficulty of selecting an adequate ANN topology and the necessary parameters for a given problem, many recent studies combine ANNs with other machine learning or optimization concepts in the following sense: The given problem is solved by different ANNs and a meta-learning algorithm observes the performance of the ANNs on the problem. The meta-learning algorithm tries to improve the ANNs by varying their topology, their parameters, etc. We will return to this point in Section 9.9.

Many finance-related ANN applications used MLP networks for the prediction of time series, e.g., daily stock returns [75], futures contracts prices [76], real estate returns [77] or three-year default probabilities of companies [78]. Refenes [79] provides a good survey on selected ANN applications for equities (e.g., modelling stock returns, testing the Efficient Markets Hypothesis), foreign exchange prices (e.g., prediction of exchange rates, development of trading rules), bond markets and other economic data.

A standard application of ANNs for classification purposes in finance are bankruptcy prediction and related tasks like bond rating, obligor classification into rating categories, etc. Besides other finance-related ANN topics, Trippi and Turban [80] offer a collection of such classification applications. Other examples are Odom and Sharda [81], Coleman et al. [82], McLeod et al. [83], Baetge and Krause [84], Wilson and Sharda [85]. Most of these studies reported better empirical classification performance of Neural Networks in comparison to standard models like discriminant analysis (for a description of this method see e.g., [86]) or linear regression, except for the study of Altman et al. [87]. A better performance of ANNs is not surprising from a theoretical view if we consider the non-linear approximation abilities of the ANNs (see Theorem 2 above) compared to the limited capabilities of linear models.

The learning capabilities of ANNs can be used for modelling adaptive traders in artificial markets to study the macro-level behaviour of such markets, e.g., measured by asset prices, depending on the micro-level decision strategies and learning capabilities of the traders (cf. also our remarks in the GA section). See e.g., Beltratti et al. [88] for examples of such models.

A number of articles studied the approximation capabilities of MLP networks for option valuation purposes, e.g., Mallaris and Salchenberger [89], Hutchinson et al. [90], Lajbcygier et al. [91], Hanke [92], Herrmann and Narr [93]. Lajbcygier [94] gives a review of related work. An interesting observation in [90] was that a small MLP network using only four hidden neurons in one hidden layer and backpropagation training yielded a good approximation of the results obtained by applying the Black/Scholes formula (cf. [95], [96]) to given option pricing parameters (e.g., the mean R^2 over different option pricing parameters was $R^2 = .9948$). Furthermore, a narrative conclusion of all the above studies is that the results from the training of ANNs using market data can yield a significantly better approximation of option prices compared to closed form option pricing formulas.

Locarek-Junge and Prinzler [97] applied MLPs and backpropagation to market risk (here: *Value-at-Risk*) estimation where the goal of the calculation is to estimate a certain α-percentile (e.g., $\alpha = .99$) of the distribution of future losses from a portfolio whose value depends on a fixed number of market risk factors. They used a Gaussian Mixture Density Network (cf. [98] for details) defined by the following equations:

$$f_{X|Z=z}(x) = \sum_{i=1}^{m} a_i(z)h_i(x|z),$$ (9.4)

$$h_i(x|z) = (2\pi)^{l/2}\sigma_i^{-l}(z)e^{-\frac{\|x-\mu_i(z)\|^2}{2\sigma_i^2(z)}},$$ (9.5)

where $a_i \geq 0, \sum_i a_i = 1$ were the mixture parameters to be estimated by an MLP network backpropagation training that minimised the maximum likelihood error function of the specified model given historical observations of market risk factor changes Z and historical target function values (market risk returns) X. The results from the Mixture Density Network estimation for a USD/DEM exchange rate dependent sample portfolio were compared to the results from a historical simulation (see e.g., Jorion [99]) and from a variance-covariance approach proposed by the *RiskMetrics*^{TM} model [100]. The Neural Network based approach yielded a better estimate particularly for the $\alpha = .99$ percentile if at least two different a_i had to be determined by the learning process, i.e., if there were two or more densities to be included in the mixture.

Naim et al. [101] combined MLP networks as a forecasting tool for financial markets with ideas from asset allocation for portfolio management. Their considerations rely on the predictability of real-world asset returns by ANNs. An interesting remark pointed out by Naim et al. is the explicit difference between a two-step approach that separates the prediction of the necessary asset parameters by ANNs from the port-folio choice problem, and an integrated approach that optimizes both the prediction error and the portfolio choice alternatives in one backpropagation training process. The results of an empirical study simulating a strategic asset allocation problem us-ing data from G7 countries' capital markets indicated that the two-step approach was dominated by the integrated approach. The annualised return was higher and the standard deviation was even smaller for the portfolio calculated by the latter ap-proach compared to the portfolio resulting from the two-step approach running on the same data set. Both methods were significantly better than a standard Markowitz [18] mean-variance portfolio selection procedure based on historical estimation of parameters from the data set.

Bonilla et al. [102] used MLP nets trained by backpropagation to forecast ex-change rate volatilities of six currencies against the Spanish Peseta from historical observations and compared the results to different parametric GARCH-type models (see e.g., [103] for an overview of volatility modelling). In their study the MLP nets' forecasting accuracy was superior for three currencies and inferior for the other three currencies, but the results of the cases where the MLP nets were superior to the para-metric models were significantly better than the cases where the MLP networks were inferior.

In general, ANNs were successfully applied to many non-linear and complex finance problems, e.g., time series prediction, classification and modelling of learn-ing entities in financial simulations. This is especially true for MLPs which are the most widely used ANN computing paradigm in finance. The success of ANNs is mainly caused by their excellent approximation capabilities for non-linear depen-dencies that are even theoretically justified for many classes of functions. But be-sides the achieved results, an exact algorithm for constructing a suitable ANN for a given, arbitrary problem has not been found yet. It is still a matter of experiments and experiences to construct an adequate ANN, therefore this is considered to be a chal-lenge for hybrid approaches discussed later in this article. And it must be emphasized that many approximation capability proofs assume a large or even infinite number of neurons in the ANN resulting in a significantly larger number of weights to be adapted. The dimension of the corresponding weight optimization problem usually grows over-proportional in the number of neurons, therefore the resulting learning problem will get very complex if we have to use many neurons in our finance ap-plications. More precisely, the problem of learning in an MLP is **NP**-*complete* (see [104] for an overview of related results). Finally, it has to be kept in mind that an ANN approach is mainly a data driven problem solving method, i.e., the availabil-ity of sufficiently representative, high-quality data sets is a crucial point in all ANN applications.

9.8 Fuzzy Logic

In contrast to the approaches which have been presented so far in this survey, *Fuzzy Logic (FL)* is not a local search algorithm, but a heuristic approach invented in 1965 by Zadeh [105] that supports the use of vaguely defined variables in calculations and logical inferences.

The building block of the FL framework is a so-called *fuzzy* set which is a generalization of a set in the traditional sense of Cantor. In traditional (in Fuzzy theory also called *crisp*) sets, each element e from a universe of discourse U can either be member of a set $A \subseteq U$ (denoted by $e \in A$) or not be member of A ($e \notin A$). Using a *membership function* $m_A : U \to \{0, 1\}$ this can be modelled as follows:

$$\forall e \in U : m_A(e) = \begin{cases} 1 & e \in A, \\ 0 & \text{otherwise.} \end{cases} \tag{9.6}$$

A fuzzy set generalizes this definition by allowing arbitrary functions of the type $m_A : U \to [0, 1]$ to express the degree of membership of an element e from U to the set A. Therefore, a fuzzy set over U is defined by the pair $F_A := (U, m_A)$ where m_A is the membership function of F_A. The usual set operations were originally generalized by Zadeh [105] as follows:

Given are two fuzzy sets $F_A := (U, m_A)$ and $F_B := (U, m_B)$.

- The complement of F_A is $F_A^c := (U, m_A^c)$ satisfying
  $\forall e \in U : m_A^c(e) := 1 - m_A(e)$.
- The union $F_C = F_A \cup F_B$ where $F_C := (U, m_C)$ is obtained by setting $\forall e \in U :$
  $m_C(e) := \max\{m_A(e), m_B(e)\}$.
- The intersection $F_C = F_A \cap F_B$ where $F_C := (U, m_C)$ is defined by $\forall e \in U :$
  $m_C(e) := \min\{m_A(e), m_B(e)\}$.

Based on these properties, relations between fuzzy sets and an arithmetic for fuzzy numbers can be defined, see e.g., Klir [107, 108] for a detailed coverage of these topics.

An important goal of fuzzy modelling in applications is *computing with words*. Since computations on a machine have to be done by using accurate numbers while human problem solving methods are often more heuristically, FL is commonly used as an interface between problem solving knowledge stated in a natural language and corresponding computations in exact arithmetic to be performed by a machine. For this functionality, the concept of *linguistic variables* is essential. A linguistic variable L is a tupel $L := (V, T, U, G, M)$ where V is the name of the variable, T is the domain of linguistic terms for V, U is the domain of crisp values for V (universe of discourse), G is the syntax rule (grammar) for building the linguistic terms and M is the semantic rule that assigns a fuzzy set F_t to each $t \in T$. Figure 2 shows an example for a linguistic variable that describes asset returns.

Using such linguistic variables, many FL applications incorporate *fuzzy rules*. Assume we have defined the following linguistic variables:

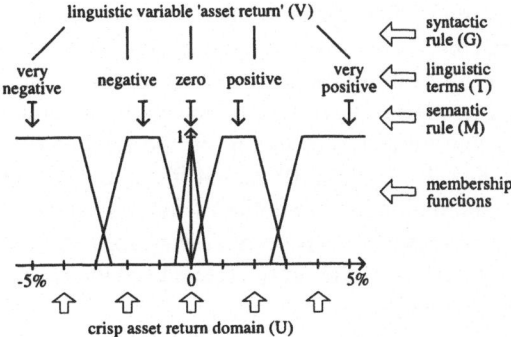

FIGURE 2. Linguistic variable for asset return

$$L_{input} := (V_{input}, T_{input}, U_{input}, G_{input}, M_{input}), \qquad (9.7)$$

$$L_{output} := (V_{output}, T_{output}, U_{output}, G_{output}, M_{output}). \qquad (9.8)$$

Then we can formulate rules of the following form:

$$R_i : \text{ IF } V_{input} = t_1 \text{ THEN } V_{output} = t_2 \qquad (9.9)$$

where $t_1 \in T_{input}$ and $t_2 \in T_{output}$. The input (independent) variable is described by the linguistic variable L_{input}, and the output (dependent) variable is defined using the linguistic variable L_{output}. A *fuzzy system* uses a rule base RB containing $k \in \mathbb{N}$ such rules $R_1, \ldots, R_k$. Note that both the input and the output variables in (9.9) can be aggregations of different linguistic variables using the fuzzy set operators (negation, intersection, union etc.) for combining the fuzzy values of linguistic variables. To use the fuzzy system e.g., for the approximation of the crisp output $g(e)$ that belongs to a given crisp input $e \in U$ for a function $g : U \to \mathbb{R}$, the rule base has to contain rules describing the functional dependence between the input and the output using fuzzy rules like (9.9). These rules are usually heuristic interpretations of the dependence between the crisp input and the crisp output variables. They can either be maintained by a human expert, or be generated and adapted for instance by a local search algorithm.

Algorithm 6. *Fuzzy System Computation*

Input: crisp value $e \in U$, fuzzy rule base RB
Fuzzificate e
$\forall R_i \in RB$: Aggregation of IF condition to determine rule fulfilment of R_i
$\forall R_i \in RB$: Activation of R_i to calculate output activation (THEN part)
Accumulation of the output of all $R_i \in RB$ to obtain output fuzzy set
Defuzzification of the output fuzzy set to obtain crisp output (e)
Output: output (e)

Assuming the presence of adequate rules in the rule base, the fuzzy system computes the crisp output $output(e)$ which approximates $g(e)$ for a given arbitrary, but fixed crisp input $e \in U$ using the scheme shown in Algorithm 6 (cf. Nelles [10], p. 304). There are a number of aggregation, activation, accumulation and defuzzification schemes in the literature depending on the type of fuzzy system and the application context. See also Nelles [10], p. 304 ff. for an overview.

Concerning the approximation capability of fuzzy systems, Wang [109] proved the following interesting result (cf. his original work for a more precise formulation that includes the specification of the fuzzy system used in the proof, and also Kosko [110] for further approximation results):

Theorem 3. *Given is an arbitrary function $g : U \to \mathbb{R}$ where U is a (crisp) compact subset of $\mathbb{R}^n$, and a real number $\epsilon > 0$. Then there is a fuzzy system such that $sup_{e \in U} |output(e) - g(e)| < \epsilon$.*

A natural field for FL applications in finance are rule-based systems which have to deal with vague, non-quantitative or uncertain inputs. The book by von Altrock [111] covers the basic fuzzy system methodology which is necessary for real-world implementations, and a variety of applications concerning the creditworthiness check of obligors, fraud detection etc.

Rommelfanger [112] describes a fuzzy system for checking the credit solvency of small companies. It is based on lingustic variables e.g., for inputs like market share, market growth, rate of innovation and contains rules like 'IF MarketShare = positive AND MarketGrowth = medium AND RateOfInnovation = medium THEN Sales = positive' which were provided by human experts. The system is used by a German bank to evaluate potential obligors in a credit rating process. Similar applications were reported e.g., by Weber [113], and in the book by Ruan et al. [114]. This book contains a selection of other recent FL applications in finance and risk management, e.g., several articles concerning the interesting difference between using probabilistic decision theory or making fuzzy decisions in risky situations, an application of FL in electricity market pricing, and a contribution that proposes a Discounted Fuzzy Cash Flow model for capital budgeting.

In the same volume, Korolev et al. [115] extended Merton's model [116] of the valuation of a premium for bank deposit insurance which had originally used the Black/Scholes option pricing formula [95, 96] to determine the risk-based premium. In their study the Black/Scholes valuation framework is extended by considering a so-called *Fuzzy Valued Asset* as the underlying, i.e., the crisp stock price in the Black/Scholes framework is replaced using a fuzzy set on the payoff scale as the universe of discourse. According to Korolev et al., the use of the fuzzy underlying particularly avoids the criticism against the original Merton model which used a rather unrealistic crisp payoff structure generated by the crisp stock price, and makes the model more useful for real-world applications.

9.9 Hybrid Approaches

All problem solving approaches discussed in the preceding sections (and of course, also other methods which were not mentioned above) exhibit certain strengths and certain weaknesses. For instance, HC is a very fast local search method that causes low computational cost but suffers from the danger of getting stuck in sub-optimal solutions. In contrast to this trade-off, randomized algorithms like SA, TA and EC use heuristic mechanisms to reduce the risk of early convergence to sub-optimal solutions at the price of slower convergence speed and potentially higher computational cost to find reasonably good solutions. This observation suggests the combination of different problem solving methods to emphasize the overall benefits of the resulting, so-called *hybrid approach* and obtain less weakness of the combined approach compared to the sum of individual weaknesses of its ingredients. Of course, the combination should use the minimum number of different methods which leads to the desired properties of the hybrid approach, otherwise it will be difficult to analyse and predict its behaviour. See e.g., Fogel & Michaelwicz [8], p. 391 ff. for a discussion of these issues.

Moreover, all problem solving methods presented in the preceding sections require an adequate representation of the parameters of the problem to be solved as well as initial parameter values which lead to a desired performance of the respective problem solving method. An appropriate combination of methods can be used to derive both the necessary parameters to obtain good solutions and the solutions themselves, which will provide strong support to the user who has to determine the parameters manually, otherwise.

For an overview of different general aspects of hybridization as well as a classification of different approaches, see Goonatilake & Khebbal [117]. The book by Abraham & Koeppen [118] contains a selection of recent developments in the area of hybrid systems, while Rutkowska's recent book [119] particularly covers the fusion of ANN and FL. This topic is also addressed by Jin [120], as well as the integration of EC and FL. While these and many other surveys concentrate on non-finance problems, we will now consider some financial applications.

Since the mid- and late 1990s, many hybrid systems for the support of asset trading in financial markets have been created and integrated into commercial trading platforms. Besides the huge number of commercial products and software in this area there are also many academic publications which report successful applications of hybrid methods. For instance, Herrmann et al. [123] created a combination of a GA and a fuzzy system to support equity traders. The GA worked on a population of individuals, each of which represented an FL rule base consisting of rules like 'IF PriceEarningsRatio = high AND AverageTradingVolume = high THEN buy'. For each individual, both the rules themselves and the fuzzy membership functions used in the linguistic variables were optimized by the GA using daily stock price information of the 30 companies listed in the German DAX index over a period of seven years. The hybrid system e.g., yielded a positive excess return over the DAX index performance in 66% to 75% of the fuzzy rule applications.

Siekmann et al. [124] used a combination of ANN and FL (*NeuroFuzzy system*) to predict the sign of the daily returns of the German DAX index. The fuzzy system rule base contained rules based on different technical indicators in the 'IF'-condition which were evaluated using historical stock price information, and the fuzzy conclusion in the 'THEN' part was either 'NextDailyDAXReturn = positive', 'NextDaily-DAXReturn = neutral' or 'NextDailyDAXReturn = negative'. An MLP network was used to represent the fuzzy system rules, which were initially provided by a human expert, and this network was trained using daily DAX returns. The goal of the MLP training was to remove inappropriate rules from the fuzzy system by pruning their representing nodes from the MLP network, and to adapt the membership functions of the linguistic variables in the remaining fuzzy rules. After the training, the remaining rules were applied to an additional test set containing daily DAX returns not used in the training, and the trading rules based on the prediction by the NeuroFuzzy system e.g., yielded a much higher daily return than naive trading rules based on the previous change of the DAX or than the fuzzy system rules originally provided by the human expert.

For another study that has a similar focus, but discusses the hybridisation of ANN and GA to support trading decisions in the US 30-year T-bond future market, see Harland [125].

In a series of recent papers we proposed a hybrid approach that combined GA and HC to compute risk-return efficient portfolio structures for a discrete set of credit portfolio investment alternatives under constraints. The hybrid approach was implemented and empirically tested both using a single objective function that related the total net risk adjusted return to the total unexpected loss (measured by downside risk) of a given credit portfolio (cf. Schlottmann & Seese [126]), and using two separate objective functions for risk and return (cf. Schlottmann & Seese [127])). For instance, in the latter study the hybrid approach required three minutes to find an approximation for a constrained, global Pareto-efficient set based on 20 assets on a standard Personal Computer, while the upper computational bound for this problem determined by a complete enumeration of the search space was 72 minutes. Moreover, the hybrid approach showed higher convergence speed towards feasible, optimal solutions while consuming low additional computational resources compared to its GA counterpart without the additional HC component.

The above examples of successful hybridisation in financial contexts are the basis for some conclusions and possible future research directions which are derived in the final section below.

9.10   Conclusions

In the preceding sections we have discussed different heuristic approaches which were successfully applied to a variety of complex financial problems. The respective results underline the fact that heuristic approaches are an interesting alternative to other problem solving algorithms e.g., if a problem is computationally very hard, if the problem's parameters cannot be defined using exact bounds or if the functional dependency between input and output and/or the set of input parameters is a priori unknown.

However, we have also pointed out that besides the advantages, there are certain design problems when choosing a heuristic approach. The representation of the exogeneous variables (e.g., decision variables), the parameters of the heuristic algorithm, etc. have to be chosen carefully to obtain good results, and this is itself a non-trivial and in many cases complex task. Beyond that, the early success of certain heuristic approaches, e.g., ANNs, caused too high expectations towards the results, which simply had to create disappointment due to the complexity and sometimes the dynamics of the problems to be solved, for example in stock market analysis and prediction.

Moreover, none of the heuristic approaches fits into all problem contexts and some were especially useful in certain problem settings. For example, the stochastic search heuristics like Simulated Annealing, Threshold Accepting, Tabu Search as well as the different methods from the Evolutionary Computation paradigm were particularly successful in combinatorial problems while Artificial Neural Networks mainly yielded good results for function approximation, e.g., in non-parametric, non-linear regression. Evolutionary Computation is also a natural approach to modelling evolving and learning financial entities, and Fuzzy Logic is the first choice for problems which cannot be modelled well using crisp numbers or crisp sets.

Much recent academic effort in the area of heuristics has been spent on hybrid methods, which try to combine the strengths of different approaches. These methods seem to be promising for future studies developing modern heuristics in the finance context since many of them incorporate an estimation mechanism that determines necessary algorithmic parameters instead of relying on trial-and-error experiments by the user. We think that another crucial point for future success is the explicit exploitation of problem-specific financial knowledge in the algorithms instead of applying the standard heuristic algorithm scheme to the problem under consideration. The flexibility of many of the heuristic approaches discussed in our survey concerning the integration of problem-specific knowledge is the true strength and one of the best justifications for choosing such methods.

References

[1] M. Garey, D. Johnson, *Computers and Intractability*, New York, W. H. Freeman & Company, 1979.

[2] C. Papadimitriou, *Computational Complexity*, Reading, Addison-Wesley, 1994.

[3] D. Seese, F. Schlottmann, The building blocks of complexity: a unified criterion and selected applications in economics and finance, presented at *Sydney Financial Mathematics Workshop 2002*, http://www.qgroup.org.au/SFMW

[4] G. Ausiello, P. Crescenzi, G. Gambosi, V. Kann, A. Marchetti-Spaccamela, M. Protasi, *Complexity and Approximation*, Springer, Heidelberg, 1999.

[5] C. Reeves (ed.), *Modern Heuristic Techniques for Combinatorial Problems*, Oxford, Blackwell Scientific Publishers, 1993.

[6] I. Osman, J. Kelly (eds.), *Meta-heuristics: Theory and Applications*, Dordrecht, Kluwer, 1996.

[7] E. Aarts and J. Lenstra (eds.), *Local Search in Combinatorial Optimization*, Chichester, John Wiley & Sons, 1997.

[8] D. Fogel, Z. Michalewicz, *How to Solve it — Modern Heuristics*, Springer, Heidelberg, 2000.

[9] D. Pham, D. Karaboga, *Intelligent Optimization Techniques*, Springer, London, 2000.

[10] O. Nelles, *Nonlinear System Identification*, Springer, Heidelberg, 2001.

[11] S. Chen (ed.), *Evolutionary Computation in Economics and Finance*, Springer, Heidelberg, 2002.

[12] S. Kirkpatrick, C. Gelatt and M. Vecchi, Optimization by simulated annealing, *Science* **220** (1983), 671–680.

[13] V. Cerny, Thermodynamical approach to the travelling salesman problem: an efficient simulation algorithm, *Journal of Optimization Theory and Applications* **45** (1985), 41–51.

[14] W. Metropolis, A. Rosenbluth, M. Rosenbluth, A. Teller, and E. Teller, Equation of the state calculations by fast computing machines, *Journal of Chemical Physics* **21** (1953), 1087–1092.

[15] E. Aarts and J. Korst, Simulated annealing and Boltzmann machines: a stochastic approach to combinatorial optimization and neural computing, Chichester, John Wiley & Sons, 1989.

[16] E. Aarts, J. Korst and P. van Laarhoven, Simulated annealing, in: E. Aarts and J. Lenstra, Local search in combinatorial optimization, Chichester, John Wiley & Sons, 1997, 91–120.

[17] T. Chang, N. Meade, J. Beasley, Y. Sharaiha, Heuristics for cardinality constrained portfolio optimization, *Computers & Operations Research* **27** (2000), 1271–1302.

[18] H. Markowitz, *Portfolio Selection: Efficient Diversification of Investments*, John Wiley & Sons, New York, 1959.

[19] G. Dueck and T. Scheurer, Threshold accepting: A general purpose algorithm appearing superior to simulated annealing, *Journal of Computational Physics* **90** (1990), 161–175.

[20] G. Dueck and P. Winker, New concepts and algorithms for portfolio choice, *Applied Stochastic Models and Data Analysis* **8** (1992), 159–178.

[21] M. Gilli and E. Kellezi, Portfolio optimization with VaR and expected Shortfall, in: E. Kontoghoirghes, B. Rustem and S. Siokos (eds.), *Computational Methods in Decision-making, Economics and Finance*, Kluwer, Dordrecht, 2002.

[22] M. Gilli and E. Kellezi, Threshold accepting for index tracking, Research paper, University of Geneva, http://www.unige.ch/ses/metri/gilli/portfolio/Yale-2001-IT.pdf.

[23] P. Winker, *Optimization Heuristics in Econometrics*, John Wiley & Sons, Chichester, 2001.

[24] F. Glover, Future paths for integer programming and links to artificial intelligence, *Computers and Operations Research* **13** (1986), 533–549.

[25] P. Hansen, The steepest ascent mildest descent heuristic for combinatorial programming, presented at *Congress on Numerical Methods in Combinatorial Optimization*, Capri, 1986.

[26] F. Glover, M. Laguna, *Tabu Search*, Kluwer, Dordrecht, 1997.

[27] A. Hertz, E. Taillard, and D. de Werra, Tabu search, in: E. Aarts and J. Lenstra, *Local Search in Combinatorial Optimization*, John Wiley & Sons, Chichester, 1997, 121–136.

[28] F. Glover, J. Mulvey, and K. Hoyland, Solving dynamic stochastic control problems in finance using tabu search with variable scaling, in: H. Osman and J. Kelly (eds.), *Meta-heuristics: Theory and Applications*, Kluwer, Dordrecht, 1996, 429–448.

[29] K. DeJong, D. Fogel, and H. Schwefel, A history of evolutionary computation, in: T. Baeck, D. Fogel and Z. Michalewicz (eds.), *Evolutionary Computation 1*, Bristol, IOP Publishing, 2000, 40–58.

[30] L. Fogel, A. Owens, and M. Walsh, *Artificial Intelligence through Simulated Evolution*, John Wiley & Sons, New York, 1966.

[31] J. Koza, *Genetic Programming*, MIT Press, Cambridge, MA, 1992.

[32] J. Koza, *Genetic Programming II*, MIT Press, Cambridge, MA, 1994.

[33] J. Koza, F. Bennett, D. Andre, and M. Keane, *Genetic Programming III*, Morgan Kaufmann, San Francisco, 1999.

[34] J. Holland, *Adaptation in Natural and Artificial Systems*, Michigan University Press, Ann Arbor, 1975.

[35] I. Rechenberg, Cybernetic solution path of an experimental problem, *Royal Aircraft Establishment Library Translation* **1122**, 1965.

[36] H. Schwefel, *Evolution and Optimum Seeking*, John Wiley & Sons, Chichester, 1995.

[37] T. Baeck, D. Fogel, Z. Michalewicz (eds.), *Evolutionary Computation 1*, Bristol, IOP Publishing, 2000.

[38] T. Baeck, D. Fogel, Z. Michalewicz (eds.), *Evolutionary Computation 2*, Bristol, IOP Publishing, 2000.

[39] W. Banzhaf, J. Daida, A. Eiben, M. Garzon, V. Honavar, M. Jakiela, R. Smith (eds.), *Proc. of the Genetic and Evolutionary Computation Conference*, Morgan Kaufmann, San Francisco, 1999.

[40] G. Rudolph, Finite Markov chain results in evolutionary computation: A tour d'horizon, *Fundamentae Informaticae*, 1998, 1–22.

[41] H. Muehlenbein, Genetic Algorithms, in: E. Aarts and J. Lenstra (eds.), *Local Search in Combinatorial Optimization*, John Wiley & Sons, Chichester, 1997, 137–172.

[42] S. Droste, T. Janses and I. Wegener, Perhaps not a free lunch but at least a free appetiser, in: W. Banzaf et al. (eds.), Proceedings of First Genetic and Evolutionary Computation Conference, San Francisco, Morgan Kaufmann, 1999, 833–839.

[43] M. Vose, *The Simple Genetic Algorithm*, MIT Press, Cambridge, MA, 1999.

[44] I. Wegener, On the expected runtime and the success probability of Evolutionary Algorithms, *Lecture Notes in Computer Science* **1928**, Springer, Heidelberg, 2000.

[45] T. Riechmann, *Learning in Economics*, Physica, Heidelberg, 2001.

[46] C. Rieck, Evoluationary simulation of asset trading strategies, in: E. Hillebrand, J. Stender (eds.): *Many-agent Simulation and Artificial Life*, IOS Press, 1994, 112–136.

[47] P. Tayler, Modelling artificial stock markets using genetic algorithms, in: S. Goonatilake, P. Treleaven (eds.), *Intelligent Systems for Finance and Business*, John Wiley & Sons, New York, 1995, 271–287.

[48] B. LeBaron, W. Arthur, R. Palmer, Time series properties of an artificial stock market, *Journal of Economic Dynamics & Control* **23** (1999), 1487–1516.

[49] J. Coche, An evolutionary approach to the examination of capital market efficiency, *Evolutionary Economics* **8**, 357–382.

[50] J. Farmer, A. Lo, Frontiers of finance: Evolution and efficient markets, Santa Fe Institute, 1999, http://www.santafe.edu/~jdf.

[51] R. Walker, E. Haasdijk, M. Gerrets, Credit evaluation using a genetic algorithm; in: S. Goonatilake, P. Treleaven (eds.), *Intelligent Systems for Finance and Business*, John Wiley & Sons, New York, 1995, 39–59.

[52] S. Mott, Insider dealing detection at the Toronto Stock Exchange Modelling artificial stock markets using genetic algorithms, in: S. Goonatilake, P. Treleaven (eds.), Intelligent systems for finance and business, John Wiley & Sons, New York, 1995, 135–144.

[53] A. Frick, R. Herrmann, M. Kreidler, A. Narr, D. Seese, A genetic based approach for the derivation of trading strategies on the German stock market, in: Proceedings ICONIP '96, Springer, Heidelberg, 1996, 766–770.

[54] R. Bauer, *Genetic Algorithms and Investment Strategies*, John Wiley & Sons, New York, 1994.

[55] J. Kingdon, *Intelligent Systems and Financial Forecasting*, Springer, Heidelberg, 1997.

[56] R. Tsang, P. Lajbcygier, Optimization of technical trading strategy using split search Genetic Algorithms, in: Y. Abu-Mostafa, B. LeBaron, A. Lo, A. Weigend (eds.), *Computational Finance 1999*, MIT Press, Cambridge, MA, 2000, 690–703.

[57] W. Langdon, R. Poli, *Foundations of Genetic Programming*, Springer, Heidelberg, 2002.

[58] C. Keber, Option valuation with the Genetic Programming approach, in: Y. Abu-Mostafa, B. LeBaron, A. Lo, A. Weigend, *Computational finance 1999*, MIT Press, Cambridge, MA, 2000, 370–386.

[59] M. Brennan, E. Schwarz, The valuation of American put options, *Journal of Finance* 32 (1977), 449–462.

[60] J. Cox, S. Ross, M. Rubinstein, Option pricing: a simplified approach, *Journal of Financial Economics* 7 (1979), 229–263.

[61] J. Li and E. Tsang, Reducing failures in investment recommendations using Genetic Programming, presented at *6th Conference on Computing in Economics and Finance, Barcelona*, 2000.

[62] S. Baglioni, C. da Costa Pereira, D. Sorbello and A. Tettamanzi, An evolutionary approach to multiperiod asset allocation, in: R. Poli, W. Banzhaf, W. Langdon, J. Miller, P. Nordin and T. Fogarty (eds.), Genetic Programming, Proceedings of EuroGP 2000, Springer, Heidelberg, 2000, 225–236.

[63] S. Chen, T. Kuo, Towards an agent-based foundation of financial econometrics: An approach based on Genetic-Programming financial markets, in: W. Banzhaf et al. (eds.), *Proc. of the Genetic and Evolutionary Computation Conference*, Morgan Kaufmann, San Francisco, 1999, 966–973.

[64] W. McCulloch, W. Pitts, A logical calculus of the ideas immanent in nervous activity, *Bulletin of Mathematical Biophysics* 5 (1943), 115–133

[65] R. Schalkoff, *Artificial Neural Networks*, New York, McGraw-Hill, 1997.

[66] M. Arbib, *The Handbook of Brain Theory and Neural Networks*, MIT Press, Cambridge, MA, 1995.

[67] T. Kohonen, *Self-organising Maps*, Springer, Heidelberg, 1995.

[68] G. Deboeck and T. Kohonen, *Visual Explorations in Finance*, Springer, Heidelberg, 1998.

[69] U. Seiffert, L. Jain (eds.), *Self-organising Neural Networks*, Springer, Heidelberg, 2002.

[70] D. Rumelhart, J. McClelland, *Parallel Distributed Processing: Explorations in the Microstructure of Cognition, Vol. 1: Foundations*, MIT Press, Cambridge, MA, 1986.

[71] R. Hecht-Nielsen, *Neurocomputing*, Addison-Wesley, Reading, MA, 1990.

[72] K. Hornik, Approximation capabilities of multilayer feedforward networks, *Neural Networks* 4 (1991), 251–257.

[73] M. Anthony, P. Bartlett, *Learning in Neural Networks*, University Press, Cambridge, UK, 1999.

[74] A. Zapranis, P. Refenes, *Priciples of Neural Model Identification*, Springer, London, 1999.

[75] D. Witkowska, Neural Networks application to analysis of daily stock returns at the largest stock markets, in: P. Szczepaniak (ed.), *Computational Intelligence and Applications*, Heidelberg, Physica, 1999, 351–364.

[76] M. Azoff, *Neural Network Time Series Forecasting of Financial Markets*, John Wiley & Sons, New York, 994.

[77] R. Bharati, V. Desai, M. Gupta, Predicting real estate returns using Neural Networks, *Journal of Computational Intelligence in Finance* 7 (1999) 1, 5–15.

[78] J. Baetge, A. Jerschensky, Measurement of the probability of insolvency with Mixture-of-Expert Networks, in: W. Gaul, H. Locarek-Junge (eds.), *Classification in the Information Age*, Springer, Heidelberg, 1999, 421–429.

[79] A. Refenes, *Neural Networks in the Capital Markets*, John Wiley & Sons, Chichester, 1995.

[80] R. Trippi, E. Turban, *Neural Networks in Finance and Investing*, Probus Publishing, Chicago, 1993.

[81] M. Odom, R. Sharda, A Neural Network model for bankruptcy prediction, *Proceedings of the IEEE International Joint Conference an Neural Networks*, Vol. 2, 1990, 163–167.

[82] K. Coleman, T. Graettinger and W. Lawrence, Neural Networks for bankruptcy prediction: The power to solve financial problems, in: *AI review* (1991) 4, 48–50.

[83] R. McLeod, D. Malhotra and R. Malhotra, Predicting credit risk, A Neural Network Approach, *Journal of Retail Banking* (1993) 3, 37–44.

[84] J. Baetge and C. Krause, The classification of companies by means of Neural Networks, *Journal of Information Science and Technology* 3 (1993) 1, 96–112.

[85] R. Wilson, R. Sharda, Bankruptcy prediction using Neural Networks, *Decision Support Systems* 11 (1994), 545–557.

[86] E. Altman, Financial ratios, discriminant analysis and the prediction of corporate bankruptcy, *Journal of Finance* 23 (1968), 189–209.

[87] E. Altman, G. Marco and F. Varetto, Corporate distress diagnosis: Comparisions using linear discriminant analysis and Neural Networks, *Journal of Banking and Finance* 18 (1994) 3, 505–529.

[88] A. Beltratti, S. Margarita and P. Terna,*Neural Networks for Economic and Financial Modelling*, International Thomson Computer Press, London, 1994.

[89] M. Malliaris and L. Salchenberger, Beating the best: A Neural Network challenges the Black-Scholes formula, *Applied Intelligence* 3 (1993) 3, 193–206.

[90] J. Hutchinson, A. Lo, and T. Poggio, A nonparametric approach to pricing and hedging derivative securities, *Journal of Finance* 49 (1994) 3, 851–889.

[91] P. Lajbcygier, A. Flitman, A. Swan, and R. Hyndman, The pricing and trading of options using a hybrid Neural Network model with historical volatility, *NeuroVest Journal* 5 (1997) 1, 27–41.

[92] M. Hanke, Neural Network approximation of analytically intractable option pricing models, *Journal of Computational Intelligence in Finance* 5 (1997) 5, 20–27.

[93] R. Herrmann, A. Narr, Risk neutrality, *Risk* (1997) 8.

[94] P. Lajbcygier, Literature review: The non-parametric models, *Journal of Computational Intelligence in Finance* 7 (1999) 6, 6–18.

[95] F. Black, M. Scholes, The valuation of option contracts and a test of market efficiency, *Journal of Finance* 27 (1972), 399–417.

[96] F. Black, M. Scholes, The pricing of options and corporate liabilities, *Journal of Political Economy* 81 (1973), 637–654.

[97] H. Locarek-Junge and R. Prinzler, Estimating Value-at-Risk using Artificial Neural Networks, in: C. Weinhardt, H. Meyer zu Selhausen and M. Morlock (eds.), *Informationssysteme in der Finanzwirtschaft*, Springer, Heidelberg, 1998, 385–399.

[98] C. Bishop, *Neural Networks for Pattern Recognition*, Clarendon Press, Oxford, 1995.

[99] P. Jorion, Value-at-Risk: The new benchmark for controlling market risk, Irwin, Chicago, 1997.

[100] J. P. Morgan and Reuters, *RiskMetrics*™ Technical Document, New York, 1996, http://www.rmg.com.

[101] P. Naim, P. Herve, and H. Zimmermann, Advanced adaptive architectures for asset allocation, in: C. Dunis (ed.), *Advances in Quantitative Asset Management*, Kluwer Academic Publishers, Norwell, MA, 2000, 89–112.

[102] M. Bonilla, P. Marco, I. Olmeda, Forecasting exchange rate volatilities using Artificial Neural Networks, in: M. Bonilla, T. Casasus and R. Sala, *Financial Modelling*, Physica, Heidelberg, 2000, 57–68.

[103] C. Alexander, Volatility and correlation: Measurement, models and applications, in: C. Alexander (ed.), *Risk Management and Analysis, Vol. 1: Measuring and Modelling Financial Risk*, John Wiley & Sons, New York, 1998, 125–171.

[104] S. Judd, Time complexity of learning, in: M. Arbib (ed.), *Handbook of Brain Theory and Neural Networks*, MIT Press, Cambridge, MA, 1995, 984–990.

[105] L. Zadeh, Fuzzy sets, *Information and Control* **8**, (1965) 338–352.

[106] L. Zadeh, Outline of a new approach to the analysis of complex systems and decision processes, *IEEE Transactions on Systems, Man and Cybernetics*, **SMC-3** (1973) 1, 28–44.

[107] G. Klir, B. Yuan, *Fuzzy Sets and Fuzzy Logic: Theory and Applications*, Prentice-Hall, Upper Saddle River, NJ, 1995.

[108] G. Klir , B. Yuan (eds.), *Fuzzy Sets, Fuzzy Logic and Fuzzy Systems*, Singapore, World Scientific, 1995.

[109] L. Wang, Fuzzy systems are universal approximators, in: *Proceedings of the First IEEE International Conference on Fuzzy Systems*, San Diego, 1992, 1163–1169.

[110] B. Kosko, Fuzzy systems as universal approximators, in: *IEEE Transactions on Computers*, **43** (1994) 9, 1329–1333.

[111] C. von Altrock, *Fuzzy Logic and NeuroFuzzy Applications in Business and Finance*, Prentice-Hall, Upper Saddle River, NJ, 1997.

[112] H. Rommelfanger, Fuzzy logic based systems for checking credit solvency of small business firms, in: R. Ribeiro, H.-J. Zimmermann, R. Yager and J. Kacprzyk (eds.), *Soft Computing in Financial Engineering*, Physica, Heidelberg, 1999, 371–387.

[113] R. Weber, Applications of Fuzzy logic for credit worthiness evaluation, in: R. Ribeiro, H.-J. Zimmermann, R. Yager and J. Kacprzyk (eds.), *Soft Computing in Financial Engineering*, Physica, Heidelberg, 1999, 388–401.

[114] D. Ruan, J. Kacprzyk, M. Fedrizzi, *Soft Computing for Risk Evaluation and Management*, Physica, Heidelberg, 2001, 375–409.

[115] K. Korolev, K. Leifert, and H. Rommelfanger, Fuzzy logic based risk management in financial intermediation, in: D. Ruan, J. Kacprzyk, M. Fedrizzi, *Soft Computing for Risk Evaluation and Management*, Physica, Heidelberg, 2001, 447–471.

[116] R. Merton, An analytic derivation of the cost of deposit insurance and loan guarantees, *Journal of Banking in Finance* (1977) 1, 3–11.

[117] S. Goonatilake, S. Khebbal (eds.), *Intelligent Hybrid Systems*, John Wiley & Sons, Chichester, 1995.

[118] A. Abraham, M. Koeppen (eds.), *Hybrid Information Systems*, Springer, Heidelberg, 2002.

[119] D. Rutkowska, *Neuro-fuzzy Architectures and Hybrid Learning*, Springer, Heidelberg, 2002.

[120] Y. Jin, *Advanced Fuzzy Systems Design and Applications*, Springer, Heidelberg, 2002.

[121] J. Balicki, Evolutionary Neural Networks for solving multiobjective optimization problems, in: P. Szczepaniak, *Computational Intelligence and Applications*, Springer, Heidelberg, 1999, 108–199.

[122] M. Gupta, Fuzzy neural computing, in: P. Szczepaniak, *Computational Intelligence and Applications*, Springer, Heidelberg, 1999, 34–41.

[123] R. Herrmann, M. Kreidler, D. Seese and K. Zabel, A fuzzy-hybrid approach to stock trading, in: S. Usui, T. Omori (eds.), Proceedings ICONIP '98, Amsterdam, IOS Press, 1998, 1028–1032.

[124] S. Siekmann, R. Neuneier, H.-J. Zimmermann and R. Kruse, Neuro-Fuzzy methods applied to the German stock index DAX, in: R. Ribeiro, H.-J. Zimmermann, R. Yager and J. Kacprzyk (eds.), *Soft Computing in Financial Engineering*, Physica, Heidelberg, 1999, 186–203.

[125] Z. Harland, Using nonlinear Neurogenetic models with profit related objective functions to trade the US T-Bond future, in: Y. Abu-Mostafa, B. LeBaron, A. Lo, A. Weigend (eds.), *Computational Finance 1999*, MIT Press, Cambridge, MA, 2000, 327–343.

[126] F. Schlottmann, D. Seese, A hybrid genetic-quantitative method for risk-return optimization of credit portfolios, *Proc. QMF'2001 (abstracts)*, Sydney, 2001, http://www.business.uts.edu.au/resources/qmf2001/ F_Schlottmann.pdf.

[127] F. Schlottmann, D. Seese, Finding Constrained Downside Risk-Return Efficient Credit Portfolio Structures Using Hybrid Multi-Objective Evolutionary Computation, in: G. Bol, G. Nakhaeizadeh, S. Rachev, T. Ridder, K.-H. Vollmer (eds.), Credit Risk, Heidelberg, Springer, 2003, 231–265.

10

On Relation Betweeen Expected Regret and Conditional Value-at-Risk

Carlos E. Testuri

Stanislav Uryasev

ABSTRACT   The paper compares portfolio optimization approaches with expected regret and Conditional Value-at-Risk (CVaR) performance functions. The expected regret is defined as an average portfolio underperformance comparing to a fixed target or some benchmark portfolio. For continuous distributions, CVaR is defined as the expected loss exceeding α-Value-at Risk (VaR), i.e., the mean of the worst (1-α)100% losses in a specified time period. However, generally, CVaR is the weighted average of VaR and losses exceeding VaR. Optimization of CVaR can be performed using linear programming. We formally prove that a portfolio with a continuous loss distribution, which minimizes CVaR, can be obtained by doing a line search with respect to the threshold in the expected regret. An optimal portfolio in CVaR sense is also optimal in the expected regret sense for some threshold in the regret function. The inverse statement is also valid, i.e., if a portfolio minimizes the expected regret, this portfolio can be found by doing a line search with respect to the CVaR confidence level. A portfolio, optimal in expected regret sense, is also optimal in CVaR sense for some confidence level. The relation of the expected regret and CVaR minimization approaches is explained with a numerical example.

10.1   Introduction

Modern portfolio optimization theory was originated by Markowitz (1952), who demonstrated that quadratic programming can be used for constructing efficient portfolios. Relatively recently, linear programming techniques, which have superior performance compared to quadratic programming, became popular in finance applications: the mean absolute deviation approach, Konno and Yamazaki (1991), the regret optimization approach, Dembo and King (1992), Dembo and Rosen (1999), and the minimax approach, Young (1998). A reader interested in applications of optimization techniques in finance can find many relevant papers in Ziemba and Mulvey (1998) and in Zenios (1996).

The *expected regret* (see, Dembo and King (1992), Dembo and Rosen (1999)) which is also called the *low partial moment* (see, Harlow (1991)) is defined as the average portfolio underperformance compared to a fixed target or some benchmark portfolio. Although both terms are quite popular, in this paper we use the term "expected regret". A similar concept to the expected regret was utilized by Cariño and Ziemba (1998) in the Russell–Yasuda Kasai financial planning model. In this application, several target thresholds were used and portfolio underperformance was penalized with different coefficients for various thresholds. Probabilistic performance measures similar to the expected regret, such as the conditional expectation constraints and integrated chance constraints described in Prekopa (1995) have been successfully used in various engineering applications outside of financial context. High numerical efficiency of the expected regret approach in portfolio optimization is related to using state-of-the-art linear programming techniques.

This paper establishes relation of the expected regret with *Conditional Value-at-Risk* (CVaR) risk measure. The CVaR risk measure is closely related to *Value-at-Risk* (VaR) performance measure, which is the percentile of the loss distribution. A description of various methodologies for the modeling of VaR can be seen, along with related resources, at URL http://www.gloriamundi.org/. The term Conditional Value-at-Risk was introduced by Rockafellar and Uryasev (2000). For continuous distributions, CVaR is defined as the conditional expected loss under the condition that it exceeds VaR, see Rockafellar and Uryasev (2000). For continuous distributions, this risk measure also is known as Mean Excess Loss, Mean Shortfall, or Tail Value-at-Risk. For continuous distributions, Hurliman (2001) presents ten equivalent definitions of CVaR which were used in different forms in reliability, actuarial science, finance and economics. However, for general distributions, including discrete distributions, CVaR has been defined only recently by Rockafellar and Uryasev (2002) as a weighted average of VaR and losses strictly exceeding VaR. Also, Acerbi et al. (2001), Acerbi and Tasche (2001) redefined expected shortfall similar to CVaR. For general distributions, CVaR, which is a quite similar to the VaR measure of risk has more attractive properties than VaR. CVaR is sub-additive and convex for general distributions, see Rockafellar and Uryasev (2002). Moreover, CVaR is a *coherent* measure of risk in the sense of Artzner et al. (1997). Coherency of CVaR for general distributions was first proved by Pflug (2000); see also Rockafellar and Uryasev (2002), Acerbi et al. (2001), Acerbi and Tasche (2001).

Rockafellar and Uryasev (2000,2002) demonstrated that optimization of CVaR can be performed using linear programming. Several case studies showed that risk optimization with the CVaR performance function and constraints can be done with relatively small computational resources, see Rockafellar and Uryasev (2000,2002), Krokhmal, Palmquist, and Uryasev (1999), Andersson et al. (2001), Bogentoft, Romeijn and Uryasev (2001), Jobst and Zenios (2001).

This paper compares portfolio optimization approaches with expected regret and CVaR utility functions. In order to allow the comparison, regret is specified with ℓ_1-norm and it is assumed that expectations are calculated using a density function. We show for continuous distributions that an optimal portfolio in the CVaR sense is also optimal in the expected regret sense for some threshold in the regret function.

The portfolio which minimizes CVaR can be obtained by adjusting the threshold in the expected regret function and minimizing the expected regret function. An inverse statement is also valid, i.e., a portfolio optimal in the expected regret sense, is also optimal in CVaR sense for some confidence level. If a portfolio minimizes the expected regret, it can be found by solving a one-dimensional minimization problem with respect to CVaR confidence level. We formally prove the statement on the relation of the expected regret and CVaR minimization approaches and explain statements with an example.

10.2 Comparison of expected regret and CVaR

10.2.1 Assumptions and notation

Let $f(\mathbf{x}, \mathbf{y})$ be a loss function, i.e., $-f(\mathbf{x}, \mathbf{y})$ defines the utility of return on investments, associated with a decision vector $\mathbf{x}$ and a random vector $\mathbf{y}$. The decision vector $\mathbf{x}$ can be interpreted in various ways; for instance, it is a portfolio consisting of n instruments with positions belonging to a feasible set $X \subseteq \Re^n$. The random vector $\mathbf{y} \in \Re^m$ accounts for uncertainties in the loss function. To simplify formal analysis, it is supposed that the random vector $\mathbf{y}$ is drawn from a joint density function $p(\mathbf{y})$. For each $\mathbf{x}$, $f(\mathbf{x}, \mathbf{y})$ is a random variable, since it is a function of the random vector $\mathbf{y}$. The distribution function of $f(\mathbf{x}, \mathbf{y})$ for a given $\mathbf{x}$,

$$\Psi(\mathbf{x}, \zeta) \stackrel{\Delta}{=} \int_{f(\mathbf{x},\mathbf{y}) \leq \zeta} p(\mathbf{y}) \, d\mathbf{y},$$

measures the probability of the event that the losses will not exceed a given level of losses, ζ. $\Psi(\mathbf{x}, \zeta)$ is a cumulative probability function of $\mathbf{x}$ which is nondecreasing and right-continuous with respect to ζ in the general case. We assume that $\Psi(\mathbf{x}, \zeta)$ is continuous with respect to ζ; this is accomplished for a given $\mathbf{x}$ if the probability measure $\int_{f(\mathbf{x},\mathbf{y})=\zeta} p(\mathbf{y}) \, d\mathbf{y}$ is zero for all ζ. This assumption as well as the assumption on the existence of density $p(\mathbf{y})$ is imposed to establish continuity of the function $\Psi(\mathbf{x}, \zeta)$ with respect to decision vector $\mathbf{x}$. In some common situations, the required continuity follows from properties of loss $f(\mathbf{x}, \mathbf{y})$ and the density $p(\mathbf{y})$; see Uryasev (1995). We denote the ℓ_1-norm regret function for a portfolio $\mathbf{x}$ as

$$G_\zeta(\mathbf{x}) \stackrel{\Delta}{=} \int_{\mathbf{y} \in \Re^m} [f(\mathbf{x}, \mathbf{y}) - \zeta]^+ \, p(\mathbf{y}) \, d\mathbf{y}, \tag{10.1}$$

where the integrand may be interpreted as a measure of underperformance of the portfolio with respect to a given benchmark ζ; and the positive part operator, $[\cdot]^+$, is defined as $\max(0, \cdot)$; see Harlow (1991), Dembo and King (1992). VaR for a confidence level $\alpha \in (0, 1)$, α-VaR, is defined as a minimal value of losses such that the potential losses do not exceed ζ with probability α,

$$\zeta_\alpha(\mathbf{x}) \stackrel{\Delta}{=} \min\{\zeta \in \Re : \Psi(\mathbf{x}, \zeta) \geq \alpha\}. \tag{10.2}$$

Since $\Psi(\mathbf{x}, \zeta)$ is continuous by assumption and nondecreasing with respect to ζ, there could exist more than one ζ such that $\Psi(\mathbf{x}, \zeta) = \alpha$; this is the reason for the use of the minimum operator. For continuous distributions considered in this paper, CVaR, with a given confidence level α (α-CVaR), is defined as the conditional expectation of losses exceeding VaR,

$$\phi_\alpha(\mathbf{x}) \triangleq (1 - \alpha)^{-1} \int_{f(\mathbf{x}, \mathbf{y}) \geq \zeta_\alpha(\mathbf{x})} f(\mathbf{x}, \mathbf{y}) p(\mathbf{y}) \, d\mathbf{y}, \tag{10.3}$$

where the probability that $f(\mathbf{x}, \mathbf{y}) \geq \zeta_\alpha(\mathbf{x})$ is $1 - \alpha$. However, for general distributions, including discrete distributions, CVaR may not be equal to the conditional expectation defined in (10.3). For general distributions, CVaR is defined as a weighted average of VaR and conditional expectation of losses strictly exceeding VaR.

10.2.2   Formal statements

The minimum expected regret problem on a feasible set X is stated as

$$\min_{\mathbf{x} \in X} G_\zeta(\mathbf{x}). \tag{10.4}$$

Similarly, we formulate the CVaR minimization problem

$$\min_{\mathbf{x} \in X} \phi_\alpha(\mathbf{x}), \tag{10.5}$$

to obtain a decision that minimizes risk with probability level α.

Rockafellar and Uryasev (2000) showed that α-VaR and α-CVaR can be characterized in terms of the function

$$F_\alpha(\mathbf{x}, \zeta) \triangleq \zeta + (1 - \alpha)^{-1} G_\zeta(\mathbf{x}), \tag{10.6}$$

which is convex and continuously differentiable with respect to ζ under considered assumptions. They deduced that α-VaR is a minimizer of function $F_\alpha(\mathbf{x}, \zeta)$ with respect to (w.r.t.) ζ and α-VaR satisfies the equation

$$\zeta_\alpha(\mathbf{x}) = \inf A_\alpha(\mathbf{x}), \tag{10.7}$$

where

$$A_\alpha(\mathbf{x}) \triangleq \operatorname*{Arg\,min}_{\zeta \in \Re} F_\alpha(\mathbf{x}, \zeta) \tag{10.8}$$

is a nonempty, closed, and bounded interval (since $\Psi(\mathbf{x}, \zeta)$ is continuous and nondecreasing with respect to ζ). The fact that α-VaR is a minimizer of function $F_\alpha(\mathbf{x}, \zeta)$ can be directly proved by differentiating the function $F_\alpha(\mathbf{x}, \zeta)$ and equating the derivative to zero. Indeed,

$$\frac{\partial}{\partial \zeta} F_\alpha(\mathbf{x}, \zeta) = 1 + (1 - \alpha)^{-1} \frac{\partial}{\partial \zeta} G_\zeta(\mathbf{x}).$$

However, as it was shown by Rockafellar and Uryasev (2000)

$$\frac{\partial}{\partial \zeta} G(\zeta) = \Psi(\mathbf{x}, \zeta) - 1 .$$

Therefore,

$$\frac{\partial}{\partial \zeta} F_\alpha(\mathbf{x}, \zeta) = 1 + (1 - \alpha)^{-1}(\Psi(\mathbf{x}, \zeta) - 1) = (1 - \alpha)^{-1}(\Psi(\mathbf{x}, \zeta) - \alpha) .$$

The derivative $\frac{\partial}{\partial \zeta} F_\alpha(\mathbf{x}, \zeta)$ equals zero when $\Psi(\mathbf{x}, \zeta) = \alpha$; i.e., α-VaR is a minimizer of the function $F_\alpha(\mathbf{x}, \zeta)$. This implies that α-CVaR of the losses associated with $\mathbf{x} \in X$ can be determined from

$$\phi_\alpha(\mathbf{x}) = \min_{\zeta \in \Re} F_\alpha(\mathbf{x}, \zeta). \tag{10.9}$$

Indeed,

$$\phi_\alpha(\mathbf{x}) = (1 - \alpha)^{-1} \int_{f(\mathbf{x},\mathbf{y}) \geq \zeta_\alpha(\mathbf{x})} f(\mathbf{x}, \mathbf{y}) p(\mathbf{y}) \, d\mathbf{y} = \zeta_\alpha(\mathbf{x}) + (1 - \alpha)^{-1} G_{\zeta_\alpha}(\mathbf{x})$$

$$= F_\alpha(\mathbf{x}, \zeta_\alpha) = \min_{\zeta \in \Re} F_\alpha(\mathbf{x}, \zeta) .$$

This allows us to calculate α-CVaR without having to calculate α-VaR on which its definition depends. Also, Rockafellar and Uryasev (2000) showed that the CVaR minimization problem (10.5) may be solved by minimizing function $F_\alpha(\mathbf{x}, \zeta)$ simultaneously with respect to both arguments,

$$\min_{\mathbf{x} \in X} \phi_\alpha(\mathbf{x}) = \min_{\mathbf{x} \in X} \min_{\zeta \in \Re} F_\alpha(\mathbf{x}, \zeta) = \min_{(\mathbf{x}, \zeta) \in X \times \Re} F_\alpha(\mathbf{x}, \zeta). \tag{10.10}$$

Let us denote by $S_\alpha \overset{\Delta}{=} \text{Arg min}_{(\mathbf{x}, \zeta) \in X \times \Re} F_\alpha(\mathbf{x}, \zeta)$ a solution set of the second optimization problem in (10.10), a solution set of the CVaR optimization problem by $X_\alpha^C \overset{\Delta}{=} \text{Arg min}_{\mathbf{x} \in X} \phi_\alpha(\mathbf{x})$, a solution set of the minimum regret problem by $X_\zeta^R \overset{\Delta}{=} \text{Arg min}_{\mathbf{x} \in X} G_\zeta(\mathbf{x})$, and a projection of S_α on ζ line by $A_\alpha \overset{\Delta}{=} \{\zeta :$ there exist $\mathbf{x}$ such that $(\mathbf{x}, \zeta) \in S_\alpha\}$

In order to explain the algebraic notation, we consider a geometric visualization of the solution sets. For a two-dimensional example ($n = 2$), the relations between the defined sets are illustrated in Figure 1. Let us consider that

$$X \overset{\Delta}{=} \{x_1 + x_2 = 1, \ x_1 \geq 0, \ x_2 \geq 0\}.$$

Figure 1 displays X and the solution set S_α (the shaded region) belonging to $X \times \Re$. It is shown that A_α is a projection of S_α on ζ line, and X_α^C is a projection of S_α on X. Also, it is shown that the solution set X_ζ^R of the minimum regret problem (10.4) for some $\bar{\zeta} \in A_\alpha$ is contained in X_α^C, and $\mathbf{x}^*$ and its associated $A_\alpha(\mathbf{x}^*)$.

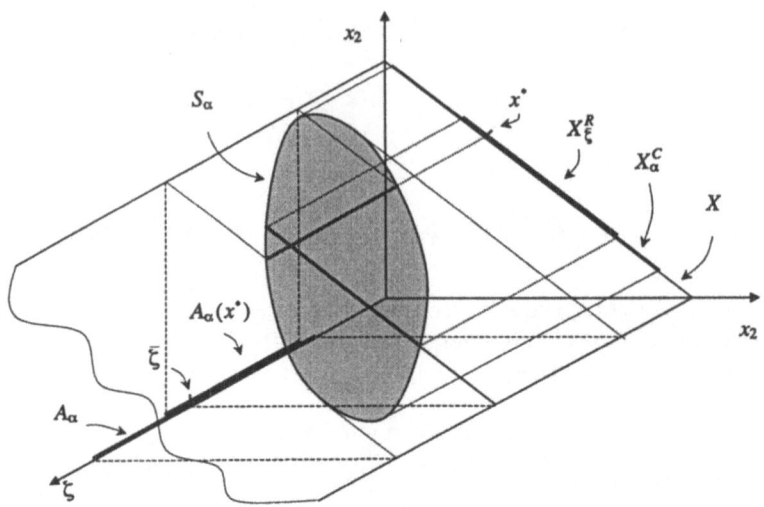

FIGURE 1. Two-dimensional example illustrating solution sets relationship.

Further, we formulate a theorem stating that for each CVaR optimization problem (10.5), there is a regret optimization problem (10.4) having the same set of portfolio solutions.

Theorem 1 (CVaR $\Rightarrow$ Regret). *For any $\alpha \in (0, 1)$ and $\mathbf{x}^* \in X_\alpha^C$ there exists a pair $(\mathbf{x}^*, \zeta^*) \in S_\alpha$ such that $\mathbf{x}^* \in X_{\zeta^*}^R$.*

Proof. Equality (10.10) implies (see Theorem 2 in Rockafellar and Uryasev (2000)) that for any $x^* \in X_\alpha^C$ there exists $(\mathbf{x}^*, \zeta^*) \in S_\alpha$; therefore, we need to prove only that $\mathbf{x}^* \in X_{\zeta^*}^R$. Indeed, if

$$(\mathbf{x}^*, \zeta^*) \in \text{Arg} \min_{(\mathbf{x}, \zeta) \in X \times \Re} F_\alpha(\mathbf{x}, \zeta),$$

then

$$\mathbf{x}^* \in \text{Arg} \min_{\mathbf{x} \in X} F_\alpha(\mathbf{x}, \zeta^*) = \text{Arg} \min_{\mathbf{x} \in X} \{\zeta^* + (1 - \alpha)^{-1} G_{\zeta^*}(\mathbf{x})\}$$

$$= \text{Arg} \min_{\mathbf{x} \in X} G_{\zeta^*}(\mathbf{x}) = X_{\zeta^*}^R.$$

The theorem is proved. $\square$

The next theorem proves an inverse statement to Theorem 1, i.e., for each regret optimization problem (10.4) there exists an equivalent CVaR optimization problem (10.5).

Theorem 2 (Regret $\Rightarrow$ CVaR). *For any $\zeta \in \Re$ and $\mathbf{x}^* \in X_\zeta^R$ there exists a unique $\alpha \in (0, 1)$ such that $\zeta \in A_\alpha(\mathbf{x}^*)$, $(\mathbf{x}^*, \zeta) \in S_\alpha$, and $\mathbf{x}^* \in X_\alpha^C$.*

Proof. Since $\Psi(\mathbf{x}^*, \zeta)$ is continuous with respect to ζ, there exists $\alpha \in (0, 1)$ such that $\Psi(\mathbf{x}^*, \zeta) = \alpha$. The derivative of the function $F_\alpha(\mathbf{x}^*, \zeta)$ with respect to ζ equals (see proof of Theorem 1 in Rockafellar and Uryasev (2000))

$$\frac{\partial}{\partial \zeta} F_\alpha(\mathbf{x}^*, \zeta) = (1 - \alpha)^{-1}(\Psi(\mathbf{x}^*, \zeta) - \alpha) = 0$$

The function $F_\alpha(\mathbf{x}^*, \zeta)$ is convex with respect to ζ (see Theorem 2 in Rockafellar and Uryasev (2000)). Therefore,

$$\zeta \in A_\alpha(\mathbf{x}^*) = \text{Arg} \min_{\tau \in \Re} F_\alpha(\mathbf{x}^*, \tau), \tag{10.11}$$

and the first statement of the theorem is proved.

Also, since $\mathbf{x}^* \in X_\zeta^R$, we obtain

$$\mathbf{x}^* \in \text{Arg} \min_{\mathbf{x} \in X} G_\zeta(\mathbf{x}) = \text{Arg} \min_{\mathbf{x} \in X} \{\zeta + (1 - \alpha)^{-1} G_\zeta(\mathbf{x})\} = \text{Arg} \min_{\mathbf{x} \in X} F_\alpha(\mathbf{x}, \zeta). \tag{10.12}$$

Inclusions (10.11) and (10.12) imply that $(\mathbf{x}^*, \zeta) \in \text{Arg} \min_{(\mathbf{x}, \tau) \in X \times \Re} F_\alpha(\mathbf{x}, \tau) = S_\alpha$, and the second statement of the theorem is proved. Finally, the last statement $\mathbf{x}^* \in X_\alpha^C$ follows from Theorem 2 in Rockafellar and Uryasev (2000). □

The following corollary considers the case when $S_\alpha = X_\alpha^C \times A_\alpha$. For instance, this is valid when the set A_α consists only of one VaR point $\zeta_\alpha(\mathbf{x})$.

Corollary 1. *If in addition to conditions of Theorem 2, $S_\alpha = X_\alpha^C \times A_\alpha$, then $X_\alpha^C = X_\zeta^R$ for any $\zeta^* \in A_\alpha$*

Proof. The statement follows from (10.12). □

10.3   Numerical example

In this section, we illustrate the formal statements with a numerical example demonstrating the equivalence of the expected regret and CVaR approaches. For numerical calculations, we used the implementation framework described by Rockafellar and Uryasev (2000).

We considered a universe of $n = 1792$ listed stocks for the portfolio optimization problem. The decision vector $\mathbf{x}$ consists of stock positions in the portfolio. Components of the vector $\mathbf{y} \in \Re^m$ are random returns of instruments, hence $m = n$. Distribution of the vector $\mathbf{y} \in \Re^m$ is modeled by $s = 156$ historical weekly scenario returns, $\{\mathbf{y}^1, \ldots, \mathbf{y}^s\}$, with equal discrete probability, $1/s$. The feasible set X is a convex polytope given by the following constraints:

$$s^{-1} \sum_{j=1}^{s} \mathbf{x}^T \mathbf{y}^j \geq R \quad \text{(lower bound on expected return)},$$
$$\sum_{i=1}^{n} x_i = 1 \quad \text{(normalization constraint)},$$
$$x_i \geq 0, \quad i = 1, ..., n \text{ (no short positions)}.$$

The loss function for a scenario j is given by $f(\mathbf{x}, \mathbf{y}^j) = \mathbf{x}^T(-\mathbf{y}^j)$, which is the negative return on a portfolio $\mathbf{x}$. Generally, the integrals in (10.1) and (10.6) cannot be obtained analytically; therefore, numerical procedures are used to estimate them. For this case, we use a statistical approximation method with historical scenarios. The corresponding approximations to $G_\zeta(\mathbf{x})$ and $F_\alpha(\mathbf{x}, \zeta)$ are

$$\widetilde{G}_\zeta(\mathbf{x}) \triangleq \frac{1}{s} \sum_{j=1}^{s} [-\mathbf{x}^T \mathbf{y}^j - \zeta]^+,$$

and

$$\widetilde{F}_\alpha(\mathbf{x}, \zeta) \triangleq \zeta + (1 - \alpha)^{-1} \widetilde{G}_\zeta(\mathbf{x}).$$

Then, the problems of minimizing $\widetilde{G}_\zeta(\mathbf{x})$ on X and minimizing $\widetilde{F}_\alpha(\mathbf{x}, \zeta)$ on $X \times \Re$ are convex programming problems, since $f(\mathbf{x}, \mathbf{y}^j)$ is convex with respect to $\mathbf{x}$. Both problems can be reduced to linear programming problems using additional variables.

Distribution characteristics of the instrument returns (for the population of instruments) are shown in Table 1.

TABLE 1: POPULATION OF INSTRUMENTS*

Minimum	Maximum	Mean	Std. Dev.	Skewness	Kurtosis
-0.30684	0.327555	0.003048	0.078163	-0.259866	9.568777

* Distribution characteristics of average instrument returns in 156 weeks (minimum of mean instrument returns, maximum of mean instrument returns, mean of instrument returns, standard deviation of instrument returns, skewness of average instrument returns, and kurtosis of average instrument returns).

For the considered portfolio of stocks, we conducted numerical experiments comparing minimum regret and minimum CVaR approaches. We specified a set of minimum regret problems by making a grid in parameter ζ (fifty values). Further, for value $\bar{\zeta}$ in this grid a sensitivity analysis with respect to α was performed to find a matching CVaR problem with $\min_{\zeta \in \Re} |\zeta - \bar{\zeta}|$. The target weekly return was set to $R = 0.003$. The fifty $\bar{\zeta}$ values produced α ranging approximately between 0.75 and 0.98. The minimum regret and CVaR models were solved using linear programming by previously reformulating the piece-wise linear convex functions $\widetilde{G}_\zeta(\mathbf{x})$ and $\widetilde{F}_\alpha(\mathbf{x}, \zeta)$ into equivalent linear functions with the additional auxiliary variables. The search procedure ($\zeta^* = \arg \min |\zeta - \bar{\zeta}|$) and the linear programming problem were implemented with GAMS, Brooke et al. (1992), and solved with CPLEX's mathematical programming library, ILOG (1997), in a personal computer with a Pentium-II

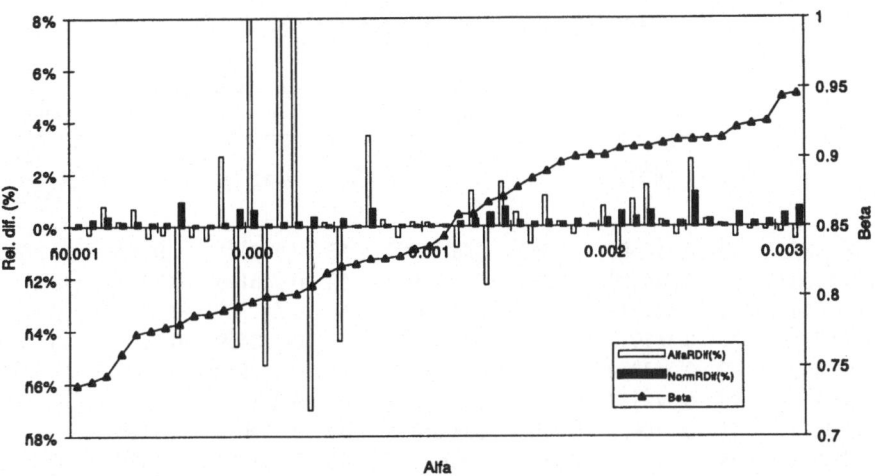

FIGURE 2. Zeta and solution norm relative differences between minimum regret and CVaR for different $\bar{\zeta}$ values and associated α.

300 MHz processor and 128 MB memory. Table 2 depicts a summary of numerical experiments. The numerical results indicated that the expected regret solution can be found quite precisely by solving a minimization problem with respect to α in the CVaR minimization approach. The relative differences of portfolio solution norms between the two considered approaches were less than 1%. Also, there is a close correspondence of ζ^* and $\bar{\zeta}$ values: the relative difference is less than 5% or the absolute difference is less than 0.0002. Figure 2 depicts a summary of numerical comparisons of minimum regret and CVaR approaches.

10.4   Conclusion

This paper demonstrated that the minimum expected regret and the minimum CVaR approaches are closely related. For the case with ℓ_1-norm and constant target value, a portfolio optimal in expected regret sense can be obtained by solving a minimization problem with respect to the confidence parameter in the minimum CVaR approach. Also, the inverse statement is valid, i.e., a portfolio optimal in CVaR sense can be obtained by solving a minimization problem with respect to the target value in the minimum expected regret approach. Numerical experiments confirmed the formal mathematical statements. Also, numerical experiments demonstrated that both approaches, minimum regret and minimum CVaR, can be very efficiently implemented using small computational resources, such as a Pentium-II 300 MHz computer.

TABLE 2: COMPARISON OF THE MINIMUM EXPECTED REGRET AND CVAR
SOLUTIONS.*

Regret				CVaR			Comparison	
$\bar{\zeta}$	Sol. size	$\|x_{\zeta}^{R}\|_2$	α	ζ^*	Sol. size	$\|x_{\alpha}^{C}\|_2$	$\frac{\zeta^*-\bar{\zeta}}{|\bar{\zeta}|}$	$\frac{\|x_{\alpha}^{C}-x_{\zeta}^{R}\|_2}{\|x_{\zeta}^{R}\|_2}$
-0.0011884	54	0.546686	0.7363	-0.0011891	53	0.546485	-0.056%	0.083%
-0.0011033	53	0.545907	0.7393	-0.0011066	52	0.546145	-0.298%	0.212%
-0.0010182	51	0.541216	0.7435	-0.0010102	50	0.541307	0.780%	0.327%
-0.0009330	49	0.541990	0.7589	-0.0009313	49	0.542439	0.182%	0.121%
-0.0008479	47	0.543070	0.7728	-0.0008423	46	0.542642	0.664%	0.147%
-0.0007628	48	0.542182	0.7754	-0.0007661	47	0.542092	-0.433%	0.107%
-0.0006777	48	0.540022	0.7781	-0.0006798	47	0.539613	-0.310%	0.110%
-0.0005925	49	0.549551	0.7803	-0.0006175	48	0.548881	-4.206%	0.881%
-0.0005074	47	0.551212	0.7864	-0.0005093	46	0.551060	-0.365%	0.047%
-0.0004223	47	0.555201	0.7871	-0.0004244	47	0.555241	-0.512%	0.032%
-0.0003372	48	0.556890	0.7901	-0.0003281	48	0.557170	2.702%	0.118%
-0.0002520	48	0.560760	0.7930	-0.0002636	47	0.559461	-4.579%	0.630%
-0.0001669	49	0.565039	0.7958	-0.0001496	49	0.566291	10.34%	0.599%
-0.0000818	49	0.569552	0.7996	-0.0000861	49	0.569453	-5.282%	0.062%
0.0000033	49	0.570694	0.7999	0.0000104	48	0.570732	211.0%	0.130%
0.0000885	49	0.573080	0.8013	0.0000956	48	0.573292	8.015%	0.137%
0.0001736	47	0.571145	0.8073	0.0001614	46	0.571929	-6.994%	0.332%
0.0002587	47	0.570329	0.8164	0.0002591	46	0.570296	0.162%	0.037%
0.0003438	46	0.570130	0.8213	0.0003288	46	0.570573	-4.380%	0.269%
0.0004290	46	0.568194	0.8230	0.0004290	46	0.568193	0.019%	0.002%
0.0005141	45	0.572725	0.8267	0.0005320	45	0.574391	3.492%	0.646%
0.0005992	46	0.572726	0.8268	0.0006008	46	0.572689	0.262%	0.042%
0.0006843	48	0.566529	0.8286	0.0006815	48	0.566545	-0.412%	0.045%
0.0007695	46	0.566841	0.8335	0.0007707	45	0.566861	0.168%	0.023%
0.0008546	44	0.567263	0.8358	0.0008558	43	0.567181	0.138%	0.026%
0.0009397	43	0.572998	0.8430	0.0009403	43	0.573061	0.060%	0.023%
0.0010248	43	0.569597	0.8584	0.0010167	42	0.569538	-0.796%	0.139%
0.0011100	44	0.570064	0.8586	0.0011250	44	0.570147	1.359%	0.245%
0.0011951	43	0.573454	0.8672	0.0011681	43	0.572051	-2.256%	0.469%
0.0012802	43	0.570857	0.8713	0.0013021	43	0.568413	1.708%	0.695%
0.0013653	44	0.566464	0.8784	0.0013723	44	0.566491	0.513%	0.181%
0.0014504	42	0.566328	0.8846	0.0014407	42	0.566368	-0.670%	0.109%
0.0015356	45	0.566256	0.8898	0.0015535	45	0.566520	1.170%	0.194%
0.0016207	43	0.563752	0.8965	0.0016232	42	0.563500	0.155%	0.124%
0.0017058	42	0.555593	0.9004	0.0017006	41	0.556432	-0.307%	0.212%
0.0017909	41	0.547912	0.9014	0.0017907	41	0.547814	-0.014%	0.037%
0.0018761	42	0.544711	0.9017	0.0018901	41	0.544074	0.750%	0.283%
0.0019612	42	0.536406	0.9063	0.0019404	41	0.536946	-1.060%	0.532%
0.0020463	42	0.532997	0.9074	0.0020672	42	0.532065	1.021%	0.342%
0.0021314	42	0.529311	0.9076	0.0021650	41	0.527889	1.576%	0.560%
0.0022166	41	0.525621	0.9101	0.0022211	41	0.525570	0.205%	0.131%

0.0023017	40	0.520152	0.9127	0.0022944	39	0.520477	-0.318%	0.149%
0.0023868	40	0.516449	0.9127	0.0024482	40	0.513859	2.571%	1.255%
0.0024719	41	0.515979	0.9134	0.0024780	41	0.516526	0.245%	0.249%
0.0025571	39	0.509568	0.9141	0.0025592	38	0.509469	0.083%	0.042%
0.0026422	38	0.511696	0.9214	0.0026316	37	0.511280	-0.402%	0.479%
0.0027273	37	0.514012	0.9238	0.0027237	37	0.514044	-0.132%	0.135%
0.0028124	37	0.519918	0.9258	0.0028086	36	0.519317	-0.135%	0.198%
0.0028976	38	0.514950	0.9434	0.0028908	37	0.514028	-0.232%	0.436%
0.0029827	38	0.515775	0.9453	0.0029683	37	0.517958	-0.483%	0.691%
0.0030678	38	0.503796	0.9688	0.0030676	37	0.503834	-0.005%	0.023%

* Regret heading: $\bar{\zeta}$ value, solution size (number of non-zero components in the solution), solution norm; CVaR heading: α value, ζ solution value, solution size, and solution norm; Comparison heading: relative zeta difference and relative solution norm difference.

References

[1] Acerbi, C., C. Nordio , and C. Sirtori (2001): *Expected shortfall as a tool for financial risk management.* Working paper, http://www.gloriamundi.org.

[2] Acerbi, C., and D. Tasche (2001): *On the coherence of expected shortfall.* Working paper, http://www.gloriamundi.org.

[3] Andersson, F., H. Mausser, D. Rosen, and S. Uryasev (2001): Credit Risk Optimization with Conditional Value-at-Risk Criterion. *Mathematical Programming*, Series B 89, 2001, 273–291.

[4] Artzner, P., F. Delbaen, J.M. Eber, and D. Heat (1997): Thinking Coherently, *Risk*, Vol. 10, 68–71.

[5] Bogentoft E., H.E. Romeijn, and S. Uryasev (2001): Asset/Liability Management for Pension Funds Using CVaR Constraints. *The Journal of Risk Finance.* Vol. 3, No. 1, 57–71.

[6] Brooke, A., D. Kendrick, A. Meeraus, and R. Rosenthal (1992): GAMS, *A User's Guide.* *Redwood City*, CA: Scientific Press.

[7] Cariño, D.R. and W.T. Ziemba (1998): Formulation of the Russell- Yasuda Kasai Financial Planning Model, *Operations Research.* Vol. 46, No. 4, 443–449.

[8] Dembo, R.S. and A.J. King (1992): Tracking Models and the Optimal Regret Distribution in Asset Location, *Applied Stochastic Models and Data Analysis*, Vol. 8, 151–157.

[9] Dembo, R.S. and D. Rosen (1999): The Practice of Portfolio Replication: A Practical Overview of Forward and Inverse Problems. *Annals of Operations Research.* Vol. 85, 267–284.

[10] Embrechts, P. (1999): Extreme Value Theory as a Risk Management Tool, *North American Actuarial Journal*, vol. 3.

[11] ILOG (1997): CPLEX, 6.0 ed. Mountain View, CA.

[12] Jobst, N.J. and S.A. Zenios (2001): The Tail That Wags The Dog: Integrating Credit Risk In Asset Portfolios. *The Journal of Risk Finance.*, Vol. 3, No. 1, 31–44.

[13] Harlow, W.V. (1991): Asser Allocation in a Downside-Risk Framework. *Fin. Anal. J.*, Vol. 47, No. 5, 28–40.

[14] Hurliman, W. (2001): Conditional Value-at-Risk bounds for Compound Poisson Risks and Normal Approximation. *MPS: Applied mathematics/0201009*, Working Paper, www.mathpreprints.com/math/Preprint/werner.huerlimann/ 20020111/1/.

[15] Konno, H. and H. Yamazaki (1991): Mean Absolute Deviation Portfolio Optimization Model and Its Application to Tokyo Stock Market. *Management Science,* 37, 519–531.

[16] Krokhmal, P., J. Palmquist, and S. Uryasev (2002): Portfolio Optimization with Conditional Value-at-Risk Objective and Constraints. *The Journal of Risk*, Vol. 4, No. 2.

[17] Markowitz, H.M. (1952): Portfolio Selection. *Journal of Finance*. Vol. 7, No. 1, 77–91.

[18] Krokhmal. P., J. Palmquist, and S. Uryasev (2002): Portfolio Optimization with Conditional Value-At-Risk Objective and Constraints. *The Journal of Risk*, Vol. 4, No. 2, 2002.

[19] Pflug, G. Ch. (2000): Some Remarks on the Value-at-Risk and the Conditional Value-at-Risk. In. Uryasev S. (Ed.) *Probabilistic Constrained Optimization: Methodology and Applications*, Kluwer Academic Publishers.

[20] Prekopa, A. (1995): *Stochastic Programming*, Kluwer Academic Publishers.

[21] Rockafellar, R.T. and S. Uryasev (2000): Optimization of Conditional Value-at-Risk. *The Journal of Risk*, Vol. 2, No. 3.

[22] Rockafellar R.T. and S. Uryasev (2002): Conditional Value-at-Risk for General Loss Distributions. *The Journal of Banking and Finance*, Vol. 26, No. 7, 1443–1471.

[23] Uryasev, S. (1995). Derivatives of Probability Functions and Some Applications. *Annals of Operations Research*, Vol. 56, 287–311.

[24] Young, M.R. (1998): A Minimax Portfolio Selection Rule with Linear Programming Solution. *Management Science*. Vol. 44, No. 5, 673–683.

[25] Ziemba, W.T. and J.M. Mulvey (Eds.) (1998): *Worldwide Asset and Liability Modeling*, Cambridge Univ. Press.

[26] Zenios, S.A. (Eds.) (1996): *Financial Optimization*, Cambridge University Press, Cambridge, U.K.

11

Estimation, Adjustment and Application of Transition Matrices in Credit Risk Models

Stefan Trück

Emrah Özturkmen

ABSTRACT   The paper gives a survey on recent developments on the use of numerical methods in rating based Credit Risk Models. Generally such models use transition matrices to describe probabilities from moving from one rating state to the other and to calculate Value-at-Risk figures for portfolios. We show how numerical methods can be used to find so-called true generator matrices in the continuous-time approach, adjust transition matrices or estimate confidence bounds for default and transition probabilities.

11.1   Introduction

Borrowing and lending money have been two of the oldest financial transactions. They are the core of the modern world's sophisticated economy, providing funds to corporations and income to households. Nowadays, corporations and sovereign entities borrow either on the financial markets via bonds and bond-type products or they directly borrow money at financial institutions such as banks and savings associations. The lenders, e.g., banks, private investors, insurance companies or fund managers are faced with the risk that they might lose a portion or even all of their money in case the borrower cannot meet the promised payment requirements.

In recent years, to manage and evaluate Credit Risk for a portfolio especially, so-called rating based systems have gained more and more popularity. These systems use the rating of a company as the decisive variable but not — like the formerly used so-called structural models of the value of the firm — when it comes to evaluating the default risk of a bond or loan. The popularity is due to the straightforwardness of the approach but also to the upcoming "New Capital Accord" of the Basel Committee on Banking Supervision, a regulatory body under the Bank of International Settlements, publicly known as Basel II. Basel II allows banks to base their capital requirement on internal as well as external rating systems. Thus, sophisticated credit risk models are being developed or demanded by banks to assess the risk of their credit portfolio better by recognizing the different underlying sources of risk. Default probabilities for certain rating categories, as well as the probabilities for moving from one rating state to another, are important issues in such Credit Risk Models. We will start with a brief description of the main ideas of rating based Credit Risk Models and then give a survey of numerical methods that can be applied when it comes to estimating continous time transition matrices or adjusting transition probabilities.

11.2   Rating Based Credit Risk Models

11.2.1   Reduced Form Models — the beginning

In 1994, Jerome S. Fons, was the first to develop a so-called reduced form model and to derive credit spreads using historical default rates and recovery rate estimates. Before that most models were based on the so-called structural approach developed by Merton (1974) calculating default probabilities and credit spreads using the value of the company as the decisive variable. However, in the model developed by Fons, in calculating the price of credit risk the decisive variables were the rating of a company and historical default probabilities for rating classes and not the value of the firm. Although his approach is rather simple in relation to other theoretical models of credit risk, the predicted credit spreads derived by Fons showed strong similarity towards real market data.

However, not only the 'worst case' event of default has influence on the price of a bond, but also a change in the rating of a company or an issued bond. Therefore, refining Fons' model, Jarrow, Lando, Turnbull (JLT)[1] introduced a discrete-time Markovian Model to estimate changes in the price of loans and bonds. They incorporate possible rating upgrade, stable rating and rating downgrade (with default as a special event) in the reduced form approach. To determine the price of credit risk, both historical default rates and a transition matrix (an exemplary historical transition matrix is shown in Table 1) is used.

11.2.2   Rating Models — the JLT model

Deterioration or improvement in the credit quality of the issuer is highly important for example if someone wants to calculate VaR figures for a portfolio or evaluate credit derivatives like credit spread options whose payouts depend on the yield spreads that are influenced by such changes. One common way to express these changes in the credit quality of market participants is to consider the ratings given by agencies like Standard & Poor and Moody's. Downgrades or upgrades by the rating

---

[1] JLT

TABLE 1. An example for Moody's one-year transition matrix

Initial Rating	Rating at year-end (%)							
	Aaa	Aa	A	Baa	Ba	B	Caa	Default
Aaa	93.40	5.94	0.64	0	0.02	0	0	0
Aa	1.61	90.55	7.64	0.26	0.09	0.01	0	0.02
A	0.07	2.28	92.44	4.63	0.45	0.12	0.01	0
Baa	0.05	0.26	5.51	88.48	4.76	0.71	0.08	0.15
Ba	0.02	0.05	0.42	5.16	86.91	5.91	0.24	1.29
B	0	0.04	0.13	0.54	6.35	84.22	1.91	6.81
Caa	0	0	0	0.62	2.05	4.08	69.20	24.06

agencies are taken very seriously by market players to price bonds and loans, thus effecting the risk premium and the yield spreads.

Jarrow/Lando/Turnbull (JLT) in 1997 constructed a model that considers different credit classes characterized by their ratings and allows moving within these classes. In this section, we will describe the basic idea of the the JLT model and also some of the extensions by Lando (1998).

To ease the understanding and interpret the outcomes better, we first focus on the theoretical framework in the discrete time before going to the continuous time case.

11.2.3 The Discrete Time Case

The JLT models default and transition probabilities by using a discrete time, time-homogenous Markov chain on a finite state space S={1,......,K}. The state space S represents the different rating classes. While 1 is for the best credit rating, K represents the default case. Hence, the (KxK) one-period transition matrix is:

$$P = \begin{pmatrix} p_{11} & p_{12} & \cdots & p_{1K} \\ p_{21} & p_{12} & \cdots & p_{2K} \\ \cdots & \cdots & \cdots\cdots \\ p_{K-1,1} & p_{K-1,2} & \cdots & p_{K-1,K} \\ 0 & 0 & \cdots & 1 \end{pmatrix} \quad (11.1)$$

where $p_{ij} \geq 0$ for all i, j, i $\neq$ j, and $p_{ii} \equiv 1 - \sum_{\substack{j=1 \\ j\neq i}}^{K} p_{ij}$ for all i. p_{ij} represents the actual probability of going to state j from state i in one time step.

JLT makes the following key assumptions underlying their model:[2]

- The interest rates and the default process are independent under the martingale measure Q.

[2] The assumptions of complete markets with no arbitrage opportunities, thus existence and uniqueness of the martingale measure $\tilde{Q}$, is a key assumption of all intensity models that will not be mentioned further during the rest of the chapter.

- The transition matrix is time-homogenous, i.e., we use the same matrix for each point in time.
- The default state is an absorbing state, represented by $p_{Ki} = 0$ for $i = 1, \ldots, K - 1$ and $p_{KK} = 1$.

The multi-period transition matrix is:

$$P_{0,n} = P^n.$$

Under the martingale measure the one-period transition matrix is:

$$\tilde{Q}_{t,t+1} = \begin{pmatrix} \tilde{q}_{11}(t, t+1) & \tilde{q}_{12}(t, t+1) & \cdots \tilde{q}_{1K}(t, t+1) \\ \tilde{q}_{21}(t, t+1) & \tilde{q}_{22}(t, t+1) & \cdots \tilde{q}_{2K}(t, t+1) \\ \cdots & \cdots & \cdots \cdots \\ \tilde{q}_{K-1,1}(t, t+1) & \tilde{q}_{K-1,2}(t, t+1) & \cdots \tilde{q}_{K-1,K}(t, t+1) \\ 0 & 0 & \cdots 1 \end{pmatrix} \quad (11.2)$$

where $\tilde{q}_{ij}(t, t+1) \geq 0$, for all i,j, $i \neq j$, $\tilde{q}_{ij}(t, t+1) \equiv 1 - \sum_{\substack{j=1 \\ j \neq i}}^{K} \tilde{q}_{ij}$, and $\tilde{q}_{ij}(t, t+1) > 0$ if and only if $q_{ij} > 0$ for $0 \leq t \leq \tau - 1$.

JLT argue that, without additional restrictions, the martingale probabilities $\tilde{q}_{ij}(t, t + 1)$ can depend on the entire history up to now, i.e., the Markov property is not satisfied anymore. Therefore, they assume that the martingale probabilities $\tilde{q}_{ij}(t, t + 1)$ satisfy the equation

$$\tilde{q}_{ij}(t, t+1) = \pi_i(t) p_{ij} \quad (11.3)$$

for all i,j, $i \neq j$ where $\pi_i(t)$ is a deterministic function of time.

In other words, given a current state i, the martingale probabilities to move from state i to j are proportional to the actual probabilities and the proportionality factor depends on time and the current state i, but not on the next state j. JLT call this factor the risk premium.

We can write the equation (3.16) also in matrix form:

$$\tilde{Q}_{t,t+1} - \mathcal{I} = \Pi(t)[P - \mathcal{I}] \quad (11.4)$$

where $\mathcal{I}$ is the (KxK) identity matrix, $\Pi(t) = \text{diag}(\pi_1(t), \ldots, \pi_{K-1}(t), 1)$ is a (KxK) diagonal matrix. Given the time dependence of the risk premium, one gets a time-inhomogenous Markov chain under the martingale measure.

Often in practice market prices of defaultable claims do not reflect historical default or transition probabilities. We will see later how with the help of numerical methods, risk premiums can be calculated so historical transition matrices can be transformed into e.g., a one-year transition matrix under the martingale measure.

But first, having outlined the basic ideas of the JLT model, we will now describe the continuous-time case. We will see later that using the continuous approach to calculate transition and default probabilities has some advantages.

11.2.4 The Continuous-Time Case

A continuous-time, time-homogenous Markov chain is specified via the (KxK) generator matrix:

$$\Lambda = \begin{pmatrix} \lambda_{11} & \lambda_{12} & \cdots \lambda_{1K} \\ \lambda_{21} & \lambda_{22} & \cdots \lambda_{2K} \\ \cdots & \cdots & \cdots\cdots \\ \lambda_{K-1,1} & \lambda_{K-1,2} & \cdots \lambda_{K-1,K} \\ 0 & 0 & \cdots 0 \end{pmatrix} \tag{11.5}$$

where $\lambda_{ij} \geq 0$, for all i, j and $\lambda_{ii} = -\sum_{\substack{j=1 \\ j\neq i}}^{K} \lambda_{ij}$, for i = 1,.....,K. The off-diagonal elements represent the intensities of jumping to rating j from rating i. The default state K is an absorbing one.

The (KxK) t-period transition matrix under the actual probabilities is then given by

$$P(t) = e^{t\Lambda} = \sum_{k=0}^{\infty} \frac{(t\Lambda)^k}{k!} = \mathcal{I} + (t\Lambda) + \frac{(t\Lambda)^2}{2!} + \frac{(t\Lambda^3)}{3!} + \cdots . \tag{11.6}$$

For example consider the transition matrix:

P	A	B	D
A	0.90	0.08	0.02
B	0.1	0.80	0.1
D	0	0	1

Then the corresponding generator matrix is

Λ	A	B	D
A	-0.1107	0.0946	0.0162
B	0.1182	-0.2289	0.1107
C	0	0	0

Similar to their discrete-time framework, they transform the empirical generator matrix to a risk-neutral generator matrix by multiplying it with the risk premium matrix, i.e., adjustments of risk to transform the actual probabilities into the risk-neutral probabilities for valuation purposes:

$$\tilde{\Lambda}(t) \equiv \mathcal{U}(t)\Lambda(t) \tag{11.7}$$

where $U(t) = \text{diag}(\mu_1(t), \ldots , \mu_{K-1}(t), 1)$ is a $(K \times K)$ diagonal matrix whose first $K - 1$ entries are strictly positive deterministic functions of t.

Thus, JLT define a methodology to value risky bonds as well as credit derivatives based on ratings allowing changes in credit quality before default. In the following section we will give a brief outline of the advantages of continuous modeling versus discrete time models.

11.2.5   Continuous vs. discrete time modeling

Lando/Skodeberg (2000) and Lando/Christensen (2002) focus in their papers on the advantages of the continuous-time modeling over the discrete-time approach used by rating agencies to analyze rating transition data. Generally, transition probabilities are estimated by rating agencies using the multinomial method by computing

$$\hat{p}_{ij} = \frac{N_{ij}}{N_i} \tag{11.8}$$

for $j \neq i$ and where N_i is the number of firms in rating class i at the beginning of the year and N_{ij} is the number of firms that migrated to rating class j.

The authors argue that these transition probabilities do not capture rare events such as a transition from AAA to default as they may not be observed. However, it is possible that a firm reaches default through subsequent downgrades from AAA — even within one year, i.e., the probability of moving from AAA to default must be non-zero. Therefore, they apply the continuous-time method by first estimating the generator matrix via the maximum-likelihood estimator:[3]

$$\hat{\lambda}_{ij} = \frac{N_{ij}(T)}{\int_0^T Y_i(s)ds} \tag{11.9}$$

where $Y_i(s)$ is the number of firms in rating class i at time s and $N_{ij}(T)$ is the total number of transitions over the period from i to j, where $i \neq j$. Under the assumption of time-homogeneity, the transition matrix can be computed by the formula $P(t) = e^{t\Lambda}$.

Consider the following example taken from Lando/Skodeberg. There are three rating classes A, B and D for default. Assume that at the beginning of the year there are 10 firms in A and 10 in B and none in default. Suppose that one A-rated company is downgraded to B after one month and stays there for the rest of the year; a B-rated company is upgraded to A after two months to remain there for the rest of the period and a B-rated company defaults after six months. The discrete-time multinomial method then estimates the following one-year transition matrix:

P	A	B	D
A	0.90	0.10	0
B	0.1	0.80	0.1
D	0	0	1

The maximum-likelihood estimator for the continuous-time approach computes the following generator matrix first:

$\hat{\Lambda}$	A	B	D
A	-0.10084	0.10084	0
B	0.10909	-0.21818	0.10909
D	0	0	0

---

[3] See Küchler and Sorensen [1997] for details.

For instance, the non-diagonal element λ_{AB} is:

$$\hat{\lambda}_{AB} = \frac{N_{AB}(1)}{\int_0^1 Y_A(s)ds} = \frac{1}{9 + \frac{1}{12} + \frac{10}{12}} = 0.10084$$

The diagonal element $\hat{\lambda}_{AA}$ is computed so that the row sums to zero. Exponentiating the generator gives the following one-year transition matrix:

P	A	B	D
A	0.90887	0.08618	0.00495
B	0.09323	0.80858	0.09819
D	0	0	1

Thus, we have a strictly positive default probability for class A although there have been no observations from A to default in this period. However, this makes sense as the migration probability from A to B and from B to default are non-zero. Thus, it could happen within one year that a company is downgraded from rating state A to B and then defaults.

Before we go to the time-inhomogenous case let us briefly summarize the key advantages of the continuous-time approach:[4]

- We can get non-zero estimates for probabilities of rare events which the multinomial method estimates to be zero.
- We can obtain transition matrices for arbitrary time horizons.
- We have a good method to assess the proposal of Basel-II to impose a minimum probability of 0.03% for events that are estimated with a binomial or multinomial method to have a zero probability.
- In the continous-time approach we do not have to worry which yearly periods we consider. Using a discrete-time approach may lead to quite different results depending on the starting point of our consideration.
- The continuous framework permits of generating confidence sets as discussed in later sections.
- The dependence on covariates can be tested and business cycles effects can be quantified.

For the time-inhomogeneous case, the Nelson–Aalen estimator is used:

$$\hat{\lambda}_{hj}(t) = \sum_{\{k:T_{hjk} \leq t\}} \frac{1}{Y_h(T_{hjk})}$$

where $T_{hj1} < T_{hj2} < \cdots$ are the observed times of transitions from h to j and $Y_h(t)$ counts the number of firms in rating class h at time just prior to t. The transition matrix is then computed as follows:

---

[4] See Lando/Skodeberg [2000] p.2ff. and Lando/Christensen [2002] p.2ff.

$$\hat{P}(s,t) = \prod_{i=1}^{m}(\mathcal{I} + \Delta\hat{\Lambda}(T)) \tag{11.10}$$

where $\mathcal{I}$ is the identity matrix, T_i is a jump time in the interval $]s,t]$ and

$$\Delta\hat{\Lambda}(T) = \begin{pmatrix} -\frac{\Delta N_1(T_i)}{Y_1(T_i)} & -\frac{\Delta N_{12}(T_i)}{Y_1(T_i)} & \cdots & -\frac{\Delta N_{1K}(T_i)}{Y_1(T_i)} \\ -\frac{\Delta N_{21}(T_i)}{Y_2(T_i)} & -\frac{\Delta N_2(T_i)}{Y_2(T_i)} & \cdots & -\frac{\Delta N_{2K}(T_i)}{Y_2(T_i)} \\ \cdots & \cdots & \cdots\cdots \\ -\frac{\Delta N_{K-1,1}(T_i)}{Y_{K-1}(T_i)} & -\frac{\Delta N_{K-1,2}(T_i)}{Y_{K-1}(T_i)} & \cdots & -\frac{\Delta N_{K-1,K}(T_i)}{Y_{K-1}(T_i)} \\ 0 & 0 & \cdots 0 \end{pmatrix}$$

The diagonal elements count the total number of transitions away from class i divided by the number of exposed firms and the off-diagonal elements count the number of migrations to the corresponding class, again divided by the number of firms exposed. This estimator can be interpreted as a cohort method applied to very short intervals.

Implementing this approach to our example we obtain the following matrices:

$$\Delta\Lambda(T_{\frac{1}{12}}) = \begin{pmatrix} -0.1 & 0.1 & 0 \\ 0 & 0 & 0 \\ 0 & 0 & 0 \end{pmatrix}$$

$$\Delta\Lambda(T_{\frac{2}{12}}) = \begin{pmatrix} 0 & 0 & 0 \\ \frac{1}{11} & -\frac{1}{11} & 0 \\ 0 & 0 & 0 \end{pmatrix}$$

$$\Delta\Lambda(T_{\frac{6}{12}}) = \begin{pmatrix} 0 & 0 & 0 \\ 0 & -0.1 & 0.1 \\ 0 & 0 & 0 \end{pmatrix}$$

Using equation 11.10 we get

$$\hat{P}(0,1) = \begin{pmatrix} 0.90909 & 0.08181 & 0.00909 \\ 0.09091 & 0.81818 & 0.09091 \\ 0 & 0 & 1 \end{pmatrix}$$

As in the time-homogeneous case we obtain strictly positive default probability for class A although the entries in the matrix above are slightly different than in the time-homogenous case. However, Lando and Skodeberg indicate that in most cases the results are not dramatically different for large data sets. However, compared to the estimator for the transition matrix using the discrete multinomial estimator we get quite different results.

11.3 Finding Generator Matrices for Markov chains

So far we described the basic ideas of rating based credit risk evaluation methods and the advantages of continuous-time transition modeling over the discrete-time case.

Despite these advantages of continuous modeling, there are also some problems to deal with like the existence, uniqueness or adjustment of the generator matrix to the corresponding discrete transition matrix.

An important issue when dealing with the continuous-time approach is whether for a given discrete one-year transition matrix a so-called true generator exists. For some discrete transition matrices there does not exist a generator matrix, while for some though there exists a generator, it has negative off-diagonal elements. This would mean that, considering short time intervals, transition probabilities may be negative, which is from a practical point of view not acceptable. Examining this question and suggesting numerical methods for finding true generators or approximations for true generators, we will follow an approach by Israel/Rosenthal/Wei (2000).

In their paper they first identify conditions under which a true generator does or does not exist. Then they provide a numerical method for finding the true generator once its existence is proved and how to obtain an approximate one in case of the absence of a true generator.

The authors define two issues with transition matrices: *Embeddability,* which is to determine if an empirical transition matrix is compatible with a true generator or a Markov process; and *identification,* which is to search for the true generator once its existence is known.

11.3.1 Computing the Generator Matrix to a given Transition Matrix

Given the one-year $N \times N$ transition matrix P we are interested in finding a generator matrix Λ such that:

$$P = e^{\Lambda} = \sum_{k=0}^{\infty} \frac{\Lambda^k}{k!} = \mathcal{I} + \Lambda + \frac{\Lambda^2}{2!} + \frac{\Lambda^3}{3!} + \cdots . \tag{11.11}$$

Dealing with the question whether there exists a generator matrix for a given transition matrix P, the first task is to calculate S with

$$S = max\{(a - 1)^2 + b^2; a + bi \text{ is an eigenvalue of } P, \ a, b \in R\}$$

where all the (possibly complex) eigenvalues of P are examined by computing the absolute square of the eigenvalue minus 1 and taking the maximum of these.

It can be shown that if the condition $S < 1$ holds,

$$\Lambda^* = (P - \mathcal{I}) - \frac{(P - \mathcal{I})^2}{2} + \frac{(P - \mathcal{I})^3}{3} - \frac{(P - \mathcal{I})^4}{4} + \cdots \tag{11.12}$$

converges geometrically quickly and is an $N \times N$ matrix having row-sums of zero and satisfying $P = e^{\Lambda^*}$ exactly.[5]

It is important to note that even if the series Λ^* does not converge or converges to a matrix that cannot be a true generator, P may still have a true generator.

There is also a simpler way to check if $S < 1$, namely if the transition matrix consists of diagonal elements that are greater than 0.5. Then S is less than 1 and its convergence guaranteed. For most transition matrices in practice this will be true, so we can assume that generators having row-sums of zero and satisfying $P = e^{\Lambda^*}$ can be found.

However, often there remains one problem: The main disadvantage of series (11.12) is that Λ^* may converge but does not have to be a generator matrix, particularly it is possible that some off-diagonal elements are negative.

We will illustrate this with an example. Consider the one-year transition matrix:

P	A	B	C	D
A	0.9	0.08	0.0199	0.0001
B	0.050	0.850	0.090	0.010
C	0.010	0.090	0.800	0.100
D	0	0	0	1

Calculating the generator that exactly matches $P = e^{\Lambda^*}$ we get with (11.12)

Λ	A	B	C	D
A	-0.1080	0.0907	0.0185	-0.0013
B	0.0569	-0.1710	0.1091	0.0051
C	0.0087	0.1092	-0.2293	0.1114
D	0	0	0	0

having a negative entry in λ_{AD}.

From the economic viewpoint this is not acceptable because a negative entry in the generator may lead for very short time intervals to negative transition probabilities. Israel/Rosenthal/Wei (IRW) show that it is possible that sometimes there exist more than one generator. They provide conditions for the existence or non-existence of a valid generator matrix and a numerical algorithm for finding a valid generator.

11.3.2   Conditions for the Existence of a valid Generator

Conditions for the non-existence of a generator

We will first define the conditions to conclude the non-existence of a generator. If

- $det(P) \leq 0$; or
- $det(P) > \prod_i p_{ii}$; or
- there are states i and j such that j is accessible from i, but $p_{ij} = 0$,

then there is not a generator matrix for P.

---

[5] For proof see Appendix 1 on p.23 in Israel et al.

Conditions for the uniqueness of a generator

Beside the problem of the existence of the generator matrix, there is also the problem of its uniqueness, as P sometimes has more than one valid generator. As it is unlikely for a firm to migrate to a rating "far" from its current rating, the authors suggest to choose among valid generators the one with the lowest value of

$$J = \sum_{i,j} |j - i||\lambda_{ij}|$$

which ensures that the chance of jumping too far is minimized.

A theorem for the uniqueness of the generator is as follows:

- If $det(P) > 0.5$, then P has at most one generator.
- If $det(P) > 0.5$ and $||P - \mathcal{I}|| < 0.5$, then the only possible generator for P is $ln(P)$.
- If P has distinct eigenvalues and $det(P) > e^{-\pi}$, then the only possible generator is $log(P)$.

For the proofs of the above conditions and further material we refer to the original article.

Conditions for the non-existence of a valid generator

Investigating the existence or non-existence of a valid generator matrix with only positive off-diagonal elements, we start with another result obtained by Singer and Spilerman (1976):

Let P be a transition matrix that has real distinct eigenvalues.

- If all eigenvalues of P are positive, then $log(P)$ is the only real matrix Λ such that $exp(\Lambda) = P$.
- If P has any negative eigenvalues, then there is no real matrix Λ such that $exp(\Lambda) = P$.

Using the conditions above we can conclude a condition for the non-existence of a valid generator.

Let P be a transition matrix such that at least one of the following three conditions hold:

- $det(P) > 1/2$ and $|P - I| < 1/2$ or
- P has distinct eigenvalues and $det(P) > e^{-\pi}$ or
- P has distinct real eigenvalues.

Suppose further that the series (11.12) converges to a matrix Λ with negative off-diagonal entries. Then there does not exist a valid generator for P.

In our example we find for

$$P = \begin{pmatrix} 0.9 & 0.08 & 0.0199 & 0.0001 \\ 0.050 & 0.850 & 0.090 & 0.010 \\ 0.010 & 0.090 & 0.800 & 0.100 \\ 0 & 0 & 0 & 1 \end{pmatrix}$$

- $det(P) = 0.6015$ and $|P - I| = 0 \leq 1/2$,
- P has the distinct positive eigenvalues 0.9702, 0.8529, 0.7269 and 1.0000 and $det(P) = 0.6015 > 0.0432 = e^{-\pi}$,
- P has distinct real eigenvalues.

So all three conditions hold — however, to show that one of the conditions holds would have been enough — and since the series (11.12) converges to

$$\Lambda = \begin{pmatrix} -0.1080 & 0.0907 & 0.0185 & -0.0013 \\ 0.0569 & -0.1710 & 0.1091 & 0.0051 \\ 0.0087 & 0.1092 & -0.2293 & 0.1114 \\ 0 & 0 & 0 & 0 \end{pmatrix}$$

we find that there exists no true generator.

An algorithm to search for a valid generator

Israel et al. also provide a search algorithm for a valid generator if the series (11.12) fails to converge, or converges to a matrix that has some negative off-diagonal terms while none of the three conditions above holds. We stated already that it is still possible that a generator exists even if (11.12) fails to converge.

Israel et al. use Lagrange interpolation in their search algorithm. They assume that P is $n \times n$ with distinct eigenvalues $e_1, e_2, ..., e_n$ and that there exists an arbitrary function f analytic in a neighborhood of each eigenvalue with $f(P) = g(P)$. Here g is the polynomial of degree $n - 1$ such that $g(e_j) = f(e_j)$ for each j. Using the Lagrange interpolation formula leads to

$$g(x) = \sum_j \prod_{k \neq j} \frac{x - e_k}{e_j - e_k} f(e_j). \tag{11.13}$$

Obviously the product is over all eigenvalues e_k except e_j and the sum is over all eigenvalues e_j.

To search for generators, the values $f(e_j)$ must satisfy

$$f(e_j) = log|e_j| + i(arg(e_j) + 2\pi k_j \tag{11.14}$$

where k_j is an arbitrary integer and e_j may be a complex number and $|e_j| = \sqrt{Re(e_j)^2 + Im(e_j)^2}$ and $arg(e_j) = arctan(Im(e_j)/Re(e_j))$. In theory, it is possible to find a generator by checking all branches of the logarithm of e_j and using Lagrange interpolation to compute $f(P)$ for each possible choice of $k_1, k_2, ..., k_n$.

Each time, it is checked if the non-negativity condition of the off-diagonal elements is satisfied. If P has distinct eigenvalues, then this search is finite, as the following theorem shows.

If P has distinct eigenvalues and if λ is a generator for P, then each eigenvalue e of Q satisfies that $|Im(e)| \leq |ln(det(P))|$. In particular, there are only a finite number of possible branch values of $log(P)$ which could possibly be generators of P since, if $det(P) \leq 0$, then no generator exists.

Thus, due to the finite number of possible branch values of $log(P)$ whenever P has distinct eigenvalues, it is possible to construct a finite algorithm to search for all possible generators of P.

The necessary algorithm can be described as follows:

- Step 1. Compute the eigenvalues $e_1, e_2, ..., e_n$ of P, and verify that they are all distinct.
- Step 2. For each eigenvalue e_j, choose an integer k_j such that $|Im\lambda| = |arg(e_j)+ 2\pi k_j| \leq |ln(det(P))|$.
- Step 3. For the collection of integers $k_1, k_2, ..., k_n$ set $f(e_j) = log|e_j| + i(arg(e_j) + 2\pi k_j)$. Let then g(x) be the function given in (11.13).
- Step 4. For g(x) compute the matrix $Q = g(P)$, and see if it is a valid generator.
- Step 5. If Q is not a valid generator return to Step 2, modifying one or more of the integers k_j. Continue until all allowable collections $k_1, k_2, ..., k_n$ have been considered.

For the rare cases where repeated eigenvalues are present, the authors refer to Singer and Spilerman 1976.

11.3.3   Methods for finding Approximations of the Generator

As we saw in the previous section, it is not unlikely that the series (11.12) converges to a generator with negative off-diagonal elements and there exists no generator without off-diagonal elements. Thus, either the calculated generator has to be adjusted or we have to use different methods to calculate the generator matrix. In the following section different methods will be discussed.

If we find a generator matrix with negative entries in a row, we will have to correct this. The result may lead to a generator not providing exactly $P = e^{\Lambda^*}$ but only an approximation, though ensuring that from the economic viewpoint the necessary condition that all off-diagonal row entries in the generator are non-negative is guaranteed.

The literature suggests different methods to deal with this problem if calculating Λ leads to a generator matrix with negative row entries:

The Jarrow, Lando, Turnbull Method

Jarrow et al. (1997) suggest a method where every firm is assumed to have made either zero or one transition throughout the year. Under this hypothesis it can be shown that for $\lambda_i \neq 0$ for $i = 1, \ldots, K - 1$:

$$exp(\Lambda) = \begin{pmatrix} e^{\lambda_1} & \frac{\lambda_{12}(e^{\lambda_1}-1)}{\lambda_1} & \cdots & \frac{\lambda_{1K}(e^{\lambda_1}-1)}{\lambda_1} \\ \frac{\lambda_{21}(e^{\lambda_2}-1)}{\lambda_2} & e^{\lambda_2} & \cdots & \frac{\lambda_{2K}(e^{\lambda_2}-1)}{\lambda_2} \\ \cdots & \cdots & \cdots & \cdots \\ \frac{\lambda_{K-1,1}(e^{\lambda_{K-1}}-1)}{\lambda_{K-1}} & \frac{\lambda_{K-1,2}(e^{\lambda_{K-1}}-1)}{\lambda_{K-1}} & \cdots & \frac{\lambda_{K-1,K}(e^{\lambda_{K-1}}-1)}{\lambda_{K-1}} \\ 0 & 0 & \cdots & 1 \end{pmatrix} \quad (11.15)$$

The estimates of $\hat{\Lambda}$ can be obtained by solving the system:

$$\hat{q}_{ii} = e^{\hat{\lambda}_i} \text{ for } i = 1, \ldots, K - 1 \text{ and}$$

$$\hat{q}_{ij} = \hat{\lambda}_{ij}(e^{\hat{\lambda}_i} - 1) \text{ for } i, j = 1, \ldots, K - 1.$$

JLT provide the solution to this system as

$$\hat{\lambda}_i = log(\hat{q}_{ii}) \text{ for } i = 1, \ldots, K - 1 \text{ and}$$

$$\hat{\lambda}_{ij} = \hat{q}_{ij} \cdot \frac{log(\hat{q}_{ii})}{(\hat{q}_{ii} - 1)} \text{ for } i \neq j \text{ and } i, j \ldots, K - 1.$$

This leads only to an approximate generator matrix, however it is guaranteed that the generator will have no non-negative entries except the diagonal elements.

For our example, the JLT method gives the associated approximate generator

Λ	A	B	C	D
A	-0.1054	0.0843	0.0210	0.0001
B	0.0542	-0.1625	0.0975	0.0108
C	0.0112	0.1004	-0.2231	0.1116
D	0	0	0	0

with non-negative entries, however, $exp(\Lambda)$ is only close to the original transition matrix P:

P_{JLT}	A	B	C	D
A	0.9021	0.0748	0.0213	0.0017
B	0.0480	0.8561	0.0811	0.0148
C	0.0118	0.0834	0.8041	0.1006
D	0	0	0	1

Especially in the last column high deviations (from 0.0001 to 0.0017 in the first row or 0.0148 instead of 0.010 in the second) for low default probabilities have to be considered as a rather rough approximation. We conclude that the method suggested by JLT in 1997 solves the problem of negative entries in the generator matrix, though we get an approximation that is not really close enough to the 'real' transition matrix.

Methods suggested by Israel, Rosenthal and Wei

Due to the deficiencies of the method suggested by JLT, in their 2000 paper IRW suggest a different approach to finding an approximate true generator. They suggest, using (11.12) to calculate the associated generator and then adjusting this matrix using one of the following methods:

- Replace the negative entries by zero and add the appropriate value back in the corresponding diagonal entry to guarantee that row-sums are zero. Mathematically,

$$\lambda_{ij} = max(\lambda_{ij}, 0), j \neq i; \quad \lambda_{ii} = \lambda_{ii} + \sum_{j \neq i} min(\lambda_{ij}, 0).$$

The new matrix will not exactly satisfy $P = e^{\Lambda^*}$.

- Replace the negative entries by zero and add the appropriate value back into *all* entries of the corresponding row proportional to their absolute values. Let G_i be the sum of the absolute values of the diagonal and non-negative off-diagonal elements and B_i the sum of the absolute values of the negative off-diagonal elements:

$$G_i = |\lambda_{ii}| + \sum_{j \neq i} max(\lambda_{ij}, 0); \quad B_i = \sum_{j \neq i} max(-\lambda_{ij}, 0).$$

Then set the modified entries

$$\lambda_{ij} = \begin{cases} 0, & i \neq j \text{ and } \lambda_{ij} < 0, \\ \lambda_{ij} - \frac{B_i|\lambda_{ij}|}{G_i} & \text{otherwise if } G_i > 0, \\ \lambda_{ij}, & \text{otherwise if } G_i = 0. \end{cases}$$

In our example where the associated generator was

Λ	A	B	C	D
A	-0.1080	0.0907	0.0185	-0.0013
B	0.0569	-0.1710	0.1091	0.0051
C	0.0087	0.1092	-0.2293	0.1114
D	0	0	0	0

applying the first method and setting λ_{AD} to zero and adding -0.0013 to the diagonal element λ_{AA} we would get for the adjusted generator matrix Λ^*

Λ^*	A	B	C	D
A	-0.1093	0.0907	0.0185	0
B	0.0569	-0.1710	0.1091	0.0051
C	0.0087	0.1092	-0.2293	0.1114
D	0	0	0	0

which gives us for the approximate one-year transition matrix:

P_{IRW1}	A	B	C	D
A	0.8989	0.0799	0.0199	0.0013
B	0.0500	0.8500	0.0900	0.0100
C	0.0100	0.0900	0.8000	0.1000
D	0	0	0	1

Obviously the transition matrix P_{IRW1} is much closer to the 'real' one-year transition than using the JLT method. Especially for the second and third row we get almost exactly the same transition probabilities as for the 'real' transition matrix. Also the deviation for the critical default probability λ_{AD} is clearly reduced compared to the JLT method described above.

Applying the second suggested method and again replacing the negative entries by zero but 'redistributing' the appropriate value to *all* entries of the corresponding row proportional to their absolute values gives us the adjusted generator

Λ^*	A	B	C	D
A	-0.1086	0.0902	0.0184	0
B	0.0569	-0.1710	0.1091	0.0051
C	0.0087	0.1092	-0.2293	0.1114
D	0	0	0	0

and the associated one-year transition matrix

P_{IRW1}	A	B	C	D
A	0.8994	0.0795	0.0198	0.0013
B	0.0500	0.8500	0.0900	0.0100
C	0.0100	0.0900	0.8000	0.1000
D	0	0	0	1

Again we get results that are very similar to the ones using the first method by Israel et al. The authors state that generally by testing various matrices they found similar results. To compare the goodness of the approximation they used different distance matrix norms.

While the approximation of the method suggested by JLT in their 1997 seminal paper is rather rough, the methods suggested by IRW give better approximations of the true transition matrix. In the next section we will now deal with the question how to adjust transition matrices according to some conditions derived from market prices or macro-economic forecasts.

11.4 Modifying Transition Matrices

11.4.1 The Idea

Another issue when dealing with transition matrices is to adjust, re-estimate or change the transition matrix due to some economic reason. In this section we will describe numerical methods that can be applied to modify e.g., historical transition matrices. As we stated before, the idea of time-homogenous transition matrices is often not realistic. Due to changes in the business cycles, like recession or expansion of the economy, average transition matrices estimated out of historical data are modified. A well-known approach where transition matrices have to be adjusted according to the macro-economic situation is the *CreditPortfolio View* model by the McKinsey Company. A more thorough description of the ideas behind the model can be found in Wilson. Another issue e.g., described in JLT (1997) is for example to match transition matrices with default probabilities implied in bond prices observed in the market. We will now describe some of the methods mentioned by the literature so far.

11.4.2 Methods suggested by Lando

Again we consider the continuous-time case where the time-homogenous Markov chain is specified via the $(K \times K)$ generator matrix:

$$\Lambda = \begin{pmatrix} \lambda_{11} & \lambda_{12} & \cdots \lambda_{1K} \\ \lambda_{21} & \lambda_{22} & \cdots \lambda_{2K} \\ \cdots & \cdots & \cdots \cdots \\ \lambda_{K-1,1} & \lambda_{K-1,2} & \cdots \lambda_{K-1,K} \\ 0 & 0 & \cdots 0 \end{pmatrix} \tag{11.16}$$

where

$$\lambda_{ij} \geq 0 \text{ for all i, j and}$$

$$\lambda_{ii} = -\sum_{\substack{j=1 \\ j \neq i}}^{K} \lambda_{ij} \text{ for i} = 1,.....K.$$

The off-diagonal elements represent the intensities of jumping to rating j from rating i. The default state K is an absorbing one. As we already showed in the previous section using the generator matrix, we get the $K \times K$ t-period transition matrix by

$$P(t) = exp(t\Lambda) = \sum_{k=0}^{\infty} \frac{(t\Lambda)^k}{k!}. \tag{11.17}$$

In his paper Lando (1999) extends the JLT approach and describes three different methods to modify the transition matrices so that the implied default probabilities are matched. Assuming a known recovery rate φ we can derive the implied survival probabilities from the market prices of risky bonds for each rating class

$$\tilde{Q}_0^i(\tau > t) = \frac{v^i(0, t) - \varphi p(0, t)}{(1 - \varphi) p(0, t)} \qquad \text{for i = 1,......K-1} \qquad (11.18)$$

where $v^i(0, t)$ is the price of a risky bond with rating and i, $p(0, t)$ is the price of a riskless bond. We can then calculate the one-year implied default probabilities as

$$p_{iK} = 1 - \tilde{Q}_0^i(\tau > 1). \qquad (11.19)$$

Based on this prework Lando aims to create a family of transition matrices $(\tilde{Q}(0, t))_{t>1}$ so that the implied default probabilities for each maturity match the corresponding entries in the last column of $\tilde{Q}(0, t)$. Assuming a given one-year transition matrix P and an associated generator matrix Λ with $P = e^\Lambda$, he proposes the following procedure:

1. Let $\tilde{Q}(0, 0) = \mathcal{I}$.
2. Given $\tilde{Q}(0, t)$, choose $\tilde{Q}(t, t + 1)$ such that

$$\tilde{Q}(0, t)\tilde{Q}(t, t + 1) = \tilde{Q}(0, t + 1), \quad \text{where } (\tilde{Q}(0, t + 1))_{iK} = 1 - \tilde{Q}^i(\tau > t + 1)$$

for $i = 1, ...K - 1$.
3. Go to step 2.

The difference between the methods is how step 2 is performed. It is important to note that this procedure only matches the implied default probabilities. To match the implied migration probabilities, one needs suitable contracts such as credit swaps against downgrade risk with the full maturity spectrum.

In each step the generator matrix Λ is modified such that

$$\tilde{Q}(0, 1) = e^{\Lambda(0)},$$
$$\tilde{Q}(1, 2) = e^{\Lambda(1)},$$

and they satisfy

$$\tilde{Q}(0, 1)\tilde{Q}(1, 2) = \tilde{Q}(0, 2)$$

where $\Lambda(0)$ is a modification of Λ depending on $\Pi_1 = (\Pi_{11}, \Pi_{12}, \Pi_{13})$ and $\Lambda(1)$ is a modification of Λ depending on $\Pi_2 = (\Pi_{21}, \Pi_{22}, \Pi_{23})$ where Π_1 and Π_2 have to satisfy

$$p_{iK}(0, 1) = 1 - \tilde{Q}(0, 1),$$
$$p_{iK}(0, 2) = 1 - \tilde{Q}(0, 2).$$

Note that the generator matrix has to be checked if it still fulfills the criteria, namely non-negative off-diagonal elements and row sums of zero for each row. Having the generator matrix, all we have to do is to exponentiate it to get the transition matrix.

To illustrate the methods we will always give an example based on a hypothetical transition matrix with three possible rating categories $\{A, B, C\}$ and a default state $\{D\}$ of the following form. Let's assume our historical yearly transition matrix has the form:

P	A	B	C	D
A	0.900	0.080	0.017	0.003
B	0.050	0.850	0.090	0.010
C	0.010	0.090	0.800	0.100
D	0	0	0	1

Thus, the associated generator is:

Λ	A	B	C	D
A	-0.1080	0.0909	0.0151	0.0020
B	0.0569	-0.1710	0.1092	0.0050
C	0.0087	0.1092	-0.2293	0.1114
D	0	0	0	0

We further assume that we forecasted (or calculated implied default probabilities using bond prices) the following one-year default probabilities for Rating classes $\{A, B, C\}$:

	A	B	C
$\widetilde{p}_{iK}$	0.006	0.03	0.2

In the following subsections we will refer to these transition matrices and default probabilities and show adjustment methods for the generator and probability transition matrix.

Modifying default intensities

The first method we describe modifies the default column of the generator matrix and simultaneously modifies the diagonal element of the generator according to:

$$\widetilde{\lambda}_{1K} = \pi_1 \cdot \lambda_{1K} \text{ and } \widetilde{\lambda}_{11} = \lambda_{11} - (\pi_1 - 1) \cdot \lambda_{1K},$$
$$\widetilde{\lambda}_{2K} = \pi_2 \cdot \lambda_{2K} \text{ and } \widetilde{\lambda}_{22} = \lambda_{22} - (\pi_2 - 1) \cdot \lambda_{2K},$$
$$\dots \qquad\qquad\qquad \dots$$

and for row $K - 1$:

$$\widetilde{\lambda}_{K-1,K} = \pi_{K-1} \cdot \lambda_{K-1,K} \text{ and}$$

$$\widetilde{\lambda}_{K-1,K-1} = \lambda_{K-1,K-1} - (\pi_{K-1} - 1) \cdot \lambda_{K-1,K}$$

such that for the new transition matrix $\widetilde{P}$ with

$$\widetilde{P}(t) = exp(t\widetilde{\Lambda}) = \sum_{k=0}^{\infty} \frac{(t\widetilde{\Lambda})^k}{k!} \tag{11.20}$$

the last column equals the risk-neutral default probabilities estimated from the market.

$$\widetilde{P}(:, K) = \begin{pmatrix} \widetilde{p}_{1K} \\ \widetilde{p}_{2K} \\ \cdots \\ \widetilde{p}_{K-1,K} \\ 1 \end{pmatrix} \tag{11.21}$$

Thus, with these modifications $\widetilde{\Lambda}$ is also a generator with rows summing to zero. The modifications are done numerically such that all conditions are matched simultaneously.

Using numerical solution, we get $\pi_1 = 1.7443$, $\pi_2 = 4.1823$, $\pi_3 = 2.1170$ and thus, for the modified generator matrix

$\widetilde{\Lambda}$	A	B	C	D
A	-0.1095	0.0909	0.0151	0.0034
B	0.0569	-0.1869	0.1092	0.0209
C	0.0087	0.1092	-0.3537	0.2358
D	0	0	0	0

and the associated probability transition matrix is:

$\widetilde{P}$	A	B	C	D
A	0.8987	0.0793	0.0161	0.0060
B	0.0496	0.8365	0.0840	0.0300
C	0.0094	0.0840	0.7066	0.2000
D	0	0	0	1

We find that due to the fact that the changes in the generator only take place in the last column and in the diagonal elements, for the new probability transition matrix most of the probability mass is shifted from the default probability to the diagonal element — especially when the new (calculated) default probability is significantly higher. Still, if a jump occurs, interpreting $-\frac{\lambda_{ij}}{\lambda_{ii}}$ as the probability for a jump into the new rating class j also these probabilities slightly change since λ_{ii} is modified. In our case we find that for example for rating class A the conditional probability for a default has increased from 1.8% to more than 3% the other conditional probabilities slightly decrease from 84% to 83% and from 14% to 13.8%. These results are confirmed by taking a look at the other rows and also by the associated new one-year transition matrix. We conclude that the main changes only happen in the diagonal element and in the default transition.

Modifying the rows of the generator Matrix

This method goes back to Jarrow, Lando and Turnbull (1997) where again a numerical approximation is used to solve the equations. The idea is not only to apply the last column and the diagonal elements of the generator matrix but to multiply each row by a factor such that the calculated or forecasted default probabilities are matched.

Thus, we get:

$$
\widetilde{\Lambda} = \begin{pmatrix}
\pi_1 \cdot \lambda_{11} & \pi_1 \cdot \lambda_{12} & \cdots \pi_1 \cdot \lambda_{1K} \\
\pi_2 \cdot \lambda_{21} & \pi_2 \cdot \lambda_{22} & \cdots \pi_2 \cdot \lambda_{2K} \\
\cdots & \cdots & \cdots \cdots \\
\pi_{K-1} \cdot \lambda_{K-1,1} & \pi_{K-1} \cdot \lambda_{K-1,2} & \cdots \pi_{K-1} \cdot \lambda_{K-1,K} \\
0 & 0 & \cdots 0
\end{pmatrix}
\tag{11.22}
$$

Applying this method and solving

$$
\widetilde{P}(t) = exp(t\widetilde{\Lambda}) = \sum_{k=0}^{\infty} \frac{(t\widetilde{\Lambda})^k}{k!}
\tag{11.23}
$$

subject to $\widetilde{P}(:; K) = (\widetilde{p}_{1K}, \widetilde{p}_{2K}, \ldots, \widetilde{p}_{K-1,K}, 1)$ numerically we get for

$\widetilde{\Lambda}$	A	B	C	D
A	-0.1455	0.1225	0.0204	0.0027
B	0.1149	-0.3457	0.2207	0.0101
C	0.0198	0.2482	-0.5212	0.2532
D	0	0	0	0

and the associated probability transition matrix is:

$\widetilde{P}$	A	B	C	D
A	0.8706	0.0988	0.0246	0.0060
B	0.0926	0.7316	0.1458	0.0300
C	0.0247	0.1639	0.6114	0.2000
D	0	0	0	1

In this case due to the different method more probability mass is shifted from the diagonal element of the transition matrix to the other's row entries. Considering for example the new transition matrix we find that for rating state *B* the probability for staying in the same rating category decreases from 0.85 to approximately 0.73 while for all the other row entries the probability significantly increases — e.g., from 0.05 to 0.09 for moving from rating state *B* to rating state *A*. These results were confirmed by applying the method to different transition matrices, so we conclude that the method that modifies the complete row of the generator spreads clearly more probability mass from the diagonal element to the other elements than the first method does. It could be used when the transition matrix should be adjusted to an economy in a rather unstable situation.

Modifying eigenvalues of the transition probability matrix

The third method described here modifies the generator and transition matrix by modifying the eigenvalues of the transition probability matrix. It is assumed that the transition matrix and thus, also the generator, are diagonalizable. Let then M be a matrix of eigenvectors of the transition matrix P and D a diagonal matrix of eigenvalues of P. Then the generator matrix is changed by numerically modifying the eigenvalues with

$$\widetilde{\Lambda} = M \, \Pi(0)D \, M^{-1}$$

where $\Pi(0)$ is a diagonal matrix with diagonal elements $(\pi_{11}, \pi_{12}, \pi_{13}, 0)$ such that for $\widetilde{\Lambda}$ we get:

$$\widetilde{P}(t) = exp(t\widetilde{\Lambda}) = \sum_{k=0}^{\infty} \frac{(t\widetilde{\Lambda})^k}{k!} \tag{11.24}$$

with

$$\widetilde{P}(:; K) = \begin{pmatrix} \widetilde{p}_{1K} \\ \widetilde{p}_{2K} \\ \cdots \\ \widetilde{p}_{K-1,K} \\ 1 \end{pmatrix} \tag{11.25}$$

Using this method we get for the modified generator matrix:

$\widetilde{\Lambda}$	A	B	C	D
A	-0.2459	0.2108	0.0347	0.0003
B	0.1319	-0.3925	0.2537	0.0069
C	0.0200	0.2538	-0.5278	0.2540
D	0	0	0	0

The associated probability transition matrix is:

$\widetilde{P}$	A	B	C	D
A	0.7930	0.1587	0.0423	0.0060
B	0.0991	0.7065	0.1644	0.0300
C	0.0253	0.1643	0.6104	0.2000
D	0	0	0	1

We find that in this case the third method shifts the most of the probability mass from the diagonal elements to the other elements of the row. The results are more similar to those when we modified the complete row of the generator than to those where only the default intensities were modified. For other transition matrices that were examined we found similar results. Clearly it can be stated that the methods modifying the eigenvalues and the complete rows of the generator should be used if rather grave changes in transition probabilities are expected, while the method that modifies the default intensities changes the transition probabilities more cautiously. It should be noted that further investigation on how the chosen method changes the adjusted transition matrix should be conducted.

After providing some methods for how generator and transition matrices can be adjusted due to business cycle effects or calculated risk neutral default probabilities, we will now have a look at how we can use generator matrices and simulation methods for calculating confidence sets for transition or default probabilities.

11.5   Confidence sets for continous time rating transitions

11.5.1   The idea

Lando and Christensen (2002) estimate the transition and default probabilities and set confidence intervals for default probabilities for each rating class by using the multinomial method and continuous-time approaches. They show that especially for higher rating classes where defaults are rare or non-existent, a bootstrap method using the continuous-time approach provides better confidence intervals than the binomial or multinomial approach to rating transitions. They also find that for speculative grades both methods provide more similar results.

The authors first start with the binomial method before describing their methodology. Consider a binomial random variable $X \sim B(p, N)$ where p is the probability of default. For the rating classes where we did not observe any defaults ($X = 0$) we could compute the largest default probability for a given confidence level α that cannot be rejected by solving the following equation:

$$(1 - p)^N = \alpha,$$

$$p^{max}(N, \alpha) = 1 - \sqrt[N]{\alpha}.$$

We can use the same approach to calculate two-sided confidence intervals for remaining classes where we observed transition to defaults. What we need is the number of firms with a rating or default at the beginning of the year t that have also a rating or default recorded at the beginning of the year $t + 1$. For a given confidence level α we have to solve the following equation to calculate the lower end of the interval

$$P(X \le \tilde{X}_i - 1 | p_i = p_i^{min}) = 1 - \frac{\alpha}{2}$$

where $\tilde{X}_i$ is the number of observed defaults in rating class i. To compute the upper end of the interval we solve

$$P(X \le \tilde{X}_i | p_i = p_i^{max}) = \frac{\alpha}{2}.$$

Assume for example that we have 200 companies in rating class *Aaa* and about 1000 companies in rating class *Aa*. Using a discrete approach in both rating classes we did not observe any default. Then solving equation

$$p^{max}(N, \alpha) = 1 - \sqrt[N]{\alpha}$$

gives us the following confidence intervals for the rating classes *Aaa* and *Aa*.

	N_i	X_i	$p_i^{min}(N_i, 0.01)$	$p_i^{max}(N_i, 0.01)$
Aaa	200	0	0	0.02276
Aa	1000	0	0	0.00459

This example already shows the main disadvantage of the binomial method — the confidence intervals are mainly dependent on the number of firms N, i.e., the lower N is the wider the confidence intervals are which is unrealistic for upper classes. Thus, if we do not observe defaults in higher rating classes the confidence interval simply depends on the number of firms in the observed rating class.

To overcome this disadvantage in estimating confidence sets, Lando and Christensen follow a different approach. Using the continous-time approach and a so-called bootstrapping method[6] one can calculate confidence sets for the generator matrix and thus, also for transition matrices for arbitrary time intervals. Their methodology can be described as follows:[7]

1. Using equation (11.9), estimate the generator matrix Λ^* for the considered period. This is the point estimator for the generator matrix - it is called the 'true generator'.
2. Exponentiate the true generator and record the corresponding 'true' one-year default probabilities for each rating category.
3. Simulate histories for the credit portfolio by using a start configuration of rating states for the portfolio and the true generator Λ^*.
4. For each history, compute the estimates of the generator via equation (11.9) and exponentiate them to obtain a distribution of the maximum-likelihood estimator of the one-year default probabilities. Repeating 3) and 4) N times we obtain N generator estimates and N transition matrices for any time period.
5. Compute the relevant quantiles e.g., for default or transition probabilities.

The main advantage of the bootstrap-method based on continuous-time rating transitions is that one gets narrower intervals in general and that it is possible to calculate confidence sets for the upper rating classes at all. This was not possible using the multinomial method. For lower rated classes, both methods have similar results. In addition to that the default probability estimated using the 'true' generator is approximately the mean of the simulated distributions. The binomial approach, as mentioned earlier, suffers from its dependence on the number of firms.

---

[6] An introduction to bootstrapping can be found in Efron/Tibshirani [1993].

[7] See Lando/Christensen [2002] p. 8ff.

TABLE 2. The 'true' generator estimated for the period 1997–2002

	01	02	03	04	05	06	07	08	D
01	-0.4821	0.1722	0.0517	0.1205	0.0689	0.0517	0.0172	0.0000	0.0000
02	0.0098	-0.4902	0.3625	0.0756	0.0265	0.0059	0.0039	0.0020	0.0039
03	0.0038	0.0941	-0.5694	0.3667	0.0643	0.0111	0.0085	0.0021	0.0187
04	0.0026	0.0154	0.2550	-0.5658	0.2250	0.0240	0.0157	0.00225	0.0258
05	0.0036	0.0090	0.0626	0.2935	-0.5972	0.1008	0.0656	0.0286	0.0334
06	0.0018	0.0055	0.0238	0.0880	0.2492	-0.5626	0.1375	0.0257	0.0312
07	0.0047	0.0093	0.0279	0.0652	0.1513	0.1257	-0.5538	0.0954	0.0745
08	0.0066	0.0000	0.0132	0.0197	0.0728	0.09923	0.1324	-0.4766	0.1324
D	0.0000	0.0000	0.0000	0.0000	0.0000	0.0000	0.0000	0.0000	0.0000

11.5.2   Some empirical results

In this section we will present some results from implementing the Lando/Christensen algorithm on the rating data of a German commercial bank.

The rating system consists of eight rating classes and the considered period was from April 1996 to May 2002. The first task was to compute the generator matrix from the entire given information about rating transition in the loan portfolio. The estimated generator matrix is considered to be 'true'; it is the point estimator for the real empirical generator. Using the maximum-likelihood estimator as in (11.9) we computed the generator matrix in Table 2 that gives us the one-year transition matrix in Table 3.

Our considered portfolio consists of 1160 companies with the following rating structure. This structure was used as starting population for the simulation. Using the estimated generator, for the population of firms in the portfolio, we simulated 20,000 histories for the portfolio. As the waiting time for leaving state i has an exponential distribution with the mean $\frac{1}{\lambda_{ii}}$ we draw an exponentially-distributed random variable t_1 with the density function

TABLE 3. The estimated one-year transition probability matrix based on the 'true' generator (in%)

	01	02	03	04	05	06	07	08	D
01	62.80	9.70	7.44	8.72	5.78	3.26	1.58	0.26	0.46
02	0.62	64.06	18.96	10.83	3.27	0.74	0.53	0.21	0.79
03	0.30	4.99	64.86	17.99	7.30	1.32	0.97	0.28	1.99
04	0.23	2.09	12.57	66.63	11.14	2.48	1.74	0.55	2.58
05	0.26	1.00	6.89	14.75	62.38	5.42	4.08	1.87	3.37
06	0.19	0.61	3.19	8.58	13.26	60.91	7.50	2.31	3.45
07	0.32	0.77	2.90	6.37	9.51	7.22	61.01	5.27	6.63
08	0.42	0.24	1.39	3.11	6.02	6.26	7.72	63.81	11.03
D	0.00	0.00	0.00	0.00	0.00	0.00	0.00	0.00	100

$$f(t_1) = \lambda_{ii} e^{-\lambda_{ii} t_1}$$

for each company with initial rating i. If we get $t_1 > 6$, the company stays in its current class during the entire period of six years. If we get $t_1 < 6$, we have to determine which rating class the company migrates to.

TABLE 4. The distribution of ratings of the 1160 companies

01	02	03	04	05	06	07	08
11	106	260	299	241	95	99	49

For this, we divide the interval [0,1] into sub-intervals according to the migration intensities calculated via $\frac{\lambda_{ij}}{\lambda_{ii}}$ and draw a uniform distributed random variable between 0 and 1. Depending on which sub-interval the random variable lies in we determine the new rating class j. Then we have to check again whether the company stays in the new rating class or migrates — we draw again from an exponentially-distributed random variable t_2 with parameter λ_{jj} from the generator matrix. If we find that $t_1 + t_2 > 6$ it stays in the new rating, the simulation is completed for this firm, and if it does not we have to determine the new rating class. The procedure is repeated until we get $\sum t_k > 6$ or the company migrates to default state.

This procedure is carried out for all 1160 firms in the portfolio. Thus, we get 1160 transition histories that are used to estimate the generator matrix and calculate the one-year transition matrix using exponential series.

Since we simulated 20,000 histories we obtained 20,000 default probabilities for each rating class and could use the 'empirical distribution' of default probabilities to compute e.g., 95% confidence intervals for the one-year default probabilities of the rating states.

The results are similar to those of Lando and Christensen. For almost all rating classes (also for the upper classes) we obtain nicely shaped probability distributions for the simulated default probabilities. For the upper rating classes (see e.g., Figure 1) the distribution is asymmetric and slightly skewed to the right. For the lower rating categories the distribution becomes more and more symmetric (see Figures 2 and 3) and are close to the normal distribution.

Comparing the calculated default probabilities using the continuous bootstrap approach and the binomial approach, we find that confidence intervals are, especially for the higher rating classes, much smaller with the numerical simulation approach. For rating class 1 we get an interval of length 0.004887 using the bootstrap method compared to an interval of length 0.047924 using the binomial approach — this is about a factor of 10. But also for lower rating classes in most cases the intervals we get with the numerical method are more precise. We conclude that, especially when we are interested in (credit) Value-at-Risk figures, using the simulation method described above is a useful tool for deriving 95% or 99% confidence sets.

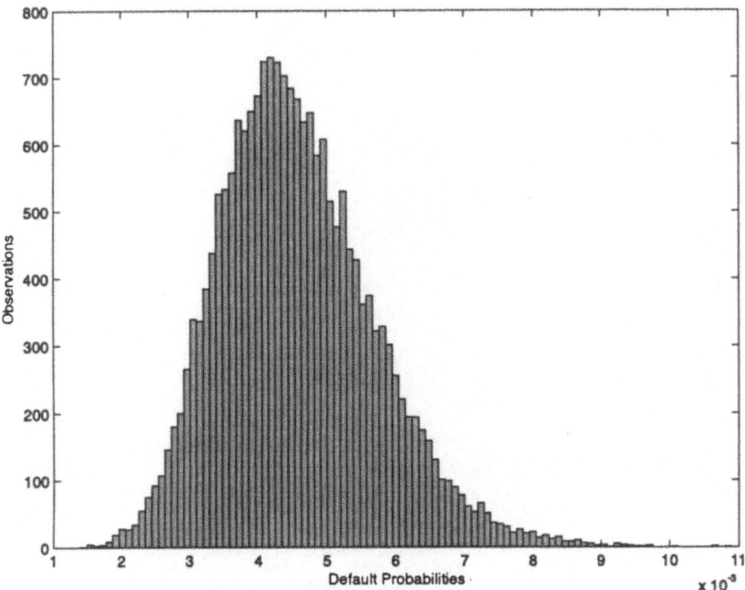

FIGURE 1. Histogram of simulated default probabilities for rating class 01

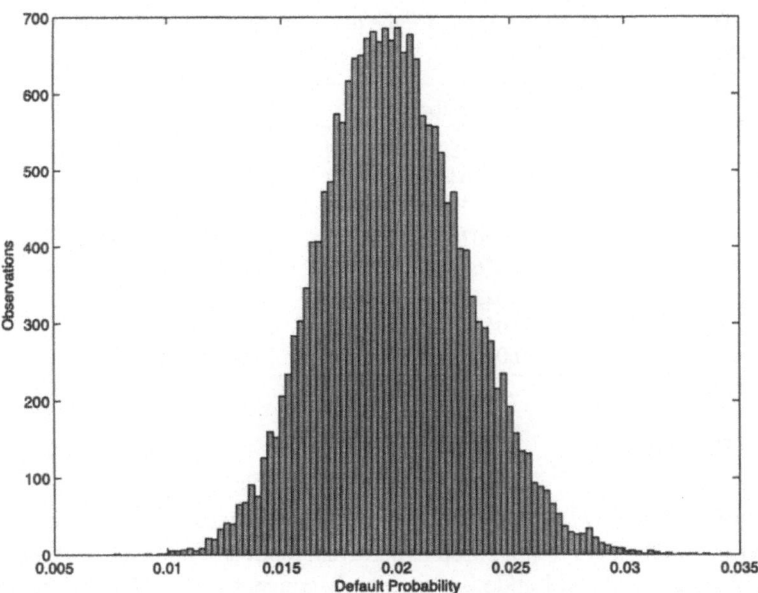

FIGURE 2. Histogram of simulated default probabilities for rating class 03

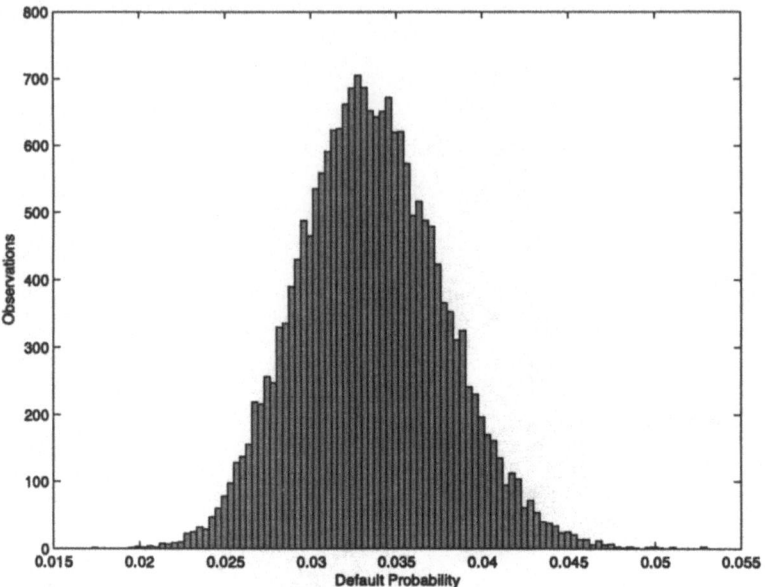

FIGURE 3. Histogram of simulated default probabilities for rating class 05

TABLE 5. The 95% - confidence interval for the default probability of each rating class using the generator and binomial method

Rating class	Generator method	Binomial Approach
01	[0.002659, 0.007546]	[0, 0.047924]
02	[0.004049, 0.012930]	[0.001036, 0.009697]
03	[0.014794, 0.025426]	[0.013346, 0.024524]
04	[0.020285, 0.031131]	[0.019039, 0.030789]
05	[0.026485, 0.040820]	[0.022834, 0.038941]
06	[0.023988, 0.046128]	[0.017074, 0.044922]
07	[0.049405, 0.084044]	[0.042718, 0.086257]
08	[0.079200, 0.143780]	[0.066564, 0.160339]

11.6   Summary

The aim of this paper was to provide an overview of recent developments in the use of numerical methods in rating based Credit Risk models. The main work in the area so far was done by David Lando, so we mainly concentrated on his results and suggestions. After providing insights on why using a continuous-time approach can be more favourable than a discrete-time approach, we found that finding an associated generator to a transition matrix can lead to some difficulties. Especially when for higher rating classes the probabilities for default in the transition are rather small, we may get a generator with negative off-diagonal elements. This matrix cannot be considered as a 'correct' continuous-time transition matrix, since for short time periods we may get negative transition probabilities. Thus, we provided some methods for how a valid generator can be found or how the calculated generator matrix can be adjusted to get a valid generator. In the next section we summarized some numerical solution methods for the adjustment of transition matrices due to changes in the business cycle or calculated risk-neutral default probabilities. We concluded that the results depend significantly on the chosen method and that further research on how the chosen method changes the adjusted transition matrix should be conducted. Finally we described how the continuous-time approach and a bootstrap simulation method can be used to calculate confidence bounds for default and transition probabilities. In an empirical study we found that confidence intervals especially for default probabilities in higher rating classes were much narrower than when a discrete binomial approach was used. We conclude that the use of a continuous-time approach in rating systems and the associated numerical methods may give a great benefit when transition matrices have to be adjusted or when confidence bounds for default and transition probabilities have to be calculated. In view of internal-rating based systems and Basel II the benefits of these methods should not be neglected.

References

[1] Bomfim, A.N. (2001). *Understanding Credit Derivatives and their potential to Synthesize Riskless Assets.* The Federal Reserve Board. Working paper.

[2] Carty, L., Lieberman, D., Fons, J.S. (1995). *Corporate Bond Defaults and Default Rates 1970–1994.* Special Report. Moody's Investors Service.

[3] Carty, L. and Lieberman, D. (1996). *Corporate Bond Defaults and Default Rates 1938–1995.* Special Report. Moody's Investors Service.

[4] Carty, L. (1997). *Moody's Rating Migration and Credit Quality Correlation, 1920–1996.* Special Comment. Moody's Investors Service.

[5] Crosby, P. (1998). *Modelling Default Risk.* in Credit Derivatives: Trading & Management of Credit & Default Risk. John Wiley & Sons. Singapore.

[6] Crouhy, M., Galai, D., Mark, R. (2000). *A Comparative Analysis of Current Credit Risk Models.* Journal of Banking and Finance (24).

[7] Das, S.R. (1998). *Credit Derivatives — Instruments.* in Credit Derivatives: Trading & Management of Credit & Default Risk. John Wiley & Sons. Singapore.

[8] Duffie, D. and Singleton, K.J. (1999). *Modeling Term Structures of Defaultable Bonds.* Review of Financial Studies (12).

[9] Efron, B. and Tibshirani, R.J (1993). *An Introduction to the Bootstrap.* Chapman & Hall, New York.

[10] Fons, J.S. (1994). *Using Default Rates to Model the Term Structure of Credit Risk.* Financial Analysts Journal, 25–32.

[11] Fons, J.S., Cantor, R., Mahoney, C. (2002). *Understanding Moody's Corporate Bond Ratings and Rating Process.* Special Comment. Moody's Investors Service.

[12] Gupton, G. (1998). *CreditMetrics: Assessing the Marginal Risk Contribution of Credit.* in Credit Derivatives: Trading & Management of Credit & Default Risk. John Wiley & Sons. Singapore.

[13] Israel, R.B, Rosenthal, J.S., Wei, J.Z. (2000). *Finding Generators for Markov Chains via Empirical Transition Matrices, with Application to Credit Ratings.* Working Paper.

[14] Jarrow, R.A. and Turnbull, S.M. (1995). *Pricing Derivatives on Financial Securities Subject to Credit Risk.* Journal of Finance (1).

[15] Jarrow, R.A., Lando, D., Turnbull, S.M. (1997). *A Markov Model for the Term Structure of Credit Risk Spreads.* Review of Financial Studies (10).

[16] Küchler, U. and Sørensen, M. (1997). *Exponential Families of Stochastic Processes.* Springer-Verlag, New York.

[17] Lando, D. (1998). *On Cox Processes and Credit Risky Securities.* University of Copenhagen. Working Paper.

[18] Lando, D. (1999). *Some Elements of Rating-Based Credit Risk Modeling.* University of Copenhagen. Working Paper.

[19] Lando, D. and Skødeberg, T. (2000). *Analyzing Rating Transitions and Rating Drift with Continuous Observations.* University of Copenhagen. Working Paper.

[20] Lando, D. and Christensen, J. (2002). *Confidence Sets for Continuous-time Rating Transition Probabilities.* University of Copenhagen. Working Paper.

[21] Madan, D.B. and Ünal, H. (1998). *Pricing the Risks of Default.* Review of Derivatives Research (2).

[22] Merton, R. (1974). *On the pricing of corporate debt: The Risk Structure of Interest Rates.* Journal of Finance (29).

[23] Saunders, A. and Allen, L. (2002). *Credit Risk Measurement.* John Wiley.

[24] Singer, B. and Spilerman, S. (1976). *The representation of Social Processes by Markov models.* American Journal of Sociology (82).

[25] Schönbucher, P.J. (2000). *The Pricing of Credit Risk and Credit Risk Derivatives.* University of Bonn. Working Paper.

[26] Trück, S. and Peppel J. (2003). *Credit Risk Models in Practice – A Review.* Physica-Verlag, Heidelberg.

[27] Uhrig-Homburg M. (2002). *Valuation of Defaultable Claims – A Survey.* Schmalenbach Business Review (54).

[28] Wilson, T.C. (1997). Portfolio Credit Risk I. Risk (10–9).

[29] Wilson, T.C. (1997). Portfolio Credit Risk II. Risk (10–10).

[30] Wilson, T.C. (1997). Measuring and Managing Credit Portfolio Risk. McKinsey & Company.

12

Numerical Analysis of Stochastic Differential Systems and its Applications in Finance

Ziyu Zheng

ABSTRACT   In this note, we provide a survey of recent results on numerical analysis of stochastic differential systems and its applications in Finance.

12.1   Introduction

The past three decades have witnessed an unprecedented series of theoretical and empirical advances in our understanding of financial markets. These breakthroughs in financial technology have led to an explosion in the variety and volume of new financial products and services that require increasingly sophisticated quantitative tools to properly assess and manage their risks and returns. Such a demand gave birth to a new interdisciplinary branch among engineering, finance, mathematics and statistics: Numerical methods in Finance.

For the numerous contributions from both academic researchers and practitioners before 1997, we refer the readers to Rogers and Talay [123] and the references therein. From the point of view expressed there, it seems to us that the following aspects of Numerical methods in Finance are of eminent importance:

- **Numerical methods for partial differential equations (PDEs).** This classical art initiated first from mechanics and physics found its new playground in the financial world. In particular, the finite difference methods for viscosity solutions (cf. Barles [15], Barles and Souganidis [16], Akian et al. [4]) appears to be the suitable technique due to the high degeneracy presented in the PDEs from Finance modelling, e.g., the Black & Scholes equations from European option pricing, the variational inequalities from American option pricing, the boundary value problems from Barrier option and Lookback option, etc.

- **Numerical methods for stochastic differential equations (SDEs), stochastic optimal control problems and stochastic differential games.**[1] The recognized procedures for stochastic system simulation include time discretization schemes (Euler scheme, Milshtein scheme, Runge–Kutta scheme, Romberg extrapolations) & Monte Carlo/quasi-Monte Carlo method (cf. Asmussen et al. [6], Boyle [27], Boyle et al. [28], Clark [36], Duffie and Glynn [48], Jacod and Protter [72], Kloeden and Platen [77], Kohatsu-Higa and Ogawa [79], Kohatsu-Higa and Protter [80], Kurtz and Protter [84], Milshtein [103], [104], Newton [110], [111], [112], Niederreiter [113], [114], [115], Talay [126], [127], Talay and Tubaro [129]), the Markov chain approximation method (cf. Kushner and Dupuis [87]), the lattice binomial & multinomial approximation (cf. Cox–Ross–Rubinstein [37]), the forward shooting grid method (cf. Barraquand and Pudet [19]), etc. On one hand, these methods provide probabilistic algorithms for deterministic nonlinear PDEs (cf. Kushner [86], [85], Talay and Tubaro [128], Talay and Zheng [131]), on the other hand, they provide direct approaches to simulate the sophisticated stochastic Finance models (cf. Rogers and Talay [123], Gobet and Temam [68]) and to access some quantities which are impossible or difficult to achieve via deterministic algorithms, e.g., the numerical computation of Value-at-Risk (cf. Talay and Zheng [130], [133]). An important advantage of such probabilistic algorithms is that they themselves can be interpreted as natural discrete models of the same financial situation, and thus have an inherent value independent of their use as a numerical method. In addition, as a well-known merit of the Monte Carlo method, probabilistic algorithms are less dimension-sensitive than their deterministic counterparts, which is especially important for large scale Financial Engineering. Finally we would like to point out the recent successful applications of Malliavin calculus in the numerical methods for stochastic systems (cf. Bally and Talay [13], [14], Protter and Talay [121], Fournié et al [58], [59], Gobet [66], Kohatsu-Higa [78]).
- **Statistical procedures and Filtering to identify model structures** (cf. Barndorff-Nielsen and Shephard [17], Del Moral et al. [105], Florens-Zmirou [56], Fournie and Talay [60], Kutoyants [91], [92], Viens [138]).
- **Numerical methods for backward stochastic differential equations (BSDEs).** The notion of backward stochastic differential equations (BSDEs) was introduced by Pardoux and Peng [118]. Independently, Duffie and Epstein [46], [47] introduced stochastic differential utilities in economics models, as solutions to certain BSDEs. From then on, BSDEs found numerous applications in mathematical economics and finance (cf. Duffie et al. [49], El Karoui et al. [53]). The numerical methods for BSDEs were studied by Briand et al. [29], Chevance [34], Douglas et al. [44], Ma et al. [95], [94], Zhang [140].

The latter three aspects can be classified into numerical analysis of stochastic differential systems. In this note, we make an endeavor to give a (far away from

---

[1] The book by Bouleau & Lepingle [26] presents various mathematical tools necessary to construct and analyze the numerical methods for the approximation of a wide class of stochastic processes.

complete and obviously biased) survey of the new results on numerical analysis of stochastic differential systems and its applications in finance since 1997 up to date. We also provide a list of interesting problems from our point of view.

12.2 Models and Challenges

From the finance point of view, the numerical resolutions of the following models are of eminent importance:

Model A. $(X_t(x))$ is a d-dimensional diffusion, solution to

$$X_t(x) = x + \int_0^t A_0(s, X_s(x))ds + \sum_{i=1}^r \int_0^t A_i(s, X_s(x))dW_s^i,$$

where (W_s) is an r-dimensional Brownian motion.

Model B. $(X_t(x))$ is a d-dimensional process, solution to

$$X_t = X_0 + \int_0^t B(s, X_{s-})dZ_s,$$

where X_0 is an $\mathbb{R}^d$-valued random variable and (Z_t) is an r-dimensional Lévy process, null at time 0.

Model C. We are given a set $Ad(u)$ of admissible controls. For each admissible control $u(\cdot) \in Ad(u)$, $(X_t^{x,u})$ is a d-dimensional controlled diffusion, solution to

$$X_t^{x,u} = x + \int_0^t b(X_s^{x,u}, u_s)ds + \int_0^t \sigma(X_s^{x,u}, u_s)dW_s. \tag{12.1}$$

System (12.1) can be generalized to include optimal stopping problems, Dynkin games, stochastic differential games, mixed control problems, etc.

In order to illustrate the motivations and the difficulties for these models, we consider five disciplines of problems originated from finance, namely, option hedging & pricing, numerical methods for ARCH/stochastic volatility models, Value-at-Risk type analysis, asset portfolio management and simulation problems raised from interest rate term structure modelling.

12.2.1 Option hedging & pricing

The stock price (S_s), $0 \le s \le T$, is modelled by a stochastic process and the riskless interest rate is denoted by r. The pricing of various options is reduced to the computation of the expectation of certain functionals of (S_s), e.g.,

- European call option:

$$u(S, t) = \mathbb{E}\left[\exp(-r(T - t))(S_T - K)^+ | S_t = S\right].$$

- American put option:

$$u(S, t) = \sup_{t \leq \theta \leq T} \mathbb{E}\left[\exp(-r(\theta - t))(K - S_\theta)^+ | S_t = S\right].$$

- Barrier option (In the money knock-out call):

$$u(S, t) = \mathbb{E}\left[\exp(-r(T - t))(S_T - K)^+ \mathbb{I}_{\sup_{t \leq s \leq T} S_s < B} | S_t = S\right].$$

- Lookback put option:

$$u(S, t) = \mathbb{E}\left[\exp(-r(T - t)) \sup_{t \leq \theta \leq T} (S_\theta - S_T)^+ | S_t = S\right].$$

- Asian call option:

$$u(S, t) = \mathbb{E}\left[\exp(-r(T - t))(\frac{1}{T - t} \int_t^T S_s ds - K)^+ | S_t = S\right].$$

- Passport call option:

$$u(S, t) = \sup_{|q_s| \leq C} \exp(-r(T - t))\mathbb{E}\left[\psi_T(q) | S_t = S\right],$$

where (q_s) is a trading strategy chosen by the option buyer and $\psi(q)$ denote the trading account.

From the above list we noticed that more and more sophisticated path-dependent exotic derivatives were introduced into the over-the-counter market according to the demands from the financial clients and then became standard components in the Exchanges. Such a trend made it harder to obtain the closed-form hedging & pricing solutions. In addition, in practice, one also has to take constraints including transaction costs, short-selling prohibition, etc. into consideration.

On the other hand, most closed-form formulas are based on the facts that, in the Black & Scholes framework, the logreturn of stock price is supposed to be normal and the volatility and expected return rate are supposed to be constant, which poorly fits the empirical evidence. It is to this end that various alternative models are introduced in order to catch better the market data characteristics, e.g., ARCH type models, illiquid market models, jump-diffusions models, large investor models, pure jump models, regime-switching volatility models, stochastic volatility models, etc.

The complexity of the payoff functions and financial models makes numerical solution mandatory in numerous practical situations. Option hedging & pricing mainly leads to the numerical resolutions of $\mathbb{E}[F(X.)]$, where F is a functional of a process (X_t) which models the price of the underlying, and (X_t) is the solution to Models (A) or (B) with the following main difficulties[2]:

---

[2] Another often mentioned difficulty is the irregularity of the system coefficients. However, we are not convinced that it is an inherent difficulty because of the discrete nature of financial data: the Model only needs to fit into data on finite discrete observations.

- The heavy and complicated path-dependence of the payoff function.
- High degeneracy.
- High nonlinearity.
- Non-homogeneous diffusion.

It is natural to apply time discretization schemes and Monte Carlo methods to compute $\mathbb{E}[F(X_{.})]$. *It is often preferable to use the simplest Euler scheme rather than more complicated ones* due to simulation difficulties, cf., Talay [127]. Another important viewpoint in Talay [127] is that *it is often not useful and even clumsy to try to approximate the diffusion process on the space of trajectories, when one wants to compute a quantity which depends on its law*: approximate processes efficient for simulations may not converge almost surely to the considered diffusion process, in short, weak convergence is often more preferable than strong convergence. Moreover, an important technique developed in Talay & Tubaro [129] gave error expansion for the Euler scheme. Basing on such technique, it is possible to obtain higher order precision by a Romberg extrapolation of Euler schemes with different step-sizes. This technique is extremely valuable in practice. Jacod and Protter [73] obtained the asymptotic error distribution for the Euler scheme for SDE driven by continuous semi-martingales by normalizing the error process.

We list several concrete problems concerning option hedging & pricing, on which significant progress was made recently, whereas certain inherent difficulties still remain standing and to be completely resolved.

- **Monte Carlo method for American contingent claims.** Monte Carlo method for American contingent claims (American option, American swaption, shout option, American Asian option, American game option, etc.) has remained an active and challenging direction for a long time. The importance comes from the Monte Carlo method being the only practical method for systems with more than one underlying security. The difficulty comes from the cash flows of American contingent claims being dependent both on the price path of the underlying asset and the decisions of the owner. Until the early 1990s, the valuation of American contingent claims was widely considered outside the scope of Monte Carlo. Thereafter, the continual endeavor in the last decade results in the following three classes of approaches:
  - First breakthrough to attack this difficulty via approximating optimal exercise policy makes a long (incomplete) list including Tilley [135], Barraquand and Martineau [18], Broadie and Glasserman [30], Boyle et al. [28], Broadie et al. [31], Carr [32], Longstaff and Schwartz [93], Fu et al. [62], etc. and we now could regard such approaches as classical ones. These approaches provide lower bounds for the price of American options and approximating optimal exercise policy, they are thus useful for the option buyer. The drawback of this class of methods is that they require *the determination of optimal decisions* which leads to the heavy complexity of the algorithm, bias of the estimators, lack of clear convergence rate error estimate, etc.
  - Rogers [122] applied a duality approach to obtain the upper bound for the price of American options and thus is useful for the option writer.

- Recently, El Karoui et al. [52] introduced BSDEs with reflection (RBSDEs), that is, to a setting with an additional continuous, increasing process added in the standard BSDE; the function of this additional process is to keep the solution above a certain prescribed lower-boundary process (obstacle) and to do so in a minimal fashion. These authors also made the crucial observation that the solution is the value function of an optimal stopping problem and provides the viscosity solution of related parabolic variational inequalities within the Markovian framework. Therefore, RBSDEs provide an alternative approach to study the American option. Cvitanić and Karatzas [39] generalized these results to the RBSDEs with two-sided obstacles. They proved that the solution is the value of a Dynkin game. Cvitanić and Ma [41] applied these results to study the American game option. Chevance [35], Bally et al. [10], [11], [12], Ma and Zhang [97], [96], [98] studied the numerical method for RBSDEs and its applications to American options. We emphasize that there is not explicitly an optimal stopping time in the setting of RBSDEs, thus it is possible to *avoid the simulation and the determination of optimal decisions* and provide a completely different Monte Carlo type approach for American contingent claims. Moreover, the optimal stopping policy can be retrieved from the hitting time of the solution to RBSDE to the obstacle and then can be approximated. Finally this approach is based on the BSDE theory and thus can be easily generalized to nonlinear underlying security models and general obstacle processes. The drawback of this class of methods is that they often need artificial boundary conditions and lead to space discretization which suffers from the curse of dimension. In addition, the coding for such methods is much more complicated than its counterparts.

- **Numerical methods for hypoelliptic non-homogeneous diffusions.** As we pointed out in the introduction, Malliavin calculus proved itself a powerful tool in the numerical methods for stochastic systems. We now further explain the reasons by considering the numerical resolution of Model (A). The Euler scheme is defined by

$$X^n_{(p+1)T/n}(x) = X^n_{pT/n}(x) + A_0(pT/n, X^n_{pT/n}(x))\frac{T}{n}$$

$$+ \sum_{i=1}^{r} A_i(pT/n, X^n_{pT/n}(x))(W^i_{(p+1)T/n} - W^i_{pT/n})$$

for $p = 0, \ldots, n - 1$. When the coefficients of Model (A) is assumed Lipschitz, it can be easily shown that $\mathbb{E}[f(X_T) - f(X^n_T)]$ is of order $\frac{1}{\sqrt{n}}$ when f is regular. Such a result is not very satisfactory because:

- One is always interested in the distribution function and density function of X_T, which are merely measurable or even Dirac function of X_T.

- Taking the number of simulations of the Monte Carlo method N into consideration, the numerical cost is very expensive.

In order to study the case when f is irregular, a natural assumption is that the diffusion should have certain non-degeneracy. It is well-known that Malliavin calculus is the sharpest weapon up to date to provide a sense of non-degeneracy to a widest class of diffusions. Bally and Talay [13] first observed this point and successfully studied the case of measurable f, obtained error estimate with order $\frac{1}{n}$ and error expansion under the assumption that (X_t) is hypoelliptic.

Unfortunately, the stochastic systems from finance modelling present strong degeneracy and time dependence and do not generally satisfy the uniformly hypoelliptic and homogeneous hypotheses which were standard assumptions in the classical literatures in order to obtain the proper estimates for the Malliavin covariance matrix (cf. Kusuoka and Stroock [88], [89], [90], etc.), e.g., consider the simplest model for single stock price:

$$dS_t = rS_t dt + \sigma(t, S_t) S_t dW_t,$$

which is neither hypoelliptic nor homogeneous. New classes of hypotheses thus must be designed in order to meet the new characteristics from finance modelling. The recent research of Cattiaux and Mesnager [33] on hypoelliptic non-homogeneous diffusions and Talay and Zheng [130], [133] on partially non-degenerate non-homogeneous diffusions (i.e., diffusions satisfying a so-called Condition (M)) introduced new conditions which fit the characteristics from finance modelling better than the classical ones. We thus expect that such new conditions and computation techniques can find numerous applications in finance.

For example, Talay and Zheng [130], [133] studied Euler schemes for Model (A). Let $(X_s^t(x'), 0 \le s \le T - t)$ be a smooth version of the flow solution to

$$X_s^t(x') = x' + \int_0^s A_0(t + \theta, X_\theta^t(x'))d\theta + \sum_{i=1}^r \int_0^s A_i(t + \theta, X_\theta^t(x'))dW_{t+\theta}^i.$$

Denote by $M(t, s, x')$ the Malliavin covariance matrix of $X_s^t(x')$. Condition (M) is stated as:

(M) For all $p \ge 1$ there exist a non-decreasing function K, a positive real number r, and a positive Borel measurable function Ψ such that

$$\left\| \frac{1}{M_d^d(t, s, x')} \right\|_p \le \frac{K(T)}{s^r} \Psi(t, x')$$

for all t in $[0, T)$ and s in $(0, T - t]$. In addition, Ψ satisfies: for all $\lambda \ge 1$, there exists a function Ψ_λ such that

$$\sup_{t \in [0,T]} \mathbb{E}[\Psi(t, X_t(x))^\lambda] < \Psi_\lambda(x),$$

and

$$\sup_{n > 0} \sup_{t \in [0,T]} \mathbb{E}[\Psi(t, X_t^n(x))^\lambda] < \Psi_\lambda(x).$$

We emphasize that the condition (M) is related to the Malliavin covariance of a component, not on the Malliavin covariance matrix of $X_T(x)$ which, thus, may be degenerate. This point is essential in view of financial applications. Talay and Zheng [133] listed various financial models which satisfy Condition (M). In spite of the fact that Condition (M) is much weaker than the usually assumed conditions on the Malliavin covariance matrix, one can obtain the convergence rate estimates on the marginal laws of the Euler scheme and the convergence rate estimates on the quantile of $X_T^d(x)$ of the Euler scheme under Condition (M). On the other hand, most of the financial computations may be reduced to the computation of one marginal law of the solution to stochastic models, which presents the Profit & Loss of the investor/trader.

Cattiaux and Mesnager [33] relaxed the hypoelliptic non-homogeneous diffusions to another direction, that is, the coefficients A_0, A_i are allowed to be only Hölder continuous w.r.t. the time variable. No direct application to finance has been published basing on their technique yet, however, it is obvious that one may apply their technique to study nonlinear filtering theory.

Natural next steps consist in extending Condition (M) to Model (B), seeking middle ground between Condition (M) and Cattiaux and Mesnager's technique, etc.

- **Numerical methods for Skorohod Integral.** The integral by parts formula founded the power of the Malliavin calculus (cf. Nualart [117], [116]). In numerical methods for stochastic systems, the integral by parts formula showed more and more importance, e.g., Bally and Talay [13], [14] applied the integral by parts formula to obtain the error estimate and expansion for distribution function and density function of the Euler scheme, Gobet [66], Gobet and Temam [68], Gobet [67] applied the integral by parts formula to approximate killed diffusion and applied such techniques to study Barrier option, Kohastu-Higa [78] applied the integral by parts formula to study the approximation of stochastic differential equations with boundary conditions, etc.

Unfortunately, except in a few examples, one does not know much about how to compute the divergence operator (also called Skorohod integral, anticipating integral) δ which is deduced from the integral by parts formula. Such a fact brought a dilemma for some applications of Malliavin calculus in numerical methods. For example, Fournié et al [58] compute the Greeks via integral by parts formula to eliminate the unwanted derivative operation. The Greeks are defined by

$$\partial_\alpha \mathbb{E}\left[F(S(\alpha))\right] = \mathbb{E}\left[F'(S(\alpha))J\right],$$

where F models the payoff functions of derivatives, S models the underlying security which is sensitive w.r.t. the system parameter α, e.g., initial value, volatility, etc., and J models the Jacobian of S w.r.t. the system parameter α. Formally, via the integral by parts formula one has

$$\mathbb{E}\left[F'(S(\alpha))J\right] = \mathbb{E}\left[F(S(\alpha))\delta(S(\alpha), J)\right],$$

where $\delta(S(\alpha), J)$ represents a Skorohod integral. With such a transform one significantly accelerates the computation by removing the derivative term *in case that the distribution of the deduced Skorohod integral part is known to us*. However, in the simplest examples when such terms are known, in fact one may obtain the closed-form or nearly closed-form solutions of the Greeks and then numerical solution lost its true value, while for the complicated models such terms are unknown for us.

To our knowledge, the numerical methods for the Skorohod integral were only studied in Ahn and Kohatsu-Higa [3] and Cvitanić et al [42]. Ahn and Kohatsu-Higa [3] studied the Euler scheme for solutions of anticipating SDEs where the anticipation is due to the initial random variable, where the condition on the coefficients and the error are comparable to the ones in the usual adapted case. Cvitanić et al [42] studied the hedging of digital options where the diffusion is assumed to be non-degenerate. The technique therein is based on the representation theorem for BSDEs, cf., Ma and Zhang [97], [96]. The numerical results therein indicated that, equipped with the present computation technique for Skorohod Integral, it is not clear if the Malliavin calculus based Greek computation method is superior to the classical ones including the delta method, RVM method, etc.

Therefore a deep and systematical study of the numerical methods for the Skorohod integral appears to be essentially important and provide a large number of new problems and applications (not restricted in finance).

- **Numerical methods for jump-diffusion models.** A modelling argument was made that the standard model of a diffusion is incorrect for a variety of reasons, and that one needs a model that has a large number of small jumps. In the stock market, for example, prices are not continuous but change by units of 12.5 cents; the stock market closes overnight and on weekends, and opening prices often have jumps. Indeed, the New York Stock Exchange employs "specialists" to try to smooth out inherently unstable or "jumpy" stock prices. Aside from this, there occur with regularity external shocks, both predictable and totally inaccessible. Predictable ones include earnings announcements, going ex-dividend, scheduled meetings of the Federal Reserve Board to adjust interest rates, etc. Inaccessible ones include unexpected events such as political assassinations, currency collapses (such as the Mexican peso recently), and national disasters.

    In the government security market alone there are often substantial jumps related either to central bank intervention or to the release of significant macroeconomic information. Analogous considerations apply in the foreign exchange currency markets and require models with jumps, cf. Navas [108]. Finally, as regards Finance Theory, we note that the idea of including Lévy process driven security prices is not new, but goes back at least to 1963 when Mandelbrot [99] and Fama [55] deduced that one needed models with infinite variances; had modern tools been available, a likely construction would have been SDEs driven by symmetric stable processes. All these arguments lead to the implementation of Model (B) in finance.

On the other hand, the recent research of Avesani and Bertrand [8], Fournier [61], Jacod [71], Mikulevicius and Platen [102], Platen [119], Protter and Talay [121], Zhang [141], [143], [142] gave us strong evidence how to further investigate the numerical methods for Model (B). We also would like to point out that the Malliavin calculus approach for jump processes was recently developed by Jacod, Fournier, etc., therefore we might expect that various Malliavin calculus based results for the diffusion Models (A) can be extended in parallel to the jump-diffusion Models (B) in the near future.

However, it seems to us that the most important problem concerning jump-diffusion models is: *do jump-diffusion models really outperform diffusion models and are they implementable?* From our point of view, though it is theoretically valuable to model the totally inaccessible jumps, it is hard to implement such models in practice. The main interest thus should be focused on the predictable ones. The first part of the question is then reduced to: to model a given phenomenon, between jumps and the sharp movements of diffusions, can empirical research distinguish which one is superior in case that the observation is actually discrete? If jump does outperform (e.g., it seems the case for the opening price), the second part of the question is reduced to:

– How to estimate the model parameters, in particular, to separate the parameters related to jump components from the ones related to the diffusion components? It seems that the robust statistics technique might be helpful for this part.

– Once concrete jump-diffusion models are available, one needs to establish the pricing and hedging framework for such models. We refer the readers to the pioneering work of Jarrow and Rosenfeld [76], Navas [108], Jarrow and Madan [74], [75].

12.2.2   Numerical methods for ARCH/stochastic volatility models

Econometric models of changing volatility were first fitted to data by the introduction of the autoregressive conditionally heteroscedastic (ARCH) model by Engle [54] and Bollerslev [23]. The volatility of an ARCH type model depends on the variability of *past* observations, e.g., in the original ARCH(p) model of Engle, the volatility σ_t^2 is driven by a linear combination of p lagged square error terms

$$\sigma_t^2 = \omega + \sum_{i=1}^{p} \alpha_i \epsilon_{t-i}^2$$

where $\alpha_i \geq 0$. The ARCH type model therefore belongs to the so-called time-lag models and its success is because the model can modify itself according to the past observations. ARCH type models play such a prominent role in various arena of financial econometrics that even a modest list of references can not be given here, e.g., Engle [54] has been cited more than a thousand times.

Meanwhile, an alternative modelling methodology was introduced by Taylor [134] in which the volatility is driven by *unobserved* components, e.g., in Ball and

Roma [9], the author explained the well-known smile effect via the following model:

$$
\begin{cases}
dS(t) & = \mu dt + \sigma(t)S(t)dW_t^1, \\
dv(t) & = \varphi(t)dt + \psi(t)dW_t^\sigma,
\end{cases}
$$

where v is a monotone function of σ. This class of models are called continuous-time stochastic volatility models. Stochastic volatility models have also been intensively studied in the last decade.

As a matter of fact, the discrete time model outperforms the continuous time model in estimation and calibration due to the discrete nature of market observations; whereas the continuous time model outperforms the discrete time model in pricing and hedging thanks to the well-established Black & Scholes framework and the sharp weapons provided from the continuous-time stochastic analysis arsenal.

Therefore, how to naturally bridge the gap between the empirically convincing discrete-time ARCH type model and the flexible continuous-time stochastic volatility model became one of the central problems of finance research. A breakthrough happened in Nelson [109], where the author proved that ARCH processes converge in law to some stochastic volatility models, based on the techniques from Stroock and Varadhan [125]. Success in this direction leads to the viewpoint that "one could regard the ARCH model as merely a device which can be used to perform filtering or smoothing estimation of unobserved volatilities" as in Bollerslev and Rossi [24] and a number of books. We refer the readers to Mele and Fornari [100] and the references therein for a long list of publications concerning the ARCH type model (whose number is still increasing steadily), the stochastic volatility model and the connections between them.

In spite of the significant success of the ARCH/stochastic volatility framework, in this subsection, we would like to point out some dilemmas therein. We also propose some possible remedies to cover such flaws.

We first observe that:

- The convergence argument of Nelson is based on an assumption that the sampling frequency can be higher and higher. In practice one actually has only discrete observations of the system instead of the commonly assumed continuous observations. *The sampling frequency has an upper limit.*

- Although the convergence result is proven, one still can not explain clearly why the *time-lag* characteristic of the ARCH type model disappears and a new *hidden process* comes into being. This question is even more fatal when the limitation of the sampling frequency is taken into consideration. We emphasize that these two characteristics are the basic spirits of these two classes of models respectively; both clear financial and strict mathematical interpretations must be given if both models are to be good models for the real world.

- In stochastic volatility models, one has only *partial* observations of the system, whereas in ARCH type models one has *complete* observations of the system.

In order to understand the deep nature inside the above three phenomena, we return to the original convergence study in Nelson [109], where the author proved that a series of *consistent* GARCH(1,1) or exponential ARCH models converge weakly to a stochastic volatility model when the sampling frequency tends to infinity. We now consider a simplest example which was studied in Engle [54], i.e., an ARCH(1) model such that

$$\sigma_t = \epsilon_t h_t^{1/2} \; ; \; h_t = \alpha_0 + \alpha_1 \sigma_{t-1}^2,$$

whose sampling interval is one day. This model predicts that the volatility at next noon follows a conditional normal with a conditional variance which depends on and only on the volatility at this noon. *We also suppose that this model is a good approximation to a given market.* Now let us double the sampling frequency and build a corresponding new ARCH model, whose sampling interval is half a day. In order to make this new ARCH model also a good approximation to the same market (and actually should be a better model because of the higher sampling frequency), *it must coincide with or outperform the first model.* There are thus two natural choices: either the new model is of type ARCH(2) such that

$$\sigma_t = \epsilon_t h_t^{1/2} \; ; \; h_t = \alpha_0' + \alpha_1' \sigma_{t-0.5}^2 + \alpha_2' \sigma_{t-1}^2,$$

in order to take the observations at this noon and at this midnight into consideration; or the new model is still ARCH(1) such that

$$\sigma_t = \epsilon_t h_t^{1/2} \; ; \; h_t = \alpha_0'' + \alpha_1'' \sigma_{t-0.5}^2,$$

if we assume that the information at this noon can be completely contained in the information at this midnight. The latter was chosen in Nelson [109] and actually leads to a Markovian setting if we raise the sampling frequency higher and higher and assume a consistent condition. The deep investigation in Stroock and Varadhan [125] told us that the "consistent" assumption suffices to guarantee that a sequence of well-behaved discrete time Markov processes converges to an Itô process. But here is the source of all the troubles: *a Markovian setting discards the time-lag characteristics of ARCH type models because a Markovian assumption claims that all the information from the past can be contained in the present state; and the power of the ARCH type model comes from the fact that it is autoregressive conditioned on the past observations.* A simplest example is, the NASDAQ index is around 3,000 both at the end of 1999 and at the beginning of 2001. The ARCH type model can tell the difference between them because the history before these two same states are different; a Markovian model can not because a Markovian model "forgets" the history before the present state. Another example is the well-known leverage effect first observed by Black that volatility tends to grow in reaction to bad news and to fall in response to good news, which has been explained by the EGARCH model. Such an effect also can not be explained in a Markovian setting. *If we do believe that the past has influence on the present, we must raise the number of the parameters in the corresponding ARCH type models when we raise the sampling frequency.*

We now provide another viewpoint to understand the nature of the convergence. Under consistent assumptions, the limiting process of a sequence of ARCH processes with sampling intervals h_n, $h_n \downarrow 0$, is an Itô process; then without loss of generality, we can assume that it is the solution to a certain Model (A). Consider the Euler scheme with step size h_n for this Model (A), it coincides with the corresponding ARCH process with sampling interval h_n *with only a higher order error which is omitted under the consistency assumption*. It is to this end that the time-lag characteristics of the ARCH models disappeared in the limiting model. It is also because we omitted the past information that we have only partial information for the limiting model and a "hidden" process comes into being.

Now a natural question is: if the time-lag is *in* the limiting model, what kind of sequence of ARCH type models should we choose?

It is to this end that we would like to highlight the recent working papers by Kuchler and Platen [81], [82] on the weak and strong discrete-time approximation of stochastic differential equations with time delays.

Let $(X_t(x))$ be a d-dimensional diffusion, solution to

$$X_t(x) = x + \int_0^t A_0(X_s(x), X_{s-r}(x))ds + \sum_{i=1}^d \int_0^t A_i(X_s(x), X_{s-r}(x))dW_s^i,$$

$$(12.2)$$

with initial value

$$X(u) = \xi(u), \quad u \in [-r, 0].$$

One approximates the above stochastic differential equation with time delays by an Euler scheme with $n = ML$ steps and with step size $\Delta t = \frac{r}{L}$. The Euler scheme is thus

$$X_{(p+1)T/n}^n(x) = X_{pT/n}^n(x) + A_0(X_{pT/n}^n(x), X_{(p-L)T/n}^n(x))\frac{T}{n}$$

$$+ \sum_{i=1}^d A_i(X_{pT/n}^n(x), X_{(p-L)T/n}^n(x))(W_{(p+1)T/n}^i - W_{pT/n}^i)$$

for $p = 0, \ldots, n - 1$. Kuchler and Platen [81], [82] proved the strong and weak convergence results of the Euler scheme under proper assumptions.

It is easy to observe that there is a surprising coincidence between the time discretization of (12.2) and nonlinear ARCH(2) models where the observations come from $\frac{r}{L}$ time ago (which can be regarded as present for large L) and r time ago. In addition, the time delays in the continuous SDE (continuous limiting model) and Euler schemes with different step sizes (discrete model with different sampling interval) *remain the same*. Therefore the convergence results of Kuchler and Platen actually provided another limiting continuous model of ARCH(2) models once some proper corresponding consistency assumptions were designed.

But one can do much more in this direction! Let us consider the discrete-time model

$$X^n_{(p+1)T/n}(x) = X^n_{pT/n}(x)$$

$$+ A^n_0(X^n_{pT/n}(x), X^n_{(p-1)T/n}(x), ..., X^n_{(p-L)T/n}(x))\frac{T}{n}$$

$$+ A^n_1(X^n_{pT/n}(x), X^n_{(p-1)T/n}(x), ..., X^n_{(p-L)T/n}(x))\epsilon_{p,\frac{r}{L}} \quad (12.3)$$

for $p = 0, \dots, n - 1$, where A^n_0, A^n_1 can be defined in the same manner as ARCH(L+1), GARCH(L+1,m), $1 \leq m \leq L + 1$, EGARCH, etc., and $\epsilon_{p,\frac{r}{L}}$ are i.i.d normals with variance $\frac{r}{L}$. The limiting continuous model, if existing, would be a time-lag model taking the observations in the past r time interval into account. We here make a wild guess on the form of the limiting model:

$$X_t(x) = x + \int_0^t A_0\left(X_s(x), \int_{s-r}^s \alpha_\theta X_\theta(x)d\theta, g(X_\theta(x), s - r \leq \theta \leq s)\right) ds$$

$$+ \int_0^t A_1\left(X_s(x), \int_{s-r}^s \beta_\theta X_\theta(x)d\theta, g(X_\theta(x), s - r \leq \theta \leq s)\right) dW_s, \quad (12.4)$$

with initial value

$$X(u) = \xi(u), \quad u \in [-r, 0].$$

The three terms inside A_0, A_i are corresponding to the latest observation, the weighted average of observations in the last r interval, and the trend or pattern of the observations in the last r interval respectively. These terms may vary according to the actual ARCH type model we choose, e.g., a possible choice of $g(X_\theta(x), s - r \leq \theta \leq s)$ could be simply $X_s(x) - X_{s-r}(x)$; if we construct (12.3) as an ARCH(L+1), then the limiting model may contain only the first two terms or even only the second term under proper assumptions.

Unfortunately, there does not exist a well-established theory for (12.4) comparable to the case of Stroock and Varadhan [125] for the Markovian diffusions. We thus propose a step-by-step procedure toward a Theory of time-lag continuous models in finance:

- The existence and uniqueness of (12.4). It is expected when proper regular conditions are assumed.
- The numerical resolutions of (12.4).
- Design the consistency condition.
- Under the consistency condition, prove the weak convergence for (12.3).
- Keeping in mind that the sampling frequency has an upper limit, one then studies the convergence rate estimate of (12.3) to (12.4) in order to estimate the coefficients of (12.4) from (12.3).
- Study the pricing and hedging problem under Models (12.3) and (12.4) respectively and compare the outputs in order to justify the coincidence between the ARCH type model and its "limiting" continuous-time model.
- Establish the connection between (12.4) and stochastic volatility models and give financial interpretation to the "hidden" process.

Besides, there are some other aspects concerning ARCH/stochastic volatility framework deserving notice:

- One fundamental assumption for the ARCH type model is that the unconditional variance of the process is constant; it seems that some alternative assumptions such as the unconditional variance of the process is periodic deserves investigation.
- Estimation technique via simulation for the stochastic volatility model has been intensively studied, e.g., Barndorff-Nielsen and Shephard [17] studied the case where the spot volatility is supposed to be stationary and stochastically independent from the underlying Brownian motion. Unfortunately, without the stationary assumption, we do not know much how to estimate the coefficients of stochastic model efficiently because one has to face a single-sample problem in such cases.
- The recent work of Del Moral et al. [105] provides an exciting new access for the discrete-time partial observation filtering problems: let $X = (X_t)_{t>0}$ be an $\mathbb{R}^d$-valued stochastic process governed by the stochastic differential equation. One aims to compute the conditional expectations

$$\pi_{Y,N} f = E(f(X_N)|Y_1, \ldots, Y_N)$$

based on noisy observations $Y_1, \ldots, Y_N$ for all reasonable functions on $\mathbb{R}^d$ in case that no explicit form for the transition semigroup $(Q_t)_{t \geq 0}$ of the Markov process X_t is known. These authors proposed to approximate this filter with the help of a Monte Carlo simulation technique. It deserves noticing that their methodology fits well the structure of the stochastic volatility model. It is stated in Del Moral et al. [105] that their estimates are not optimal according to the results from numerical experiments. Therefore it should be possible to sharpen the theoretical results in their paper. On the other hand, one should be able to improve the performance of the stochastic volatility model by applying the filtering technique with discrete observation, cf., Viens [138] applied their results to portfolio optimization under partially observed stochastic volatility.

12.2.3   Value-at-Risk type analysis

New methods of measuring and managing risk have evolved in parallel with the growth of the OTC derivatives market. During the last decade, one of these measures, known as Value-at-Risk, has become especially prominent, and now serves as the basis for the most recent BIS market risk-based capital requirement. Value-at-Risk prominence has grown because of its conceptual simplicity and flexibility. It can be used to measure the risk of an individual instrument, or the risk of an entire portfolio. Value-at-Risk is also potentially useful for the risk management of a firm. In addition, the importance to measure the trading risk exposures correctly for regulators is reflected by the Basle Committee on Banking Supervision [1], [2].

Value-at-Risk is usually defined by the largest loss in portfolio value that would be expected to occur due to change in market prices over a given period of time in all but a small percentage of circumstances, i.e., a quantile of a random variable which models the future loss.

The importance of quantile is not only limited in Value-at-Risk, but involves various aspects of finance including the Risk measurements (cf. Artzner et al [5]), the development of the Value-at-Risk based risk management (cf. Basak and Shapiro [20]), the quantile hedging (cf. Föllmer and Leukert [57]), etc. The motivation of the quantile-based financial strategies can be explained as follows: even in a complete market many investors *do not want a perfect hedging because it takes away completely the opportunity to make a profit together with the risk of a loss*, cf. Föllmer and Leukert [57]. The investor actually looks for the most efficient allocation of capital to pursue the profit while keeping the total business risk under control. A theory building on such a spirit therefore might outperform the non-arbitrage based theory in fitting the reality.

While quantile is conceptually simple and flexible, for a quantile figure to be useful, it needs to be reasonably accurate, and needs to be available on a timely enough basis that it can be used for risk management. Pritsker [120] investigates the trade-off between accuracy and the computational time for six alternative Value-at-Risk computation methods. Among them, the delta-gamma Monte Carlo method provides the best trade-off but still leads to significant errors in the Value-at-Risk figures, especially for deeply out of the money options.

It is in Talay and Zheng [130], [133] that a new numerical approach has been proposed and *the global error estimates* for general quantile computation is for the first time available. This numerical approach is a combination of the Monte Carlo method and the Euler scheme. The global error on the quantile figure is of order

$$\mathcal{O}\left(\frac{1}{\overline{p}_T^d(\rho(x, \delta))n}\right) + \mathcal{O}\left(\frac{1}{p_T^{n,d}(\rho(x, \delta))\sqrt{N}}\right), \tag{12.5}$$

where n is the number of steps of the discretization in the Euler scheme, N is the number of simulations of the Monte Carlo method, $p_T^{n,d}(\cdot)$ denotes the density at time T of the d-*th* component of the Euler scheme, and, denoting by $p_T^d(x, y)$ the density of $X_T^d(x)$, we have set

$$\overline{p}_T^d(\rho(x, \delta)) := \inf_{y \in (\rho(x, \delta)-1, \rho(x, \delta)+1)} p_T^d(x, y).$$

The popular delta Monte Carlo method (resp. delta-gamma Monte Carlo method) can be regarded as a one step Euler scheme and Monte Carlo method (resp. one step Milstein scheme and Monte Carlo method), that is, $n = 1$. Therefore however large the number of simulations N one chooses, one can not reduce the discretization error. From the finance point of view, the one step scheme leads to normal distributions whereas multi-steps scheme may capture the heavy-tailed distribution being observed in the financial market. This new numerical approach therefore includes these popular approaches and outperforms them due to the above reasons. We also emphasize that the existence of a positive marginal density seems necessary for any quantile computation since if not, the quantile figure may present jumps and thus out of the scope of numerical approximation.

We also should highlight the work of Glasserman, Heidelberger and Shahabuddin [65] where the authors modelled the heavy-tail phenomenon by using t-distribution, the work of Duffie and Pan [50] where the authors used the delta-gamma Monte Carlo method for classical multi-factor jump-diffusion for default intensities and asset returns under which between-jump returns are correlated Brownian motion with return jumps at Poisson arrivals that are jointly normally distributed.

Interesting and challenging next step questions include:

- Quantitatively study the Value-at-Risk based risk management and quantile hedging problem.
- Seek the optimal condition where Estimate (12.5) holds, and most importantly, find out for which class of models the quantile figure *is not* reliable.
- Variance reduction. The variance reduction for quantiles appears to be of a different nature from the classical variance reduction problem, few references are available, cf., Glasserman, Heidelberger and Shahabuddin [64]. New ideas and techniques are in need.

12.2.4 Asset portfolio management

A typical model for the asset prices and the portfolio value is

$$
\begin{cases}
X_t^j(x) & = x^j + \int_0^t \sigma_0^j(s, X_s(x))X_s^j(x)ds \\
& \quad + \sum_{i=1}^r \int_0^t \sigma_i^j(s, X_s(x))X_s^j(x)dW_s^i, \quad j = 1, \ldots, d-1, \quad (12.6) \\
X_t^d(x) & = \sum_{j=1}^{d-1} \pi^j(s, X_s(x))X_s^j(x),
\end{cases}
$$

where $(X_t^d(x))$ represents the portfolio process, $(X_t^j(x), 1 \le j \le d-1)$ represents the $d-1$ financial assets, and $\pi^j(t, X_t(x))$ represents the investing strategies of the portfolio. Asset portfolio management is of essential importance in finance (cf. Duffie [45]). The central problem therein is to measure and manage the financial risk carried by the portfolio and thus can be reduced to various systems of Class (C). The fact that the market might be incomplete and illiquid leads to the heavy nonlinearity of the coefficients in (12.6); more importantly, various constraints from the real market must be taken into consideration, e.g., transaction costs, short-selling prohibition, gain process constraints, etc. Thus one needs to study the numerical methods for constrained nonlinear stochastic systems. Problems from this discipline are of high variety, e.g., Benth et al. [22] modelled optimal portfolio selection with consumption as a nonlinear integro-differential equation with gradient constraint and studied it via a viscosity solution approach, Delbaen et al. [43] studied exponential hedging and entropic penalties, to mention a few.

The final step in asset portfolio management, is always solving a Model (C) analytically or numerically. The standard technique for solving Model (C) numerically, is the Markov chain approximation method. The main idea of this method is to approximate the original controlled process by an appropriate controlled Markov chain on a finite state space, and to approximate the original cost function by a function which is appropriate for the approximating chain. The state space is a discretization of the original state space and the approximation Markov chain is chosen such that local consistency properties (that is mean and mean square change per step) of the approximation chain are similar to those of the original controlled process. The notion of local consistency of an approximation chain is analogous to the notion of consistency for a finite difference approximation, except that here it is the controlled process which is approximated, not a partial differential equation. One then has to solve the optimal control problem for the approximation controlled chain and to prove that the solutions to the approximate problem actually converge to the exact value as the approximation parameter goes to zero. The convergence proofs for the Markov chain approximation method are purely probabilistic and based on the theory of weak convergence of probability measures. It can be used for either controlled or uncontrolled stochastic processes. We refer the readers to Kushner and Dupuis [87] for full discussions[3].

We here pick two specific topics from asset portfolio management problems, namely, asset portfolio management with partial information and model risk measurement and management problems.

Nagai [106], Nagai and Peng [107] studied the optimal investment problem as a risk-sensitive stochastic control problem, where the mean returns of individual securities are explicitly affected by economic factors defined as Gaussian processes. As an improvement, the investment strategies are assumed to depend only on the history of security prices. The problem turns out to be a stochastic control problem with partial information. Results in both finite and infinite time cases are presented. We refer the readers to Bensoussan [21] for full details in this direction.

On the other hand, the fact that investors have only partial information also leads to the notion of Model Risk. Consider (12.6), suppose that a trader precisely *knows* the model followed by the real market, and this model can be described by a system of stochastic differential equations. Then, in a complete market, the trader can apply the non-arbitrage pricing theory to construct perfectly replicating self-financed strategies in order to hedge options. When the trader has only imperfect information on the model (that is actually this case in practice), the trader's strategies of course may deviate from the target and therefore lead to financial risk. Such risk caused by misspecified models is called Model Risk. When one has a rather precise information on the model of the market, then one can take advantage of the robustness of formulae of Black and Scholes type (see, e.g., Avellaneda et al [7], El Karoui, Jeanblanc–Picqué & Shreve [51] and Romagnoli & Vargiolu [124], another approach is super-replication techniques developed in, e.g., Touzi [136] and references therein). When one has only vague information on the model of the mar-

---

[3] An alternative deterministic approach is based on the theory of numerical analysis for possibly degenerate PDEs, whose proof is based on the strong comparison theorem for viscosity solutions to PDEs, see Barles and Souganidis [16]. Nevertheless, it leads to similar coding as from the Markov chain approximation method. Both methods could be regarded as an advanced version of the classical finite difference method. The basic spirit is to keep the monotone condition valid.

ket, the problem is much more complicated, we refer the readers to Gibson [63] for various contributions in this direction.

Talay and Zheng [132] studied model risk management problems. The authors studied a strategy for the trader which, in a sense, guarantees good performances whatever is the unknown model for the assets of his/her portfolio. The trader chooses trading strategies to decrease the risk and therefore acts as a minimizer; the market systematically acts against the interest of the trader, so that we consider it acts as a maximizer. Thus one may consider the model risk control problem as a two players (Trader versus Market) zero-sum stochastic differential game problem. Therefore this construction corresponds to a 'worst case' worry and, in this sense, can be viewed as a continuous-time extension of discrete-time strategies based upon prescriptions issued from Value-at-Risk analyses at the beginning of each period. In addition, the initial value of the optimal portfolio can be seen as the minimal amount of money which is needed to face the worst possible damage. The authors gave a proper mathematical statement for such a game problem and proved that the value function of this game problem is the unique viscosity solution to a Hamilton–Jacobi–Bellman–Isaacs (HJBI) equation, and satisfies the Dynamic Programming Principle. The authors then answered a part of the open problems listed in the Conclusion in Cvitanić and Karatzas [40]. Talay and Zheng [131] numerically studied a concrete model risk control problem originated from the European bond options hedging and computed the value function and optimal strategies for the trader, and illustrate their financial interpretations.

As mentioned above, problems concerning portfolio management presents high variety. We finish this subsection by giving one problem from another branch in this discipline: to maximize the probability of a portfolio of reaching a given ceiling before reaching a floor by a given deadline T. This is a long-standing open problem in portfolio optimization. We refer the readers to Kulldorff [83] for known techniques.

12.2.5   Simulation problems arising from interest rate term structure modelling

In order to characterize the random evolution of the term structure of interest rates, models with one factor, generally chosen as the short term rate, have been developed because they are easy to implement (see, e.g., Merton [101], Vasicek [137], Cox, Ingersoll and Ross [38], Hull and White [70], etc.), even if most empirical studies using a principal component analysis have decomposed the motion of the interest rate term structure into three independent and non-correlated factors, which respectively capture the level shift in the term structure, the twist in opposite direction of short and long term rates, and the butterfly factor that captures the fact that the intermediate rate moves in the opposite direction of the short and the long term rates. The multi-factor models do significantly better than single-factor models in explaining the dynamics and the shape of the entire term structure, but the latter provide analytical expressions for the prices of simple interest rates contingent claims, whereas a multi-factor model generally leads to numerically or quasi-analytically solve partial differential equations in a higher dimension to obtain prices and hedge ratios for the interest rate-contingent claims.

We consider two typical stochastic systems from the interest rate term structure modelling:

- In the Cox–Ingersoll–Ross (CIR) model, the short term rate satisfies

$$dr_t = k(\theta - r_t)dt + \sigma_{CIR}\sqrt{r_t}dW_t,$$

  where the speed of mean reversion coefficient k, the long run mean interest rate level θ and the volatility σ_{CIR} are positive constants.

- In the Heath–Jarrow–Morton (HJM) [69] model, for all time T^*, the instantaneous forward rate $f(t, T^*)$ satisfies the stochastic differential equation

$$f(t, T^*) = f(0, T^*) + \int_0^t \alpha(s, T^*)ds + \int_0^t \sigma(s, T^*)dW_s,$$

  for a given Borel measurable function $f(0, \cdot) : \mathbb{R}^+ \to \mathbb{R}$, and given random maps $\alpha : C \times \Omega \to \mathbb{R}$, $\sigma : C \times \Omega \to \mathbb{R}$, where $C := ((s, t) \mid 0 \le s \le t \le T^*)$. We suppose that $\alpha(\cdot, T^*)$ and $\sigma(\cdot, T^*)$ are adapted processes such that

$$\int_0^{T^*} |\alpha(s, T^*)|ds + \int_0^{T^*} |\sigma(s, T^*)|^2 ds < \infty \quad \mathbb{P} - a.s.$$

The advantage of such an infinite dimensional model is that if at time 0, one can set the theoretical forward rate $f(0, T)$ equal to the observed one $f^*(0, T)$, then one has a perfect fitting of the whole current term structure and the problem of inverting the yield curve to calibrate is avoided.

The new difficulties arising in the simulation of the interest rate process are:

- For the models of CIR type, a coefficient structure like $\sqrt{r_t}$ arises, which satisfies neither the classical Lipschitz assumption nor the non-degeneracy assumption used in the numerical methods for stochastic systems.

Progress in this direction was made very recently. Yan [139] studied the weak convergence of the Euler scheme for stochastic differential equations whose coefficients may be discontinuous on a set of Lebesgue measure zero. The rate of convergence is estimated when coefficients are Hölder continuous. Bossy et al. [25] studied the discretization of

$$dX_t = b(X_t)dt + \sigma\sqrt{X_t}dW_t.$$

We emphasize that the usual Euler scheme is even not well defined for such CIR models because X_t^n might be negative. The authors overcame the difficulties that the derivatives of coefficients may explode near 0 and that the system is not strongly elliptic, and obtained an almost optimal convergence rate for the approximation of $\mathbb{E}[f(X_T)]$ where f may be irregular. The technique therein, without surprise, is again Malliavin calculus. We finally would like to point out the possible connection between such approximation techniques and the convergence of ARCH models we discussed before.

The above results indicate that we shall soon be able to understand the simulation problem of stochastic systems with Hölder continuous coefficients in a new level.

• For the infinite-factor models of HJM class, it is obviously impossible to simulate the complete term structure. One must make a tradeoff between the tractability and computation complexity. To our knowledge, this direction is still completely open up to date.

It is impossible even to give an incomplete list of interesting new problems presented in the direction of Numerical methods in Finance. We expect that the following years will witness an even more prospective growth of contribution in this arena, and we wish that this note would help to attract more researchers into this field.

References

[1] Framework for supervisory information about derivatives activities of banks and securities firms, 1995. Basle Committee on Banking Supervision, manuscript, Bank for Internal Settlements.

[2] Principles for the management of interest rate risk, 1997. Basle Committee on Banking Supervision, manuscript, Bank for Internal Settlements.

[3] H. Ahn, and A. Kohatsu-Higa, Numerical Solutions of Anticipating Stochastic Differential Equations. *Stochastics and Stochastic Reports*, 54(3-4), 1995.

[4] M. Akian, J-L. Menaldi, and A. Sulem, On an investment–consumption model with transaction costs. *SIAM J. Control and Optim.*, 34(1), 1996.

[5] P. Artzner, F. Delbaen, J.-M. Eber, and D. Heath, Coherent measures of risk. *Math. Finance*, 9(3):203–228, 1999.

[6] S. Asmussen, P. Glynn, and J. Pitman, Discretization error in simulation of one-dimensional reflecting Brownian motion. *Ann. Appl. Probab.*, 5(4), 1995.

[7] M. Avellaneda, A. Lévy, and A. Paras, Pricing and hedging derivative securities in markets with uncertain volatilities. *Applied Mathematical Finance*, 2, 1995.

[8] R. Avesani and P. Bertrand, Does volatility jump or just diffuse? A statistical approach. In L.C.G. Rogers and D. Talay, editors, *Numerical Methods in Finance*, pages 270–289. Cambridge University Press, 1997.

[9] C.A. Ball and A. Roma, Stochastic Volatility Option pricing. *Journal of Financial and Quantitative Analysis*, 29(4), 1994.

[10] V. Bally and G. Pagès, A quantization algorithm for solving multi-dimensional optimal stopping problems, 2000. Prepublications du laboratoire Probabilites et Modeles Aleatoires Paris 6.

[11] V. Bally and G. Pagès, Error analysis of the quantization algorithm for obstacle problems, 2001. Prepublications du laboratoire Probabilites et Modeles Aleatoires Paris 6.

[12] V. Bally, G. Pagès, and J. Printems, First order schemes in the numerical quantization method, 2002. Prepublications du laboratoire Probabilites et Modeles Aleatoires Paris 6.

[13] V. Bally and D. Talay, The law of the Euler scheme for stochastic differential equations (I) : convergence rate of the distribution function. *Probability Theory and Related Fields*, 104(1), 1996.

[14] V. Bally and D. Talay, The law of the Euler scheme for stochastic differential equations (II) : convergence rate of the density. *Monte Carlo Methods and Applications*, 2:93–128, 1996.

[15] G. Barles, Convergence of numerical schemes for degenerate parabolic equations arising in finance theory. In L.C.G. Rogers and D. Talay, editors, *Numerical Methods in Finance*, Publications of the Newton Institute, pages 1–21. Cambridge University Press, 1997.

[16] G. Barles and P.E. Souganidis, Convergence of approximation schemes for fully nonlinear second-order equations. *Asymptotic Analysis*, 4:271–283, 1991.

[17] O.E. Barndorff-Nielsen and N. Shephard, Econometric analysis of realized volatility and its use in estimating stochastic volatility models. *J. R. Stat. Soc. Ser. B Stat. Methodol.*, 64(2), 2002.

[18] J. Barraquand and D. Martineau, Numerical valuation of high dimensional multivariate American securities. *Journal of Financial and Quantitative Analysis*, 30(3), 1995.

[19] J. Barraquand and T. Pudet, Pricing of American path-dependent contingent claims. *Math. Finance*, 6(1), 1996.

[20] S. Basak and A. Shapiro, Value-at-Risk Based Risk Management: Optimal Policies and Asset Prices. *Review of Financial Studies*, 14(2), 2001.

[21] A. Bensoussan, *Stochastic control of partially observable systems*. Cambridge University Press, Cambridge, 1992.

[22] F. Benth, K.H. Karlsen, and K. Reikvam, Optimal portfolio selection with consumption and nonlinear integro-differential equations with gradient constraint: a viscosity solution approach. *Finance and Stochastics*, 5(3), 2001.

[23] T. Bollerslev, Generalized Autoregressive Conditional Heteroscedasticity. *Journal of econometrics*, 31:307–327, 1986.

[24] T. Bollerslev and P.E. Rossi, Introduction: Modelling Stock Market Volatility - Bridging the GAP to Continuous Time, 1996.

[25] M. Bossy, A. Diop, and D. Talay, A numerical scheme for CIR type models, 2002.

[26] N. Bouleau and D. Lepingle, *Numerical Methods for Stochastic Processes*. J. Wiley, 1993.

[27] P. Boyle, Options: a Monte Carlo approach. *Journal of Financial Economics*, 4, 1977.

[28] P. Boyle, M. Broadie, and P. Glasserman, Monte Carlo methods for security pricing. Computational financial modelling. *J. Econom. Dynam. Control*, 21(8-9), 1997.

[29] P. Briand, B. Delyon, and J. Mémin, Donsker-type theorem for BSDEs. *Electron. Comm. Probab.*, 6:1–14, 2001.

[30] M. Broadie and P. Glasserman, Pricing American-style securities using simulation. *J. Econom. Dynam. Control*, 21(8-9), 1997.

[31] M. Broadie, P. Glasserman, and S. Kou, A continuity correction for discrete barrier options. *Math. Finance*, 7(4), 1997.

[32] P. Carr, Randomization and the American put. *Review of Financial Studies*, 11, 1998.

[33] P. Cattiaux and L. Mesnager, Hypoelliptic non-homogeneous diffusions. *Probability Theory and Related Fields*, 123(4), 453–483, 2002.

[34] D. Chevance, Numerical methods for backward stochastic differential equations. In L.C.G. Rogers and D. Talay, editors, *Numerical Methods in Finance*, Publications of the Newton Institute. Cambridge University Press, 1997.

[35] D. Chevance, *Résolution Numérique des Èquations Différentielles Stochastiques Rétrogrades*. PhD thesis, Université de Provence, 1997.

[36] J.M.C. Clark, An efficient approximation for a class of stochastic differential equations. In W. Fleming and L. Gorostiza, editors, *Advances in Filtering and Optimal Stochastic Control*, volume 42 of *Lecture Notes in Control and Information Sciences*. Proceedings of the IFIP Working Conference, Cocoyoc, Mexico, 1982, Springer-Verlag, 1982.

[37] J. Cox, S. Ross, and M. Rubinstein, Option pricing: a simplified approach. *J. of Economics*, January 1978.

[38] J.C. Cox, J.E. Ingersoll, and S.A. Ross, A theory of the term structure of interest rates. *Econometrica*, 53:385–407, 1985.

[39] J. Cvitanić and I. Karatzas, Backward sde's with reflection and Dynkin games. *Annals of Probability*, 24:2024–2056, 1996.

[40] J. Cvitanić and I. Karatzas, On dynamic measures of risk. *Finance & Stochastics*, 3(4):451–482, 1999.

[41] J. Cvitanić and J. Ma, Reflected forward-backward sde's and obstacle problems with boundary conditions. *Journal of Appl. Math. Stoch. Anal.*, 14(2), 2001.

[42] J. Cvitanić, J. Ma, and J. Zhang, Efficient Computation of Hedging Portfolios for Options with Discontinuous Payoffs. *Math. Finance.*

[43] F. Delbaen, P. Grandits, T. Rheinländer, D. Samperi, M. Schweizer, and C. Stricker. Exponential hedging and entropic penalties. *Math. Finance*, 12(2), 2002.

[44] J.J. Douglas, J. Ma, and P. Protter, Numerical methods for forward-backward stochastic differential equations. *Ann. Appl. Probab.*, 6(3), 1996.

[45] D. Duffie, *Dynamic Asset Pricing Theory*. Princeton University Press, 1992.

[46] D. Duffie and L. Epstein, Asset pricing with stochastic differential utility. *Review of Financial Studies*, 5:411–436, 1992.

[47] D. Duffie and L. Epstein, Stochastic differential utility. *Econometrica*, 60(2):353–394, 1992.

[48] D. Duffie and P. Glynn, Efficient monte carlo simulation of security prices. *Ann. Appl. Prob.*, 5(4):897–905, 1995.

[49] D. Duffie, J. Ma, and J. Yong, Black's consol rate conjecture. *Annals of Applied Probability*, 5:356–382, 1995.

[50] D. Duffie and J. Pan, Analytical value-at-risk with jumps and credit risk. *Finance and Stochastics*, 5(2), 2001.

[51] N. El Karoui, M. Jeanblanc-Picqué, and S.E. Shreve, Robustness of the Black and Scholes formula. *Mathematical Finance*, 8(2):93–126, 1998.

[52] N. El Karoui, C. Kapoudjian, E. Pardoux, and S. Peng and M.C. Quenez, Reflected solutions of backward sde's, and related obstacle problems for pde's. *Ann. Probab.*, 25:702–737, 1997.

[53] N. El Karoui, S. Peng, and M.C. Qnenez, Backward stochastic differential equations in finance. *Mathematical Finance*, 7(1):1–71, Jan 1997.

[54] R.F. Engle, Autoregressive Conditional Heteroscedasticity with Estimates of the Variance of United Kingdom Inflation. *Econometrica*, 50(1), 1982.

[55] E. Fama, Behavior of stock market prices. *J. Business*, 38:34–105, 1965.

[56] D. Florens-Zmirou, On estimating the diffusion coefficient from discrete observations. *J. Appl. Prob.*, 30:790–804, 1993.

[57] H. Föllmer and P. Leukert, Quantile Hedging . *Finance and Stochastics*, 3(3), 1999.

[58] E. Fournié, J-M. Lasry, J. Lebuchoux, and P-L. Lions, Applications of Malliavin calculus to Monte Carlo methods in finance. *Finance Stoch.*, 3(4):391–412, 1999.

[59] E. Fournié, J-M. Lasry, J. Lebuchoux, and P-L. Lions. Applications of Malliavin calculus to Monte Carlo methods in finance II. *Finance Stoch.*, 5(2):201–236, 2001.

[60] E. Fournié and D. Talay, Application de la statistique des diffusions à un modèle de taux d'intérêt. *Finance*, 12(2):79–111, 1991.

[61] N. Fournier, *Calcul des variations stochastiques sur l'espace de Poisson, applications des EDPS paraboliques avec sauts et certaines quations de Boltzmann*. PhD thesis, Paris 6, 1999.

[62] M. Fu, C. Laprise, S.B. Madan, D.B. Su, and R. Wu, Pricing American options: a comparison of Monte Carlo simulation approaches. *Journal of Computational Finance*, 4, 2001.

[63] R. Gibson, editor, *Model Risk: Concepts, Calibration and Pricing*. Risk Books. 2000.

[64] P. Glasserman, P. Heidelberger, and P. Shahabuddin, Variance Reduction Techniques for Value-at-Risk With Heavy-Tailed Risk Factors. In *Proceedings of the 1999 simulation conference on Winter simulation: Simulation-a bridge to the future*, 2000.

[65] P. Glasserman, P. Heidelberger, and P. Shahabuddin. Portfolio value-at-risk with heavy-tailed risk factors. *Math. Finance*, 12(3), 2002.

[66] E. Gobet, Weak approximation of killed diffusion using Euler schemes. *Stochastic Process. Appl*, 87(2), 2000.

[67] E. Gobet, Monte Carlo evaluation of Greeks for multidimensional barrier and lookback options. *Mathematical Finance*, 2003.

[68] E. Gobet and E. Temam, Discrete time hedging errors for options with irregular payoffs. *Finance Stoch*, 5(3), 2001.

[69] D. Heath, R. Jarrow, and A. Morton, Bond pricing and the term structure of interest rates: a new methodology for contingent claim valuation. *Econometrica*, 60:77–105, 1992.

[70] J. Hull and A. White, Pricing interest rate derivative securities. *Review of Financial Studies*, 3:573–592, 1990.

[71] J. Jacod, The Euler scheme for Lvy driven by stochastic differential equations: Limit theorems, 2002. Prepublications du laboratoire Probabilites et Modeles Aleatoires Paris 6.

[72] J. Jacod and P. Protter, Une remarque sur les équations différentielles stochastiques à solutions Markoviennes. In *Séminaire de Probabilités XXV*, volume 1485 of *Lecture Notes in Mathematics*, pages 138–139. Springer-Verlag, 1991.

[73] J. Jacod and P. Protter, Asymptotic error distributions for the Euler method for stochastic differential equations. *Ann. Probab.*, 26(1), 1998.

[74] R. Jarrow and D. Madan, Valuing and hedging contingent claims on semimartingales, 1994. Working paper.

[75] R. Jarrow and D. Madan, Option pricing using the term structure of interest rates to hedge systematic discontinuities in asset returns. *Math. Finance*, 5:311–336, 1995.

[76] R. Jarrow and E.R. Rosenfeld, Jump risks and the intertemporal capital asset pricing model. *J. of Business*, 57:337–351, 1984.

[77] P.E. Kloeden and E. Platen, *Numerical Solution of Stochastic Differential Equations*. Springer–Verlag, 1992.

[78] A. Kohatsu-Higa, Weak approximations: A Malliavin calculus approach . *Mathematics of Computation*, 70, 2001.

[79] A. Kohatsu-Higa, and S. Ogawa, Weak rate of convergence for an euler scheme of nonlinear sde's. *Monte Carlo Methods and its Applications*, 3:327–345, 1997.

[80] A. Kohatsu-Higa, and P. Protter, The Euler scheme for SDEs driven by semimartingales. In H. Kunita and H.H. Kuo, editors, *Stochastic Analysis on Infinite Dimensional Spaces*, pages 141–151. Pitman, 1994.

[81] U. Kuchler and E. Platen, Strong Discrete Time Approximation of Stochastic Differential Equations with Time Delay, 2000.

[82] U. Kuchler and E. Platen, Weak Discrete Time Approximation of Stochastic Differential Equations with Time Delay, 2001.

[83] M. Kulldorff, Optimal control of favorable games with a time limit. *SIAM J. Control Optim.*, 31(1), 1993.

[84] T. Kurtz and P. Protter, Wong–Zakai corrections, random evolutions and numerical schemes for S.D.E.'s. In *Stochastic Analysis: Liber Amicorum for Moshe Zakai*, pages 331–346, 1991.

[85] H. Kushner, Numerical methods for variance control, with applications to optimization in finance, 1998.

[86] H.J. Kushner, *Probability Methods for Approximations in Stochastic Control and for Elliptic Equations*. Academic Press, New York, 1977.

[87] H.J. Kushner and P. Dupuis, *Numerical Methods for Stochastic Control Problems in Continuous Time*. Springer Verlag, 1992.

[88] S. Kusuoka and D. Stroock, Applications of the Malliavin Calculus, part I. In *Taniguchi Symp. SA*, pages 271–306, 1982.

[89] S. Kusuoka and D. Stroock, Applications of the Malliavin Calculus, part II. *J. Fac. Sci. Univ. Tokyo*, 32:1–76, 1985.

[90] S. Kusuoka and D. Stroock, Applications of the Malliavin Calculus, part III. *J. Fac. Sci. Univ. Tokyo*, 34:391–442, 1987.

[91] Y.A. Kutoyants, *Parameter Estimation for Stochastic Processes*. Heldermann Verlag, 1984.

[92] Y.A. Kutoyants, *Identification of Dynamical Systems with Small Noise*, volume 300 of *Mathematics and its Applications*. Kluwer Academic Publishers, Dordrecht, 1994.

[93] F.A. Longstaff and E.A. Schwartz, Valuing American options by simulation. *Review of Financial Studies*, 14, 2001.

[94] J. Ma, P. Protter, J. San Martin, and S. Torres, Numerical Method for Backward SDE's. *Ann. Appl. Probab.*, 12, 2002.

[95] J. Ma, P. Protter, and J. Yong, Solving forward-backward stochastic differential equations explicitly – A four step scheme. *Probab. Theory Relat. Fields*, 98:339–359, 1994.

[96] J. Ma and J. Zhang, Representation Theorems for Backward Stochastic Differential Equations. *Annals of Applied Probability*.

[97] J. Ma and J. Zhang, Path Regularity for Solutions to Backward SDE's. *Probability Theory and Related Fields*, 2002.

[98] J. Ma and J. Zhang, Representation and Regularity of Solutions to Backward Stochastic Differential Equations with Reflections, 2002.

[99] B. Mandelbrot, The valuation of certain speculative prices. *J. Business*, 36:394–419, 1963.

[100] A. Mele and F. Fornari, *Stochastic volatility in Financial Markets: crossing the bridge to continuous time*. Kluwer, 2000.

[101] R.C. Merton, Theory of rational option pricing. *Bell J. of Econom. and Management Sci.*, 4:141–183, 1973.

[102] R. Mikulevicius and E. Platen, Time discrete Taylor approximations for Ito processes with jump component. *Math. Nachr.*, 138:93–104, 1988.

[103] G.N. Milshtein, Approximate integration of stochastic differential equations. *Theory Probab. Appl.*, 19:557–562, 1974.

[104] G.N. Milshtein, A method of second-order accuracy integration of stochastic differential equations. *Theory Probab. Appl.*, 23:396–401, 1976.

[105] P. Del Moral, J. Jacod, and P. Protter, The Monte-Carlo method for filtering with discrete-time observations. *Probab. Theory Related Fields*, 120(3), 2001.

[106] H. Nagai, Risk-sensitive dynamic asset management with partial information. In *Stochastics in finite and infinite dimensions*. Birkhäuser Boston, 2001.

[107] H. Nagai and S.G. PENG, Risk-sensitive dynamic portfolio optimization with partial information on infinite time horizon. *Ann. Appl. Probab.*, 12(1), 2002.

[108] J. Navas, *Valuation of Foreign Currency Options under Stochastic Interest Rates and Systematic Jumps Using the Martingale Approach*. PhD thesis, Purdue University, 1994.

[109] D.B. Nelson, ARCH models as diffusion approximations. *Journal of Econometrics*, 45:7–38, 1990.

[110] N.J. Newton, An asymptotically efficient difference formula for solving stochastic differential equations. *Stochastics*, 19:175–206, 1986.

[111] N.J. Newton, An efficient approximation for stochastic differential equations on the partition of symmetrical first passage times. *Stochastics and Stochastic Reports*, 29:227–258, 1990.

[112] N.J. Newton, Asymptotically efficient Runge-Kutta methods for a class of Ito and Stratonovich equations. *SIAM Journal of Applied Mathematics*, 51(2):542–567, 1991.

[113] H. Niederreiter, Quasi-Monte Carlo methods and pseudo-random numbers (survey). *Bull. of the American Math. Soc.*, 84:957–1041, 1978.

[114] H. Niederreiter, *Random Number Generation and Quasi–Monte Carlo Methods*. CBMS-NSF Regional Conference Series in Appl. Math. SIAM, 1992.

[115] H. Niederreiter, New developments in uniform pseudorandom number and vector generation. In P.J.-S. Shiue and H. Niederreiter, editors, *Monte Carlo and Quasi-Monte Carlo Methods in Scientific Computing*, volume 106 of *Lecture Notes in Statistics*, pages 87–120, Heidelberg, New York, 1995. Springer Verlag.

[116] D. Nualart, *Lectures on probability theory and statistics. École d'Été de probabilités de Saint-Flour XXV*, chapter Analysis on Wiener space and anticipating stochastic calculus. Berlin: Springer, 1995.

[117] D. Nualart, *Malliavin Calculus and Related Topics*. Probability and its Applications. Springer-Verlag, 1995.

[118] E. Pardoux and S. Peng, Adapted solution of a backward stochastic differential equation. *Systems and Control Letters*, 14:55–61, 1990.

[119] E. Platen, An approximation method for a class of Ito processes with jump component. *Lietuvos Matemematikos Rinkiniys*, 124–136, 1982.

[120] M. Pritsker, Evaluating Value-at-Risk methodologies: Accuracy versus computational time. In *Model Risk*. Risk books, 2000.

[121] P. Protter and D. Talay, The Euler scheme for Lévy driven stochastic differential equations. *Ann. Probab.*, 25(1):393–423, 1997.

[122] L.C.G. Rogers, Monte Carlo Valuation of American Options. *Math. Finance*, (3), 2002.

[123] L.C.G. Rogers and D. Talay, editors, *Numerical Methods in Finance*. Publications of the Newton Institute. Cambridge University Press, 1997.

[124] S. Romagnoli and T. Vargiolu, Robustness of the Black–Scholes approach in the case of options on several assets. *Finance & Stochastics*, 4:325–341, 2000.

[125] D. Stroock and S. Varadhan, *Multidimensional Diffusion Processes*. Springer, 1979.

[126] D. Talay, Discrétisation d'une e.d.s. et calcul approché d' espérances de fonctionnelles de la solution. *Mathematical Modelling and Numerical Analysis*, 20(1):141–179, 1986.

[127] D. Talay, Simulation and numerical analysis of stochastic differential systems: a review. In P. Krée and W. Wedig, editors, *Probabilistic Methods in Applied Physics*, volume 451 of *Lecture Notes in Physics*, chapter 3, pages 54–96. Springer-Verlag, 1995.

[128] D. Talay, and L. Tubaro, Probabilistic Numerical Methods for Partial Differential Equations. Book in preparation.

[129] D. Talay, and L. Tubaro, Romberg extrapolations for numerical schemes solving stochastic differential equations. *Structural Safety*, 8, 1990.

[130] D. Talay, and Z. Zheng, Approximation of quantiles of components of diffusion processes. Submitted to Stochastic Processes and their Applications.

[131] D. Talay, and Z. Zheng, A Hamilton Jacobi Bellman Isaacs equation for a financial risk model. In J-L. MENALDI, E. ROFMAN, and A. SULEM, editors, *Optimal control and PDE - Innovations and applications*. IOS Press, 2000.

[132] D. Talay, and Z. Zheng, Worst case model risk management. *Finance and Stochastics*, 6(4):517–537, 2002.

[133] D. Talay, and Z. Zheng, Quantiles of the Euler scheme for diffusion processes and financial applications. *Mathematical Finance*, 13(1):187–199, 2003.

[134] S. Taylor, *Modeling Financial Time Series*. UK: Wiley, 1986.

[135] J.A. Tilley, Valuing American options in a path simulation model. *Transactions of the Society of Actuaries*, 45, 1993.

[136] N. Touzi, Direct characterization of the value of super–replication under stochastic volatility and portfolio constraints. *Stoch. Proc. Appl*, 88:305–328, 2000.

[137] O. Vasicek, An equilibrium characterization of the term structure. *Journal of Financial Economics*, 5:177–188, 1977.

[138] F. Viens, Portfolio optimization under partially observed stochastic volatility, 2001.

[139] Liqing Yan, The Euler scheme with irregular coefficients. *Annals of Probab.*, 30(3):1172 –1194, 2002.

[140] J. Zhang. A numerical scheme for backward stochastic differential equations, 2002. Submitted.

[141] X.L. Zhang, Options américaines et modèles de diffusion avec sauts. *Note aux Comptes-Rendus de l'Académie des Sciences*, 317:857–862, 1993.

[142] X.L. Zhang, Numerical analysis of American option pricing in a jump-diffusion model. *Math. Oper. Research*, 22(3):668–690, 1997.

[143] X.L. Zhang, Valuation of American option in a jump diffusion model. In L.C.G. Rogers and D. Talay, editors, *Numerical Methods in Finance*, Publications of the Newton Institute, pages 93–114. Cambridge University Press, 1997.

List of Contributors

Keyvan Amir-Atefi
Assistant Vice President
Market Risk Management
HSBC Securities USA
452 Fifth Avenue
New York, NY 10018 USA

Zauresh Atakhanova
Kazakhstan Institute of Management, Economics
    and Strategic Research (KIMEP)
4 Abai Avenue
Almaty, 480100, Kazakhstan
zaat@kimep.kz

Almira Biglova
Institut für Statistik und Mathematische Wirtschaftstheorie
Universität Karlsruhe(TH), Postfach 6980
76128 Karlsruhe, Germany
and
Ufa State Aviation Technical University
450000, Ufa, K. Marx Str., 12, Russia

Oliver J. Blaskowitz
Center for Applied Statistics and Economics (CASE)
Humboldt–Universität zu Berlin
Wirtschaftswissenschaftliche Fakultät
Spandauer Str. 1, 10178 Berlin, Germany
blaskowitz@wiwi.hu-berlin.de

Dylan D'Souza
Vice President
Credit Risk Management
HSBC Bank USA
452 Fifth Avenue
New York, NY 10018, USA
dylan.m.d'souza@us.hsbc.com; dylan@econ.ucsb.edu

Wolfgang K. Härdle
Center for Applied Statistics and Economics (CASE)
Humboldt–Universität zu Berlin
Wirtschaftswissenschaftliche Fakultät
Spandauer Str. 1, 10178 Berlin, Germany

Isabella Huber
Institut für Statistik und Mathematische Wirtschaftstheorie
Universität Karlsruhe (TH)
Postfach 6980
76128 Karlsruhe
Germany
huber@statistik.uni-karlsruhe.de

Irina Khindanova
Colorado School of Mines
Golden, CO 80401, USA
ikhindan@mines.edu

Arturo Kohatsu-Higa
Department of Economics, Universitat Pompeu Fabra
Ramón Trias Fargas, 25-27, 08005 Barcelona, Spain
kohatsu@upf.es

Piotr S. Kokoszka
Mathematics and Statistics
Utah State University
3900 Old Main Hill
Logan, Utah 84322-3900, USA
piotr@stat.usu.edu

Miquel Montero
Departament de Física Fonamental
Universitat de Barcelona
Diagonal 647, 08028 Barcelona, Spain

Emrah Özturkmen
Institut für Statistik und Mathematische Wirtschaftstheorie
Universität Karlsruhe (TH)
Postfach 6980
76128 Karlsruhe, Germany

Sergio Ortobelli Lozza
Department MSIA, University of Bergamo
Via dei Caniana, 2, 24127 Bergamo, Italy
sergio.ortobelli@unibg.it

Gilles Pagès
Laboratoire de Probabilités et Modèles Aléatoires
CNRS, UMR 7599
Université Paris 6
4 Place Jussieu
75451 Paris Cedex 05, France

Andrejus Parfionovas
Mathematics and Statistics
Utah State University
3900 Old Main Hill
Logan, Utah 84322-3900 USA
andrej@cc.usu.edu

Huyên Pham
Laboratoire de Probabilités et
Modèles Aléatoires
CNRS, UMR 7599, Université Paris 6
2 Place Jussieu
75251 Paris Cedex 05, France
pham@gauss.math.jussieu.fr

Jacques Printems
Centre de Mathématiques
Faculté de Sciences et Technologie
CNRS, UMR 8050
Université Paris 12
Val de Marne , 61, Av du Général de Gaulle
94010 Créteil Cedex , France
printems@univ-paris12.fr

Svetlozar T. Rachev
Institut für Statistik und Mathematische Wirtschaftstheorie
Universität Karlsruhe
Postfach 6980, 76128 Karlsruhe, Germany
    and
Department of Statistics and Applied Probability
University of California, Santa Barbara
CA 93106-3110, USA
rachev@statistik.uni-karlsruhe.de

Borjana Racheva-Jotova
Managing Director (FinAnalytica, Seattle, USA)
    and Faculty of Economics and Business
Sofia University, Bulgaria
65 Milin Kamak Str. Sofia 1421, Bulgaria
borjana.racheva@finanalytica.com

Frank Schlottmann
GILLARDON AG Financial Software Research Dept.
Alte Wilhelmstr. 4
D-75015 Bretten, Germany
frank.schlottmann@gillardon.de
frank.schlottmann@gillardon.de

Peter Schmidt
Quantitative Analyst
Equities, Asset Management Research
Bankgesellschaft Berlin AG
Alexanderplatz 2, 10178 Berlin, Germany
peter.schmidt@ib.bankgesellschaft.de

Detlef Seese
Institut AIFB
Universität Karlsruhe (TH)
Postfach 6980
D-76128 Karlsruhe, Germany
seese@aifb.uni-karlsruhe.de

Stoyan Stoyanov
Chief Financial Analyst (FinAnalytica, Seattle, USA)
440-F Camino del Remedio
Santa Barbara, CA 93110, USA
and
Faculty of Mathematics and Informatics
Sofia University, Sofia, Bulgaria
stoyan.stoyanov@finanalytica.com

Carlos E. Testuri
Computer Science Institute
Universidad de Uruguay
J. Herrera y Reissig 565
Montevideo 11300, Uruguay
ctesturi@fing.edu.uy

Stefan Trück
Institut für Statistik und Mathematische Wirtschaftstheorie
Universität Karlsruhe (TH)
Postfach 6980, 76128 Karlsruhe, Germany
stefan@statistik.uni-karlsruhe.de

Stanislav Uryasev
Department of Industrial and Systems Engineering
University of Florida
PO Box 116595, 303 Weil Hall
Gainesville, FL 32611-6595, USA
uryasev@ise.ufl.edu

Ziyu Zheng
Department of Mathematical Sciences
University of Wisconsin-Milwaukee
Milwaukee, WI 53201-0413, USA
ziyu@uwm.edu